ADHESIVE TECHNOLOGY

668·3

LIBRARY
No. B 7197

B 7197

ROCKET PROPULSION ESTABLISHMENT
LIBRARY

18/11

Please return this publication, or request a renewal, by the date stamped below.

Name	Date
~~D YOUNG~~ B426	12/1/8

(4/64) L.23964 442077 Wt29280 D7061 10/64 10M T&Co G871.   R.P.E. Form 243

ADHESIVE TECHNOLOGY

Developments Since 1977

Edited by S. Torrey

NOYES DATA CORPORATION

Park Ridge, New Jersey, U.S.A.

1980

Sole distribution in the UK by:
Gothard House Publications
Gothard House
Henley-on-Thames, Oxon
RG9 1AJ   Tel. 049 12 3602

Copyright © 1980 by Noyes Data Corporation
    No part of this book may be reproduced in any form
    without permission in writing from the Publisher.
Library of Congress Catalog Card Number: 79-25936
ISBN: 0-8155-0787-9
Printed in the United States

Published in the United States of America by
Noyes Data Corporation
Noyes Building, Park Ridge, New Jersey 07656

Library of Congress Cataloging in Publication Data

Torrey, S
    Adhesive technology.

    (Chemical technology review ; no. 148)
    Updates the 1978 ed. published under title: Adhesives
technology annual.
    Includes indexes.
    1. Adhesives--Patents. I. Noyes Data Corporation.
II. Adhesives technology annual. III. Title.
IV. Series.
TP967.T67 1980   668'.3'0272   79-25936
ISBN 0-8155-0787-9

Foreword

This book is a data-based publication, providing information retrieved and made available from the U.S. patent literature. It thus serves a double purpose in that it supplies detailed technical information and can be used as a guide to the patent literature in this field. By indicating all the information that is significant, and eliminating legal jargon and juristic phraseology, this book presents an advanced, commercially oriented review of adhesive technology based on about 280 U.S. patents issued since January 1978. This title contains all new developments since our previous title *Adhesives Technology Annual,* published in 1978.

As was its predecessor, this is a practical, useful manual. It reflects the efforts and skills of many talented inventors. Its continuing purpose is to present the necessary chemistry, as well as changing technology and applications, notably the replacement of organic solvents by less toxic and less polluting carriers, to serve the varied interests of the makers and users of adhesives.

While the information is taken from U.S. patents, the coverage is actually worldwide in scope, as nowadays over 35% of all processes patented in the U.S. are developed by foreign investigators from practically every industrial nation.

The U.S. patent literature is the largest and most comprehensive collection of technical information in the world. There is more practical, commercial, timely process information assembled here than is available from any other source. The technical information obtained from a patent is extremely reliable and comprehensive; sufficient information must be included to avoid rejection for "insufficient disclosure." These patents include practically all of those issued on the subject in the United States during the period under review; there has been no bias in the selection of patents for inclusion.

The patent literature covers a substantial amount of information not available in the journal literature. The patent literature is a prime source of basic commercially useful information. This information is overlooked by those who rely primarily on the periodical journal literature. It is realized that there is a lag between a patent application on a new process development and the granting of a patent, but it is felt that this may roughly parallel or even anticipate the lag in putting that development into commercial practice.

Many of these patents are being utilized commercially. Whether used or not, they offer opportunities for technological transfer. Also, a major purpose of this book is to describe the number of technical possibilities available, which may open up profitable areas of research and development. The information contained in this book will allow you to establish a sound background before launching into research in this field.

v

Advanced composition and production methods developed by Noyes Data are employed to bring these durably bound books to you in a minimum of time. Special techniques are used to close the gap between "manuscript" and "completed book." Industrial technology is progressing so rapidly that time-honored, conventional typesetting, binding and shipping methods are no longer suitable. We have bypassed the delays in the conventional book publishing cycle and provide the user with an effective and convenient means of reviewing up-to-date information in depth.

The table of contents is organized in such a way as to serve as a subject index. Other indexes by company, inventor and patent number help in providing easy access to the information contained in this book.

15 Reasons Why the U.S. Patent Office Literature is Important to You

1.   The U.S. patent literature is the largest and most comprehensive collection of technical information in the world. There is more practical commercial process information assembled here than is available from any other source.
2.   The technical information obtained from the patent literature is extremely comprehensive; sufficient information must be included to avoid rejection for "insufficient disclosure."
3.   The patent literature is a prime source of basic commercially utilizable information. This information is overlooked by those who rely primarily on the periodical journal literature.
4.   An important feature of the patent literature is that it can serve to avoid duplication of research and development.
5.   Patents, unlike periodical literature, are bound by definition to contain new information, data and ideas.
6.   It can serve as a source of new ideas in a different but related field, and may be outside the patent protection offered the original invention.
7.   Since claims are narrowly defined, much valuable information is included that may be outside the legal protection afforded by the claims.
8.   Patents discuss the difficulties associated with previous research, development or production techniques, and offer a specific method of overcoming problems. This gives clues to current process information that has not been published in periodicals or books.
9.   Can aid in process design by providing a selection of alternate techniques. A powerful research and engineering tool.
10.   Obtain licenses—many U.S. chemical patents have not been developed commercially.
11.   Patents provide an excellent starting point for the next investigator.
12.   Frequently, innovations derived from research are first disclosed in the patent literature, prior to coverage in the periodical literature.
13.   Patents offer a most valuable method of keeping abreast of latest technologies, serving an individual's own "current awareness" program.
14.   Copies of U.S. patents are easily obtained from the U.S. Patent Office at 50¢ a copy.
15.   It is a creative source of ideas for those with imagination.

Contents and Subject Index

Hot Melt Adhesives

POLYOLEFIN

Having High Strength at Elevated Temperatures

Hot melt adhesives produce a bond by mere cooling as distinguished from cross-linking or other chemical reactions. Prior to heating, the hot melt adhesives are solids that can be prepared in bulk or pellet form for ease of handling. Upon heating, the hot melt adhesive composition melts and flows freely for application to a substrate. If the hot melt adhesives are thermoplastic rather than thermosetting, and thus remeltable, they can be applied to a particular substrate and later remelted to form a hot melt bond between this substrate and another substrate.

Hot melt adhesives are useful for bonding various substrates together such as wood, paper, plastics, and textiles, as well as other materials. One use for which they are well suited is the fabrication of corrugated paperboard, and for cardboard case sealing and closing. Hot melt adhesives useful for producing corrugated paperboard must have high bond strength under conditions of shock, stress, high humidity and extremes of temperature encountered in transportation and storage. In addition, the melt point, wetting time, initial tack, setting time, pot life and general handling qualities on automatic corrugated board machinery are essential considerations.

Hot melt adhesives are widely used by industry in the construction of various packaging containers. For example, one application is the use of hot melt adhesives as the side seam adhesive in the preparation of cans from fibrous materials which may have a metal or plastic foil laminate.

The aim of *D.A. Godfrey; U.S. Patent 4,076,670; February 28, 1978; assigned*

1

to Eastman Kodak Company was to provide polyolefin-based hot melt adhesives that have high strength at elevated temperatures and pressures and resistance to creep.

The process shows that a blend comprising low density polyethylene, tackifying resin, and crystalline polypropylene or a crystalline propylene-containing copolymer provides a hot melt adhesive having this combination of properties.

The low density polyethylenes useful in the process are well known in the art and can be prepared by the high pressure polymerization of ethylene in the presence of free radical catalysts. These polyethylenes have a melt index at 190°C of about 1 to 50, and most preferably about 20. These polyethylenes have a density of about 0.910 to 0.940 g/cm^3, and most preferably about 0.925 g/cm^3. The polyethylenes can be used in amounts of about 30 to 70% by wt, preferably about 35 to 45% by wt.

Useful tackifying resins are hydrocarbon resins, synthetic polyterpenes, rosin esters and the like. Commercially available resins include Eastman's Resin H-130, Goodyear's Wingtack resins, and the Sta-Tac, Nirez and Betaprene H resins from Reichhold. These tackifying resins, which preferably have softening points of at least 100°C and preferably 120°C, can be used in amounts of about 10 to 50% by wt of the adhesive composition, preferably about 20 to about 40% by wt and most preferably about 30% by wt.

The crystalline, hexane-insoluble polypropylenes or propylene-containing copolymers useful in the adhesive composition of this process are produced by the polymerization of propylene or propylene and ethylene in the presence of stereospecific catalysts. They have crystalline ethylene/propylene copolymers containing no more than 20% by wt ethylene, and having flow rates at 230°C of about 30 to 180, preferably about 120 to about 180. These crystalline polypropylenes and propylene-containing copolymers can be used in amounts of 10 to 40% by wt of the adhesive composition, preferably about 25 to 35% by wt, and most preferably about 30% by wt.

The adhesive compositions are prepared by blending together the components in the melt at a temperature of about 180° to 230°C until a homogeneous blend is obtained, approximately 2 hours. Various methods of blending materials of this type are known to the art and any method that produces a homogeneous blend is satisfactory.

In addition to the above listed components, it is desirable for the hot melt adhesive composition to contain about 0.1 to 1.5% by wt, preferably about 0.25 to 1.0% by wt, of one or more stabilizers or antioxidants.

A particularly effective antioxidant is Irganox 1010 used in combination with Cyanox 1212. Adhesive properties were determined as follows. Viscosity of the adhesives was determined at 190°C on an F.F. Slocomb Melt Viscometer.

Creep Resistance: A sheet of fiberboard substrate was wound around a mandrel having a diameter of about 2.5 inches into the shape of a cylinder with an overlap of 0.3 inch. The cylinder was lap shear bonded with the adhesive to be evaluated. This sample was then marked with a pencil along the bond. The sample was put into a cold box at 35°F for 24 hr at which time the bond was inspected for creep. If the bond edge still matched up with the pencil mark, no bond creep had been experienced. Samples prepared in this way were also used to test for bond failure at 0°F after 1 hr exposure.

Bond Strength at Elevated Temperature: Samples (½ inch wide) of the fiberboard substrates were bonded with the adhesive by applying it with a heated spatula. An overlapping bond of 0.3 inch was made. Thus, the bond area was 0.5 inch x 0.3 inch. Bonds were allowed to condition at room temperature for 24 hr. The bond strength of the adhesive was determined by separating the bond at 0.1 in/min on an Instron tensile tester. A jaw separation distance of approximately 0.7 inch was used. A temperature cabinet was used to determine bond strength at various temperatures.

Example: The following adhesive composition was prepared.

	Percent by Weight
DAC-B hydrocarbon resin having a Softening Point of 100°C, and a viscosity of 140 cp at 190°C and a Ring and Ball Softening Point of 138°C (Resin H-138)	29.55
Crystalline polypropylene having a melt flow rate at 230°C of 160 and a density of 0.905 (Tenite 428S)	30
Polyethylene having a melt index at 190°C of 20 and a density of 0.925 g/cm³ (Tenite 18BO)	40.0
Tetrakis[methylene-3-(3',5'-di-tert-butyl-4'-hydroxy-phenyl)propionate] methane (Irganox 1010)	0.10
Laurylstearylthiodipropionate (Cyanox 1212)	0.35

This composition provided an adhesive composition with the following properties.

Melt viscosity at 190°C	120,000 cp
Ring and Ball Softening Point	159°C
Creep at 35°F after 24 hr	ND
Bond strength at	
73°F	88 psi
105°F	77 psi
140°F	57 psi

Note: ND is none detectable.

This shows that this adhesive has excellent resistance to elevated temperature and pressure.

In an effort to produce a polyethylene-containing hot melt adhesive with a reduced melt viscosity, shorter hot tack time, and improved elevated temperature properties *D.A. Godfrey; U.S. Patent 4,127,619; November 28, 1978; assigned to Eastman Kodak Company* developed an adhesive blend comprising a polyethylene, a tackifying resin, and a high-density, low viscosity polyethylene wax having a molecular weight of about 1,000.

The polyethylenes can be any of the normally solid polyethylenes having a melt viscosity of 500 to 20,000 cp at 190°C, preferably 1,000 to 6,000, and a density greater than 0.90 to about 0.97 g/cc. The tackifying resins are the same as those suggested in the previous patent.

The high density, low viscosity polyethylene waxes useful in the blend are widely available articles of commerce and can be prepared by known techniques. These waxes are made by degrading high density polyethylene to the desired viscosity. Polyethylene waxes normally have, and should have, for this process, Ring and Ball softening points of 110° to 120°C and preferably of 113° to 117°C. This wax has a molecular weight of about 800 to 1,200, preferably about 1,000, a penetration hardness at 23°C of about 1.0, and a melt viscosity of about 25 cp at 150°C. Such a wax is, e.g., Bareco 1000 polywax. The polyethylene wax can be used in amounts of about 7 to 15% by wt of the adhesive composition, preferably about 10% by wt.

Example 1: Low-molecular weight polyethylene (210 g) having a melt viscosity of 3,075 cp at 177°C, density of 0.906 g/cc, and Ring and Ball softening point of 105°C, 89.4 g of Eastman Resin H-130 hydrocarbon resin tackifier having a density of 1.028 g/cc at 70°F, melt viscosity of 2,150 cp at 177°C, and Ring and Ball softening point of 129°C, and 0.6 g of tetrakis[methylene-3-(3',5'-di-tert-butyl-4-hydroxyphenyl)propionate] methane, are placed in a 500 ml round bottom resin pot fitted with a mechanical stirrer and nitrogen purge line. The system is purged with nitrogen and heated to 200°C using a metal bath.

The polymer blend is stirred for one hr after melting to insure that the blend is homogeneous. The molten polymer blend is poured in the form of a thin cake, cooled and chopped into small pieces for use as an adhesive. This composition has a melt viscosity of 2,040 cp at 177°C. The adhesive properties of this material are tested in the following manner using corrugated board and an adhesive tester that simulates a case sealing line.

Corrugated board specimens 6.5 inches long by 2.5 inches wide are placed in the grips of the tester. The bottom specimen is cut so that the flutes are parallel to the adhesive bead with the unglazed side being bonded. All top specimens have the flutes perpendicular to the bead with the machine-glazed side being bonded since this is the manner in which a box is bonded in actual practice. The bottom specimen is then moved forward at a constant speed under the melt applicator for the application of the adhesive bead and is stopped directly under the top specimen.

After a predetermined open time, the top specimen is moved down to contact the lower specimen. Contact is maintained for a given compression time and force, after which the top substrate is separated from the bottom substrate using a constant force (40 psig). The separation force required is measured by a pressure-transducer and recorded on a digital peak load meter. The adhesive bead width (before compression) is adjusted by nitrogen pressure on the hot-melt adhesive applicator to give a width of 70 mils. Ten tests are carried out at each compression time under the following conditions.

Open time	0.75 sec
Down pressure	16.5 psig
Up pressure	40.0 psig
Compression times of 0.75	4.0 sec

The bond separation force is measured in pounds to separate a bond having a constant length of 6.5 inches. The percent fiber tear is determined as the fraction of the bond length showing fiber tear. The hot tack time (defined as the compression time required to give a bond separation force of 30 lb) of this adhesive is 1.9 seconds. After a compression time of 4.0 seconds, 60% of the bond shows fiber tear.

Example 2: Low-molecular weight maleated polyethylene (186.7 g) having a saponification number of 5, a melt viscosity of 3,350 cp at 190°C, Ring and Ball softening point of 106°C, and a density of 0.908 g/cc, 78.65 g of Eastman Resin H-130 hydrocarbon resin tackifier having a density of 1.028 g/cc at 70°F, melt viscosity of 2,150 cp at 177°C, and Ring and Ball softening point of 129°C, 33.3 g of Bareco 1000 polywax having a melting point of 94° to 99°C, 1.05 g of lauryl stearyl ester of thiodipropionic acid, and 0.30 g of tetrakis[methylene-3-(3',5'-di-tert-butyl-4-hydroxyphenyl)propionate] methane are blended according to the procedure of Example 1.

The addition of the Bareco 1000 polywax to this composition reduced the melt viscosity from 1,860 to 1,050 cp at 177°C. The effect of the addition of the Bareco 1000 polywax and the maleated polyethylene on bond separation force is reflected in the very short hot tack time of 0.85 second obtained with this material. The percent fiber tear obtained is significantly higher than that obtained with the composition described in Example 1, even at a compression time of 0.75 second.

To solve the problems previously encountered in using polyolefin-based hot melt adhesives for bonding nonporous substrates, such as foil-to-foil and foil-to-paper, *D.A. Godfrey; U.S. Patent 4,146,521; March 27, 1979; assigned to Eastman Kodak Company* worked out a formulation comprising at least one polyethylene resin, an ethylene/lower alkyl acrylate copolymer, a tackifying resin, and a high density, low viscosity polyethylene wax having a molecular weight of about 1,000, which provides a hot melt adhesive which can be applied without solvents, has excellent bonding to nonporous substrates, good bond aging properties and excellent elevated temperature properties.

The polyethylene can be any of the normally solid polyethylenes having a melt index of 100 to 5,000 at 190°C, preferably 2,000 to 2,500, and a density greater than 0.90 to 0.93 g/cc. The tackifying resins and polyethylene waxes used in the formulation are those described in the previous patent.

The ethylene/lower alkyl acrylate copolymer suitable for use in the adhesive is, for example, an ethylene/ethyl acrylate copolymer having a melt index of 2 to 30, preferably 6 to 20, a density of 0.915 to 0.95 and an ethyl acrylate content of 10 to 25%, preferably 18 to 23% by wt. A typical material is Bakelite.

The polyethylene wax can be used in amounts of about 7 to 15% by wt of the adhesive composition, preferably about 10 to 12% by wt. The hydrocarbon resins and polyterpene tackifying resins can be used either alone or in combination. These tackifying resins can be used in amounts of about 13 to 40% by wt of the adhesive composition, preferably about 25 to 30% by wt. The polyethylene or a blend of such polyethylenes can be used in amounts of 30 to 40% by wt in the adhesive formulation, preferably 32 to 35% by wt. The ethylene/lower alkyl acrylate copolymer can be used in an amount of 20 to 40% by wt, preferably 26 to 32% by wt.

Example: Low molecular weight polyethylene (104 g) having a melt viscosity of 3,075 cp at 177°C, density of 0.906 g/cc, and Ring and Ball softening point of 105°C; 75 g ethylene/ethyl acrylate copolymer containing 23 wt % ethyl acrylate and a melt index of 20; 70 g of Eastman Resin H-130 hydrocarbon resin tackifier having a density of 1.028 g/cc at 70°F, melt viscosity of about 1,000 cp at 190°C, and Ring and Ball softening point of 129°C; 30 g of Bareco 1000 polywax having a melting point of 113° to 117°C and a melt viscosity of 4.2 cp at 177°C; and 0.6 g of Irganox 1010 are placed in a 500 ml round bottom resin pot fitted with a mechanical stirrer and nitrogen purge line.

The system is purged with nitrogen and heated to 200°C using a metal bath. The polymer blend is stirred for one hr after melting to insure that the blend is homogenous. The molten adhesive is applied as a thin film to aluminum foil and bonded to another aluminum foil by applying pressure and heat to the foils. This adhesive bonded the aluminum foils which on cooling formed an excellent bond between the foils.

A similar formulation, but omitting the Bakelite (ethylene/ethyl acrylate copolymer) produced an adhesive which did not form a satisfactory bond.

Binding of Fibrous or Porous Materials

M. Abe, T. Murata, M. Furuichi, N. Shiraishi, S. Wada, S. Tai and T. Maeda; U.S. Patent 4,081,414; March 28, 1978; assigned to Japan Synthetic Rubber Co., Ltd., Japan have developed a hot melt adhesive for bonding fibrous or porous materials, such as woven and nonwoven fabrics, carpets and rugs, paper and boards, plastic and synthetic rubber foams, and wood and wood products.

The hot melt adhesive composition has a melt viscosity of 50 to 50,000 poises within a temperature of 100° to 150°C, and is prepared by mixing 100 parts by weight of a polybutadiene having a 1,2-bond structure content of not less than 70% and a crystallinity of not less than 10% with from 50 to 1,000 parts by weight of an inorganic filler, from 10 to 300 parts by weight of a tackifier, a plasticizer, and/or a flow improving agent.

The 1,2-polybutadiene must also have an intrinsic viscosity (η) of from 0.7 to 3.0, preferably 1.0 to 2.0 as measured in toluene at 30°C. Such a 1,2-polybutadiene can be made as described in U.S. Patents 3,498,963 or 3,522,332.

The inorganic filler can be calcium carbonate, aluminum hydroxide, talc, or clay, having a particle size of from 0.01 to 100 μ.

The tackifier is added to the composition in order to increase the tackiness of the adhesive and provide an excellent bond strength to the surface of the product. Examples of useful tackifiers are rosin and rosin derivatives, pinene polymers, dipentene polymers, coumarone-indene resin, petroleum resin, and atactic polypropylene.

The plasticizer is used to confer increased processability, flexibility, and low temperature resistance on the hot melt adhesives. Usually it is a liquid organic substance of low volatility, for example, an ester of a phthalic acid derivative, naphthenic oil, or aromatic oil.

The flow improving agent provides added processability, increased surface hardness, and resistance to blocking and stringiness. Among useful fluidizers are paraffin wax, microcrystalline wax, and low molecular weight polyethylene.

Examples 1 through 3: To 100 parts by weight of a 1,2-polybutadiene having a 1,2-bond structure content of 90%, crystallinity of 17%, and intrinsic viscosity of 1.19 were added varying parts by weight as shown in the table of an inorganic filler (calcium carbonate), tackifiers (petroleum resin Toho Hi-Resin No. 60 and atactic polypropylene), plasticizer (naphthenic oil), and a flow improving agent (145°F paraffin wax).

The compositions were kneaded by mixing roll at from 70° to 80°C for 15 min and then pressed at 130°C for 3 min to form 200 μ thick sheets. Each sheet was sandwiched between two sheets of 500 μ thick fabric of natural fiber (cotton), and the sandwich combination was preheated to 110°C on a hot press and was hot melt bonded with a pressure of 0.5 kg/cm^2 for 1 sec. The bond strength and bleeding of the adhesive from each test piece were determined, and the results were compiled in the table. It will be seen that the 1,2-polybutadiene based compositions produce excellent bonds with no danger of bleeding.

	. . . . . . .Example		
Ingredients	1	2	3
Calcium carbonate, pbw	100	200	300
Petroleum resin, pbw	30	40	20

(continued)

| | . . . . . . .Example . . . . | | |
	1	2	3
Paraffin wax, pbw	20	20	40
Naphthenic oil, pbw	20	10	10
Atactic polypropylene, pbw	6	8	6
1,2-Polybutadiene, pbw	100	100	100
Properties			
Apparent melt viscosity, poise	5,300	14,000	25,000
Peeling strength, g/cm	2,150	810	720
Bleeding	no	no	no

Tackifier of Petroleum Resin

Tackifiers for use in hot melt polyolefin adhesive formulations have normally been naturally occurring resins such as rosin or terpenes which are objectionably costly. *S. Matubara and S. Iwai; U.S. Patent 4,081,415; March 28, 1978; assigned to Nippon Oil Co., Ltd., Japan* have developed an improved hot melt adhesive composition which includes a tackifier made of a petroleum resin of a character which is comparable, or even superior, to naturally occurring resins.

The hot melt composition essentially comprises an ethylene copolymer, a tackifier and a wax, the tackifier being a thermally cracked petroleum fraction which has a boiling point in the range of 140° to 220°C and in which the content of conjugated diolefin is 0.7 wt % or less, the ratio of conjugated diolefin content to the polymerizable component is 3% or less, the total content of indene and its alkyl derivatives is 2 wt % or less, and the rate of indene content to the polymerizable component is 8% or less, and the whole being subjected to polymerization at −30° to +60°C in the presence of a Friedel-Crafts catalyst added in amounts of 0.01 to 5 wt % based on the starting fraction. The resulting reaction product is characterized by being an aromatic hydrocarbon resin.

The ethylene copolymer is a copolymer of ethylene and monocarboxylic acid vinylester and/or acrylic acid ester. The content of this copolymer in the ultimate adhesive composition is preferably 15 to 45 wt %. Its melt index is in the range of 1 to 500 as measured by the ASTM Procedure D-1238.

The method of preparing the tackifying resin employed is fully described in U.S. Patent No. 3,778,421.

The class of waxes which may be blended to suit a particular application includes paraffin wax, microcrystalline wax, naturally occurring waxes, polyethylene wax, polypropylene wax, and atactic polypropylene wax.

The tackifying resin which forms an important part of the adhesive composition may be blended with ethylene copolymers and any of the above-listed waxes over a relatively wide range of ratios. When the adhesive is to be applied to substrates, such as paper, plastic film, aluminum foil and other packaging sheets, the ratio of ethylene copolymer blends should be rather conservative. It may be applied by roll coating, gravure coating, dip coating or other types of coating (including impregnation).

For these applications, the tackifying resin and the waxes may be blended in amounts of 20 to 300 and 10 to 400 parts by weight, respectively, per 100 parts by weight of the ethylene copolymers, with satisfactory results. For use in woodcraft, bookbinding or other applications where relatively high adhesiveness is required, the ratio of ethylene copolymers should be increased. For example, for 100 weight parts of ethylene copolymers, there may be 20 to 200 weight parts of tackifying resin and 10 to 100 weight parts of waxes.

Graft Copolymer of Polyethylene

The objective of *M. Shida, J. Machonis, Jr., S. Schmukler and R.J. Zeitlin; U.S. Patents 4,087,588; May 2, 1978; 4,087,587; May 2, 1978; both assigned to Chemplex Company* is to provide greatly enhanced adhesion to substrates such as polar polymers, metals, paper, glass, wood, etc. through the use of blends of copolymers of ethylene and ethylenically unsaturated esters with graft copolymers of high density polyethylene and its copolymers. These resins can be applied in any conventional manner and typical application processes are lamination, extrusion coating, coextrusion, powder coating, blow molding, etc.

It is well known that laminates of polyethylene with dissimilar substrates have many desirable characteristics. However, it is often difficult to bond polyethylenes to dissimilar substrates because of the differences in physical and chemical structures.

The blends of graft copolymer and ethylene copolymers of this formulation have improvements over previous systems including: eliminating the need for additional adhesive layers when bonding unmodified polyethylenes to dissimilar substrates; economic advantages due to eliminating the need to use costly highly polar copolymers of ethylenes; excellent bond strength; and moisture and chemical insensitivity of the adhesive bond between the blends of this formulation and various substrates.

By grafting suitable fused ring, unsaturated carboxylic acid anhydrides to high density polyethylene and ethylene copolymers and blending the resultant graft copolymers with an ethylene-ester copolymer, composites have been obtained with excellent adhesive strength to various substrates including polar polymers, metals, glass, paper, wood and the like. Furthermore, the adhesive bond formed is not easily affected by moisture or chemicals. Surprisingly, the adhesive strength of the blends is synergistically better than that of either component when tested alone. This occurs despite the fact that the concentration of fused ring carboxylic acid anhydride in the blends is reduced by dilution with the ungrafted resin component.

When ethylene-ester copolymers are used as the blending resin, they show dramatically increased adhesive strength compared to other resins. Ethylene-vinyl acetate copolymers are very effective as blending resins. Also, there is a preferred range of ester group content of the blending resin for maximum adhesion.

By ethylene-ester copolymers are meant copolymers of ethylene with ethylenically unsaturated monomers which contain an ester grouping. The major classes of these monomers are the vinyl esters, acrylate esters and methacrylate esters. Such esters have the general formulae:

$$CH_2{=}CH{-}O{-}\underset{\underset{O}{\|}}{C}{-}R, \qquad R'{-}\underset{\underset{CH_2}{\|}}{C}{-}\underset{\underset{O}{\|}}{C}{-}O{-}R''$$

wherein R, R' and R'' are organic radicals or hydrogen.

The term "high density polyethylene" used herein for the grafting backbone includes polymers of ethylene and copolymers with butene and other unsaturated aliphatic hydrocarbons. These high density polyethylenes and copolymers are usually prepared using transitional metal catalysts and often are referred to as low or medium pressure polyethylenes as opposed to low density polyethylenes which often involve high pressure and free radical initiators. Preferably such high density polyethylenes have a density of about 0.930 to 0.970. Also, it is preferable sometimes to graft to blends of two or more of the above homopolymers and copolymers.

Preferred ethylene-ester copolymers are ethylene-vinyl acetate copolymers.

The unsaturated, fused ring carboxylic acid anhydrides and mixtures of these as graft monomers include compounds which contain one or more acyclic, polycyclic, carbocyclic and/or heterocyclic moieties not including the anhydride ring.

Fused ring is defined in the *International Encyclopedia of Chemical Science*, D. Van Nostrand Co., Inc., Princeton, New Jersey, 1964, as "a structural element in the formula of a chemical compound consisting of two rings that are joined by having two atoms in common."

The rings may be simple, bridged, carbocyclic, heterocyclic, polycyclic or complex.

The method of making the graft copolymer of these blends in general consists of heating a mixture of the polymer or polymers and the monomer to be grafted in a solvent or above the melting point of the polyethylene with or without an initiator. Thus, the grafting occurs in the presence of air, hydroperoxides, other free radical catalysts or, in the essential absence of these materials, where the mixture is maintained at elevated temperatures and preferably under high shear.

The resulting graft copolymers used in the blends of this invention are found to consist of about 70 to 99.999 wt % of polyethylene and about 30 to 0.001 wt % of the unsaturated fused ring carboxylic acid anhydride or mixtures and these graft copolymers are capable of blending with a wide variety of ethylene-ester resins to produce the adhesive compositions of this process. The preferred range for the anhydride in the graft copolymer is about 0.001 to 5 wt %.

Excellent monomers in the graft copolymers of this process include 4-methyl-cyclohex-4-ene-1,2-dicarboxylic acid anhydride, tetrahydrophthalic acid anhydride, x-methylnorborn-5-ene-2,3-dicarboxylic acid anhydride, norborn-5-ene-2,3-dicarboxylic acid anhydride, maleo-pimaric acid and bicyclo[2.2.2]oct-5-ene-2,3-dicarboxylic acid anhydride.

The graft monomer used in the examples can be represented by the following structure:

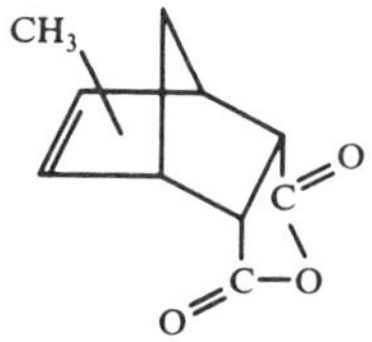

Bridged, carbocyclic mixture of isomers
x-Methylbicyclo[2.2.1] hept-5-ene-2,3-dicarboxylic
acid anhydride (XMNA)

Example 1: XMNA is reacted with high density polyethylene homopolymer resin in a twin-screw extruder to give a graft copolymer resin with 1.0 wt % XMNA incorporation and a melt index of 0.8 g/10 min. The graft copolymer is blended in varying amounts with an ethylene-vinyl acetate copolymer resin of a melt index of 1.0 g/10 min, a density of 0.929 g/cc and vinyl acetate content of 4.7 wt %. The blends as well as the graft copolymer resin itself and the ethylene-vinyl acetate copolymer resin itself are tested for adhesion to nylon 6 film. The results obtained are summarized below.

Graft Copolymer in Blend (wt %)	Adhesion to Nylon 6 Number of Weights ($1/16''$ strip)
0	<1
1	8
5	>11
40	>11
50	8
75	3
90	2
100	<1

As shown by the table, the adhesion of the blends of graft copolymer with ethylene-ester copolymer is better than that of either component when tested alone.

Example 2: The graft copolymer resin described in Example 1 is blended at the 3 wt % level with a polyethylene homopolymer resin of a melt index of 6.6 g/10 min and a density of 0.917 g/cc. The same graft copolymer resin is also blended at the 3 wt % level with an ethylene-vinyl acetate copolymer resin of a melt index of 1.0 g/10 min, a density of 0.929 g/cc and vinyl acetate content of 4.7 wt %. The results obtained are summarized below.

Blending Resin Type	Adhesion to Nylon 6 Number of Weights ($^1/_{16}$" strip)
Polyethylene homopolymer	2
Ethylene-vinyl acetate copolymer	>11

As shown by the above data, when an ethylene-ester copolymer is used as the blending resin, better adhesive strength is obtained as compared to a similar blend made with a polyethylene homopolymer as the blending resin.

Example 3: The graft copolymer described in Example 1 is blended at the 3 wt % level with an ethylene-methyl acrylate copolymer resin of a melt index of approximately 2 g/10 min, a density of approximately 0.94 g/cc and methyl acrylate content of approximately 15 wt %. The adhesion to nylon 6 film of this blend is compared to that of the polyethylene homopolymer blend, described in Example 2, in the table below.

Blending Resin Type	Adhesion to Nylon 6 Number of Weights ($^1/_{16}$" strip)
Polyethylene homopolymer	2
Ethylene-methyl acrylate copolymer	7

As shown above, ethylene-methyl acrylate copolymers may be used as the blending resin with improved results over a polyethylene homopolymer.

With High Resistance to Prolonged Immersion in Water

There is the need for a hot melt adhesive for attachment of plastic, particularly polyethylene, films to the inner surface of metal pipes for the prevention of corrosion. *J. Hobes and W. Payer; U.S. Patent 4,129,472; December 12, 1978; assigned to Ruhrchemie AG, Germany* have found that such an adhesive can be made of an olefin copolymer comprising 70 to 95% by wt ethylene, 0.5 to 10% by wt of an amide of an unsaturated carboxylic acid, and 0.5 to 15% by wt of an unsaturated carboxylic acid. Preferably, the fusion adhesive additionally contains 0.5 to 15% by wt of an ester of an unsaturated carboxylic acid with a saturated alcohol or a vinyl ester of a saturated carboxylic acid.

Preferred amides for use in the formulation are acrylamide, methacrylamide and the amide of crotonic acid. The unsaturated carboxylic acid may be acrylic, methacrylic, ethacrylic, itaconic, fumaric or maleic acid. The particularly preferred esters for use in the adhesive are those of acrylic or methacrylic acid with tert-butyl or tert-amyl alcohol. Vinyl acetate may also be used.

Manufacture of the copolymers may be carried out by common polymerization of the comonomers under high pressure in autoclaves or tube reactors. Working is usually at pressures of 1,500 to 2,500 bars and temperatures of 220° to 290°C. The polymerization is carried out in the presence of catalytic quantities of free radical initiators.

The copolymers can be manufactured with consistent product quality in this way. They are preferably used as fusion adhesives for sticking plastics and/or other metals (e.g., polyurethane, aluminum and steel) onto metals.

The copolymers, which occur in varying grain size or shape of granule, were pressed on a laboratory press (Wickert steam press) into films of approximately 0.2 mm thickness. In addition, approximately 10 g of the product was distributed uniformly between two glass fiber strengthened Teflon covers which are placed between 4 mm thick steel plates. Under optimal processing conditions of 20 kp/cm^2 for 2 min at 180°C films which were well plasticized throughout were obtained. Pieces of approximately 70 x 100 mm^2 in size were cut from these pressed films and were pressed between two aluminum or steel plates, also under the above conditions, to produce respective composite systems.

The plates (100 x 100 mm^2) were purified with CCl_4 before pressing and bent at right angles, so that a pure pressing surface of 70 x 100 mm^2 remained. The thickness of the plates was approximately 0.1 to 0.2 mm. Resistance to scaling, measured on aluminum plates coated with polyethylene, is the force per tension length required to peel off a specified strip of coating. A distinction is made between initial and further cracking force. The latter is of primary importance regarding system adhesive strength.

Determination of resistance to scaling was effected with a Zwick tension testing machine (vertical type of construction), in whose clamping jaws the surfaces (30 x 100 mm^2) of the composite system, which had been bent through 90°, were held. The actual bonding surface (70 x 100 mm^2) was perpendicular to the clamping jaws. The tension speed was 50 mm/min. The stripping or scaling forces occurring were registered by means of a recording instrument.

In order to estimate conclusively the bonding behavior, the products were subjected to an ageing test and were stored for 72 hr in water at 60°C and then tested as described above.

Example: A reaction mixture composed of 96.25% by wt ethylene, 1.5% by wt methacrylic acid, 1.5% by wt tert-butyl acrylate, 0.75% by wt acrylic amide was compressed at 2,000 bars. Polymerization is initiated with 50 ppm tert-butyl peroctoate in the form of a benzene solution. The reaction temperature is 230°C; 8.7% methacrylic acid, 8.0% tert-butyl acrylate and 2.7% acrylic amide were incorporated. The fusion index of the reaction mixture was 11.2 g/10 min.

The adhesion properties measured on the composite aluminum/adhesive/aluminum system were as follows.

Before ageing		After ageing	
Initial cracking	9.5–10.2 kp/cm	Initial cracking	8.1–13.7 kp/cm
Further cracking	1.4–3.5 kp/cm	Further cracking	2.0–3.8 kp/cm

In a comparative example, where the copolymer was composed of 98.88% by wt ethylene and 1.12% acrylic acid, polymerization was initiated with 260 ml/hr of a 0.7% isobutyl peroctoate solution in benzene (83 ppm). The reaction temperature was 225°C. After a time of approximately 52 sec of direct contact of the reaction mixture in the reactor, the polymer was removed. It contained 95% by wt ethylene and 5% acrylic acid in polymer bound form. The fusing index (MFI 190/2) was 5.3 g/10 min. The bonding properties of the product were determined by use of the aluminum/adhesive/aluminum system in accordance with the process described above.

The scaling resistance values were:

	Kiloponds per Centimeter
Before ageing	
Initial cracking	3.4–5.0
Further cracking	0.9–1.1
After ageing	
Initial cracking	0.5–1.5
Further cracking	0.1–0.3

Even worse results were obtained for corresponding ethylene-acrylester and ethylene-amide copolymers whose adhesion values were already 0.1 to 0.3 before the ageing test.

Having Improved High Temperature Shear Properties

Hot-melt adhesives are useful for bonding various substrates together, such as wood, paper, plastics, and textiles, as well as other materials. One use for which they are well suited is the fabrication of corrugated paperboard, and for cardboard case sealing and closing. Hot-melt adhesives useful for producing corrugated paperboard must have high bond strength under conditions of high elevated temperature shear and are particularly useful where extremes of temperature are encountered in transportation and storage, particularly in warehouses.

W.A. Ames; U.S. Patent 4,159,287; June 26, 1979; assigned to Eastman Kodak Company has found that such adhesive formulations can be made by grafting amorphous polyolefins (such as Eastman's Eastobond M-5 polyolefins) with unsaturated alkyl carboxylic acid components in the presence of free radical initiators to provide modified amorphous polyolefins. These modified amorphous polyolefins are partially neutralized with a tetraalkyl ammonium hydroxide to form a partial salt of the modified amorphous polyolefin. The partially neutralized modified amorphous polyolefin is then reacted with pivalolactone in an amount of from about 5 to 40% by wt to provide a grafted amorphous polyolefin based hot melt adhesive.

The grafting of the unsaturated alkyl carboxylic acid component to the amorphous polyolefin is carried out by grafting the unsaturated acid component to the amorphous polymers in the presence of a free radical initiator. The preferred initiators are the alkyl and acyl peroxides, such as di-tert-butyl peroxide,

2,5-dimethyl-2,5-di(tert-butylperoxy)hexane, benzoyl peroxide, lauroyl peroxide, and the like, which are efficient in the abstraction of hydrogen from the polymer chain.

The amorphous polyolefin can conveniently be polypropylene with a saponification number of 10 to 25, a melt viscosity at 190°C of 1,500 to 15,000 cp, a Ring and Ball softening point of 95 to 105, and a saponification number of 10 to 25. The unsaturated alkyl carboxylic acid can be crotonic acid; and the tetraalkyl ammonium hydroxide can be tetrabutyl ammonium hydroxide.

Example: Grafting Crotonic Acid onto Amorphous Polypropylene – A 2 liter resin flask fitted with an anchor stirrer, metal thermometer and N_2 bubbler was charged with 1,087 g of amorphous polypropylene having a melt viscosity of about 1,700 cp at 190°C and a Ring and Ball softening point of about 100°C. The polymer was melted and heated with stirring under N_2 to 180°C, 100 g of crotonic acid was added in one charge. The temperature decreased to 160°C. 35 g of di-tert-butyl peroxide was added dropwise over 3 hr with rapid stirring. The polymer was stripped with N_2 for 3 hr to remove unreacted crotonic acid. It had a melt viscosity at 190°C of 663 cp and an acid number of 11.7.

Grafting Pivalolactone onto Crotonic Acid-Modified Amorphous Polypropylene – 150 g of 11.7 acid number crotonic acid-grafted amorphous polypropylene was dissolved in 1,200 ml of benzene in a 3 liter flask fitted with stirrer, thermowell, condenser, N_2 bubbler, and a Dean-Stark trap.

18 ml of 0.87 N-tetrabutylammonium hydroxide in methanol was added at 60°C to the reaction vessel to neutralize 50% of the acid groups. The temperature was increased to reflux to azeotrope water and remove methanol. The contents of the flask were cooled to 60°C. 400 ml of tetrahydrofuran was added to solvate the tetrabutylammonium carboxylate groups. 37.4 g of pivalolactone monomer was added.

The contents of the flask were heated to reflux. The viscosity of the reaction mixture increased significantly as the polymerization proceeded. The reaction mixture was cooled and 5 ml of concentrated HCl in 50 ml tetrahydrofuran was added to the reaction vessel to convert the acid salts to carboxylic acid groups. The polymer was precipitated by adding the reaction mixture to excess methanol. The product was filtered, washed three times with methanol in a blender, stabilized with 0.5% Irganox 1010, and dried overnight in a vacuum oven.

The graft copolymer was obtained in 99% yield. It had a DSC melting point of 181°C, a Ring and Ball softening point of 181°C, and a melt viscosity at 190°C of 1,100 cp.

The compositions are useful hot melt adhesives for packaging and product assembly, especially where high elevated temperature shear is desirable. They would also be useful as palletizing adhesives, particularly for use in hot warehouses.

POLYESTER

Containing Spheroidal Metal Powder

In the manufacture and repair of metal bodies, such as automobiles and appliances, solder compositions containing lead are frequently used to fill cavities and voids. These lead solders are a health hazard which mandates special handling to protect workers and are also extremely dense. Conventional hot melt adhesives are not satisfactory for such cavity and void filling applications because they cannot be sanded rapidly at assembly line speed. They do not readily accept paint because they bleed through, and they do not withstand the curing temperatures for the paint.

A need therefore exists for a cavity or void forming composition which is less dense and toxic than lead solder, forms a strong bond to metal substrate, withstands extremes of humidity and temperature, is readily applied, rapidly sets to a sandable state, is easily sanded smooth, and accepts paint without bleeding through.

To fill these needs, *J.G. Martins and D.D. Donermeyer; U.S. Patents 4,097,445; June 27, 1978; and 4,073,973; February 14, 1978; both assigned to Monsanto Company* have developed a method for filling cavities with a hot melt adhesive composition. The composition comprises:

(a) From about 70 to 30 parts by weight of a polyester-amide block copolymer melting at about 155° to 225°C, having from about 30 to 70% by wt of crystalline polyester segments derived from at least one aliphatic or alicyclic diol having from 2 to 10 carbon atoms and at least one alicyclic or aromatic dicarboxylic acid having from 8 to 20 carbon atoms, and from about 70 to 30% by wt of amorphous polyamide segments derived from an aliphatic polycarboxylic acid containing at least 40 wt % of a C_{18-54} polycarboxylic acid and an aliphatic or alicyclic primary diamine containing 2 to 10 carbon atoms; and

(b) From about 30 to 70 parts by weight of finely divided spheroidal metal powder selected from the group consisting of aluminum, iron, mild steel, stainless steel and zinc.

The polyester-amide composition of the formulation is prepared by reacting a crystalline polyester with an amorphous polyamide or with the component polycarboxylic acid and diamine by the one step or two step method set forth in U.S. Patent 3,650,999.

The second component of the adhesive composition is a finely divided spheroidal metal powder selected from the group consisting of aluminum, iron, mild steel, stainless steel and zinc. The metal powder is substantially uniformly dispersed in the polyester-amide. It may be of number average particle size of 0.2 to 150 μ and is preferably of number average particle size of 4 to 100 μ. The preferred metal filler is atomized aluminum, particularly when the adhesive composition is used for cavity filling, since it allows the hot melt composition to be readily smoothed and burnished when it is sanded.

In general, platelike, acicular, or multifaceted granular powdered metals are unsatisfactory, surprisingly causing high viscosity in the hot melt and "blinding" or filling and occlusion of sandpaper when the filled composition is sanded.

The amount of metal powder which is dispersed in the polyester-amide is sufficient to improve the high temperature creep resistance without causing unmanageable rheology. It is preferable in about 30 to 70 parts by weight of metal powder dispersed in about 70 to 30 parts by weight of polyester-amide. The melt viscosity of the hot melt composition containing the metal filler is preferably less than 150,000 cp at a temperature of 232°C and a shear rate of 3 to 4 sec^{-1} measured in a Brookfield Thermocel Unit Model HBT.

In addition to improving the creep resistance of the polyester-amide component, the metallic component improves the rate of melting of the adhesive composition, allows the composition to be applied and spread more easily with less pressure, imparts longer "open" time between application of the hot melt and closing of the bond and higher "green" strength or faster onset of bond strength, and reduces the degree of shrinkage of the adhesive composition when it is cooled from the hot melt temperature to ambient temperature.

The cost of the composition is also considerably reduced. When the composition is used to fill cavities, it can be readily sanded, as discussed above, withstands extremes of temperature and humidity, is exceptionally solvent resistant and is readily painted without absorbing solvent, swelling, and blistering.

These hot melt adhesive compositions find widespread utility wherever hot melt adhesives are used. They are especially valuable in those applications where resistance to creep at elevated temperatures is a necessary requirement. They may be used to great advantage to bond a variety of substrates, including metal, glass, synthetic and natural textiles, leathers, synthetic polymeric sheet material, wood, paper, etc.

Example: A block copolymer which is approximately 60% by wt crystalline polyethylene terephthalate segments and 40% by wt amorphous polyamide made from dimer acid and hexamethylenediamine is prepared in two steps. In the first step 157.5 parts (0.272 mol) of a C_{36} dibasic acid and 30.8 parts (0.266 mol) of 1,6-hexanediamine are charged to a reaction vessel and heated with agitation at about 215°C for 1 hr to form a polyamide resin.

During the first 30 min the pressure rises to 264 g/cm^2 after which time the reaction vessel is vented to reduce the pressure to 158 g/cm^2. At the end of 1 hr the pressure is released and 269 parts of a crystalline polyethylene terephthalate (MP = 260°C/inherent viscosity 0.147) and 5.9 parts (0.095 mol) of ethylene glycol are charged to the vessel along with a minor amount of an antioxidant. The vessel is flushed with nitrogen and the mixture is heated to about 280°C while maintaining a nitrogen pressure of 70 g/cm^2. After 0.5 hr the vessel is vented and vacuum applied and the reaction is continued under full vacuum (0.1 to 5 mm of mercury) for 2 hr. At the end of this time the resulting molten polyester-amide is discharged under pressure into a water bath to quench the material.

The polymer obtained melts at 185°C and the inherent viscosity is 0.50. To a stainless steel reactor fitted with an anchor agitator and a jacketed hot oil heating system is added 100 parts by weight of the polyester-amide and heating is begun. When the contents have reached 250°C, agitation is begun at 60 rpm and 100 parts by weight dry aluminum powder (Alcoa Atomized Powder 123) is fed into the mass at a rate of 10 parts by weight per minute. The agitation is continued and the temperature raised to 266°C under a nitrogen blanket. Agitation is continued for 15 min after the second addition is completed, and the molten mass is discharged under slight N$_2$ pressure (70 g/cm^2) quenched in a bath, ground and redried.

This material is used as a hot melt to fill dents and orifices in large metal structures. After application it is cooled to room temperature, sanded smooth with 80 grit tungsten carbide abrasive and painted with an automotive topcoat. No "telegraphing" is observed.

For Polyester Fibers, Films and Other Shaped Articles

Materials, such as polyesters and polyvinyl fluoride, can be extruded to form shaped articles, such as fibers, films, and builtup or structural articles. It is desirable to shape and fasten these shaped articles together and a hot melt adhesive is particularly desirable for this purpose. *M.V. Kulkarni and J. Von Kamp; U.S. Patent 4,097,548; June 27, 1978; assigned to The Goodyear Tire & Rubber Company* describe compositions useful for forming laminates of various polyester materials.

The hot melt adhesive preferably should have a softening point of 215° to 225°C, a Shore D hardness of 50 to 60, and a Brookfield Thermoseal melt viscosity at 245°C of 1,000 to 1,500 poises to permit films of polyester and/or polyvinyl fluoride to be adhered together to form useful laminates.

A composition having the above attributes as a hot melt adhesive is composed on a weight percent basis of about 9 to 30% of a copolymer of ethylene-alkyl acrylate or methacrylate having an alkyl radical of C$_{1-4}$ carbon atoms and containing 18 to 23% of alkylacrylate or methacrylate, about 40 to 45% of a copolyester terephthalate of ethylene and neopentyl glycol, about 50 to 55% of a terpolyester of the structure obtained by condensing ethylene glycol with a mixture of terephthalic acid, isophthalic acid and azelaic acid, or their anhydrides, and an effective amount of an antioxidant.

The copolyester terephthalates are condensation products of terephthalic acid or its anhydride with a mixture of the glycols of 2 to 5 carbon atoms with the amount of ethylene glycol preferably being from 60 to 70%, and the neopentyl glycol being from 40 to 30%, having an intrinsic viscosity at 25°C of 0.55 to 0.66 and a glass transition temperature of 70°C.

The terpolyesters are obtained by condensing ethylene glycol with from 50 to 57 mol % of terephthalic acid or its anhydride, 1 to 7 mol % isophthalic acid or anhydride, and 30 to 36 mol % of azelaic acid or its anhydride, to give a composition having an intrinsic viscosity at 25°C of 0.66 to 0.76 and a glass transition temperature of about −10°C.

The organophosphonates are particularly effective antioxidants in these compositions. The organo radical can be alkyl, aryl or cycloalkyl radical, preferably of 6 to 20 carbon atoms, such as octyl, phenyl, nonyl and cyclohexyl.

Example: A blend containing 9.9% ethylene/ethyl acrylate (23% ethyl acrylate and a melt index of 20) and 90.1% of the mixture of the following composition was prepared, in which percentages are by weight. The blend had a viscosity of 145,000 cp at 475°F. It yielded satisfactory bonds to the polyester (Mylar) and polyvinyl fluoride (Tedlar).

57/7/36	Ethylene-terephthalate/isophthalate/azelate	54.94
70/30	Ethyl/neopentyl-terephthalate	43.96
	Phosphonate, antioxidant	1.10

Thermoplastic Copolyesters Made Under Mild Conditions

The object of the adhesive formulation developed by *P.C. Georgoudis; U.S. Patent 4,124,571; November 7, 1978; assigned to National Starch & Chemical Corporation* was the production of thermoplastic copolyesters exhibiting good thermal stability, high heat and cold resistance, good wetting characteristics, low viscosity, good cohesive strength and a high compatibility with a large variety of tackifying resins, plasticizers, solvents, modifiers, pigments, etc., and one which could further be cured through functional groups (e.g., hydroxyl or carboxyl groups) to improve its strength, heat and solvent resistance.

Another object was to provide a method for manufacturing such thermoplastic copolyesters wherein mild conditions involving temperatures no higher than 265°C, pressures of 1 to 30 mm Hg and relatively short polycondensation times of 2 to 7 hr could be employed.

It was found that the above objects could be provided by a copolyester of at least one aromatic dicarboxylic acid member, at least one aliphatic dicarboxylic acid member and at least one C_{2-10} glycol, wherein the total dicarboxylic acid members are composed of 20 to 80 mol % in the aromatic dicarboxylic acid member and correspondingly 80 to 20% of the aliphatic dicarboxylic acid member. It has further been found that copolyesters characterized by superior performance are prepared when at least one nonpolymeric polyfunctional member is added to the base copolyester composition to function as a chain extender. Examples of the last group are glycerol and trimethylolethane.

These copolyester compositions are prepared by combining the aromatic dicar-boxylic acid member, the aliphatic dicarboxylic acid member and the glycol in a two or three stage process with the polycondensation reaction occurring at about 225° to 265°C for 2 to 7 hr at a pressure of about 1 to 30 mm Hg as will be described below.

The resulting thermoplastic copolyesters are characterized, in varying degrees depending upon the specific optional components present, by a low glass transi-tion temperature (preferably less than 0°C), a relatively high melting point, superior heat stability, as well as unusual compatibility with additives conven-tionally used in the production of adhesive formulations.

These thermoplastic copolyesters are useful by themselves as hot melt or solu-tion adhesives or may be compounded with various additives to lend themselves either to pressure sensitive adhesive formulations useful for nonvinyl wall coverings, labels, decals and the like, or to nonpressure sensitive formulations useful as structural adhesives and for packaging, edge bonding and laminating paper, cloth, wood and other nonvinyl substrates.

In the examples, the I.V. (Intrinsic Viscosity) was determined using 0.5, 1.0 and 1.5% solution in 60/40 phenol/1,1,2,2-tetrachloroethane solvent and extrap-olating to 0 concentration according to known methods. The values for Tg and Tm were determined on a DuPont Model 900 Differential Thermal Analyzer.

Example 1: This example illustrates the preparation of chain extended thermo-plastic copolyesters.

291.0 parts dimethyl terephthalate, 360 parts 1,4-butanediol, 0.51 part di-butylin oxide, 0.20 part antimony trioxide and 0.41 part 4,4'-thiobis(6,tert-butyl-m-cresol) were charged into a 2 liter flask and heated with stirring to 136° to 230°C for 2.5 hr and methanol was distilled off. 388.0 parts iso-octadecenyl succinic anhydride, 86.0 parts xylene and 5.64 parts trimethylolpropane were added to the reactor and heated with stirring for 3 hr under reflux at 188° to 211°C, water being distilled off, until the acid number was 0.50.

Thereafter the copolyester prepolymer was polycondensed at 188° to 265°C and 1.5 to 2.0 mm Hg until a highly viscous product was obtained and discharged into silicon release paper. This product, which was also crystalline, had the following properties. I.V. equals 0.75 dl/g, Tg equals −33°C, and Tm equals +90°C and +128°C.

Example 2: This example illustrates the utility of these copolyesters as hot melt adhesives for laminating wood.

50 parts copolyester from Example 1 and 50 parts Picco-L 60 (Hercules Chemi-cal Co.) were blended at 300°F until a uniform solution was obtained. A small amount of the molten adhesive composition was applied at about 300°F on a 1″ x 3″ x 1/8″ yellow birch splint with a 1″ square measured off at one end. The

composition was drawn down from the 1" line to the end of the splint. A second splint was quickly placed on top of the first one making a bond of 1 in². A 500 g wt was then placed over the laminate and hard manual pressure applied against the weight (and hence against the laminate) for 5 seconds. When set, excessive adhesive was trimmed off, and the laminate was aged for 24 hr at room temperature. The shear strength was measured on an Instron Tester having a crosshead speed of 12"/min (this being the speed at which the adhesive bond is pulled apart), a chart speed of 12"/min and a range of 0 to 1,000 psi. The shear strength was determined to 740 psi.

Amorphous Polyolefin Added for Lower Melt Viscosities

Many polyesters have useful properties as hot melt adhesives for the bonding of various substrates including metals, fabrics, and plastics. However, in order to obtain high bond strength on these substrates, it is necessary to use high molecular weight polyester materials. For example, the I.V. of polyesters used is normally about 0.5 up to about 1.4. These high molecular weight polyesters, therefore, have very high melt viscosities at the temperatures at which they must be applied to the substrates. These high viscosities are deleterious in that they make the adhesive difficult to apply and, in many cases, the polymer melt does not wet out the substrate sufficiently.

It was found by *R.L. McConnell, J.R. Trotter, and F.B. Jayner; U.S. Patent 4,146,586; March 27, 1979; assigned to Eastman Kodak Company* that a blend comprising at least one substantially amorphous olefin polymer such as amorphous polypropylene, amorphous poly-1-butene, poly(propylene-co-higher 1-olefin), and a polyester provides a hot melt adhesive which has good wet out on a wide variety of substrates, provides good bond strength, and maintains a good bond strength over a wide variety of end use conditions. The higher 1-olefin may be a C_{6-10} 1-olefin.

They provide a hot melt adhesive composition comprising a blend of an amorphous polyolefin and a partially crystalline polyester which has a heat of fusion of about 1 to 12 calories per gram, a melting point of about 80°C to 225°C, and an inherent viscosity of at least 0.5, and preferably at least 0.6, as measured at 25°C, using 0.50 g of polyester per 100 ml of a solvent consisting of 60% by volume phenol and 40% by volume tetrachloroethane. The amorphous polyolefin may be polypropylene, poly-1-butene, poly-1-pentene, or a propylene/ 1-butene copolymer.

Example: About 42.75 g of copolyester prepared from 75 mol % dimethyl terephthalate, 25 mol % adipic acid, 75 mol % ethylene glycol and 25 mol % 1,4-butanediol (0.78 I.V., Tm equals 140°C) and about 2.25 g of a substantially amorphous polypropylene (1,500 cp at 190°C; Ring and Ball softening point 105°C) are physically blended under a nitrogen blanket in a Brabender Plastograph (sigma blade mixer) at 190°C for 5 min. The blend is stabilized against thermal degradation by the addition of 0.2 wt % Eastman Inhibitor DOPC [2,6-bis(1-methylheptadecyl)p-cresol].

The resulting blend is removed from the Brabender Plastograph, allowed to cool to 23°C, then cut into small convenient pieces, and dried in a vacuum oven at 50°C. At 23°C the blend is nontacky to the touch. The melt viscosity of the blend is 45,200 cp at 220°C and 1,000⁻¹ (Sieglaff-McKelvey rheometer). Melt pressed films of the blend (0.005±0.0005" thick) are used to bond 65/35 Kodel polyester/cotton fabric (twill weave) using a Sentinel heat sealer at 425°F (218°C) and 28 psi for 2 sec. The bonded specimens have an initial T-peel strength of 19.0 lb/in width (measured at 23°C) and 2"/min peel rate 24 hr after bond formation. Similarly good results are achieved using amorphous polypropylene having a melt viscosity of 3,000 cp at 190°C instead of 1,500 cp at 190°C.

The melt viscosity of the unmodified copolyester is 90,500 cp at 220°C and 1,000 sec⁻¹. Fabric bonds that were made with films (0.005±0.0005" thick) of unmodified copolyester have an initial T-peel strength of about 22 lb/in width.

POLYAMIDE

For Garment Interlinings

Polyamides are used as hot melt adhesives for textiles to fuse interlinings to face fabrics to impart stiffening to the latter. Generally, the interlining manufacturer sells interlinings, already coated with a uniform distribution of adhesive spots, to garment manufacturers, who cut pieces of the required shape and size and adhere them to the face fabric in a heated press.

The principal properties required of the adhesives are:

A softening point less than 150°C, preferably 100° to 130°C;

Flexibility;

A viscosity sufficiently low when heated and pressed to enable the adhesive to flow without penetrating the face fabric, normally 1,000 to 20,000 poises at 150°C;

Resistance to dry cleaning fluids, e.g., perchloroethylene or trichloroethylene;

Resistance to warm water;

The capability of being produced in or ground to a free-flowing fine powder.

A polymeric amide hot melt adhesive formulation fulfilling the above requirements, as developed by *C.L. Murray and C.P. Vale; U.S. Patent 4,118,351; October 3, 1978; assigned to British Industrial Plastics, Ltd., England* comprises:

From 30 to 60% by wt of units derived from ε-caprolactam or ε-aminocaproic acid;

From 25 to 55% by wt of units derived from equimolar quantities of hexamethylenediamine and isophthalic acid and;

From 5 to 35% by wt of units derived from equimolar quantities of hexamethylenediamine and adipic acid.

Examples 1 to 9: The compositions shown in the table were produced by charging the raw materials to a polymerization vessel provided with a nitrogen blanket and heating the charge stepwise as follows. The water of condensation evolved was collected during Steps 2 to 4:

(1) Temperature raised from 25° to 160°C over 1 hr;

(2) Temperature raised from 160° to 180°C over 1 hr;

(3) Temperature raised from 180° to 200°C over 1 hr;

(4) Temperature quickly raised to 220° to 240°C and held there for 1 hr.

(5) Temperature quickly raised to 260° to 270°C and held there for 3 to 4 hr.

All the steps were effected at atmospheric pressure, but it may be desirable to carry out the final step at reduced pressure. The resultant terpolymers, which were clear and transparent, were allowed to cool and were then ground to a fine powder passing a 60 mesh sieve. These were found particularly effective as hot melt adhesives applied by a standard perforated screen method.

In the latter method the finely powdered polymer is spread onto a substrate to be bonded, e.g., a textile interlining for suiting, and fused to the substrate by heating in an oven, e.g., at 150°C. The substrate to be bonded is then pressed to the material to which it is to be bonded, e.g., a textile suiting, in a heat transfer press and heated to a rather high temperature, e.g., 170°C, to complete the bonding process. In the following table the melt flow index is measured in g/10 min at 190°C under 2.16 kg load according to ISO recommendation No. R.1133 (ISO/R1133-1969E).

Example Number	Raw Materials Composition (%)			ACA Percent on CL	Viscosity Stabilizer, Percent*	Melting Point (°C)	Melt Flow Index
	CL	IPH	AH				
1	45	35	20	3.8	stearic acid, 2.0	100–115	22.2
2	45	35	20	3.8	stearic acid, 2.3	95–110	27.8
3	45	35	20	3.8	stearic acid, 2.5	100–115	48.6
4	45	35	20	0.0	stearic acid, 2.3	105–112	32.0
5	45	35	20	3.8	benzoic acid, 1.0	95–110	26.4
6	45	35	20	3.8	adipic acid, 1.2	105–115	25.8
7	45	35	20	3.8	sebacic acid, 1.6	105–110	22.4
8	50	30	20	3.8	stearic acid, 2.3	105–115	29.6
9	45	35	20	3.8	—	125–140	1.1

NOTE: CL is caprolactam (and aminocaproic acid.) IPH is hexamethylene isophthalamide salt. AH is nylon 66 salt or hexamethylene adipamide. ACA is aminocaproic acid.

*On total of other ingredients.

Polyoxamide from Polyoxypropylene Polyamine

The utilization of the polyamide reaction products of certain aliphatic polyamines and vegetable or animal-based dimer or trimer fatty acids as thermoplastic or hot melt adhesives for the bonding of a multitude of materials, such as
leather, textiles, wood, and the like, is well known.

However, the vegetable or animal-based dimer and trimer fatty acid materials
usually employed in the preparation of polyamide based thermoplastic adhesives
are in increasingly short supply and are continuously increasing in price. Moreover, many aliphatic polyamine compounds are also used in great quantities as
curing agents for polyepoxide resins, thus making them more unobtainable.

*H. Schulze; U.S. Patent 4,119,615; October 10, 1978; assigned to Texaco Devel-
opment Corporation* has succeeded in formulating a thermoplastic adhesive
which is made from a resinous polyoxamide formed by reacting a polyoxypropylene polyamine selected from the group consisting of diamines, triamines
and mixtures thereof, and having an average molecular weight of about 190 to
3,000 with oxalic acid to form a liquid prepolymer, said prepolymer being
further reacted with a diamine having the formula NH_2-R-NH_2, where R is a
divalent hydrocarbon radical containing 2 to 18 carbon atoms, to form the
resinous polyoxamide. These compounds usually have a melting point less than
$250°C$.

Preferably, there are used polyoxypropylenediamines of the formula:

$$NH_2 +CH-CH_2O +_y CH_2-CH-NH_2$$
$$CH_3 CH_3$$

wherein y is a number from about 2 to 40, and polyoxypropylenetriamines of
the formula:

$$CH_2 +OCH_2CH(CH_3)]_x NH_2$$
$$CH_3CH_2CCH_2 +OCH_2CH(CH_3)]_y NH_2$$
$$CH_2 +OCH_2CH(CH_3)]_z NH_2$$

where x, y and z represents numbers of about 1 to 15, and the sum of x, y and
z is from 3 to about 40. The preferred polyoxypropylenediamines of the formula have average molecular weights between about 190, where y is an average of
2.0, and about 2,000 where y is an average of about 33.2. Preferred polyoxypropylenetriamines of the above formula have average molecular weights between about 200 and 3,000. These polyoxypropylenedi- and triamines are
readily available commercially in a wide variety of molecular weight ranges, such
as Jeffamine.

Example 1: To a 1 liter 3-necked flask, fitted with a stirrer, thermometer,
distilling head and nitrogen flow system, was added 292 g (2.0 mol) of
diethyl oxalate. The ester was heated up to $105°C$ under nitrogen. Then

421.5 g (1.02 mols) of a polyoxypropylenediamine having an average molecular weight of 400 (Jeffamine D-400) was slowly added over a 1 hr period as ethanol was removed by distillation. When most of the alcohol had been removed, the pot temperature had reached 125°C, and the flask was subjected to high vacuum (0.025 mm) for 1 hr. The residue in the flask was a light yellow mobile oil. The saponification number was 154, corresponding to an equivalent weight of 364. Analysis—Calculated for N: 4.80%. Found: 4.81% N.

The structure of this prepolymer is shown below.

$$EtO-\underset{\substack{\|\\O}}{C}-\underset{\substack{\|\\O}}{C}\underset{\substack{|\\CH_3}}{\left[NH\left[CH-CH_2-O\right]_y CH_2CH\right.}-\underset{\substack{|\\CH_3}}{NH}\underset{\substack{\|\\O}}{C}-\underset{\substack{\|\\O}}{C}\Big]_z OEt$$

where y is 5.5 and z is 1.25.

Example 2: The prepolymer of Example 1 was reacted further with a diamine to produce the adhesive. The procedure was as follows.

To a 500 ml flask, fitted as described in Example 1, was added 109.2 g (0.15 mol) of the diester of Example 1. The prepolymer diester was stirred under nitrogen at 70° to 75°C while 17.4 g (0.15 mol) of hexamethylenediamine was added at once. The reaction was exothermic, and the temperature reached 110°C as the mixture became a viscous semisolid. Further heating melted the advanced prepolymer and ethanol began to distill over.

With the reaction mixture at 210°C most of the condensate (12.0 g) had distilled over. High vacuum (0.2 mm) was applied for 30 min at 230° to 240°C. The thick, yellow polymer was poured into molds where it solidified to a tough, flexible plastic with a Shore A hardness of 94 to 95. The yield was 102 g. The specific viscosity (0.5% in 90% formic acid at 25°C) was 0.27.

E.L. Yeakey and H.G. Waddill; U.S. Patent 4,128,525; December 5, 1978; also assigned to Texaco Development Corporation have solved the problem of a satisfactory replacement of the vegetable or animal-based long-chain dimer or trimer fatty acids previously used to react with certain aliphatic polyamines to provide hot melt adhesives for the bonding of materials, such as leather, textiles, wood, etc., in a slightly different way. They have found that the addition of piperazine to the polyamide-dicarboxylic acid resin greatly improves the adhesive strength of the adhesives.

Their formulation, therefore, is comprised of a resinous polyamide reaction product of a polyoxypropylene polyamine selected from diamines, triamines, or mixtures thereof having an average molecular weight of from about 190 to 3,000, piperazine, and an aliphatic or aromatic dicarboxylic acid, ester or anhydride having from 4 to about 20 carbon atoms per molecule.

The thermoplastic adhesive compound is prepared by mixing and reacting the polyoxypropylene polyamine, piperazine and dicarboxylic acid material in a molar ratio of about 0.25:1.0 to 4.0:1.0 mol total amine:mol acid material, with the piperazine present in an amount of up to about 80 molar percent, based upon the total molar amount of amine present, at a temperature of about 175° to 270°C. The reaction time is preferably within the range of about 1 to 12 hr.

The improved thermoplastic compounds of the process thus produced have broad melting ranges of about 20° to 180°C and can be added to compatible plasticizers to produce adhesives and can also be formulated with other components such as a liquid polyepoxide resin having an epoxide equivalent weight of from about 150 to 600, filler and the like, to produce thermoplastic adhesive formulations having particularly desirable properties for use in tailored systems for bonding a variety of substrates. These thermoplastic adhesive compounds exhibit unexpectedly high adhesive strengths. As the polyoxypropylene polyamines are employed diamines and triamines of the same formulas given in the previous patent.

The preferred aliphatic dibasic acids used have a divalent, saturated, unsubstituted hydrocarbon group, while the preferred aromatic dibasic acids include an unsubstituted phenylene group. Examples of preferred dibasic acids and related materials include adipic, azelaic, sebacic, isophthalic, and terephthalic acids, phthalic and succinic anhydrides, and dimethylterephthalate esters, to name a few.

The resinous polyamide reaction product is prepared by mixing the polyoxypropylene polyamine and piperazine with the dibasic acid at a temperature of between about 175° and 270°C. The polyamines and acid compounds are preferably mixed in total polyamine:acid molar ratios of from about 0.8:1.0 to 1.25:1.0 with substantially equimolar ratios being especially preferred. The admixture is usually heated for several hours, i.e., from about 1 to 12 hr, at maximum temperature to complete the reaction, while byproduct water or alcohol, depending upon the particular compounds employed, is removed. Preferably, the reaction mixture is vacuum stripped by known procedures to develop optimum molecular weight.

The piperazine can be added before, during, or after the above described polyoxypropylene polyamine is added to the above described dibasic acid. It is preferred that all three components of the resin are mixed together and reacted at once. The piperazine is added in an amount preferably of about 30 to 70 molar percent, based upon the total molar amount of amine required for the formulation.

In an especially preferred embodiment, a diglycidyl ether of bisphenol A having an epoxide equivalent weight of about 175 to 190 is added to the above described polyamide reaction product in an amount of about 5 to 25 wt %, based upon total formulation weight, while the resinous polyamide component is heated at about 100° to 130°C. Upon cooling, the resulting compound does not exhibit gelation and has a melting point between about 100° to 180°C.

Examples 1 to 3: Three thermoplastic adhesive formulations were prepared employing the compounds set forth in the following table. The resinous polyamide reaction product compounds employed in the examples of the table were prepared by admixing the described polyamine components with isophthalic acid in the described molar ratios and heating the admixtures at a temperature of 200° to 240°C for 6 to 8 hr with continuous removal of water. The reaction product mixtures were vacuum stripped during the last stages of reaction to develop the optimum molecular weight.

The resinous polyamide reaction product components were then mixed at a temperature of about 120° to 140°C with the designated amounts of polyepoxide resin (diglycidyl ether of bisphenol A, epoxide equivalent weight of 182 to 189), plasticizer and fillers. Glass beads were added to each formulation to insure even coating between substrates. Each formulation was then coated between the designated substrates in a liquid state and allowed to solidify. The adhesion values are set forth in the table.

	. . . . .Example		
	1	2	3
Ingredients, wt %			
Polyamide*	100	—	—
Polyamide**	—	100	—
Polyamide***	—	—	100
Liquid DGEBA epoxy resin	15	15	15
Plasticizer (N-ethyl-o-, p-toluenesulfonamides)	45	45	15
Fumed silica (Cab-o-Sil)	1.5	1.5	1.5
Glass beads, 0.0035" diameter	0.3	0.3	0.3
Tensile shear strength, psi (ASTM D 1002)			
Aluminum-to-aluminum	2,150±190	1,520±50	2,590±420
Steel-to-steel	2,530±80	2,360±340	1,895±60
Laminate-to-laminate (Formica)	880±60	780±60	920±60
Wood-to-wood	360±40	300±40	—

*Prepared from Jeffamine D-400, a polyoxypropylenediamine (MW 400): piperazine:isophthalic acid; molar ratio 0.3:0.7:0.9.

**Prepared from Jeffamine D-230, a polyoxypropylenediamine (MW 230): isophthalic acid; molar ratio 1.01:1.0.

***Prepared from Jeffamine D-400: piperazine:isophthalic acid; molar ratio 0.5:0.5:0.9.

As illustrated in the table, these thermoplastic adhesives employing piperazine in combination with the polyoxypropylene polyamine (Examples 1 and 3) exhibit greatly improved adhesive strength in bonding metals, wood and laminates.

For Use on Metal Substrates

The hot melt adhesives described by *T. Eernstman, C.J. Auger and E.M. Devroede; U.S. Patent 4,132,690; January 2, 1979; assigned to Allied Chemical Corp.,* comprise admixtures of about 15 to 99 wt % of a conventional polyamide resin adhesive base component, which resin serves as the principal film former and tackifying constituent and, correspondingly, about 85 to 1 wt % of a normally solid homogeneous copolymer of ethylene and an unsaturated carboxylic acid having a high acid number, i.e., of at least about 100, and a number average molecular weight between 500 and 5,000. This copolymer contains at least a major portion of ethylene, by weight, is compatible with the resin and functions to impart additional tack, flexibility, and lower viscosity.

The compositions exhibit improved adhesive properties over known blends of polyamides and polyethylene derivatives and can be adhered directly to substrates by simply hot-pressing; for example, compacting the adhesive composition blend to the substrate at elevated temperatures, e.g., above about 300°F, for a short time so as to ensure uniform contact of the blend and substrate, and cooling. The compositions have excellent physical properties and surprisingly good adhesion to a variety of substrates, particularly metals.

Examples 1 to 15: Various amounts of a solid, homogeneous copolymer of ethylene and acrylic acid, prepared according to U.S. Patent 3,658,741, having a density of 0.95, a number average molecular weight of 2,500 (as determined by vapor phase osmometry), a Ring and Ball softening point of 86°C (as determined by ASTM E28), an average Brookfield viscosity of 475 cp at 140°C, and an acid number of 120 mg KOH/g were melt blended with Versalon 1155 polyamide resin (a polyamide resin having a softening point of about 155°C), Emerez UE 80237 polyamide resin, and Emerez 1553 polyamide resin.

Peel strength determinations (T-peel with angled specimens) were carried out in accordance with German Standard DIN 53 282 (September 1968), employing 5 samples each of aluminum foil strips, 3 cm wide and 17.5 cm long of 0.5 mm thickness, previously cleaned with isopropanol. The strips, supported by a polytetrafluoroethylene sheet were heated to about 180°C on a press, the hot melt blend indicated in the table below was applied in a thin stream around the edges and lengthwise along the center of each strip, and a second strip was placed upon the first followed by a polytetrafluoroethylene sheet.

After applying a pressure of about 1,000 lb/cm² through the press for 1 min at 180°C, the bonded metal strips were allowed to cool to room temperature. Peel strengths were determined after 2 hr by an Instron (Model 1026) machine using a crosshead and chart speed of 50 mm at a full scale load range of 0 to 5 kg.

Peel strength data are set forth in the table below. It is evident from the results that the blends of the process exhibit superior adhesive properties as compared with polyamide compositions conventionally employed in hot melt adhesive compositions. Furthermore, when cast into films, the blends exhibit good film-forming capability and produce transparent, haze-free, flexible products, thereby indicating a high degree of compatibility of the polyamide resins with the copolymer.

Example Number	Polyamide Resin Used	Percent of Copolymer in Blend	Initial Tear (kg/cm)	Peel (kg/cm)
1	Versalon 1155*	0	0.147	0.153
2	Versalon 1155	20	0.460	0.253
3	Versalon 1155	40	0.816	0.573
4	Versalon 1155	60	0.972	0.555
5	Versalon 1155	80	0.630	0.489
6	Emerez UE 80237*	0	0.115	0.114
7	Emerez UE 80237	20	0.445	0.332
8	Emerez UE 80237	40	1.072	0.584
9	Emerez UE 80237	60	1.196	0.675
10	Emerez UE 80237	80	0.437	0.406
11	Emerez 1553*	0	0.478	0.427
12	Emerez 1553	20	1.885	0.683
13	Emerez 1553	40	1.506	0.630
14	Emerez 1553	60	4.641	1.230
15	Emerez 1553	80	0.788	0.517

*Control.

Laminates for Auto Interiors

The process devised by *Y. Harada and Y. Ohya; U.S. Patent 4,137,366; Jan. 30, 1979; assigned to Daicel Ltd., Japan* relates to laminates for interior finish materials, i.e., laminates to be fitted to the inner walls of the ceiling, door, window frame or trunks of motorcars or to the inner walls of rooms of buildings.

The hot melt adhesive used in the process comprises a polyamide copolymer of a melting point of preferably 90° to 130°C, composed of at least three kinds of monomers, including lauryllactam or ω-aminoundecanoic acid as an indispensable component. A preferred polyamide is a terpolymer of nylon 12/6/66.

Workability can be improved by incorporating further 3 to 15 parts by weight of ethylene/vinyl acetate copolymer (vinyl acetate content: 5 to 20%) per 100 parts by weight of the polyamide copolymer. Though hot melt adhesives in the form of films are preferred, because they can be used easily and the operation steps can be rationalized, hot melt adhesives in other forms such as powders or nets can also be used.

Glass fibers, resin boards, paper boards and foamed synthetic resin products are suitable as frames.

As the covering materials used, laminates of a buffer material such as foamed polyurethane or foamed polyethylene and a polyvinyl chloride sheet, nonwoven fabrics and synthetic fiber foundations are suitable. Foamed polyurethane is preferred. In addition to polyester-type resins, polyether-type resins can also be used.

The motorcar interior finish materials are prepared by applying a hot melt adhesive to a polyurethane side of a covering material composed of a polyvinyl chloride sheet and foamed polyurethane layer of the like and thermally adhering the same to a shaped frame.

Example: Films 50 μ thick were prepared from nylon 6/66/12 (wt ratio: 1:1:1) of a melting point of 115°C, nylon 66/66/610 (CM-4,000) of a melting point of 147°C, ethylene/vinyl acetate copolymer of a vinyl acetate content of 18 wt % and a melting point of 96°C, polyethylene of a melting point of 105°C and modified polyolefin resin (Kuranbeter) of a melting point of 135°C.

By using the films, a frame and a covering material were adhered to each other. Between an upper hot plate heated to 90°C and a lower hot plate heated to 200°C, there was inserted a covering material composed of a polyvinyl chloride sheet and foamed polyurethane, an adhesive film and a frame. A pressure of 0.5 kg/cm^2 was applied thereto for 10 sec and for 20 sec to effect the adhesion. The samples thus adhered were subjected to adhesive strength test (180° peeling) according to the specification of JIS K 6744-71 and adhesive heat resistance test.

In the latter test, the adhesion lapping area was 25 cm x 25 cm. A load of 250 g was applied to the sample, and the sample was allowed to stand in an oven at 80°C for 5 min to observe the falling off of the adhered layers. The results of the measurements are shown in the following table. In the column of heat resistance test, O indicates no falling off and X indicates falling off.

Adhesive Film	Adhesion Time (sec)	Adhesive Strength (kg/25 mm^2)	Heat Resistance Test
This process	10	1.15	O
This process	20	2.10	O
Ethylene/vinyl acetate copolymer	10	0.70	X
Ethylene/vinyl acetate copolymer	20	1.00	X
Polyethylene	10	0.20	X
Polyethylene	20	0.60	O
Kuranbeter	10	0.40	X
Kuranbeter	20	0.85	O
CM-4,000	10	0.20	X
CM-4,000	20	0.50	X

Generally, adhesives to be used for motorcar interior finish materials, such as ceiling material, should satisfy both adhesive strength of higher than 1.00 kg/25 mm^2 and heat resistance test carried out at 80°C. It is evident from the test results that only the adhesives described satisfy both quality requirements. The fact that a sufficient adhesive strength can be obtained in a short adhesion time is very advantageous in view of productivity.

Formulations for Bonding at Low Temperatures

Hot melt polyamide adhesives possessing good heat- and solvent-resistance and waterproof qualities include homopolymers of nylon 11 and nylon 12. However, they can only be used with a limited number of substrates because of the high bonding temperatures they require.

K. Ando, M. Kamosaki, Y. Ohya and S. Asai; U.S. Patent 4,141,774; Feb. 27, 1979; assigned to Daicel Ltd., Japan have formulated a hot melt adhesive consisting essentially of a methoxymethylated nylon terpolymer of nylon 12:nylon 6:nylon 66, copolymerized in a weight ratio of 24-45:25-45:25-45, which has a melting point of less than 150°C and a methoxymethylation ratio of from 15 to 25%, wherein the methoxymethylation ratio is the ratio of nitrogen atoms bonded to methoxymethyl groups to the total number of nitrogen atoms contained in the nylon terpolymer.

The formula also contains from 0.5 to 8 wt %, based on the weight of the terpolymer, of a hardener selected from the following acid group: tartaric, maleic, crotonic, malonic, succinic, oxalic, adipic, citric, o-, m- and p-hydroxybenzoic, 2,4-dihydroxybenzoic, 2-hydroxy-4-methoxybenzoic, hypophosphorous, p-toluenesulfonic and chromic acid anhydride.

Example: 100 g of a ternary nylon copolymer (melting point, 130°C; relative viscosity, 1.80; melt index, 9.5) synthesized from 40 parts of caprolactam, 20 parts of nylon 66 salt and 40 parts of lauryllactam (hereinafter designated as untreated) were charged into a stainless steel autoclave, together with 40 g of paraformaldehyde and 120 g of methanol.

The charged materials were heated under agitation to completely dissolve the nylon copolymer component. A methanol solution containing 4 g of 85% phosphoric acid was then added at 60°C, and the reaction was continued for 30 min. After cooling, the liquid reaction product was added to a large quantity of water under agitation. The resulting N-methoxymethylated nylon copolymer (designated hereinafter as MM-2) had a methoxymethylation ratio of 15.6%.

N-methoxymethylated nylon copolymers with methoxymethylation ratios of 9.5% (hereinafter designated as MM-1), 20.5% (hereinafter designated as MM-3) and 34.5% (hereinafter designated as MM-4) were prepared in the same way as described above, by correspondingly changing the addition amounts of paraformaldehyde. 20 g of each of the control and MM-1 to MM-4 were added separately into 80 g of a mixed solvent consisting of methanol and trichloroethylene mixed at a weight of 70:30.

Each of the resulting mixtures was agitated at 50°C for completely dissolving the solutes. Each of the resulting solutions was coated on a polytetrafluoroethylene resin sheet so that the dry film thickness is 50 μ. Each of the resulting films was dried for 30 min at room temperature with cool air and for a further period of 30 min with warm air at 50°C to give a film of the untreated terpolymer and a film for each of the MM-1 to MM-4.

The melting behavior of the four films was plotted, and tests of the adhesiveness of each of the four films to a mixed spun fabric consisting of 65% polyester and 35% cotton were carried out by using a small size table press, using a bonding time of 5 sec, a constant bonding pressure of 0.1 kg/cm^2 and bonding temperatures of 90° to 160°C.

It was found that MM-2 and MM-3, which had lower and better defined melting points, also gave improved bonding strength at lower bonding temperatures. They also exhibited superior resistivity to laundering and dry cleaning.

In another patent *K. Shima, Y. Harada and Y. Ohya; U.S. Patent 4,148,775; April 10, 1979; assigned to Daicel Ltd., Japan* provides thermoplastic hot melt adhesive compositions which are capable of adhering articles at low temperatures (about 100° to 130°C) and imparting high bond strength to honeycombs by fully penetrating thereinto or forming fillets thereon. The compositions also give high bond strength to metal or plywood materials and have outstanding resistance to water, moisture and heat (at about 80° to 90°C).

These hot melt adhesives comprise 100 parts by weight of a polyamide co-polymer resin having a melting point of about 80° to 160°C and composed of at least three monomers, about 5 to 30 parts by weight of a terpene-phenol resin having a softening point of about 80° to 120°C, and about 5 to 10 parts by weight of a plasticizer.

Examples of such resins are those prepared selectively from ε-caprolactam, lauryllactam, ω-aminoundecanoic acid, nylon 6,6 salt, nylon 6,9 salt, nylon 6,10 salt, nylon 6,12 salt, etc. The term "at least three monomers" herein used usually means, three, four or five monomers. Preferable polyamide copolymer resins are those prepared from three or four monomers. Examples of such preferable resins are nylon 6/6,6/12, nylon 6/6,10/12, nylon 6/6,12/12, 6/6,9/12, nylon 6/6,6/6,10/12, 6/6,6/11/12, etc. The plasticizer may be p-toluene sulfonamide or benzene butyl sulfonamide.

These hot melt adhesive compositions are capable of adhering articles at low temperatures (about 100° to 130°C) and satisfactorily penetrating into work-pieces or forming fillets thereon, and are highly adhesive to metal, plywood and various other materials and resistant to water, moisture and heat (about 80° to 90°C). They are especially well suited to the adhesion of honeycombs.

Containing Caprolactam and Substituted Caproic Acids

The melt adhesives formulated by *M. Drawert, E. Griebsch and W. Imoehl; U.S. Patent 4,150,002; April 17, 1979; assigned to Schering AG, Germany* for bonding textiles to themselves and other adhesives are comprised of polyamides having melt viscosities (measured at 220°C) of 25 to 600 Pa-sec (where 1 Pa-sec = 10 poises), and contain:

(a) Dimerized fatty acids having a content of dimeric fatty acid from 70 to 100% by wt and optionally comprising mono-carboxylic acids as viscosity regulators (chain breakers), in which up to 50 equivalent percent of the total carboxyl groups of this mixture can derive from monocarboxylic acids;

(b) One or more aliphatic straight-chain codicarboxylic acids having from 6 to 13 carbon atoms, in which the ratio of carboxyl groups from component (a) to carboxyl groups from component (b) is from 0.05:1 to 5:1; the polyamide further comprises an amount substantially equivalent to the acids mentioned under (a) and (b);

(c) A C_{6-12} aliphatic straight-chain diprimary diamine; and

(d) Caprolactam and/or ϵ-aminocaproic acid, wherein from 0.5 to 1.5 mols of caprolactam and/or ϵ-aminocaproic acid are employed per mol of carboxyl groups of the acids mentioned under (a) and (b).

In preparing the polyamides, the dimerized fatty acid, component (a), is reacted with all or a part of the caprolactam and/or of the ϵ-aminocaproic acid component (d) in a first stage. This product is then reacted in a second stage with the further components.

Polyesteramides can be made by using the above formulation except that, under (d) 0.5 to 2.5 mols of caprolactam and/or ϵ-aminocaproic acid may be used plus:

(e) Caprolactone and/or ϵ-hydroxycaproic acid, wherein from 0.05 to 1.0 mol of caprolactone and/or ϵ-hydroxycaproic acid are employed per mol of the caprolactam, or of the ϵ-aminocaproic acid mentioned above under (d).

In preparing the polyesteramides, the dimerized fatty acid, component (a) may, if desired, be reacted with all or part of the caprolactam and/or of the ϵ-aminocaproic acid, component (d) in a first stage. This product may then be reacted in a second stage with the remaining components.

Preferred polyamides and polyesteramides are those in which sebacic acid is the codicarboxylic acid according to (b), hexamethylenediamine is the diamino component according to (c), in which the ratio of the carboxylic acids according to (a) to the carboxylic acids according to (b) is from 0.1:1 to 3:1.

For the polyamides, 0.8 to 1.2 mols of caprolactam and/or ϵ-aminocaproic acid are employed per mol of carboxyl groups of the acids (a) and (b). For the polyesteramides, 0.8 to 2 mols of caprolactam and/or ϵ-aminocaproic acid are employed per mol of carboxyl groups of the acids mentioned under (a) and (b). Further, from 0.1 to 0.5 mol of caprolactone and/or ϵ-hydroxycaproic acids are employed per mol of caprolactam and/or ϵ-aminocaproic acid.

For the preparation of the polyesteramides of the present process, a dimerized fatty acid according to (a) is preferred in which the content of dimeric fatty acid is more than 90% by wt.

The melt adhesives according to the process, which comprise the sufficiently available fatty acids, caprolactam or ϵ-aminocaproic acid, caprolactone or ϵ-hydroxycaproic acid, diamines, and codicarboxylic acids, show good resistance to halogenated hydrocarbons and, simultaneously, also to alkali washing baths at $60°C$ and, in part, also at $95°C$. This resistance expresses itself by:

(1) Very good initial resistance to tearing, i.e., the resistance values before stressing by a washing or cleaning process; and by

(2) Very high wet resistance to tearing, i.e., the resistance values after cleaning while the materials are in a still-moist condition; and by

(3) Outstanding values of the resistance to tearing after drying, which values are almost attained again after many washing or cleaning cycles.

Example: 22.23 kg of dimerized tall oil fatty acid and 14.81 kg of tall oil fatty acid were heated with 146.90 kg of caprolactam under nitrogen at 250°C in a reactor equipped with a stirrer, a descending condenser, and thermometer, and were kept for 4 hr at this temperature. The acid so obtained was combined with 131.40 kg of sebacic acid and 143.00 kg of 1,12-diaminododecane and condensation was completed in 4 hr at 250°C. During the last 2 hr, a vacuum of 2 mm Hg was applied.

The polyamide obtained had the following values. The softening point was 131°C. The melt viscosity (220°C) was 154.0 Pa-sec. The resistance to separation (kg/5 cm) was 3.5 (washed at 60°C and torn wet). Resistance to separation after cleaning with perchloroethylene (kg/5 cm) was 4.1 (torn wet). The initial value (kg/5 cm) was 5.7.

PRESSURE SENSITIVE

With Good Resistance to Plasticizer Migration

Pressure-sensitive adhesives that have good tack and adhere to numerous substrates are widely used by industry in various applications in the construction of floor tiles having a preapplied adhesive for the do-it-yourself market. In this application the pressure-sensitive adhesive is thinly coated onto the undersurface of floor tiles and covered with a protective release sheet. Installation of the floor tiles is accomplished by removing the release sheet and pressing the tile into place.

The pressure-sensitive adhesives presently used in this type application contain either a styrene-butadiene rubber or a styrene-isoprene rubber. Although these pressure-sensitive adhesives provide adequate adhesive properties, they have certain deficiencies that limit their usefulness, such as poor resistance to plasticizer migration. Consequently in applications such as use on floor tiles the plasticizer in the tile migrates into the adhesive, thereby causing the adhesive to soften and string excessively.

L.A. Ardemagni; U.S. Patent 4,072,735; February 7, 1978; assigned to Eastman Kodak Company has found that a blend comprising ethylene-propylene rubber, tackifying resin, polybutene, and optionally a crystalline polypropylene provides a hot melt pressure-sensitive adhesive having good resistance to plasticizer migration and good heat stability.

The ethylene-propylene rubbers useful in the process are well-known materials and can be prepared by copolymerizing ethylene and propylene. Small amounts of a third monomer, such as a nonconjugated diene, can be added to provide unsaturation. The polymerization is usually carried out at a temperature below 100°C using a coordination catalyst, such as one comprised of aluminum alkyls and alkyl aluminum chloride and vanadium oxychloride. These ethylene-propylene rubbers contain 0 to 5 mol % unsaturation, have an ethylene content of 50 to 90 wt %, preferably 65 to 80 wt %, and have a flow rate at 230°C of about 0.2 to 0.8, preferably about 0.4 to 0.6.

An example of one such commercially available ethylene-propylene rubber useful in the process is Vistalon 702 rubber. These rubbers can be used alone or in combinations in amounts of about 5 to 15% by wt of the adhesive composition, preferably about 8 to 12% by wt.

The tackifying resins useful in the adhesive compositions can be a hydrocarbon resin having a softening point of 100°C and available commercially as Resin H-100, Wingtack 95 and the Sta-Tac and Betaprene H resins.

Other suitable resins are terpene polymers and the rosin esters, which include ethylene glycol, polyethylene glycol, glycerol and pentaerythritol rosin esters, hydrogenated rosin esters or methylated rosin esters, for example, the commercially available materials "Staybelite" Ester 3, triethylene glycol ester of hydrogenated rosin, "Foral" 85 and 105, highly stabilized ester resins of pentaerythritol and rosin base.

The hydrocarbon resins, polyterpenes, and rosin esters can be used either alone or in combinations. However, in general, the better results have been obtained with the hydrocarbon resins used alone. These tackifying resins can be used in amounts of about 30 to 60% by wt of the adhesive composition, preferably about 45 to 55% by wt.

The polybutenes useful in the adhesive compositions of this process are butylene polymers composed predominantly of monoolefins having 4 carbon atoms (85 to 98% by wt), the balance being isoparaffins. The molecular weight of the polybutene should be about 300 to 3,000 (as determined by Microlab Osmometer), preferably about 600 to 1,500. Such polymers are commercially available as Indopol H-100 polymers. These polybutylenes can be used alone or in combinations of about 20 to 50% by wt of the adhesive composition, preferably about 35 to 40% by wt.

The crystalline, hexane insoluble polypropylenes useful as an optional component in the adhesive compositions are produced by the polymerization of propylene in the presence of stereospecific catalysts. One method for preparing these polypropylenes is disclosed in U.S. Patent 3,679,775. These polypropylenes have inherent viscosities (I.V.), as measured in tetralin at 145°C, preferably about 1.0 to 3.0. They can be used in amounts preferably of about 2.0 to 4.0% by wt.

The compositions are prepared by blending together the components in the melt at a temperature of about 160° to 200°C until a homogeneous blend is obtained.

Example: The following adhesive composition was prepared.

	Percent
DAC-B hydrocarbon resin, having a softening point 100°C and a viscosity of 140 cp at 190°C (Resin H-100)	50
Ethylene-propylene rubber, 70 wt % ethylene, 0.5 flow rate at 230°C (Vistalon 702 Rubber)	10
Polybutene, MW 920 and a melt viscosity of 215 cs at 210°F (indopol H-100)	39.75
Irganox 1010 Tetrakis[methylene-3-(3',5'-di-tert-butyl-4'-hydroxyphenyl)propionate]methane	0.25

This composition provided an adhesive composition with the following properties.

Viscosity at 350°F (177°C), cp	20,000
Ring and Ball softening point, °C	85
180° peel adhesion, lb/in	3.0
Quick stick, lb/in	1.4
Shear adhesion, min to failure at 73°F, 1,000 g load, 1 in² bond area	1,200

This adhesive has good pressure-sensitive adhesive properties and good aggressive tack. It shows good peel adhesion and quick stick values. This adhesive composition was also melt coated onto floor tiles and exhibited good adhesion to the vinyl asbestos tile. Adhering the coated tiles to various substrates (wood, concrete, other tiles, and the like) resulted in destruction of the tile when removal was attempted. The adhesive maintained these properties after coated tiles had aged for 2 months at 140°F. This adhesive also exhibited good adhesion to other substrates such as polyethylene, Kraft paper, various fabrics and cloths. The compositions also showed good resistance to thermal degradation and oxidation (less than 10% viscosity change after 24 hr at 350°F).

Single-Component Adhesive

Pressure-sensitive adhesive (PSA) products have experienced a rapid growth rate in recent years because of their ease of application. Typical pressure-sensitive adhesive applications include, for example, tapes (consumer, industrial and surgical), labels, decals, films, floor tile and wall and shelf coverings.

Although PSA compositions have generally been applied to the backings from solvents, there is a strong desire to switch to PSA materials which can be applied as hot melts to eliminate solvent pollution during manufacturing of the products.

It has been discovered that in order to meet the critical requirements for use as a general purpose, pressure-sensitive adhesive, a material must:

(1) Provide a peel strength of at least 2.5 lb/in, Pressure Sensitive Tape Council (PSTC Test-1);

(2) Provide shear adhesion or strength (holding power, static shear resistance) of greater than 1,000 min when tested with 1 kg load (PSTC-7);

(3) Provide a rolling ball tack value of 2" or less (PSTC-6);

(4) Have melt viscosity stability at application temperatures such that melt viscosity will change less than 20% in 100 hr;

(5) Not leave a residue when coated tapes are peeled from highly polished steel at 73°F (PSTC-1).

R.L. McConnell, D.A. Weemes and F.B. Joyner; U.S. Patent 4,072,812; Feb. 7, 1978; assigned to Eastman Kodak Company have found that a limited number of olefin copolymers will meet the tests described above.

Suitable copolymers include propylene/higher 1-olefin copolymers containing 40 to 60 mol % of at least two higher 1-olefins. Operable 1-olefin monomers include 1-hexene, 1-heptene, 1-octene, 1-nonene and 1-decene. Propylene copolymers with less than 40 mol % comonomer or mixture of comonomers have poor tack and generally poor peel strength while those with greater than 60 mol % comonomer have poor static shear strength and they leave a residue when tapes are peeled from polished steel plates.

The pressure-sensitive copolymers must have the following structure-dependent properties.

Melt viscosity at 190°C, cp	10,000–75,000
Higher 1-olefin concentration, mol %	40–60
Density, g/cc	0.85–0.86
Tg, °C	−30 to −45
Tm	*

*Not measurable by DSC.

In addition to these structure-dependent properties, these pressure-sensitive adhesives can also be characterized by functional properties which are determined on 1 mil polyethylene terephthalate films coated with 0.75 to 1.0 mil adhesive.

Rolling back tack, in	⩽2.0
Static shear strength, min	1,000–4,000
Peel strength, lb/in	2.5–5.0

In preparing these coated tapes, it is critical that the adhesive be applied to the polyethylene terephthalate film at a melt temperature of at least 350°F. The backing material may be preheated if it is polyester, cotton or paper. Heat-sensitive backing materials, such as cellulose acetate or cellophane, are generally not preheated in order to prevent distortion of the backing. The adhesive coating thickness is also critical since commercial coating thickness is generally 1 mil or less. A coating thickness of 2 to 3 mils gives erroneous data concerning the utility of the polymer as a pressure-sensitive adhesive.

The type of catalyst and the polymerization conditions required to provide copolymers having the desired structure are quite limited. In general, the best results have been achieved by using catalyst systems which provide poor stereo-regulation in the polymerization of propylene. Combinations of Et_3Al with $AATiCl_3$ with Al/Ti molar ratios ranging from about 1:1 to 5:1 have been found to be useful. It is also necessary to conduct the polymerization at high temperatures with the preferred temperature being 150° to 160°C. The operable temperature is 140° to 170°C.

Example: In a nitrogen filled dry box, 200 ml of dry mineral spirits, 125 ml of 1-hexene (distilled and dried over sodium ribbon), 155 ml of dry 1-octene, and 0.37 g $AATiCl_3$ are placed into a clean, dry, 1 liter Parr autoclave, equipped with a stirrer. The autoclave is sealed in the dry box. A clean, dry catalyst injector is loaded with 0.5 g of Et_3Al and 35 ml of dry mineral spirits in the dry box, and it is then connected to the autoclave. The molar ratio of Al to Ti in the catalyst is 2.4:1.

After removing the autoclave from the dry box, 116 ml of liquid propylene is pressured into the autoclave. This provides a monomer mixture containing 76.5 wt % of higher 1-olefin. The autoclave is heated to 140°C with stirring. The Et_3Al solution is injected into the monomer solution to initiate the copolymerization. The copolymerization is exothermic, and the temperature of the reaction mixture increases to 150°C. This temperature is maintained for 3 hr and 200 ml of isobutyl alcohol is then pumped into the autoclave to deactivate the catalyst.

The temperature is maintained at 150°C for an additional 15 min. The autoclave is then cooled to 23°C, vented, and the copolymer is placed in an excess of isobutyl alcohol. The mixture is heated to 105°C, cooled, filtered and the soft, sticky, colorless copolymer is washed with additional isobutyl alcohol to remove catalyst residues.

The copolymer is stabilized with 0.25% Irganox 1010 and dried in a vacuum oven at 70° to 80°C. The yield of copolymer is 190 g (74% conversion). It contains 50 mol % higher 1-olefin (about 25 mol %, 1-hexene and about 25 mol % 1-octene) as determined by an NMR analysis. This tacky copolymer has a melt viscosity of 26,000 cp at 190°C, a Tg of −38°C and a density of 0.85. There is no detectable crystallinity in the sample by either x-ray or DSC analysis.

The copolymer is heated to 177°C (350°F) and coated onto polyethylene terephthalate film by means of a hot doctor blade to give a uniform 1 mil coating.

This coated tape performs well as pressure-sensitive tape material. For example, the polymer coating remains permanently tacky, and it has good adhesion to paper, steel, polyethylene, polyethylene terephthalate and the like. When the tape is peeled away from a clean stainless steel surface, no polymer residue is left on the steel demonstrating that the copolymer has good cohesive strength.

The tape has a rolling ball tack value of 3.5 cm, peel strength (on steel) of 3.4 lb/in and static shear strength (measured on steel using 1,000 g weight) of 4,000 min. When a torn page is mended with this tape, the printed matter under the tape is quite legible.

Similarly good results were obtained when the hot melt, pressure-sensitive copolymer is coated on black paper, crepe paper, 60 lb Krome-Kote paper, cloth, cellophane and cellulose acetate film backing materials.

In another process *R.L. McConnell, D.A. Weemes and F.B. Joyner; U.S. Patent 4,072,813; February 7, 1978; assigned to Eastman Kodak Company* 1-butene was found to be a suitable substitute for propylene in the copolymers. Examples of the use of 1-butene are as follows.

Example 1: In a nitrogen filled dry box, 200 ml of dry mineral spirits, 250 ml (168 g) of 1-hexene (distilled and dried over sodium ribbon), and 0.37 g AATiCl$_3$ are placed into a clean, dry 1-liter Parr autoclave, equipped with a stirrer. The autoclave is sealed in the dry box. A clean, dry catalyst injector is loaded with 0.5 g of Et$_3$Al and 35 ml of dry mineral spirits in the dry box, and it is then connected to the autoclave. The molar ratio of Al to Ti in the catalyst is 2.4:1.

After removing the autoclave from the dry box, 120 ml (72 g) of liquid 1-butene is pressured into the autoclave. This provides a monomer mixture containing 70 wt % of 1-hexene. The autoclave is heated to 140°C with stirring. The Et$_3$Al solution is then injected into the monomer solution to initiate the copolymerization. The copolymerization is exothermic, and the temperature of the reaction mixture increases to 150°C. This temperature is maintained for 3 hr and 200 ml of isobutyl alcohol is then pumped into the autoclave to deactivate the catalyst.

The temperature is maintained at 150°C for an additional 15 min. The autoclave is then cooled to 23°C, vented, and the copolymer is placed in an excess of isobutyl alcohol. The mixture is heated to 105°C, cooled, filtered and the soft, sticky, colorless copolymer is washed with additional isobutyl alcohol to remove catalyst residues.

The copolymer is stabilized with 0.25% Irganox 1010 and dried in a vacuum oven at 70° to 80°C. The yield of copolymer is 190 g (79% conversion). It contains 50 mol % 1-hexene as determined by an NMR analysis. This tacky copolymer has a melt viscosity of 26,000 cp at 190°C, a Tg of −40°C and a density of 0.85. There is no detectable crystallinity in the sample by either x-ray or DSC analysis.

The copolymer is heated and coated on the film just as described in the example of the previous patent. It had a rolling ball tack value of 3.5 cm, peel strength of 3.5 lb/in and static shear strength of 4,200 min.

Example 2: The procedure of Example 1 is followed, except that 67 ml of 1-butene, 58 ml of propylene and 250 ml of 1-hexene are used. The conversion of monomer to copolymer is 75%. The copolymer has a melt viscosity of 28,000 cp at 190°C, and it contains 50 mol % 1-hexene and about 25 mol % each of 1-butene and propylene. Coated polyethylene terephthalate tape has a peel strength of 4.5 lb/in width, rolling ball tack of 2.3 cm and static shear strength of 6,500 min. This copolymer has excellent pressure-sensitive adhesive properties.

Based on Rubbery Block Copolymers

The object of the process by *K. Bronstert, V. Ladenberger, W. Druschke, W. Groh and H. Mueller; U.S. Patent 4,089,824; May 16, 1978; assigned to BASF AG, Germany* was to provide a hot melt pressure-sensitive composition exhibiting good adhesion properties, tackiness, creep resistance and aging resistance, and, in particular, distinguished by good cohesion and good peel strength.

They found that this object was achieved by a pressure-sensitive adhesive composition based on a rubbery block copolymer of a particular structure, obtained from a monovinyl-aromatic or monovinylidene-aromatic compound and a conjugated diene which possesses radial branching and of which the olefinic double bonds are selectively hydrogenated.

The pressure-sensitive adhesive comprises a mixture of

(a) 100 parts by weight of a rubbery, radial block copolymer with terminal nonelastomeric polymer blocks of a monovinyl-aromatic or monovinylidene-aromatic compound and butadiene-based elastomeric polymer blocks in the middle position, the olefinic double bonds of the branched block copolymer having been reduced to a residual content of less than 5% by selective hydrogenation and the elastomeric polybutadiene blocks having had, prior to hydrogenation, a 1,2-vinyl content of from 30 to 70% by wt, based on the blocks, and the number-average molecular weight of the branched block copolymer being from 30,000 to 300,000;

(b) From 25 to 300 parts by weight of a tackifier resin; and

(c) From 0 to 200 parts by weight of a rubber extender oil.

The rubbery radial block copolymers to be employed according to the process form the essential elastomeric constituent of the adhesive. Examples of monovinyl-aromatic and monovinylidene-aromatic compounds which may be used to synthesize the terminal nonelastomeric polymer blocks of the branched block copolymers are styrene, styrenes alkylated in the side chain, e.g., α-methylstyrene, and nuclear-substituted styrenes, e.g., vinyltoluene and ethylvinyl-benzene. Preferably, styrene is employed by itself.

The radial block copolymer in general contains preferably from 25 to 35% by wt, based on the branched block copolymer, of the monovinyl-aromatic or monovinylidene-aromatic compounds and, accordingly, from 75 to 65% by wt, based on the branched block copolymer, of butadiene, as copolymerized units. The butadiene-based polymer blocks in the middle position are hydrogenated polybutadiene blocks which had a 1,2-vinyl content preferably of from 40 to 65% by wt prior to hydrogenation.

The examples which follow illustrate the process. Unless stated otherwise, parts and percentages are by weight. The molecular weights quoted are determined by measuring the osmotic pressure in a membrane osmometer.

In the examples, 40 μm thick polyethylene glycol terephthalate films are coated with a 25 μm thick film of the pressure-sensitive adhesive. The latter may be applied to the polyethylene glycol terephthalate film from the melt or from solution (for example in toluene). If a solvent is used, it is necessary to evaporate it under reduced pressure. During evaporation, the temperature of the coated film should not be raised to above about 130°C. In order to assess the adhesive properties of the film coated with the pressure-sensitive adhesive, the material is dried at room temperature, and the surface tack is determined by means of the peel test. Furthermore, the cohesion of the adhesive layer is determined by means of the shear test.

In the peel test, 2 cm wide test strips are adhered to a chromed sheet and peeled off parallel to the adhesive layer, i.e., at an angle of 180°, the force required being measured. The pull-off speed is 300 mm/min. The measurement is carried out 24 hr after having adhered the strips to the sheet.

In the shear test, the test strips are adhered to a high gloss chromed sheet; the area of the adhesive bond being 20 x 25 mm. The sheet carrying the test strips is clamped in vertical position. The end of the adhesive strip is loaded with 1,000 g, and the time required for the bond to break under the resulting constant tensile stress is determined. The measurement is carried out at 20° and 50°C.

Example 1: (a) Manufacture of the Rubbery Radial Block Copolymer – 145 g (1.4 mols) of purified styrene are polymerized in 2,000 g of cyclohexane and 17.7 g of tetrahydrofuran, using 21 mmol of n-butyllithium, for one hour at from 45° to 50°C under an inert gas atmosphere, after which conversion is virtually complete. The molecular weight of the living polystyrene obtained is 7,000. 335 g (6.2 mols) of butadiene are then added and polymerized first at from 45° to 50°C and then for 2.5 hr at 60°C (polybutadiene block: $\overline{M}n$ is 16,000). The living polymer is coupled by adding a stoichiometric amount of epoxidized linseed oil (Edenol B 316).

The molecular weight of the polymer after coupling is 70,000 and the 1,2-vinyl content of the polybutadiene phase is 57%. Hydrogenation is carried out using a homogeneous catalyst solution comprising 1 mmol of nickel acetylacetonate, dissolved in 30 ml of toluene, and 6 mmols of aluminum triisobutyl, at from 60° to 90°C and 5 bars. After the hydrogenation, the content of olefinic double bonds (determined by Wijs titration) is less than 2%. The polymer is precipitated by pouring the solution into alcohol and is filtered off and dried.

(b) Manufacture of a Hot Melt Adhesive Mixture – 100 parts of the selectively hydrogenated, radial block copolymer, 175 parts of the glycerol ester of hydrogenated rosin, 25 parts of an extender oil (Catenex N 945) and 5 parts of zinc dibutyldithiocarbamate are homogenized in a kneader at mixing temperatures of about 200°C. To carry out this process, the rosin ester of glycerol and the antioxidant are first introduced into the kneader. The branched block copolymer, swollen with the extender oil, is then added slowly. The adhesive composition obtained is tested as described above. The test results are shown in the table.

Example 2: The procedure followed is as in Example 1, but in this case hydrogenation after the coupling reaction is dispensed with so that a corresponding nonhydrogenated radial block copolymer is employed for the manufacture of the adhesive composition. The results obtained are to be found in the table.

Example 3: The adhesive composition is manufactured as in Example 1, but, instead of the selectively hydrogenated radial block copolymer, 100 parts of a selectively hydrogenated linear block copolymer of the polystyrene/hydrogenated polybutadiene/polystyrene type are employed. This selectively hydrogenated linear block copolymer corresponds, in molecular weight, styrene content, 1,2-vinyl content and proportion of olefinic double bonds after hydrogenation, to the selectively hydrogenated radial block copolymer used in Example 1. The adhesive composition has the properties shown in the table.

	. . . . Example.		
	1	2	3
Peel strength after 24 hr (N/2 cm)	20.0 A*	20.5 A*	12 A*
Shear strength, hr			
At 20°C	>24	>24	>24
At 50°C	>24	1.5	>24

*A denotes adhesion rupture.

With Polymers at Least Partially Crosslinked

The pressure-sensitive adhesive described by *R. Vitek; U.S. Patent 4,091,195; May 23, 1978; assigned to Kores Holding Zug AG, Switzerland* is one which can be applied in molten form to a suitable carrier or backing, such as Kraft packaging tape, without the use of a solvent.

The chain of polymers on which the adhesive is based must be at least partially crosslinked. The preferred contact adhesive compound has approximately 0.1

to approximately 10% crosslinkage and a molecular weight of from about 100,000 to 1,000,000.

In essence, the hot melt adhesive product is prepared by partially transversally crosslinking polymers using a crosslinking agent and crosslinking accelerators.

The partial crosslinking can be brought about by partial vulcanization, that is, by using sulfur as a crosslinking agent. A mixture of one or several polymers, for example, polybutadiene, polystyrene, polyisoprene, natural rubber, chlorinated butyl rubber, polyethylene, polypropylene, and sulfur is homogenized preferably by mastication, in a temperature of $100°$ to $160°C$, preferably $130°$ to about $160°C$.

Subsequently, the temperature is brought to about $80°$ to about $120°C$ and preferably to $100°$ to about $110°C$. A vulcanization accelerator is then added which brings about the transversal crosslinkage with sulfur. Precipitated sulfur in quantities of about 0.01 to 10% by wt is preferably added.

Sulfur crosslinking or vulcanization accelerators which have been found particularly useful include ammonium salts of dithiocarbamic acid, sodium dithiocarbamates, zinc dithiocarbamates, thiuram, mercapto accelerators, sulfonamide accelerators, aldehydeamide accelerators, guanidine accelerators, basic amine accelerators and any of the usual vulcanization agents employed for vulcanizing butyl rubbers, such as p-benzoquinone dioxime. The preferred amount of accelerator to be added is approximately 0.01 to about 10% by wt.

Other crosslinking agents which may be used in place of sulfur may be based on mono-, di-, tri-, or polyisocyanates. Aliphatic or aromatic peroxides may also be used. A quantity of from about 0.01 to 15% by wt of the nonsulfur crosslinking agents has been found a suitable quantity. It is preferred that the nonsulfur crosslinking agents be added at a lower temperature than that at which the homogenization of the polymers takes place and preferably just before the application of the contact adhesive compound to the adhesive carrier.

As a result of the crosslinkage reaction, double bonds in the polymers partially disappear. The resulting hot melt adhesive products are, therefore, partially oxidation- and UV-resistant. Furthermore, their sensitivity or reactivity with respect to cations, in particular with respect to manganese and iron ions, is reduced.

Example 1: In a mastication device, a mixture of the following ingredients:

	Parts by Weight
Butadiene-styrene	50
Block-polystyrene	20
"Extender-oil"	28
Precipitated sulfur	1
Ageing inhibitor	0.9

was homogenized at a temperature of 105°C. The homogeneous mixture was brought to a temperature of 95°C and mixed with 0.1 part of thiuram as a vulcanization accelerator. Immediately thereafter, the mixture was applied onto a Kraft paper support or other suitable carrier.

Example 2: In a mastication or kneading device, a mixture of 25 pbw chlorobutyl rubber, 40 pbw butadiene-styrene polymer, and 35 pbw "extruder oil" was homogenized at a temperature of 140°C. The homogeneous mixture was subsequently brought to a temperature of 100°C and reacted with 4 pbw of an agent based on guanidine.

The finished adhesive compound was then utilized by immediately being applied to a carrier sheet of polyvinyl chloride.

Substantially Amorphous Polyolefins

R.R. Schmidt III and J.D. Holmes; U.S. Patent 4,143,858; March 13, 1979; assigned to Eastman Kodak Company has found that a substantially amorphous polyolefin, such as amorphous polypropylene, can be treated with an acid catalyst at 250°C or higher to provide a low viscosity, permanently tacky, hot melt pressure-sensitive adhesive composition. This treatment renders the substantially amorphous olefin permanently tacky.

Substantially amorphous polyolefins useful in the formulation include substantially amorphous homopolymers of 1-olefins containing 3 to 5 carbon atoms, substantially amorphous copolymers prepared from ethylene and 1-olefins containing 3 to 5 carbon atoms, and substantially amorphous propylene or 1-butene/higher 1-olefin copolymers containing less than 40 mol % of at least one higher 1-olefin of the group 1-hexene, 1-heptene, 1-octene, 1-nonene, and 1-decene.

The melt viscosity of the substantially amorphous polyolefin materials is generally about 100 to 200,000 cp at 190°C while the treated products generally have melt viscosities of about 30 to 50,000 cp at 190°C (determined by ASTM D1238). Thus, the acid catalyst treatment causes a decrease in the melt viscosity of the substantially amorphous polyolefin, and this treatment also imparts permanent tack to the amorphous polyolefin composition.

The properties of amorphous polypropylene are changed significantly by treatment with acid catalyst. Such catalysts include the strong Lewis acid-type catalysts such as aluminum halides, i.e., aluminum trichloride, antimony halides, i.e., antimony dichloride, boron halides, i.e., boron trifluoride, titanium halides, i.e., titanium trichloride, and silica-aluminum type catalyst, such as Davison grade 979 silica-alumina catalyst. In particular, it becomes permanently pressure sensitive, less regular, and less opaque; has a lower glass transition temperature (Tg); and contains no vinyl or vinylidene unsaturation. It should be noted that even the higher viscosity amorphous polypropylene can be made pressure sensitive by this procedure.

The process of treating the amorphous polyolefins to make them useful as pressure-sensitive adhesives comprises reacting the polyolefins with about 0.075 wt % of at least one strong Lewis acid catalyst at temperatures of at least 250°C to provide the low viscosity, permanently tacky hot melt adhesives.

One commercially available amorphous polyolefin useful in preparing the pressure-sensitive adhesives is Eastobond M-5 polyolefin.

The pressure-sensitive adhesives may be stabilized by the addition of conventional stabilizers for polyolefin materials such as dilauryl thiodipropionate, Plastanox 1212, Irganox 1010, etc.

These modified amorphous polyolefins are useful in pressure-sensitive adhesives. As pressure-sensitive adhesives, they find utility in preparing tapes by applying the adhesive to conventional hot melt procedures to a substrate, such as a film, which can be prepared from conventional film materials, such as a polyester, for example. The pressure-sensitive adhesive can also be applied onto labels, decals, floor tile as well as wall-coverings, such as wallpaper, and shelf-coverings, such as shelf paper.

Polyacrylics

The formulation for preferred hot melt adhesives developed by *J.D.Guerin, T.W. Hutton, J.J. Miller and R.E. Zdanowski; U.S. Patent 4,152,189; May 1, 1979; assigned to Rohm and Haas Company* consists of:

(A) From about 35 to 90 pbw of a copolymer having a Tg of $-20°$ to about $-65°C$ comprising:

   (1) From about 88 to 98.5 pbw of a lower alkyl acrylate (C_{1-8}); and

   (2) From about 1.5 to 12 pbw of an ethylenically unsaturated acid or amine with

(B) From about 10 to 65 pbw of a copolymer having a Tg of 35°C or more preferably from about $35°$ to 110°C, comprising:

   (1) From about 88 to 98.5 pbw of the monomer of the formula:

$$CH_2=\overset{\overset{\displaystyle R}{\displaystyle |}}{C}-CO_2R^1$$

   wherein R is hydrogen or methyl and R^1 is lower alkyl, lower cycloalkyl or isobornyl; and

   (2) From about 1.5 to 12 pbw of an ethylenically unsaturated carboxylic acid or amine.

Pressure-sensitive adhesives can best be prepared employing from about 15 to 30 pbw of Copolymer B, whereas blends containing greater than 30 pbw of Copolymer B are suitable for use as nontacky laminating adhesives.

Especially preferred are compositions comprising a blend of:

> (A) From about 60 to 85 pbw of a copolymer comprising:
>
>> (1) From about 94 to 98 pbw of a monomer selected from butyl acrylate, 2-ethylhexyl acrylate, methyl acrylate, ethyl acrylate, isobutyl acrylate, isoamyl acrylate or n-propyl acrylate; and
>>
>> (2) From about 2 to 6 pbw of dimethylaminoethyl acrylate or methacrylate, diethylaminoethyl acrylate or methacrylate or tert-butylaminoethyl methacrylate with
>
> (B) From about 15 to 40 pbw of a copolymer comprising:
>
>> (1) From about 94 to 98 pbw of isobutyl methacrylate, isobornyl acrylate, isobornyl methacrylate, methyl methacrylate and the like; and
>>
>> (2) From about 2 to 6 pbw of an acid selected from methacrylic acid, acrylic acid or itaconic acid.

The polyacrylic hot melt adhesives of this process require no nonacrylic additives, are only very lightly colored and are substantially insensitive toward oxidation and ultraviolet exposure.

RADIATION-CURED

Containing Acrylate or Methacrylate Coupling Agent

Styrene-diene block copolymers have been formulated in the past to produce a number of types of adhesive compositions. The basic patent in this field, Harlan, U.S. Patent 3,239,478, shows combinations of these block copolymers with tackifying resins and paraffinic extending oils to produce a wide spectrum of adhesives. However, two serious limitations of these adhesive compositions are their relatively low service temperatures and their poor solvent resistance.

Generally, the highest temperature at which these styrene-diene block copolymers retain useful properties and act like a vulcanized rubber is limited by the softening temperature (Tg) of the styrene end block. Depending upon the molecular weight of the end blocks and the load, these block copolymers can begin to significantly creep or flow at 120° to 180°F. Although these adhesives can withstand very short contact with common solvents, prolonged contact with aromatic solvents or blends containing aromatic, ester or lactone solvent will cause these adhesives to soften and lose adhesive strength.

For a number of applications it would be very advantageous to have higher service temperatures. For example, these adhesives would be useful in paint shops if the masking tapes produced had solvent resistance and a 225° to 250°F service temperature. They would also be useful as laminating adhesives in, for example, retortable food pouches if they could withstand boiling water temperatures and in furniture-laminating adhesives if they could bear moderate loads at 250°F.

D.R. Hansen and D.J. St. Clair; U.S. Patent 4,133,731; January 9, 1979; assigned to Shell Oil Company have formulated a cured adhesive composition possessing excellent high temperature cohesive strength along with excellent adhesion, shear strength and solvent resistance. The adhesive is prepared by the high energy radiation, especially electron beam or ultraviolet-initiated curing, of an adhesive composition comprising:

(a)   100 pbw of a block copolymer having at least two monoalkenyl arene polymer end blocks A and at least one elastomeric conjugated diene mid block B, said blocks A comprising 8 to 55% by wt of the block copolymer;

(b)   About 25 to 250 pbw of an adhesion-promoting resin compatible with block B; and

(c)   About 1 to 50 pbw of a di- to tetrafunctional acrylate or methacrylate selected from the group consisting of the acrylic and methacrylic acid esters of polyols.

Additional components may be present in the composition including, among others, plasticizers such as rubber compounding oils or liquid resins, antioxidants, and end block A compatible resins.

The block copolymers employed in the composition are thermoplastic elastomers and have at least two monoalkenyl arene polymer end blocks A and at least one elastomeric conjugated diene polymer mid block B. The number of blocks in the block copolymer is not of special importance, and the macromolecular configuration may be linear, graft, radial or star depending upon the method by which the block copolymer is formed.

Typical block copolymers of the most simple configuration would have the structure polystyrene-polyisoprene-polystyrene and polystyrene-polybutadiene-polystyrene. A typical radial or star polymer would comprise one in which the diene block has three to four branches (radial) or five or more branches (star); the tip of each branch being connected to a polystyrene block.

Other useful monoalkenyl arenes from which the thermoplastic (nonelastomeric) blocks may be formed include alpha-methyl styrene, tert-butylstyrene and other ring alkylated styrenes as well as mixtures of the same. The conjugated diene monomer preferably has 4 to 5 carbon atoms, such as butadiene and isoprene. A much preferred conjugated diene is isoprene.

The average molecular weights of each of the blocks may be varied as desired. The monoalkenyl arene polymer blocks preferably have average molecular weights between about 7,000 and 50,000. The elastomeric conjugated diene polymer blocks preferably have average molecular weights between about 25,000 and 150,000.

The weight percentage of the thermoplastic monoalkenyl arene blocks in the finished block polymer should be between about 8 and 55%, preferably between about 10 and 30% by wt.

The block copolymer by itself lacks the required adhesion. Therefore, it is necessary to add an adhesion-promoting or tackifying resin that is compatible with the elastomeric conjugated diene block. A much preferred tackifying resin is a diene-olefin copolymer of piperylene and 2-methyl-2-butene having a softening point of about 95°C. This resin is available commercially under the trade name Wingtack 95, and is prepared by the cationic polymerization of 60% piperylene, 10% isoprene, 5% cyclopentadiene, 15% 2-methylbutene and about 10% dimer.

An essential component of the formulation is the di- to tetra-functional acrylate or methacrylate coupling agent which promotes crosslinking of the block co-polymer during exposure to the radiation. The coupling agents employed herein are di-, tri-, and tetra-functional acrylates and methacrylates selected from the group consisting of the acrylic and methacrylic acid esters of polyols.

Preferred coupling agents include 1,6-hexanediol diacrylate (HDODA), 1,6-hexanediol dimethacrylate (HDODM), trimethylolpropane triacrylate (TMPTA), trimethylolpropane trimethacrylate (TMPTM), pentaerythritol tetracrylate (PTA), and pentaerythritol tetramethacrylate (PTM).

Much preferred coupling agents are TMPTA and TMPTM with TMPTM being the most preferred. The amount of coupling agent employed varies from about 1 to 50 phr, preferably about 2 to 25 phr.

These adhesives are especially suited for preparation as 100% solids hot melt adhesives since they give relatively low processing viscosities, less than several hundred thousand centipoises, and adequate pot life, up to several hours, at processing temperatures of about 150°C.

The compositions are cured by exposure to high energy radiation such as electron beam radiation or ultraviolet radiation, with electron beam being preferred.

Example: This example demonstrates the suitability of UV radiation for initiation of crosslinking. The formulations and results are shown in the table. All three formulations, the control (Formulation A) and the two examples of the type described here (Formulations B and C), show poor strippability temperature limits (STL) and dissolve in toluene before exposure to UV radiation. AFter exposure to UV radiation, the control sample shows some improvement in STL.

Its solvent resistance remains poor, and it shows very poor shear adhesion. Formulations B and C, however, show good STL and solvent resistance as well as good shear adhesion.

	A	B	C
Formulation, phr			
Kraton 1107 rubber	100	100	100
Wingtack 95	80	80	80
TMPTM	–	25	–
TMPTA	–	–	25
DEAP (diethoxyacetophenone)	–	3	3
Properties before UV exposure			
Strippability temperature limit, °F	230	200	<120
Solvent resistance	poor	poor	poor
Polyken probe tack, kg*	1.0	0.7	0.4
Shear adhesion, min**	>4,000	<1	<1
Properties after UV exposure***			
Strippability temperature limit, °F	310	>400	>400
Solvent resistance	poor	good	excellent
Polyken probe tack, kg	0.9	0.6	0.3
Shear adhesion, min	3	>3,000	>4,000

*ASTM D-2979.
**PSTC Method No. 7, ½" x ½" contact to steel, 2 kg load.
***One pass under two UV lamps at 100 fpm.

In another patent *D.J. St. Clair and D.R. Hansen; U.S. Patent 4,151,057; April 24, 1979; assigned to Shell Oil Company* a similar formulation is given, with the difference being in the structure of the block copolymer, which component in this formulation comprises an $A'B'$ block copolymer and a multiblock copolymer having at least two end blocks A and at least one mid block B wherein the A' and A blocks are monoalkenyl arene polymer blocks and the B' and B blocks are substantially completely hydrogenated conjugated diene polymer blocks.

The weight ratio of the multiblock copolymer to the $A'B'$ two block copolymer is about 100:0 to 10:90, preferably about 80:20 to 40:60.

The multiblock copolymers employed may have a variety of geometrical structures, and may be linear, radial or branched so long as each multiblock copolymer has at least two polymer end blocks A and at least one polymer mid block B as defined above. Methods for the preparation of such polymers are known in the art.

The formulation applies especially to the use of selectively hydrogenated polymers having the configuration before hydrogenation of polystyrene-polybutadiene-polystyrene (SBS), polystyrene-polyisoprene-polystyrene (SIS), poly(α-methylstyrene)-polybutadiene-poly(α-methylstyrene) and poly(α-methylstyrene)-polyisoprene-poly(α-methylstyrene).

It will be understood that both blocks A and B may be either homopolymer or random copolymer blocks as long as each block predominates in at least one class of the monomers characterizing the blocks and as long as the A blocks individually predominate in monoalkenyl arenes and the B blocks individually predominate in dienes. The term "monoalkenyl arene" will be taken to include especially styrene and its analogs and homologs including α-methylstyrene and ring substituted styrenes, particularly ring methylated styrenes. The preferred monoalkenyl arenes are styrene and α-methylstyrene, and styrene is particularly preferred.

The blocks B may comprise homopolymers of butadiene or isoprene and co-polymers of one of these two dienes with a monoalkenyl arene as long as the blocks B predominate in conjugated diene units. When the monomer employed is butadiene, it is preferred that between about 35 and 55 mol % of the condensed butadiene units in the butadiene polymer block have 1,2 configuration. Thus, when such a block is hydrogenated, the resulting product is, or resembles, a regular copolymer block of ethylene and butene-1 (EB). If the conjugated diene employed is isoprene, the resulting hydrogenated product is or resembles a regular copolymer block of ethylene and propylene (EP).

Hydrogenation of the precursor block copolymers is preferably effected by use of a catalyst comprising the reaction products of an aluminum alkyl compound with nickel or cobalt carboxylates or alkoxides under such conditions as to substantially completely hydrogenate at least 80% of the aliphatic double bonds while hydrogenating no more than about 25% of the alkenyl arene aromatic double bonds. Preferred block copolymers are those where at least 99% of the aliphatic double bonds are hydrogenated while less than 5% of the aromatic double bonds are hydrogenated.

The average molecular weights of the individual blocks may vary within certain limits. In most instances, the monoalkenyl arene blocks will have number average molecular weights preferably on the order of 6,000 to 60,000 while the conjugated diene blocks, either before or after hydrogenation, will have average molecular weights preferably on the order of 30,000 to 150,000. The total average molecular weight of the multiblock copolymer is typically from about 35,000 to 300,000.

This adhesive formulation possesses excellent shear strength, UV and solvent resistance. The compositions are cured by electron beam radiation.

ADDITIONAL FORMULATIONS

For Use at Higher Temperatures

One of the serious limitations of adhesives made with styrene diene block co-polymers is their relatively low service temperatures, since they can begin to significantly creep or flow at 120° to 180°F. For a number of applications it would be very advantageous to have higher service temperatures. For example, these adhesives would be useful in paint shops if the masking tapes produced had a 225° to 250°F service temperature.

D.R. Hansen; U.S. Patent 4,104,323; August 1, 1978; assigned to Shell Oil Company has developed a melt-blending process for preparing a polymer composition with high temperature performance characteristics which comprises:

(a) Melt blending a low molecular weight polyphenylene ether resin and a block A compatible resin selected from the group consisting of coumarone-indene resins, vinyl toluene-α-methylstyrene copolymers, and mixtures thereof at a temperature of preferably 30°C greater than the softening point of the polyphenylene ether resin in a weight ratio of polyphenylene ether resin to block A compatible resin of between about 1:10 and about 10:1, wherein the molecular weight of the polyphenylene ether resin is between about 6,000 and 25,000. The glass transition temperature of the block A compatible resin is between about 40° and 100°C, thereby forming a polyphenylene ether resin alloy; and

(b) Melt blending about 1 to 100 parts by weight of said polyphenylene ether resin alloy with 100 parts by weight of a block copolymer at a temperature below about 230°C, said block copolymer having at least two monoalkenyl arene polymer end blocks A and at least one elastomeric conjugated diene mid block B, blocks A comprising 8 to 55% by wt of the block copolymer.

By employing this process to melt blend the polyphenylene ether resin and the styrene diene block copolymer, it is possible to avoid the use of solvents while also avoiding the degradation of the block copolymer. The resulting polymer blend not only possesses a much higher service temperature, but is surprisingly stable against oxidative degradation over a long period of time.

The adhesive composition that employs this polyphenylene oxide alloy comprises:

100 parts by weight of a block copolymer having at least two monoalkenyl arene polymer end blocks A and at least one elastomeric conjugated diene midblock B, blocks A comprising 8 to 55% by wt of the block copolymer;

About 50 to 200 parts by weight of a tackifying resin compatible with block B;

About 0 to 200 parts by weight of a hydrocarbon extending oil; and

About 1 to 100 parts by weight of a melt blend of a low molecular weight polyphenylene ether resin, and a block A compatible resin selected from the group consisting of coumarone-indene resins, vinyl toluene-α-methylstyrene copolymers, and mixtures thereof, wherein the weight ratio of the polyphenylene ether resin to the block A compatible resin is between about 10:1 and 1:10. The molecular weight of the polyphenylene ether resin is between about

6,000 and 25,000. The glass transition temperature of the block A compatible resin is between about $40°$ and $100°C$.

Adhesive compositions prepared according to this process possess service temperatures significantly higher than the similar prior art adhesives not containing the melt blend of polyphenylene ether resin and block A compatible resin. It has also been found that the selection of the particular low molecular weight polyphenylene ether resin is much preferred over the commercially available higher molecular weight resin, contrary to the expected result.

Since the higher molecular weight polyphenylene ether resin has a higher softening temperature than the corresponding lower molecular weight resin, it would be expected that adhesive compositions containing the higher molecular weight resin would have higher service temperatures than would adhesive compositions containing the lower molecular weight resin. However, it has surprisingly been found that the adhesives containing the lower molecular weight resin have the higher service temperatures.

The hot melt adhesives of this formulation are preferably used in the preparation of pressure-sensitive adhesive tapes by a method such as that disclosed in U.S. Patent 3,676,202. These tapes are preferably used where high service temperatures are required.

In the examples the following materials were employed.

Block copolymer 1—a styrene-butadiene-styrene ABA block copolymer having block molecular weights of about 10,000-52,000-10,000.

Cumar LX-509—a coumarone-indene resin having a Ring & Ball softening temperature of about $155°C$ and a glass transition temperature (Tg) determined by thermal expansion of about $88°C$.

Piccotex 120—a poly(α-methylstyrene-covinyltoluene) resin having a Ring and Ball softening temperature of $120°C$ and a Tg of about $49°C$.

PPO-L—pure poly(2,6-dimethylphenylene oxide) (PPO) having a molecular weight of about 10,400 and a Tg of about $195°C$.

The adhesive compositions were evaluated by a modified Heat Distortion Temperature test (HDT). The heat distortion temperature of the polymer blends was measured with die D tensile bars (ASTM D-412) cut from plaques 0.07" thick, compression molded at 1,000 psi. A molding temperature of $140°C$ was used. The samples were subjected to a 110 psi tensile load (engineering stress) in a temperature chamber in which the temperature increased at the rate of $40°F/hr$. The heat distortion temperature was taken to be the temperature at which the sample elongated 0.5" (ca. 30% elongation assuming all of the elongation takes place in the narrowest section of the tensile bar). Zero elongation was measured at room temperature with the load on the sample.

In the example, all materials processed on a Brabender mixer contained 0.5 phr dilauryl thiodipropionate and 0.5 phr Irganox 1010 (an antioxidant).

Example: The properties of block copolymer 1 blended with two end block resins and blends of these resins with PPO-L are examined and the results presented in the following table. Run 1 shows the properties of the block copolymer before processing. In runs 2 to 11 the block copolymer and the resins were mixed in a Brabender mixer (roller type head, 50 rpm) for 5 to 10 min at the temperatures shown in the table. The alloys of PPO-L and the end block resin were prepared by mixing in a Brabender mixer at 100 rpm and 250°C for 15 min.

From the results in the table, it can be seen that the heat distortion temperature of the block copolymers was significantly increased. The alloys of PPO-L and block A compatible resins increased the HDT by 80° to 89°F. It was noted the mixing temperatures can have a large effect on the HDT. For the 50/50 LX-509/PPO-L alloy temperatures as high as 180°C had to be employed to achieve optimum HDT. The Piccotex-120 alloys, however, because of the lower softening temperature, could be incorporated at lower temperatures.

Run Number	Block Copolymer	. . . . Alloy.					
		Alloy or Resin	Ratio	Alloy or Resin (phr)	HDT (°F)*	ΔHDT (°F)	Mixing Temperature (°F)
1	1	—	—	0	120	0	**
2	1	—	—	0	118	−2	150
3	1	LX-509	—	35	160	40	150
4	1	Piccotex-120	—	35	128	8	160
5	1	LX-509/PPO-L	75/25	17	140	20	155
6	1	LX-509/PPO-L	75/25	35	190	70	150
7	1	LX-509/PPO-L	50/50	35	118	−2	160
8	1	LX-509/PPO-L	50/50	35	175	55	175
9	1	LX-509/PPO-L	50/50	35	209	89	180
10	1	Piccotex-120/ PPO-L	75/25	35	164	44	170
11	1	Piccotex-120/ PPO-L	50/50	35	200	80	160

*At 110 psi.
**Crumb, not processed.

For Substrates with Low Surface Energies

A hot melt adhesive must be fluid enough at its application temperature to wet the surface of the substrate to be adhered to. This is essential for the formation of a secure bond to the substrate. The ability to wet the surface of the substrate is determined by the surface tension of the liquid. The adhesive must have a lower critical surface tension or surface energy than the surface energy of the substrate to enable wetting of the substrate. Prior art methods of bonding to substrates with an inherently low surface energy involve either chemically or electrically etching the surface of the substrate to increase its surface energy and, where the low surface energy is due to contaminants, sandblasting the substrate surface to achieve the same effect by removal of low surface energy contaminants.

E.F. Johnson and J.S. Carlsen; U.S. Patent 4,117,027; September 26, 1978; assigned to Raychem Corporation have provided an adhesive comprising a silicon block copolymer and a tackifier. The adhesive is capable of forming a strong bond to unetched substrates having low inherent surface energies or low surface energies due to the presence of surface impurities. Furthermore, the resultant bond is hydrolytically and thermooxidatively stable and does not deteriorate even after long periods at elevated temperatures.

Bonds made with this adhesive also possess an unusually large insulation resistance at both ambient and elevated temperatures, thereby enabling its use in unusually demanding electrical applications. Furthermore, the adhesive does not absorb or readily react with water and, therefore, has unusually high insulation resistance even in high humidity and temperature environments.

The silicon block copolymer component is preferably a siloxane block copolymer such as a polyester-poly(dimethylsiloxane) preferably the polycarbonate polydimethylsiloxane having the following formula.

where Z can be independently hydrogen, C_{1-8} alkyl, or halogen, especially chlorine or bromine positioned on the ring, where x is equal to from 1 to 100, inclusive; y is equal to from about 5 to 200, inclusive; and preferably y has an average value from about 15 to 90, inclusive; while the ratio of x to y can vary from about 0.05 to 1.5, inclusive; and where y has an average value of from about 15 to 90, inclusive; the ratio of x to y is preferably from 0.10 to about 0.25, inclusive.

The polymer preferably contains 35 to 65% of polyester units and correspondingly 65 to 35% of the siloxane units. The polymer preferably has a value for n of about 1 to 1,000 and a number average molecular weight of preferably 15,000 to 40,000.

The silicone block copolymers of the process are not in and of themselves significantly superior to many other prior art adhesives, and indeed have little or no utility as adhesives. It is only when they are combined with certain particular types of tackifiers that an outstandingly effective hot melt adhesive results.

The tackifier is believed to function to decrease the surface tension of the copolymer, thereby enabling the adhesive to more effectively wet the surface of the substrate. However, a particularly unexpected aspect of the process is that only a very limited class of tackifiers have this effect. Most known prior art tackifiers are essentially useless. The tackifier is preferably at least partially soluble in the polysilicon block copolymer, at least to the extent that segregation at interfaces is minimized.

It is advantageous for the tackifier to have a permanent dipole or be capable of a high degree of polarizability. Thus, for example, the tackifiers with easily polarized carbon-halogen bonds or aromatic unsaturation are effective in increasing the bond strength of the adhesive composition, while the less easily polarizable highly aliphatic tackifiers do not increase the overall bond strength. The tackifier should also be heat stable at elevated temperatures so that it will not decompose or otherwise react harmfully with the copolymer. This precludes the use of acidic tackifiers which degrade the copolymer at bonding temperatures. Because very few tackifiers are capable of filling these requirements, it is preferred that the tackifier be one of the following.

(1) A copolymer of vinyl toluene and α-methylstyrene;

(2) Hexabromobiphenyl;

(3) Pentabromodiphenyl ether; or

(4) Chlorinated terphenyls.

The preferred tackifiers should have a molecular weight of between 500 and 2,000, and the melt viscosity of the tackifier should be below 1 poise at 165°C.

Example: Silicon polycarbonate block copolymers (General Electric Corp.) having the molecular weights shown in the table were blended with the tackifiers shown in the table by solution blending in methylene chloride and solvent cast in films approximately 10 mil thick. These films were used to adhere together the various polymers indicated using slight applied pressure at a temperature of 320°C for 30 sec. The results obtained initially and after oven ageing at 150°C for 100 hr are shown in the table below.

The table shows that the silicon polycarbonate block copolymers are not effective as adhesives without the preferred tackifiers. When one of the preferred tackifiers is blended with any of the silicon-polycarbonates, the resulting adhesive composition forms a much stronger bond with any of the substrates tested than did the silicon polycarbonate resin by itself.

Not only does the addition of preferred tackifier increase the adhesive bond strength, it also changes the mode of failure from adhesive failure to cohesive failure in most cases. The latter is the preferred method of failure. Prior art adhesives do not undergo cohesive failure under any conditions.

Run No.*	Copolymer**	Siloxane (%)	Molecular Weight Number Average	Molecular Weight Weight Average	Tackifier*** (20% by wt)	T-Peel Strength (pli) Kynar-Stilan	T-Peel Strength (pli) Kynar-Stilan After 100 Hours at 150°C	T-Peel Strength (pli) Stilan-Stilan	T-Peel Strength (pli) Ryton-Stilan	T-Peel Strength (pli) Kapton-Stilan
1	A	65	29,000	238,000	—	1.3		1.5		2.8
2	B	65	21,000	161,000	—	2.2		8.0		
3	C	65	19,500	126,000	—	4.5		5.6		
4	D	50	21,000	113,000	—	0		0.2		
5	B	65	21,000	161,000	E	2.1		7.3		
6	B	65	21,000	161,000	F	2.0		5.1	2.0	
7	A	65	29,000	238,000	E	3.8		3.6		
8	B	65	21,000	161,000	E	9.0	11.5	17.8	15.5	8.9
9	B	65	21,000	161,000	G	9.9	11.7	15.9		
10	B	65	21,000	161,000	H	8.8	12.0	12.2		
11	B	65	21,000	161,000	I	9.3	16.1	19.7		11.0
12	C	65	19,500	126,000	E	9.8		14.8		
13	D and B†	62	21,000	154,000	H	6.2	14.6			

*Runs 1 through 7 are not in accordance with this process.

**The silicon polycarbonate block copolymers are: A is XD-1; B is LR5630, blend #1; C is Batch 1081-45B; and D is XD-7.

***The tackifiers are: E is copolymer of vinyltoluene and methylstyrene (melt viscosity <1.0 poise at 165°C); F is ester of polymerized abietic acid; G is hexabromobiphenyl; H is pentabromodiphenyl ether; and I is chlorinated terphenyl.

†Blend of 15% D/85% B.

Adhesive for Holding a Workpiece

F.B. Speyer: U.S. Patent 4,132,689; January 2, 1979; assigned to TRW Inc. describes a resinous thermoplastic hot melt composition for matrix fixturing or potting which is composed of an organic and an inorganic moiety. The organic moiety is composed of hydrocarbon polymers, preferably an alkylene hydrocarbon polymer, e.g., a polyterpene; and an aromatic hydrocarbon polymer. A plasticizer, e.g., a high molecular weight unsaturated fatty acid, such as dilinoleic acid may be advantageously present.

The inorganic moiety is preferably a finely divided inorganic filler, e.g., aluminum oxide, which is intimately dispersed in the organic material. The inorganic filler material is present from 5 to 70 parts per 100 parts of the potting composition, and preferably from 50 to 70 parts per 100 parts of the composition.

The workpiece is potted in a workpiece container by casting about the workpiece supported within the container the foregoing filled resinous thermoplastic hot melt composition at an elevated temperature, cooling the assembly to a temperature below room temperature, and then warming the assembly to room temperature either by application of relatively low heat or by allowing the assembly to regain room temperature spontaneously.

After the device is mechanically operated upon, the filled resinous material may be recovered simply by elevating the temperature to a point where the filled resinous material again becomes fluid, and removing the part from the workpiece container. The filled organic resinous potting composition may then be recovered and reused indefinitely. The composition is stable, nonreactive with the workpiece and easily removed therefrom without introducing contaminants of a deleterious nature.

In the following examples, the hot melt compositions are made by sequential melt blending of the organic ingredients which are then held at 174° to 179°C for 4 to 5 hr while the inorganic filler is stirred in.

Example 1:

	Parts by Weight
Dilinoleic acid	10
C_{5-6} terpene hydrocarbon polymer	12
Aromatic hydrocarbon polymer	13
Aluminum oxide (−325 mesh)	65

The pour temperature is from 121° to 127°C and the breaking load is about 5,000 lb. For fixturing a jet engine blade, a cool down time to set of 2 to 3 min is obtained after which removal from the fixturing jig can be done with further spontaneous or forced cooling. The amount of aluminum oxide is just below the critical pigment volume concentration (CPVC) for the system.

Example 2:

	Parts by Weight
C_{16-18} dimer acids	9
C_5 terpene hydrocarbon polymer (Wingtack 95)	11
Aromatic hydrocarbon polymer (Picco-6100)	12
Aluminum oxide (−325 mesh)	68

The foregoing example is illustrative of the maximum amount of aluminum oxide which can be used satisfactorily in the particular resin combination. At over 70% concentration of aluminum oxide, the component is too weak for satisfactory use as a potting compound. This latter concentration corresponds to a CPVC of about 32%.

Example 3:

	Parts by Weight
C_{16-18} dimer fatty acids	10
Dipentene polyterpene (MW 1,100)	20
Aromatic hydrocarbon polymer (Picco 6100)	20
Aluminum oxide (−325 mesh)	50

This composition is satisfactory as a fixturing compound, but is somewhat more expensive than the composition of Example 2 above. The amount of alumina is below the CPVC for the system.

Example 4:

	Parts by Weight
Dilinoleic acid	5
Wingtack 95 (aliphatic hydrocarbon polymer)	45
Aluminum oxode (−325 mesh)	35
Iron (−325 mesh)	15

The foregoing composition illustrates a composition utilizing as the hydrocarbon polymer a nonaromatic hydrocarbon polymer. Also illustrated is an inorganic filler which is a mixture of two inorganic fillers, i.e., a ceramic oxide and a powdered metal. The CPVC of the system is not exceeded.

SPECIAL APPLICATIONS

On-Line Binding System to Make a Multipage Document

G.O. Orth, Jr.; U.S. Patent 4,105,611; August 8, 1978; assigned to Norfin, Inc. has developed a method of on-line bonding which involved a hot melt adhesive and an endless belt or the like to deliver the adhesive to the sheet edges. The collator, sorter, and similar arts in general involve the feeding of sheets into spaced compartments. It is often desirable to bind the sheets collected in each compartment together into a multipage document. "Off-line" binding refers to the process of removing the sheets from the compartments before binding. "On-line" binding refers to the process whereby the sheets are bound together while still in the compartments.

The adhesive is melted in a glue pot from which it is deposited onto the endless belt. The belt may be equipped with its own heating means. The coated portion of the belt is then moved into contact with the edges of the sheets, which sheets have been aligned and compressed together along a line spaced a short distance back from the edges to be bound, so that the edges themselves are flared somewhat apart from each other. The still liquid adhesive flows from the belt into the spaces between the flared edges of the sheets and sets to form a tough yet flexible bond.

The formulation of a hot melt adhesive suitable for such a process involves, to some extent, a compromise between ease of application and quality of the final product. The adhesive must have a low enough viscosity when hot to flow from the pot to the belt and from the belt into the spaces between the flared sheet edges without stringing or balling up on the belt. The unused adhesive should also remain fluid so that it can be returned to the glue pot for reuse.

On the other hand, the adhesive when cold must have adequate tensile strength. Further requirements for a suitable adhesive are that it be nonblocking, that it not smoke, decompose, or give off odors when held at the glue pot temperature for reasonable periods of time, and that it set rapidly upon contacting the sheets. The components of the adhesive composition must also be compatible and must not separate at the coldest temperature that the bound sheets can be expected to encounter.

The adhesive used for this process has a viscosity of less than 5,000 cp at 250°F and is comprised of 30 to 40% of an ethylene-vinyl acetate copolymer, 25 to 35% of a wax, 20 to 40% of a modified rosin, and 10 to 30% of a fluid tackifier. Such a composition has been found to have viscosity characteristics compatible with a moving belt application, to be quick setting, to provide a tough yet flexible bond when set, to be nonblocking, and to be stable against separation at below room temperature and against degradation for reasonable periods at molten temperature.

Example: An adhesive composition was formulated by heating together 33% by wt of an ethylene-vinyl acetate copolymer having a melt index of 20 grams per 10 minutes and a vinyl acetate content of 28% by wt, sold under the name "Co-Mer EVA-505", with 30% by wt of a paraffin wax having a melting point between 143° and 145°F, with 24% by wt of a pentaerythritol ester of highly stabilized rosin, having a softening point of 219°F, sold under the trademark "Foral 105", with 12% by wt of a polyterpene resin having a molecular weight of 490 and a softening point of between 50° and 59°F sold under the trademark "Wingtack 10".

The adhesive composition was placed in the glue pot of a binding apparatus similar to the one described above and applied by this apparatus to the aligned edges of sheets of paper.

The operation of the apparatus was as follows. The frame was first moved to a position so that the belt was adjacent to the discharge passageway but spaced from the sheet edges. Then, simultaneously, the valve was opened and the belt set in motion so that a portion of the belt was coated with adhesive. The valve was closed when the length of this portion corresponded to the height of the compartments containing sheets to be bound.

The coated portion was then moved around to a position adjacent to these compartments, whereupon the frame was moved so that the coated belt contacted the sheet edges. After about 2 sec the frame was moved back, withdrawing the belt from the sheets. The adhesive cooled and set within about 15 sec after the withdrawal of the belt. During the entire operation the glue pot temperature was held at 375°F, and the belt temperature was maintained at 350°F.

The application operation proceeded smoothly without any stringing or balling up of the adhesive on the belt. The adhesive did not smoke or give off odors when hot, nor did its components separate upon cooling to room temperature. The cool adhesive was found to have produced a strong, flexible bond between the sheets, and to be nonblocking at room temperature.

Pressure Sensitive Adhesives

ACRYLIC POLYMERS

Containing Polymerized Alkoxyalkyl Ester of Unsaturated Acid

F.T. Sanderson, A. Kowalski and R.H. Kottke; U.S. Patent 4,077,926; March 7, 1978; assigned to Rohm and Haas Company have found that pressure sensitive adhesives having good tack as well as shear resistance are provided by polymers largely of alkoxyalkyl esters of α,β-unsaturated carboxylic acids, and that large proportions of "soft" alkyl methacrylates are useful in such adhesives, although the latter are not necessary.

In accordance with this formulation, the adhesive contains a tacky polymer, in parts by weight, of about 5 to 99.5% of an alkoxyalkyl ester of an α,β-unsaturated acid, preferably 0.25 to 15% of an α,β-unsaturated acid, preferably 0.5 to 10% of an hydroxyalkyl ester of an α,β-unsaturated carboxylic acid having at least one hydroxyl functionality, and as additional tackifying or "soft" monomers, 0 to 94.5% of an alkyl acrylate having an average of from 2 to 12 carbon atoms in the alkyl group or an alkyl methacrylate having from 5 to 20 carbon atoms and an average of from 5 to 16, preferably from 6 to 14, carbon atoms in the alkyl group. As is well known, the Tg of homopolymers of such esters of isomeric alcohols varies with the extent of chain branching, and this must be taken into consideration. The polymer has a Tg of below $0^{\circ}C$, and preferably below $-15^{\circ}C$. The total of the monomers is based on weight and is 100%.

In a preferred embodiment, the alkoxyalkyl ester and the alkyl methacrylate make up at least 40% of the total monomers and preferably up to 65% of the total monomers; 0.5 to 10% of the hydroxyalkyl ester with hydroxyl functionality; and the unsaturated acid monomer is present in an amount between 0.25 and 15%, and less than about half of the monomer mixture is an alkyl

61

acrylate having from 2 to 12 carbon atoms in the alkyl group. In this embodiment, at least 10% of the soft methacrylic ester is used.

By alkoxyalkyl ester of an α,β-unsaturated carboxylic acid is generally meant one having the following structure:

$$CH_2{=}\overset{\overset{\displaystyle R}{|}}{C}{-}\overset{\overset{\displaystyle O}{\|}}{C}{-}O{-}\!\!\left(R_1{-}O\right)_{\!\!n}{-}R_2$$

where R is H or methyl; R_1 is a straight chain or branched chain alkylene radical containing from 1 to 4 (preferably 2) carbon atoms; R_2 is an alkyl radical containing from 1 to 4 (preferably 2) carbon atoms; and n is 1 to 4 (preferably 1). Representative monomers are methoxyethyl methacrylate (MOEMA), ethoxyethyl methacrylate (EOEMA), methoxyethyl acrylate (MOEA), ethoxyethyl acrylate (EOEA), butoxyethyl methacrylate (BOEMA), methoxybutyl acrylate (MOBMA), methoxyethoxyethyl acrylate (MOEOEA), etc. The acid moiety is preferably acrylic or methacrylic.

An important monomer of the copolymer composition is the hydroxyalkyl ester of an α,β-unsaturated carboxylic acid having at least one hydroxyl group. This hydroxyl containing monomer can be crosslinked by the crosslinking agent to a higher degree than carboxyl containing monomers, under normal curing conditions, and this improved crosslinking is reflected in the great stability in humid atmospheres that the hydroxyl containing monomer provides to the overall composition. This monomer can be either a partial or full hydroxyalkyl ester of a monobasic or polybasic α,β-unsaturated carboxylic acid with less than 36 atoms, preferably having less than about 20 atoms.

The other essential monomer is any copolymerizable ethylenically unsaturated monocarboxylic or polycarboxylic acid. Itaconic acid and the α,β-unsaturated monocarboxylic acids, particularly methacrylic acid and acrylic acid, are preferred.

The polymers are prepared in aqueous dispersion form or in organic solutions (e.g., in xylene, methyl Cellosolve, etc.) by well-known conventional means such as free radical initiation with benzoyl peroxide or the like. Solution or aqueous dispersion polymers useful in the process preferably have a molecular weight of between 10,000 and 1,000,000.

These adhesives have been found to provide exceptionally good bonds with numerous substrates, including wood, paper, Formica brand decorative sheets, other plastic materials such as methyl methacrylate polymer polyvinyl chloride, saran, polyethylene glycol terephthalate, nylon, phenol-formaldehyde resins, urea-formaldehyde resins, and other thermoset materials such as melamine-formaldehyde resins; rayon, cotton, silk, wool, fibers of the polymeric materials mentioned above, leather, linoleum, asphalt tile, vinyl tile, ceramic tile, various silicates such as glass, mineral wool, asbestos, concrete, asbestos cement, plaster, metals such as aluminum, steel, iron, copper, zinc, chromium, nickel, as well as painted or enamelled surfaces, such as painted automobile bodies, woodwork, walls, ceilings or floors.

Precemented articles such as tapes, wallpapers, or tiles for decorating walls, floors, or ceilings may be produced in which the member to be ultimately applied to a surface to be decorated, such as a tape, wallpaper, or tile is provided on its back surface with a layer of the adhesive composition and a protective backing is applied over the adhesive layer and is adapted to be stripped therefrom at the time the member is to be applied to the surface for which it is ultimately intended. The protective layer may be a thin film of polyethylene or other material which can be stripped more or less readily from the adhesive layer.

Example: An acrylic polymer solution is prepared at 43% solids in an 80% ethyl acetate/20% toluene mixture using azobisisobutyronitrile initiator and a monomer mixture comprised of 24.5 parts by weight of a mixture of long-chain alkyl methacrylates in solids, the alkyl radicals containing 12 to 15 carbons, identified as dodecyl-pentadecyl methacrylates, 30 parts of methoxyethyl acrylate, 40 parts butylacrylate, 5 parts hydroxyethyl methacrylate and 0.5 part acrylic acid. After cooling, 46 grams of this polymer solution are charged to a glass jar and treated with 4 grams n-butanol and 0.7 gram of a solution containing 60% butylated melamine-formaldehyde resin dissolved in a 50% xylol/50% butanol solvent system. These ingredients are thoroughly mixed and used in the preparation of pressure sensitive tapes. These tapes are found to have the following properties:

180° peel strength (PSTC Method No. 1), oz/in	33
shear resistance (PSTC Method No. 7), hr	>90
tack, rolling ball (PSTC Method No. 6), inches	1.3
tack, touch rating (qualitative)	excellent

The following are general descriptions of Pressure Sensitive Tape Council tests:

PSTC Method No. 1—180° Peel Strength: This test basically involves determining resistance of pressure sensitive tape/stainless steel laminates to being delaminated at a 180° angle and at a rate of 12 inches per minute. The test laminates are consistently prepared by laying the adhesive tape onto the cleaned stainless steel plates and using only the pressure from two passes of a 4.5-pound roller to complete lamination.

PSTC Method No. 7—Shear Resistance: Shear resistance is measured as the time to failure for ½ x 1 inch tape/stainless steel laminates loaded with a 1,000-gram weight and hung in an essentially vertical position. The adhesive tape/stainless steel laminates were prepared by the same method as the peel strength test laminates.

> *PSTC Method No. 6—Rolling Ball Tack:* This tack
> rating measures the distance a 5.5-gram, $^7/_{16}$-inch diam-
> eter steel ball bearing will travel on adhesive tape before
> being stopped by the adhesive. The ball is rolled onto
> the adhesive tape from a 21.5° inclined plane from a
> point 2.0 inches above the point of contact between
> tape and inclined plane.

Water-Activatable Adhesive Tape

*R.W. Monte; U.S. Patent 4,105,824; August 8, 1978; assigned to Nashua Cor-
poration* describes a synthetic water-activated adhesive which can be utilized in
conjunction with paper sheets to form a water-activated tape. This synthetic
adhesive can be utilized to replace animal glue-based adhesives for joining the
ends of paperboard boxes. The adhesive of this composition includes water,
starch, dextrin and a polyacrylamide.

When utilizing dry polyacrylamide as the source of the polyacrylamide component, an important aspect of the process is the manner in which the adhesive
is prepared. In this situation, the adhesive is prepared by mixing the dextrin
and polyacrylamide together under conditions which prevent agglomeration.
After the dextrin and resin are sufficiently mixed, the starch is added.

The starch should be a waxy-maize variety of cornstarch which contains 95
weight % amylopectin and has a molecular weight between 20,000 and 50,000
and a low viscosity so that when it is dissolved in sufficient water to produce a
mixture having 45% solids, it will result in a solution with a viscosity of 12,000
to 16,000 cp at 160°F as measured by a Brookfield viscosity No. 3 spindle at
30 rpm. It should be present in such an amount that the final dry adhesive will
have an amylopectin content between 55 and 75% dry weight. A preferred
starch is American Maize Corporation's Amaizo 839.

The dextrin is either canary corn or yellow corn dextrin with a cold water
solubility of at least 90% (77°F) and may be prepared by the acid torrefication
of starch. Suitable commercial dextrins are Stadex 128 (A.E. Staley), Nadex
771 (National Starch), Amaize 1752 Dextrine (American Maize Co.) or Canary
S8032 (Corn Products).

The polyacrylamide must have an Oswald viscosity of from 2.5 to 2.7 cp and a
Brookfield viscosity no greater than 600 cp. It also must have a low degree of
hydrolysis so that only 5% or less of the original carboxamide group in the
polyacrylamide has been hydrolyzed to the carboxylic group, and should be
employed in amounts so that the final dry adhesive will have a polyacrylamide
content between 2 and 6% dry weight. A suitable polyacrylamide for use is
Dow Resin 164.

Daxad 11 (W.R. Grace Chemical Co.) has been used to advantage in the composition of the adhesive as a dispersing agent.

The composition of the dry adhesive of the process is set forth in Table 1. The term "dry" is intended to describe the adhesive as it appears on a coated substrate.

Table 1

Ingredients	Useable	Range Preferred
	......(percent by weight)......	
Water	5–12	6–10
Starch	55–75	60–70
Dextrin	15–40	25–35
Polyacrylamide	2–6	3–5
Dispersing agent	0–10	0.5–0.8

A particular formulation for an adhesive having a composition within the range set forth in Table 1 is as follows:

Table 2

Ingredients	Amount	
Water	550	gallons
Daxad TT	50	pounds
Foamaster B	2	capfuls
Dow 164	225	pounds
Canary Dextrine	1,900	pounds
Amaizo 839	4,400	pounds

A water activated tape can be prepared as follows: The adhesive composition is applied to a fiber glass reinforced kraft paper backing by being passed in the nip formed by a coating roll and two smaller nip rolls in a "kissroll" system. During application the adhesive composition is maintained at about 160° to 170°F. After being coated, the paper is passed through an oven maintained at a temperature between 300° and 500°F to reduce the tackiness of the adhesive and to drive off all but about 10% or less of the water. A suitable coating weight is between 17 and 20 pounds per ream of paper. The adhesive-coated paper is then wound on a dispenser roll.

The most important part of the procedure is that the resin and dextrin be allowed to mix by themselves before adding the starch. Numerous tests have indicated that when utilizing dry polyacrylamide, if the starch is added to the composition along with or prior to intimate mixing of the dextrin and resin, the tackiness of the final adhesive is inferior and the viscosity is high.

The coated paper was found to have a McLaurin value of 90 to 100. The coated paper was tested in a Fipago Testor by moistening the paper, allowing an open setting time of 1½ seconds and a closed setting time of 1 second. A typical value for the coated paper was 57. In a quick-grab test, the paper was moistened and applied to a corrugated carton. A closed setting time of only 6 to 10 seconds was needed to produce a fiber-tearing bond.

Aqueous Dispersions for Bonding Polyolefin Plastics

In an attempt to obtain pressure sensitive adhesives which have high bond strength and may be used without environmental pollution and the like, *M. Mori and R. Fukata; U.S. Patent 4,110,290; August 29, 1978; assigned to Toyo Ink Manufacturing Co., Ltd., Japan* have found that an acrylate copolymer will be remarkably enhanced in adhesiveness to polyolefin films and shaped articles by mixing the copolymer with an ethylene-vinyl acetate copolymer.

These pressure sensitive adhesives comprise, by weight: (a) 60 to 95 parts as solid matter of an aqueous copolymer dispersion obtained by emulsion polymerizing in water a mixture of the following monomers:

(1) 60 to 99 parts of at least one monomeric acrylate having a C_{4-12} alkyl group;

(2) 1 to 7 parts of at least one monomer selected from the group consisting of α,β-unsaturated carboxylic acids, acrylate and methacrylate having a hydroxyl group, acrylamide and derivatives thereof, each monomer being copolymerizable with monomer (1); and,

(3) 0 to 39 parts of at least one monomer selected from the group consisting of vinylic copolymerizable monomers other than monomers (1) and (2), and vinylidene copolymerizable monomers;

and (b) 5 to 40 parts as solid matter of an aqueous dispersion of an ethylene-vinyl acetate copolymer (EVA). The aqueous copolymer dispersions (a) and (b) may preferably be a 30 to 60% dispersion and a 40 to 60% dispersion, respectively.

The monomeric alkyl acrylate may be selected from the group consisting of butyl acrylate, hexyl acrylate, octyl acrylate and 2-ethylhexyl acrylate, the α,β-unsaturated carboxylic acids are acrylic, methacrylic, crotonic, itaconic, maleic and fumaric acids; the hydroxyl group-containing acrylates and methacrylates are β-hydroxyethyl acrylate, β-hydroxypropyl acrylate, β-hydroxyethyl methacrylate and β-hydroxypropyl methacrylate; and the monomer (3) is a member selected from the group consisting of methyl acrylate, ethyl acrylate, propyl acrylate, methyl methacrylate, ethyl methacrylate, propyl methacrylate, vinyl acetate, styrene and acrylonitrile.

An emulsifier which is polyoxyethylene alkyl phenol ether sulfate, polyoxyethylene alkyl phenol ether sulfonate, an alkyl benzene sulfonate or an α olefin sulfonate or a mixture of one of those emulsifiers with polyoxyethylene alkyl phenol ether, polyoxyethylene stearate, polyoxyethylene or polyoxypropylene is added during the emulsion polymerization, as is a thickener to increase the viscosity to 6,000 to 20,000 centipoises.

Adhesives made with such formulations showed high bond strength to articles made of polyolefins.

Curing on Contact with an Amine-Aldehyde Condensation Product

The pressure sensitive adhesives as formulated by *S.D. Pastor and M.M. Grover; U.S. Patent 4,113,792; September 12, 1978; assigned to National Starch and Chemical Corporation* are particularly characterized by both aggressive tack and cohesive strength and comprise from 50 to 85% by weight of an acrylate-based pressure sensitive adhesive polymer, 4 to 24% of a chlorosulfonated polyethylene and 11 to 40% of a polymerizable vinyl monomer or unsaturated oligomer or mixtures thereof. These compositions are cured by contact with an initiator comprising the condensation reaction product of an aldehyde and a primary or secondary amine.

The preferred pressure sensitive adhesive polymer bases used in the formulation include the interpolymers of 2-ethylhexyl, octyl or butyl acrylate, vinyl acetate and acrylic acid where the acrylate ester constitutes not less than 60% of the polymer polymerized to a relatively high molecular weight with conventional free radical initiators and characterized by a Williams plasticity number of 2.1 to 3 mm.

The chlorosulfonated polyethylene components employed contain from 25 to 70 wt % chlorine and 3 to 160 mmol sulfonyl chloride moiety per 100 g of polymer; and the polyethylene from which the polymer is prepared should have a melt index of 4 to 500. Suitable chlorosulfonated polyethylene polymers are available commercially from E.I. DuPont de Nemours & Co. under the tradename Hypalon.

As a third component of the pressure sensitive adhesives there is required at least one low molecular weight vinyl monomer or unsaturated oligomer.

In general, the pressure sensitive acrylic polymer will be employed in amounts of 50 to 85%, preferably 55 to 70%, by weight of the final improved adhesive composition; the chlorosulfonated polyethylene in an amount of 4 to 24%, preferably 8 to 15%, by weight; and the low molecular weight polymerizable monomer and/or unsaturated oligomer in an amount of 11 to 40%, preferably 22 to 35%, by weight.

It is required that the condensation reaction product of a primary or secondary amine and an aldehyde be employed as an initiator in curing of the adhesive. Particularly preferred initiators are the butyraldehyde-aniline and butyraldehyde-butyl amine condensation products sold as Accelerator 808 and Accelerator 833.

Example: A pressure sensitive adhesive polymer comprising 70% by weight of 2-ethylhexyl acrylate, 24% vinyl acetate and 6% acrylic acid is prepared by conventional free radical polymerization procedures. The polymer is characterized by a Williams plasticity number (ASTM D-926) of 2.0 to 2.5 mm and has a solids content of 35% in ethyl acetate.

77 parts (anhydrous weight) of the adhesive polymer were then combined with 10.6 parts Hypalon 30 (parts are on an anhydrous basis, Hypalon provided as a 50% solids solution in ethyl acetate), 9.1 parts trimethylolpropane triacrylate and 2.3 parts α,α-dimethylbenzyl hydroperoxide.

A 5-mil dry film of the resultant composition on release paper was transfer coated to a 2-mil thick polyester film. An adhesive bond was made to stainless steel primed with an aniline-butyraldehyde condensation product (Accelerator 808) as initiator. A 4 psi hold test (Shear Adhesion test) was run on the sample and the film found to display more than 2-hour holding power at 150°C, thus showing increased high temperature performance over a 5-mil film of the starting pressure sensitive adhesive polymer which had not been treated in accordance with the described process and which exhibited only a 3-minute hold at those elevated temperatures.

Improved Stability in Contact with Iron

The pressure sensitive adhesives prepared by *R.G. Marchessault, T.P. Carter, Jr. and M.M. Williams; U.S. Patent 4,121,028; October 17, 1978; assigned to Celanese Polymer Specialties Company* are made from acrylic polymers plus chelate esters of orthotitanic acid and show improved stability when in contact with unprotected surfaces of ferrous metals. These compositions are obtained by adding a tertiary amine to an organic solution of an acrylic interpolymer containing an interpolymerized carboxylic acid and a chelate ester of orthotitanic acid, the tertiary amine being added in the amount of 0.01 to 1.5 equivalents of tertiary amine for each equivalent of the carboxylic acid component of the interpolymer.

The acrylic interpolymer is made from (1) 0.5 to 20 wt % of a polymerizable monoethylenically unsaturated carboxylic acid, (2) at least one monomer selected from the group consisting of alkyl esters of acrylic acid or methacrylic acid, or dialkyl esters of maleic or fumaric acid, where the alkyl group contains 4 to 18 carbon atoms, and (3) optionally, another monomer copolymerizable therewith. The acrylic interpolymer has a weight average molecular weight in the range of 10,000 to 500,000 and a glass transition temperature in the range of 0° to -75°C.

The chelate ester of orthotitanic acid has the formula

$$(R_1O)_n Ti \begin{bmatrix} O \!-\!\!-\!\!-\! CH \overset{\displaystyle R_2}{\diagdown} \\ \qquad\qquad C\!-\!R_4 \\ O \!-\!\!-\!\!-\! C \underset{\displaystyle R_3}{\diagdown} \end{bmatrix}_{4-n}$$

In the formula on the preceding page, n is an integer of 2 or 3; R_1 is C_{2-10} alkyl, alkenyl, substituted alkyl or substituted alkenyl group; R_2 is C_{1-6} alkyl, alkoxy, alkenyl, or alkenoxy group; R_3 is a C_{1-6} alkyl or alkenyl group or a C_{6-10} aryl group; and R_4 is hydrogen or a C_{1-6} alkyl or alkenyl group; and R_2 and R_3 may be combined as an ethylene or trimethylene group.

A preferred ethylenically unsaturated carboxylic acid for use in the interpolymer is acrylic acid, and preferred esters are isobutyl acrylate, 2-ethylhexyl acrylate, and 2-ethylhexyl methacrylate. Examples of other monomers which optionally can be used are vinyl acetate, propionate or butyrate, methyl acrylate, diethyl maleate, hydroxypropyl methacrylate, etc.

It has been found that a small amount of tertiary amine substantially prevents color formation, inhibits corrosion and retards gelation when these pressure sensitive adhesives are in contact with ferrous metal surfaces. The tertiary amine is added in the amount of 0.01 to 1.5 equivalents of amine for each carboxylic acid equivalent in the interpolymer, and, preferably, 0.05 to 0.5 equivalent.

The tertiary amines have the formula $NR_1R_2R_3$, where R_1 is an alkyl radical containing 2 to 6 carbon atoms, R_2 and R_3 are alkyl radicals containing 2 to 4 carbon atoms. Preferred amines are triethylamine and dimethylethanol amine.

Example: To a suitable reactor were added 599 parts of denatured ethanol and 995 parts of ethyl acetate. Agitation was begun and heat was applied raising the temperature to reflux, about 160°F. A monomer catalyst mixture of 121 parts of acrylic acid, 640 parts of methyl acrylate, 1,189 parts of 2-ethylhexyl acrylate and 1.43 parts of azobisisobutyronitrile was added to the reactor over a one-hour period while keeping the temperature at reflux. Heating was continued at reflux for 75 minutes, after which time a solution of 9.8 parts of azobisisobutyronitrile in 159 parts of ethyl acetate was added over a period of one hour.

Heating at reflux was continued for four hours. At the end of this time, the temperature in the reactor was lowered to below 100°F and a solution of 42 parts of triethylamine in 119 parts of n-hexane was added to the reactor. A mixture of 19.5 parts of a solution of 75 wt % titanium diisopropyl diacetylacetonate in isopropyl alcohol, 58 parts of acetylacetone, 54 parts of ethyl acetate and 217 parts of n-hexane was added over a 60-minute period. 22 parts n-hexane and 87 parts of ethyl acetate were then added to the reactor.

After thorough mixing, the polymer solution was found to have a solids content of 45% and a viscosity, at 25°C, of 1,200 to 1,400 cp. The 0° Shear (Static Shear Adhesion Test) was greater than 24 hours; the 180° Peel, in pounds per inch of width, was 2.9, and the 90° Quick Stick, in pounds per inch of width, was 0.6.

An uncoated strip of steel was inserted into the pressure sensitive adhesive to test the effect of the adhesive on the metal. After 1½ weeks at room temperature, no discoloration, corrosion, or gelling was noted. When stored for the same period at 140°F, only a slight darkening of the solution was noted. After two months at 140°F, the solution had turned dark brown in color but exhibited no change in viscosity or loss in adhesive properties. At 110°F, the adhesive solution showed no change in color after one month and a very slight darkening after two months.

The identical adhesive solution, except with no triethylamine, was tested at room temperature in the same manner as described above. The solution discolored badly in one day, the steel began to rust in two to three days and after one week, the titanate had deposited on the metal.

Alkyl Acrylate Polymer with both Random and Terminal Hydroxyls

W.E. deVry, R.S. Drake and R.T. Morrissey; U.S. Patent 4,145,514; March 20, 1979; assigned to The B.F. Goodrich Company have formulated pressure sensitive adhesives with excellent rolling ball tack, 180° peel adhesion and shear adhesion which comprises the reaction product of (a) at least one alkyl acrylate liquid polymer containing both terminal and random hydroxyl-functionality, and (b) at least one prepolymer of a polyester glycol or a polyalkylene ether glycol with an excess amount of an aromatic diisocyanate.

Random hydroxyl functionality is defined as the presence of hydroxyl groups which are pendant from a portion of the polymeric backbone other than the ends, e.g., when a hydroxyl-containing comonomer such as 2-hydroxyethyl acrylate or vinyl benzyl alcohol is used. Excellent results were obtained using 2-hydroxyethyl acrylate. Alternatively, a carboxylated comonomer such as acrylic acid or the like may be used, and after polymerization to prepare the alkyl acrylate liquid polymer is complete, some or all of the carboxyl groups may be reacted with ethylene oxide, butanediol, or the like to produce hydroxyl groups. Excellent results were obtained using ethylene oxide, which when reacted with acrylic acid units in the polymeric backbone, produced backbone units equivalent to 2-hydroxyethyl acrylate.

Terminal hydroxyl groups are included in the alkyl acrylate liquid polymers using methods known in the art. One suitable method for preparing such is as follows:

> A carboxyl-terminated alkyl acrylate liquid polymer intermediate can be produced by the method of U.S. Patent 3,285,949, i.e., by polymerizing at least one alkyl acrylate monomer in a solvent with low chain transfer potential, preferably tert-butanol, using a bis-azocyano acid initiator having the formula:

$$HOOC-(CH_2)_n-\underset{\underset{CN}{|}}{\overset{\overset{R}{|}}{C}}-N=N-\underset{\underset{CN}{|}}{\overset{\overset{R}{|}}{C}}-(CH_2)_n-COOH$$

In the formula on the preceding page, n is an integer from 1 to 6 and R is an alkyl group containing 1 to 3 carbon atoms. Azodicyanovaleric acid is a preferred initiator for production of the carboxyl-terminated liquid polymer intermediate. Terminal hydroxyl groups may be introduced by reacting the carboxyl-terminated intermediate with ethylene oxide in the presence of a tertiary amine catalyst such as trimethylamine according to the process of U.S. Patent 3,712,916.

Component (b) of the pressure sensitive adhesive of the process comprises a prepolymer of a polyester glycol or a polyalkylene ether glycol with an excess amount of an aromatic diisocyanate. Suitable polyalkylene ether glycols are produced by methods well known to the art, contain from 3 to 6 carbon atoms per alkylene group, and have a molecular weight from 400 to 2,000, such as polypropylene ether glycol, polytetramethylene ether glycol (PTMEG) and the like. Preferred polyalkylene ether glycols contain from 3 to 5 carbon atoms per alkylene group and have a molecular weight from 400 to 1,500. Excellent results were obtained using polytetramethylene ether glycol having a molecular weight of about 600. Mixtures of polyalkylene ether glycols may also be used.

A cement of a carboxyl-terminated poly(n-butyl acrylate/butadiene/acrylonitrile/acrylic acid) liquid polymer was prepared by the method described above and converted to the random- and terminal-hydroxylated ester by reaction with ethylene oxide in the presence of trimethylamine. The polymer had a Brookfield viscosity at 25°C of about 500,000 cp and a hydroxyl number of 57.8.

The hydroxylated liquid polymer and Adiprene L-167 were mixed by hand with a spatula and spread 1-mil thick between Mylar sheets and cured at 150°C for 2 minutes. Testing for rolling ball tack, 180° peel adhesion at 70°C shear adhesion with controls—one a random hydroxylated n-butyl acrylate polymer and the other a terminal hydroxylated n-butyl acrylate polymer—showed that both terminal and random hydroxyl functionalities are necessary to give an adhesive having the full range of desirable adhesive qualities.

In certain applications, such as where two clear films are to be laminated to each other using a pressure sensitive adhesive, the adhesive must have improved clarity.

J.C. Gilles; U.S. Patent 4,145,511; March 20, 1979; assigned to The B.F. Goodrich Company has found that this is made possible by the addition of from 1 to 20 weight percent N-vinyl-2-pyrrolidone to the hydroxyl-containing alkyl acrylate liquid polymers used as one of the two ingredients in an adhesive as formulated in the previous patent.

Containing Acrylic Functional Aminocarboxylic Acids and Derivatives

By the process of *S.M. Heilmann; U.S. Patent 4,157,418; June 5, 1979; assigned to Minnesota Mining and Manufacturing Company*, high performance thermoplastic pressure sensitive adhesives are prepared by the copolymerization of N-methacryloylamino acids and acid derivatives with long-chained alkyl acrylates. The copolymers can be formulated into tapes that possess an excellent three-fold balance of the necessary tape properties of tack, adhesion and cohesion.

The adhesive compositions for application at ambient temperatures, e.g., $15°$ to $30°C$, comprise copolymers consisting essentially of monomers of (a) at least 80 mol % of one or more monomeric acrylic acid esters of a primary and/or secondary alcohol containing from 4 to 8 carbon atoms, and (b) not over 20 mol % of at least one comonomer represented by the structural formula:

$$CH_2{=}\underset{\underset{\displaystyle R^4}{|}}{\overset{\overset{\displaystyle R^1}{|}}{C}}{-}\overset{\overset{\displaystyle O}{||}}{C}{-}\overset{\overset{\displaystyle R^2}{|}}{N}{-}\overset{\overset{\displaystyle R^3}{|}}{C}{-}(CH_2)_a{-}\overset{\overset{\displaystyle O}{||}}{C}{-}X$$

wherein R^1 is H or methyl; R^2 is H, alkyl of 1–4 carbons or $-CH_2CH_2CN$; R^3 is H or alkyl of 1–4 carbons; R^4 is H, alkyl of 1–4 carbons, $-CH_2C_6H_5$, $-(CH_2)_2-SCH_3$, and $-(CH_2)_bC(O)Y$ where b is 1 or 2, where Y is NH_2 or OH; a is 0, 1, 2, or 3; X is Y; and R^2 and R^3 together can be $(CH_2)_3$.

Suitable class (a) monomers include n-butyl acrylate, isobutyl acrylate, sec-butyl acrylate, n-hexyl acrylate, 2-ethylhexyl acrylate, isooctyl acrylate, iso-amyl acrylate, sec-amyl acrylate and the like. Especially preferred class (a) monomers are isooctyl acrylate and 2-ethylhexyl acrylate.

Class (b) monomers are N-acryloyl and N-methacryloyl derivatives of amino acids and particularly of natural amino acids in optically active or racemic forms. Suitable class (b) monomers include the N-acryloyl and N-methacryloyl deriva-tives of amino acids such as glycine, alanine, β-alanine, 4-aminobutyric acid, leucine, phenylglycine, cystine, aspartic acid, glutamic acid and the like; N-acryloyl and N-methacryloyl derivatives of amino acid amides such as glycin-amide, alanamide, N-butylalanamide, and the like.

Although the aforementioned compounds are generally superior to conventional comonomers for the production of pressure sensitive adhesive formulations from alkyl acrylate esters, especially preferred are the N-methacryloyl amino acids.

The class (a) and (b) monomers can be copolymerized by standard techniques, e.g., using free radical initiators in solvents or aqueous emulsion. Suitable initia-tors include persulfates, benzoyl peroxide, di-tert-butyl peroxide, azobisisobu-tyronitrile, and the like. Suitable polymerization solvents include ethyl acetate acetone, methyl ethyl ketone, benzonitrile, and the like.

One convenient procedure which has been employed is to conduct the poly-merization solution in acetone at a monomer mol fraction of 0.18, using azobis-isobutyronitrile (0.3 mol % based on total mols of monomer) and heating with agitation at $55°C$ for 21 hours followed by heating at $60°C$ for 3 hours. As is well known in the art, vinyl polymerizations can be conducted over a wide range of conditions, the solution viscosity and desired molecular weight of products determining specific conditions. Thus, suitable monomer concentrations can be a mol fraction from 0.05 to 0.50, or preferably 0.1 to 0.25; catalyst concentrations can be from 0.1 to 1.0%; temperatures can be from less than 25° to 100°C; and reaction times can be from a few minutes to several days.

RUBBER-BASED

Sulfobutyl Rubbers plus Tackifiers

C.P. O'Farrell and J.J. Higgins; U.S. Patent 4,072,648; February 7, 1978; assigned to Exxon Research & Engineering Co. describe the preparation of sulfonated olefinically unsaturated elastomers having particular sulfo-group content and unique properties by the reaction of natural or synthetic elastomers using various complexes of SO_3.

The rubbers which may be sulfonated are the olefinic and unsaturated elastomers such as butyl rubber, halogenated butyl rubber, ethylene-propylene-conjugated or nonconjugated diolefin terpolymers, polyisobutylene, styrene-butadiene rubbers, polybutadiene, polyisoprene, natural rubber and the various types of heretofore well-known rubbers containing either high or low olefinic unsaturation.

The sulfonation, using either an SO_3-complex or acyl sulfates, is carried out so as to give a final sulfo rubber containing between about 0.1 and 5.0 mol % SO_3H groups, preferably between 0.5 and 2.0 mol %. The neutralized, partially neutralized, or unneutralized elastomers are then mixed with tackifiers, i.e., phenol-formaldehyde resins, polyisobutylene, alkylphenol formaldehyde resins, petroleum resins, etc. in an amount of between about 5 and 90 phr in between about 10 and 50 wt % solids concentration in organic solvents.

These cements are used for adhering porous or nonporous materials such as textiles, wood, cloth, sheets or strips of metals, plastics, etc. to each other. High green strength, high tensile strength, high resistance to peel, even at elevated temperatures, water impermeability, etc. characterize such adhesive layers in laminates.

A preferred solvent, 90% toluene–10% isopropyl alcohol, is used in compounding the cement and preferably an organic amine, e.g., ethylamine, is used to partially or completely neutralize the sulfonated elastomer, although inorganic bases may also be used for neutralization.

The surfaces to be adhered to one another by means of the adhesive contact cement are merely sprayed, brushed, dipped or otherwise coated with the cement to a thickness of from 1 to 15 mils, after which they are contacted with one another under mild pressure for a sufficient length of time to allow the solvent of the cement to vaporize therefrom leaving the solid material in place as the intermediate layer of the laminate.

Example 1: Two butyl cements in normal hexane (3 liters) each containing 14.2 wt % solids [butyl rubber of viscosity average molecular weight (Mv) of 350,000] were each reacted with 8.24 cc and 16.48 cc, respectively, of an acetyl sulfate solution. The acetyl sulfate solution was made up by reacting 23.4 ml of acetic anhydride with 11 ml of 96% sulfuric acid. The two cements were separately neutralized with stoichiometric amounts of ethylamine (70% in H_2O) and 0.6 g of phenyl-β-naphthylamine was added as an antioxidant. The polymers were then steam stripped and hot mill dried. Polymer number 1 contained 0.7 mol % of $SO_3^-EA^+$ (ethylamine) and polymer number 2 contained 1.5 mol % of $SO_3^-EA^+$ (ethylamine).

Example 2: Polymer number 2 of Example 1, containing 1.5 mol % of ethylamine fully neutralized SO_3H groups, was dissolved at 20 wt % total solids in a 90-10 toluene-isopropyl alcohol solution. It had a Brookfield solution viscosity of 7,000 cp. A cast, dried film of this cement, 6 mils thick, exhibited a tensile strength of 3,850 psi and an elongation at break of 1,050%. This cement was used to laminate cotton duck cloth to cotton duck cloth, cotton duck cloth to steel, and polypropylene film to polypropylene film.

Adhesion of these substrates was measured by an Instron peel test measurement at a pull rate of 2 inches per minute at 72°F. In the case of the cotton duck cloth laminated to cotton duck cloth, the T-peel adhesion measured was 25 to 30 psig. In the case of the cotton duck cloth adhering to the steel panel, the 180° peel adhesion was 3 psig, and in the case of the polypropylene film laminated to polypropylene film, the T-peel adhesion was 1 psig.

Elastomers from C₅-Hydrocarbons

The process developed by *P.R. Lakshmanan, H.E. Swift and C.Y. Wu; U.S. Patent 4,072,808; February 7, 1978; assigned to Gulf Research & Development Company* has to do with elastomers synthesized from crude C_5-hydrocarbons, preferably crude C_5-hydrocarbon mixtures which are by-products in the production of ethylene. A selected portion of the crude C_5-hydrocarbons are copolymerized in the presence of a catalyst system to prepare an elastomer having the requisite properties to produce a pressure sensitive adhesive when added to a natural or synthetic tackifier. C_5-hydrocarbon mixtures suitable for use herein are preferably obtained from a feed stream comprising a mixture of hydrocarbons having the composition defined in the table below.

Component	Broad Range	Preferred Range
	. . . . (wt %).	
C_5-monoolefinic hydrocarbons	5-50	10-30
C_5-diolefinic hydrocarbons	5-50	10-30
C_5-paraffins	5-75	20-40
C_6-aliphatics	5-75	5-40
Benzene	5-75	10-50

The process for preparing the elastomer from these hydrocarbons consists of (A) heat-soaking the process hydrocarbon mixture at a temperature sufficient to dimerize cyclopentadiene; (B) separating the resulting hydrocarbon mixture from the cyclopentadiene dimer; and (C) copolymerizing the resulting hydrocarbon mixture in the presence of a catalyst which comprises an iron complex, a trialkyl aluminum and a bidentate ligand capable of both pi and sigma bonding to form an elastomer.

The heat-soaking process is carried out in an autoclave at a temperature preferably from 110° to 150°C. After the heat-soaking, the hydrocarbon mixture comprises:

Component	Weight Percent
Isoprene	20–30
Piperylene	10–30
Cyclopentadiene	0–1
C_5-paraffin	10–30
C_5-olefin	10–30

The catalyst is selected from iron(III) acetylacetonate, triethylaluminum, 2-cyanopyridine; iron(III) naphthenate, tributylaluminum, 2-cyanopyridine; or iron(III) octoate, triethylaluminum, phenyl-2-pyridylacetonitrile; and comprises from 0.1 to 5.0 weight percent of the C_5-hydrocarbon mixture. The hydrocarbon mixture is copolymerized at a temperature of about 25° to 50°C at a pressure of about 0 to 14 kg/cm² for about 2 to 6 hours.

The elastomer finally should preferably consist of at least 95% by weight of isoprene and piperylene, in a molar ratio of from 5:1 to 1:1, and have a molecular weight of from 10,000 to 1,000,000. The elastomers should have a viscosity range from 1,000 up to about 50,000 cp at 75°F when mixed at 20 weight percent with toluene. They have high extensibility as indicated by an elongation of over 500%.

In further work by *P.R. Lakshmanan, H.E. Swift and C.Y. Wu; U.S. Patent 4,096,103; June 20, 1978; assigned to Gulf Research & Development Company* a pressure sensitive adhesive composition is formulated consisting essentially of a tackifier selected from the group consisting of: (1) an isoprene-piperylene copolymer having a ring and ball softening point of 10° to 135°C, a molecular weight of 400 to 3,000, and a weight ratio of the isoprene to piperylene of 1:10 to 10:1; (2) a piperylene/2-methyl-2-butene copolymer having a ring and ball softening point of 10° to 135°C, a molecular weight of 400 to 3,000 and a weight ratio of piperylene to 2-methyl-2-butene of 1:10 to 10:1; (3) β-terpene resins having a ring and ball softening point of 10° to 135°C and a molecular weight of 300 to 3,000; or (4) rosin or rosin esters having a ring and ball softening point of 10° to 190°C and a molecular weight of 300 to 3,000; in combination with an elastomer comprising isoprene and piperylene having a molecular weight of at least 4,000, preferably at least about 10,000; and a solvent.

Several pressure sensitive adhesive compositions were prepared and evaluated. The data obtained are tabulated in the following table.

	. . . . Run Number			
	1	2	3	4
Piccolyte S-100*, wt %	10	—	—	—
Zonester 85**, wt %	—	10	—	—
Wingtac 95***, wt %	—	—	10	—
Sta-tack 100†, wt %	—	—	—	10
Elastomer, wt %	10	10	10	10
Toluene, wt %	80	80	80	80
Tack (ASTM D-2979-71), g/cm^2	261	340	393	431
Peel strength (ASTM D-903-49), g/cm	110	241	310	268

*β-terpene resin tackifier.
**Glycerol ester of tall oil rosin.
***Piperylene/2-methyl-2-butene copolymer.
†Isoprene-piperylene based copolymer having a molecular weight of about 3,000 or less.

In these compositions the elastomer was an isoprene-piperylene copolymer having a molecular weight greater than 10,000 prepared from C_5-hydrocarbons obtained from an ethylene plant feed stream by steam-heating the C_5-hydrocarbons in an autoclave under a nitrogen atmosphere at 125°C and a pressure of 300 psig for 2 hours to dimerize cyclopentadiene. The C_5-hydrocarbons were next separated from the other components by vacuum distillation. The elastomer was prepared by adding 300 ml of the treated C_5-hydrocarbon fraction, and a catalyst comprising 2.0 g of phenyl-2-pyridylacetonitrile, 3.5 g of iron(III) acetylacetonate, and 4.0 ml of triethylaluminum to a reaction vessel. The reaction mixture was stirred at room temperature for 4 hours. The catalyst was inactivated by adding 50 cc of methyl alcohol containing 5 cc of hydrochloric acid to the elastomer (e.g., a sufficient amount to inactivate the catalyst). The elastomer was recovered using conventional methods, for example, filtering the product from the other hydrocarbons, and dried under vacuum at room temperature overnight.

The pressure sensitive adhesives were formulated by preparing a 20% by weight solution of the tackifier in toluene and a 20% by weight solution of the elastomer in toluene. The solutions were mixed using conventional apparatus for 10 hours.

Silicate, Borate, or Carbonate Adhesion Promoters

C.A. Uraneck, J.E. Burleigh and O.L. Marrs; U.S. Patent 4,092,465; May 30, 1978; assigned to Phillips Petroleum Company describe a pressure sensitive adhesive having as its constituents a rubbery polymer having an alkadienol constituent and adhesion promoters chosen from among CO-, boron- and silicon-containing compounds. In a preferred embodiment the rubbery polymer has as a constituent at least one monohydroxy-alkyl-substituted conjugated diene.

The alkadienol-containing polymers for use in the adhesive formulations are generally rubbery homopolymers, copolymers or terpolymers containing at least one monohydroxyalkyl-substituted conjugated diene. Possible co- and termonomers include conjugated dienes and copolymerizable vinylidene-group-containing monomers.

Preferred alkadienols for use in the formulations are 2,4-pentadien-1-ol; 2-methylene-3-buten-1-ol; and 3,5-hexadien-1-ol, with the first-mentioned particularly preferred.

Conjugated dienes useful as co- or termonomers with the abovedescribed alkadienols are those generally containing from 4 to 12 carbon atoms per molecule and preferably from 4 to 6 carbon atoms per molecule. Butadiene and isoprene are especially preferred.

Co- or terpolymerizable vinylidene-group-containing monomers include preferably the monovinyl-substituted aromatic compounds containing from 8 to 16, and more preferably from 8 to 10, carbon atoms per molecule.

The adhesion promoters useful in the pressure sensitive adhesives include compounds of the general formula $(R''Z)_n Y$ where Y is Si, B or CO; where Z is O, S, NH or NR'' with the proviso that Z can be NH or NR'' only when Y is Si; where n is 2 if Y is CO, 3 if Y is B and 4 if Y is Si; and where the R'' groups can be the same or different and are alkyl or cycloalkyl groups containing from 1 to 18, preferably 1 to 10, carbon atoms per R'' group, with the further proviso that the adhesion promoters contain no more than 40 carbon atoms per molecule. For solvent-containing pressure sensitive adhesives and for hot melt pressure sensitive adhesives it is especially preferred that the adhesion promoters contain 1 to 3 and 5 to 10 carbon atoms per R'' group, respectively.

Specific examples of compounds useful as adhesion promoters include dimethyl carbonate, di-n-hexyl carbonate, dicyclohexyl carbonate, methyl ethyl carbonate, triethyl borate, tricyclopentyl borate, tetraethyl silicate, tetra-n-decyl silicate, trimethyl thioborate, tetramethyl thiosilicate, octaethyl silane tetraamine, and the like. Tetraethyl silicate is preferred.

Using a Thermoplastic Elastomeric Styrene-Butadiene Block Polymer

An adhesive composition having superior tackiness, adhesive strength and cohesive force is made by *H. Inoue, H. Komai and S. Okada; U.S. Patent 4,104,327; August 1, 1978; assigned to Nippon Zeon Co., Ltd., Japan* comprising:

> (A) 100 parts by weight of an elastomer consisting of (1) 30 to 100% by weight, preferably 35 to 85%, of a thermoplastic elastomer expressed by the general formula $(S-B)_n$ or $(S-B)_{n-1}-S$ in which S is substantially a polystyrene block, B is substantially a polybutadiene block, and n is an integer of from 2 to 10; and (2) 70 to 0% by weight, preferably 65 to 15% by weight, of a thermoplastic elastomer expressed

by the general formula $(S-I)_n$ or $(S-I)_{n-1}-S$ in which S
is substantially a polystyrene block, I is substantially
a polyisoprene block, and n is an integer of from 2
to 10; and,

(B) 50 to 250 parts by weight, preferably 60 to 200
parts by weight, of a hydrocarbon resin containing 40
to 95% by weight, preferably 60 to 90% by weight,
of a 1,3-pentadiene unit and 60 to 5%, preferably 40
to 10%, by weight of an α-methylstyrene unit in the
polymer chain and having a softening point of 60° to
140°C.

Both of the thermoplastic elastomers (1) and (2) used are block polymers having a styrene content preferably of 15 to 55% by weight, and an average molecular weight preferably of 40,000 to 200,000. These elastomers include "tapered
block polymers" which contain a copolymer having a small amount of a styrene unit in a proportion progressively increasing from one end of the block to
the other as a substantial polybutadiene block or a substantial polyisoprene
block, or hydrogenated block polymers in which the double bonds of the polybutadiene block or polyisoprene block are hydrogenated partly or wholly.
These block polymers can be prepared by known methods using lithium-type
initiators, for example, the methods described in U.S. Patents 3,251,905 and
3,265,765.

Example 1: Preparation of Block Polymer — 15 parts styrene, 70 parts of 1,3-
butadiene and 15 parts of styrene were successively fed, and polymerized in
the presence of cyclohexane using n-butyl lithium as a catalyst to afford a block
polymer of the S–B–S type having a molecular weight of about 148,000 and a
styrene content of 30%. The block polymer obtained was thermoplastic and
elastomeric.

Example 2: 100 parts of the block polymer of the S–B–S type obtained in
Example 1 were dissolved in 600 parts of toluene, and 100 parts of each of the
tackifying resins shown in the table below, 60 parts of a naphthenic process oil
(Shellflex 371N) and 1 part of an antioxidant were added to form a solvent-
type pressure sensitive adhesive containing nonvolatile components in a concentration of about 30%.

The solution obtained was coated to a thickness of 25 μ on a polyester film
having a thickness of 1 mm to form an adhesive tape, and its tackiness, adhesive
strength and cohesive strength were measured.

The tackiness was measured as follows: A 10 cm adhesive tape sample was
placed on the inclined surface of a stainless steel plate inclined at 30° with the
adhesive surface facing upward. 32 stainless steel balls with a diameter varying
from $\frac{1}{32}$ to 1 inch were allowed to roll over the adhesive surface of the tape
from a position 10 cm upward of the upper edge of the tape. The tackiness is
expressed by the diameter of a ball that has the largest diameter among those
which stop on the adhesive tape.

The adhesive strength was measured as follows: An adhesive tape sample, 25 mm wide and 100 mm long, is bonded onto a stainless steel plate polished with a No. 280 waterproof abrasive paper, and peeled in a direction at an angle of 180° at a speed of 200 mm/min at 25°C. The peel strength measured is the adhesive strength.

The cohesive strength was measured as follows: An adhesive tape sample is bonded to a similarly treated stainless steel plate so that its area of 25 x 100 mm makes contact with the tape surface. A load of 1 kg is exerted vertically on the lower end of the tape at 50°C, and the time required until the adhesive tape is displaced and falls off from the stainless steel plate is measured. The results are shown in the table:

Sample	Tackifying Resin	Tackiness (x $\frac{1}{32}$ in)	Adhesive Strength (g/in)	Cohesive Strength (min)
Process				
1	Polymer A	15	800	120
2	Polymer B	20	930	150
3	Polymer C	27	780	110
4	Polymer D	20	920	120
5	Polymer E	25	910	115
Control				
6	Commercial Grade A	>3	did not adhere	unmeasurable

Polymer A: 70% 1,3-pentadiene, 30% α-methylstyrene
Polymer B: 70% 1,3-pentadiene, 20% α-methylstyrene, 10% cyclopentene
Polymer C: 60% 1,3-pentadiene, 25% α-methylstyrene, 10% cyclopentene, 5% 1,3-butadiene
Polymer D: 60% 1,3-pentadiene, 25% α-methylstyrene, 10% cyclopentene, 5% dicyclopentadiene
Polymer E: 60% 1,3-pentadiene, 30% α-methylstyrene, 10% cyclopentene
Commerical Grade A: Aromatic hydrocarbon resin having softening point of 100°C.

It can be seen from the results obtained that these adhesive compositions containing a copolymer derived from 1,3-pentadiene and α-methylstyrene as a tackifying resin have very superior tackiness and cohesive strength and superior adhesive strength to those which contain aromatic hydrocarbon resins as the tackifying resin.

Treatment with Peroxides

Pressure sensitive adhesives having permanent tackiness are formulated by *D.A. Weemes and R.L. McConnell; U.S. Patent 4,105,718; August 8, 1978; assigned to Eastman Kodak Company* by treating substantially amorphous polyolefin/hydrocarbon rubber blends with peroxide.

Substantially amorphous polyolefins useful for the formulation include homopolymers of 1-olefins containing 3 to 5 carbon atoms, copolymers prepared from ethylene and 1-olefins containing 3 to 5 carbon atoms, propylene/1-butene copolymers, propylene or 1-butene/higher 1-olefin copolymers containing less than 40 mol percent of at least one higher 1-olefin of the group consisting of 1-hexene, 1-heptene, 1-octene, 1-nonene, and 1-decene. These substantially amorphous polyolefins can be used as a blend with crystalline polyolefins or blends of such substantially amorphous polyolefins. The crystallizable polymer component of these blends, however, should not exceed a concentration of about 20%. The melt viscosity of the substantially amorphous polyolefin materials is generally in the range of 100 to 200,000 cp at 190°C as measured by ASTM D-1238.

Useful rubbery hydrocarbon polymers include ethylene/propylene copolymer rubbers, ethylene/propylene/diene terpolymers or blends of such hydrocarbon rubbers. Lower molecular weight hydrocarbon rubbers are preferred. Hydrocarbon rubbers with melt viscosity as low as about 3,000 cp are highly suitable.

One substantially amorphous polyolefin useful in this process is the essentially noncrystalline hexane-soluble amorphous polypropylene formed in minor amounts during the production of crystalline polypropylene by the polymerization of propylene in the presence of stereospecific catalysts by a process as disclosed in U.S. Patent 3,679,775. One such commercially available amorphous polyolefin is the Eastobond M-5 polyolefin available from Eastman Chemical Products, Inc.

The peroxides useful in the process include, for example di-tert-butyl peroxide, 2,5-dimethyl-2,5-di(tert-butylperoxy)hexane (Lupersol 101), dicumyl peroxide, and cumene hydroperoxide. In general, the alkyl peroxides such as di-tert-butyl peroxide and the Lupersol 101 type materials are preferred in order to provide polymers having a high degree of tack. The peroxide concentrations can preferably range from 3 to 20 weight percent, based on the amorphous polymer, and treatments are generally conducted at from 175° to 225°C.

Example 1:   Enjay EPT 3509 rubber (ethylene/propylene/diene terpolymer, I.V. 2.1) is placed in a large test tube and vacuum stripped to remove the air. It is then placed into a 400°C sand bath under a nitrogen atmosphere for 25 minutes to thermally degrade the polymer. The degraded EPT has a melt viscosity of 17,000 cp at 190°C. Low viscosity amorphous polypropylene (42.75 g; Eastobond M-5W, melt viscosity 2,500 cp measured in melt indexer at 190°C) and thermally degraded Enjay EPT 3509 rubber [2.25 g (5 wt %); melt viscosity 17,000 cp at 190°C] are melted in a Brabender Plastograph at 200°C. A solution of 9.0 g (20 wt %) Lupersol 101 in 15 ml of heptane is added to the molten polymers over a 10- to 12-minute period and the heptane is allowed to flash off. After about 30 minutes processing time in the Brabender Plastograph, the peroxide treated blend has a melt viscosity of 190,000 cp at 190°C. The product is permanently tacky and rubbery.

The sample is placed between Mylar film and a stainless steel plate and placed into a heated (140°C) Wabash hydraulic press using 2 tons of pressure. The sample is removed from the press and allowed to cool to room temperature. The coated tape (polymer thickness about 1 mil) performs well as a pressure sensitive tape material. For example, when the tape is pulled away from the stainless steel surface, no polymer residue is left on the steel, demonstrating that the polymer has good cohesive strength. The product may also be coated on Mylar tape from a solvent such as toluene or xylene. For example, a 10 wt % solution of the product in xylene forms a clear, colorless dope when cooled back to 23°C.

Example 2: A master batch of Eastobond M-5W containing 50 wt % Enjay EPT 3509 (I.V. 2.1) is made in a Brabender Plastograph for 30 minutes at 180°C. Low viscosity amorphous polypropylene (40.5 g; Eastobond M-5W, melt viscisity 2,500 cp at 190°C) and 4.5 g of the abovedescribed 50/50 Eastobond M-5W/Enjay EPT 3509 master batch are melted in a Brabender Plastograph at 200°C to provide a blend which contains 5 wt % EPT rubber. A solution of 9.0 g (20 wt %) of Lupersol 101 in 10 ml of heptane is added to the molten polymers over a 10-minute period and the heptane is allowed to flash off. After about 30 minutes process time in the Brabender Plastograph, the peroxide-treated blend has a melt viscosity of 72,000 cp at 190°C and the product is permanently tacky.

The product is coated on Mylar film from the melt by using a heated doctor blade (177°C) to give a polymer thickness of about 1 mil. The product performs well as a pressure sensitive tape material. For example, when the coated tape is bonded to a clean stainless steel surface, it peels away without leaving a residue, demonstrating that the polymer has good cohesive strength.

Based on A-B-A Block Copolymers

R. Korpman; U.S. Patent 4,136,071; January 23, 1979; assigned to Johnson & Johnson has discovered that a superior balance of properties for most industrial and other applications can be attained with pressure sensitive adhesive formulations employing particular types of A-B-A and A-B block copolymers in particular proportions. More specifically, the adhesive compositions comprise a thermoplastic elastomeric component and a resin component. The thermoplastic elastomeric component consists essentially of about 50 to 90 parts, preferably 60 to 80 parts, of a linear or radial styrene-isoprene-styrene A-B-A block copolymer and about 10 to 50 parts, preferably 20 to 40 parts, of a simple styrene-isoprene A-B block copolymer. The A-blocks are derived from styrene or styrene homologues and the B-blocks are derived from isoprene, either alone or in conjunction with small proportions of other monomers, in both the A-B-A and A-B block copolymers.

These adhesive compositions include about 20 to 300 parts, preferably 50 to 150 parts, of resin component per 100 parts by weight of the thermoplastic elastomeric component. The resin component consists essentially of tackifier resins for the elastomeric component. In general any compatible conventional tackifier resin or mixture of such resins may be employed, such as hydrocarbon resins, rosin and rosin derivatives, polyterpenes, and other tackifiers.

This adhesive possesses a superior balance of properties and is particularly advantageous for application by extrusion or melt system. Some difficulty has been experienced in coating adhesives based on A-B-A block copolymers in the form of a hot extrudate in that anchorage between the adhesive layer and certain backings is not always satisfactory. Efforts to correct this have resulted in loss of other important properties. This adhesive provides excellent mass anchorage, i.e., anchorage of the adhesive layer to its backing, without sacrificing adhesive strength, tack or quickstick and holding properties. It also exhibits excellent fiber-wetting on boxboard.

Ingredients and Characteristics	. . . . Example			
	1	2	3	4
S-I-S linear copolymer	100	85	75	50
S-I simple block copolymer	0	15	25	50
Wingtack 95 tackifier resin	80	80	80	80
Zinc dibutyl dithiocarbamate	2	2	2	2
2,5-Di-tert-amyl hydroquinone	0.5	0.5	0.5	0.5
Mass weight, oz/yd^2	4	3.9	3.8	4
Anchorage, oz/in	60	108	144	192
Adhesion to steel, oz/in width	45	41	46.5	57
Quickstick	5	9	9	16
20° hold to chrome, hr	>24	>24	18.33	12.5

Note: In the examples, all proportions are expressed in parts per 100 parts by weight of the total elastomeric component unless otherwise indicated.

In the above examples, Example 1 is a control which contains no S-I (styrene-isoprene) simple block copolymer. Examples 2, 3 and 4 contain 15, 25 and 50 parts of the S-I copolymer by weight of the total block copolymers with corresponding decreases in the amount of S-I-S copolymer. It will be seen that anchorage increases drastically with the addition of S-I copolymer, i.e., 80% with 15 parts, 140% with 25 parts and 220% with 50 parts. Quickstick also increases with the addition of S-I copolymer as does fiber wetting on boxboard. Adhesion to steel also increases at higher levels of S-I copolymer. On the other hand, hold, i.e., 20° hold to chrome, while it does decrease with the addition of S-I copolymer, remains at quite a high level with up to 50 parts S-I copolymer. At higher levels of S-I copolymer hold decreases substantially.

In the examples, the S-I-S linear copolymer and the S-I simple block copolymer both are polymerized from isoprene (the I blocks) and styrene (the S blocks) to produce 15% S blocks by weight of the total copolymer. The number average molecular weight of the S-I-S copolymer is 125,000 and 110,000 is the S-I copolymer.

With Low Volatile Liquid Content

Previously, solutions in organic solvents have been generally preferred for pressure sensitive adhesive formulations since solution-based compositions generally yield the best balance of adhesive properties. Unfortunately the use of volatile organic solvents presents many difficulties, such as high cost, fire and explosion hazards, toxicity hazards, and the generation of odors; and these difficulties must be solved for the sake of safety and to avoid environmental pollution.

C. Lee; U.S. Patent 4,145,321; March 20, 1979; assigned to Revertex Limited, England has formulated a pressure sensitive adhesive which comprises a maleinized liquid polybutadiene blended with one or more solid tackifying resins and one or more curing agents, the composition containing less than 20% by weight of volatile liquids.

The liquid butadiene polymers useful in preparing these compositions preferably have an average molecular weight in the range of 500 to 20,000, more preferably in the range of 1,000 to 10,000. The viscosity range of the polymers is from 2 to 2,500 poises at 50°C, preferably in the range from 5 to 500 poises.

The reaction between liquid polybutadiene and maleic anhydride is well known. This reaction is termed "maleinization" herein for the sake of convenience, and the product of the reaction, is termed "maleinized liquid polybutadiene." In this reaction, the maleic anhydride units are added randomly to the polymer, the anhydride structure being retained intact so long as ring-opening substances are excluded.

The conditions are conventional, requiring elevated temperatures of, for example, 190°C over a period of about 4 hours or 180°C for 5 hours in the absence of oxygen. Any liquid solvent, diluent or plasticizer, which is required in the final composition in order to reduce its viscosity, may conveniently be present during maleinization in order to maintain the reactants in a fluid condition and to help exclude oxygen, but such a substance should preferably be chemically inert towards the anhydride structure during maleinization. It is usually necessary to add a small proportion, less than 1% by weight for example, of a hindered phenol in order to obviate crosslinking, and catalysts for maleinization may also be included.

The amount of maleic anhydride added is generally sufficient to provide from 1 to 6 (preferably 1 to 3) anhydride groups per polybutadiene macromolecule. After the maleinization is complete, the reactants are cooled. It is convenient to add the tackifying resin and any other inert constituents to the warm maleinized liquid polybutadiene just after the maleinization step while the viscosity is still low.

Preferably, the tackifying resin used may amount to 25 to 55% by weight of the final composition, and may be a polyterpene, a petroleum hydrocarbon, a polybutene, etc. Suitable curing agents are polyamines, polyols, polyamides, etc.

Example 1: Preparation of Liquid Polybutadiene — A narrow molecular weight distribution, liquid polybutadiene is prepared using the following recipe and procedure:

Reagents	Amount
Dry toluene	1,500 g
n-Butyl lithium, 15 wt %	185 ml
Dry tetrahydrofuran	12.3 ml
Dry butadiene	1,500 g
Glacial acetic acid	17.1 ml

The toluene is charged to a 5-liter reactor which has been purged with nitrogen and which is equipped with a stirrer, a butadiene inlet tube, a temperature probe and a condenser adapted to be cooled using a solid carbon dioxide/methanol mixture. The temperature is raised to 50° to 55°C, the tetrahydrofuran and n-butyl lithium are charged, and the addition of butadiene is commenced at a rate of about 8.5 grams per minute. By external cooling, the temperature is maintained at 50° to 55°C. After 3 hours, the addition of butadiene is stopped, the glacial acetic acid is added, and the polymerization mixture cooled to ambient temperature.

Removal by filtration of the lithium acetate which is formed, and removal of the toluene by vacuum distillation yields a clear, liquid polybutadiene having a molecular weight of 5,000, a viscosity of approximately 100 poises (25°C), and a vinyl-1,2-microstructure content of approximately 45%.

Example 2: Maleinization of Liquid Polybutadiene — 2,500 g polybutadiene prepared as described in Example 1 are charged to a 5-liter reaction vessel with 12.5 g of an alkylated aryl phosphite antioxidant and 250 g xylene, and the mixture is heated to 120°C under a nitrogen blanket with efficient stirring.

At this temperature, 125 g maleic anhydride are added, and the reaction mixture is then heated to 180°C and held at this temperature for 5 hours. The resultant adduct is a clear, amber liquid having a viscosity of 65 poises at 25°C, and an acid value of 24 mg KOH/g.

Example 3: To 51 parts by weight of the maleinized polybutadiene of Example 2, containing approximately 5% by weight of combined maleic anhydride, are added 3.2 parts by weight of a complex of methylene bisaniline with sodium chloride, of equivalent weight 219 (Caytur 21), and 0.13 part of a phenolic antioxidant, Nonox WSO.

Separately, 49 parts by weight of a polyterpene resin of ring and ball softening point approximately 85°C (Nirez 1085) are dissolved in sufficient toluene to give a solution of 20 to 25% volatiles content. This solution is mixed with the blended polybutadiene to give a fluid adhesive composition which, when coated onto a filmic base to give a coating weight of 30 to 40 g/m^2, cured in a hot-air oven at 120°C for 5 minutes and cooled, gives a tacky adhesive film which has a peel adhesion when peeled through 180° from a clean glass plate at a separation rate of 4 in/min of greater than 600 g/in width of film.

Antistatic Composition

In the manufacture of wall coverings and shelf liners, comprising vinyl films and the like, it is known to apply pressure sensitive adhesive materials as a backing, covered by a sheet of silicone-treated release paper which, when removed, exposes the adhesive surface for contact with the wall or other substrate upon which the film is to be adhered.

Among the commonly employed pressure sensitive adhesive compositions for this purpose are the copolymers of the lower acrylic esters and vinyl monomers, particularly vinyl acetate. Removal of the protective paper from the adhesive causes an accumulation of electrostatic charges thereon which makes subsequent application of the film difficult in that it will prematurely adhere to materials carrying an opposite charge or resist contact with other materials of like charge.

R.G. Dolch and R.N. Kerr, Jr.; U.S. Patent 4,145,327; March 20, 1979; assigned to The Firestone Tire & Rubber Company have found that a pressure sensitive adhesive composition with improved antistatic properties can be formulated with an aqueous latex copolymer of a vinyl ester such as vinyl acetate and an alkyl acrylate such as 2-ethylhexyl acrylate or butyl acrylate to which an adhesive-compatible ammonium salt such as a quaternary ammonium salt has been added. The preferred salt is stearamidopropyldimethyl-β-hydroxyethylammonium nitrate and is present in an amount of from 0.5 to 5.0 parts per hundred parts of the latex copolymer. This antistatic pressure sensitive adhesive is prepared by dispersing the quaternary ammonium salt directly with the latex copolymer.

Combined with Aromatic Hydrocarbon Resin

S. Matubara and S. Iwai; U.S. Patent 4,146,514; March 27, 1979; assigned to Nippon Oil Company Ltd., Japan provide pressure sensitive adhesive compositions which contain synthetic or natural rubbers in combination with aromatic hydrocarbon resins which are less costly, more abundantly available and comparable to or even exceeding the quality of the resins previously used.

The composition contains at least 50% by weight of styrene-butadiene copolymers in combination with a resin component consisting of an aromatic hydrocarbon resin resulting from the polymerization of a thermally cracked petroleum fraction at temperatures in the range of $-30°$ to $+60°C$ in the presence of a Friedel-Crafts catalyst added in amounts of from 0.01 to 5 weight percent of the fraction. The fraction as originally recovered from petroleum refining has a boiling range of $140°$ to $220°C$ and is further fractionated to separate indenes therefrom by fractionation at about $180°C$. The final reactive fraction has a conjugated diolefin content of 0.7 wt % or less, a ratio of conjugated diolefin content to total polymerizables of 3% or less, a total content of indene and its alkyl derivatives of 2 wt % or less, and a ratio of indene content to total polymerizables of 8% or less.

Example 1:   The resin component for use in the pressure sensitive adhesive compositions is prepared in the following manner: The starting cracked oil was a by-product of the steam cracking of naphtha and had a boiling point in the range of 140° to 220°C. It was analyzed by gas chromatography to reveal the following composition:

Polymerizable components	49.0 wt %
Total content of cyclopentadiene and methyl cyclopentadiene	1.6 wt %
Total content of indene and its alkyl derivatives	8.4 wt %
Content of dicyclopentadiene	0.4 wt %
Ratio of conjugated diolefin content	4.0 %
Ratio of indene content	18.5 %

The above starting cracked oil was charged via a heater into a first fractionator A-1 and thence to a second fractionator B-1, both fractionators being as specified in the following table:

	A-1	B-1
Type of tray	sieve	bubble cap
Number of trays	30	7
Feed tray, from the bottom	18	4
Feed temperature, °C	108	55
Bottom pressure, mm Hg abs	120	110
Bottom temperature, °C	145	120
Top temperature, °C	93	25
Top pressure, mm Hg abs	68	95
Residence time at bottom, hr	1.0	0.5
Reflux ratio	5.0	2.0

The operating conditions of fractionator A-1 had been chosen so that the overhead fraction contained a total content of indene and its alkyl derivatives of 2 wt % or less and a ratio of indene content of 8% or less. The overhead product of the first fractionator A-1 was thereafter introduced into the second fractionator B-1 which was operated also under the above-tabulated conditions so that there was obtained a desired starting fraction from the bottom of fractionator B-1 in which the total content of cyclopentadiene and methyl cyclopentadiene was 0.7 wt % or less and the ratio of diolefin content was 3% or less. Conjugated diolefins were removed from the top of fractionator B-1. This operation produced 58 parts of the starting fraction from 100 parts of the starting cracked oil. The resulting starting fraction consisted of the following composition:

Polymerizable components	42.60 wt %
Total content of cyclopentadiene and methyl cyclopentadiene	0.40 wt %
Total content of indene and its alkyl derivatives	1.00 wt %
Ratio of conjugated diolefin content	0.84 %
Ratio of indene content	2.51 %

To the starting fraction thus obtained was added 0.5 wt % of boron trifluoride phenol complex as catalyst, and polymerization was carried out for 3 hours at 20°C. The product was washed with an aqueous solution of sodium hydroxide to remove the catalyst and then washed with water. Unreacted oil and low polymers were removed from the product by distillation. The resulting resin had a softening point by the ring and ball method of 95°C, a bromine value of 7 (according to ASTM D-1158-57T) and a Gardner color 1– (ASTM D-1544-58T). The resin was blended in amounts of 30 parts, 50 parts, 80 parts and 100 parts by weight, respectively, per 100 parts by weight of a commercially available styrene-butadiene rubber (SBR). The blend was dissolved in toluene to make up a 20 wt % concentration.

Example 2: 50 parts by weight of the resin prepared in Example 1 were blended with 70 parts by weight of SBR and 30 parts by weight of natural rubber, and the whole was dissolved in 600 parts by weight of toluene.

Pressure sensitive compounds formulated in this manner showed an excellent balance of adhesive strength, cohesive strength, tackiness and weather resistance.

Teleblock Copolymers

L.L. Nash; U.S. Patent 4,148,771; April 10, 1979; assigned to Phillips Petroleum Company has found that the technique of charging both a monovinyl-substituted aromatic hydrocarbon and an initiator in at least two increments prior to the addition of a conjugated diene will result in a rubbery copolymer which exhibits a very special property; i.e., when the copolymer is used to produce pressure sensitive adhesives, the adhesives exhibit greatly improved holding power as compared with adhesives prepared from copolymers not prepared by this method. These adhesives also exhibit very good quickstick and good 180° peel strength. The term "rubbery" used here means that the described material retracts within 1 minute to less than 1.5 times its original length after having been stretched at room temperature (18° to 29°C) to twice its length and held for 1 minute before being released.

The rubbery teleblock copolymers of the process are represented by the general formula $(S-D)_x Y$ where S represents a block of polymerized monovinyl-substituted aromatic hydrocarbon molecules, D represents a block of polymerized conjugated diene molecules, Y represents the residue of a polyfunctional coupling agent, x is an integer of at least 2 and can be (ideally) equal to the number of reactive sites on the polyfunctional coupling agent; and where the heterogeneity index (a measure of molecular weight distribution) of the monovinyl-substituted aromatic hydrocarbon portion of the copolymer is greatly increased by this method as compared with the corresponding prior-art copolymer, whereas the heterogeneity index of the copolymer is not appreciably correspondingly increased.

It should be noted that these copolymers can be either linear or radial although the radial (or branched) copolymers are preferred because of their superior properties in adhesives. The preparation of branched copolymers from mono-lithium-terminated polymers by coupling with agents having at least three re-active sites is discussed in U.S. Patent 3,281,383.

The preparation of linear teleblock copolymers by sequential monomer poly-merization or coupling with bifunctional agents as such is also well known in the art.

The conjugated dienes useful with the monovinyl-substituted aromatic hydro-carbon in preparation of the teleblock copolymers are those having 4 to 8 car-bon atoms. Especially preferred are butadiene and isoprene. The monovinyl-substituted aromatic hydrocarbons which can be used in the preparation of the copolymer have 8 to 12 carbon atoms per molecule. Especially preferred is styrene.

The teleblock copolymers will generally contain conjugated diene and mono-vinyl-substituted aromatic hydrocarbon in amounts within the range of 90:10 to 60:40 parts by weight of conjugated diene:monovinyl-substituted aromatic hydrocarbon. About 85:15 to about 70:30 parts by weight of conjugated diene:monovinyl-substituted aromatic hydrocarbon is the especially preferred range for best adhesive properties.

Any organomonolithium initiator known in the art can be used to initiate poly-merization of the conjugated diene and monovinyl-substituted aromatic hydro-carbon to produce the teleblock copolymers. The n-alkyllithium compounds give polymers which result in adhesives with especially good creep resistance, and n-butyllithium is the especially preferred initiator.

The amount of initiator used will be varied, depending upon the desired molecu-lar weight of the resultant copolymer. The teleblock copolymers can vary from relatively low molecular weight rubbers up to those having a molecular weight of 500,000 or more. The initiator levels useful in preparing the copolymers for the adhesives are generally in the range of 0.25 to 40 g mmol per 100 g of total monomers.

Polar compounds (such as acyclic or cyclic ethers or tertiary amines) are prefer-ably employed in the polymerization of the conjugated diene and monovinyl-substituted aromatic hydrocarbon to increase the rate of polymerization.

Example: The following tables give the ingredients and polymerization con-ditions for preparation of 80/20 butadiene/styrene radial teleblock copolymers.

Into a stirred 76-liter reactor were placed cyclohexane, tetrahydrofuran and the first increment of styrene. Upon heating the solution to 66°C the first increment of n-butyllithium (NBL) was added. Following completion of polymerization of the first increment of styrene the second increments of n-butyllithium and styrene were added to the reaction mixture. Following completion of styrene polymerization, butadiene was added; and after polymerization of the butadiene, silicon tetrachloride was added. After the reaction mixture cooled to approximately ambient temperature, the stabilizers were added and the reaction mixture was steam stripped to remove volatiles. The resultant polymer was dried in a Welding Engineers twin-screw extruder at a maximum temperature of 150° to 180°C.

Components of Polymer

	Run Number	
	1	2
	(kg)	
Cyclohexane	38.7	38.9
Styrene, 1st increment	0.68	0.68
Styrene, 2nd increment	0.68	0.68
Butadiene	5.45	5.45
Tetrahydrofuran	0.068	0.068
n-Butyllithium, 1st increment*	0.39	0.28
n-Butyllithium, 2nd increment*	0.58	0.80
Silicon tetrachloride**	0.77	0.86
2,6-Di-tert-butyl-4-methylphenol	0.05	0.05
Tris(nonylphenyl)phosphite	0.10	0.10

*Weight of 1.2 wt % solution in cyclohexane.
**Weight of 1 wt % solution in cyclohexane.

Polymerization Conditions

	Run Number	
	1	2
Initial temp at NBL addition, °C	67	66
Initial pressure at NBL addition, kPa	171	171
Temp at 2nd increment addition, °C	68	66
Pressure at 2nd increment addition, kPa	178	247
Time from NBL addition to 2nd increment addition, min	14	20
Temp at butadiene addition, °C	71	64
Pressure at butadiene addition, kPa	171	415
Time from 2nd increment addition to butadiene addition, min	16	20
Temp at $SiCl_4$ addition, °C	81	93
Pressure at $SiCl_4$ addition, kPa	359	452
Time from butadiene addition to $SiCl_4$ addition, min	5	12

The polymer was employed in a hot melt pressure sensitive adhesive and showed considerable improvement over prior art formulations in creep resistance, quick-stick and peel strength.

Unmilled Chloroprene Polymer Preparation

The process devised by *R.W. Keown, J.W. McDonald and J.K. Weise; U.S. Patent 4,156,671; May 29, 1979; assigned to E.I. DuPont de Nemours and Company* is one for the production of an organic solution of a chloroprene polymer in which the polymer is not milled to affect its properties, and yet the solution can be employed in a solvent adhesive and the adhesive will have properties substantially as good as or better than the properties of an adhesive prepared from a milled chloroprene polymer.

The process comprises preparing a solution of about 10 to 30 weight percent of a chloroprene polymer in a compatible organic solvent, heating the solution to a temperature in the range of 40° to 100°C, preferably 50° to 80°C, contacting the solution with a molecular oxygen containing gas, adding to the solution maintained at about 40° to 100°C a free-radical catalyst. The catalyst should have a half-life of 5 to 15 hours at a temperature of 40° to 100°C. The reaction time is usually about 15 minutes to 4 hours.

After the reaction has proceeded to the desired extent, usually at least about a 10% reduction in viscosity, up to about a 90% reduction in viscosity, the reaction may be terminated by adding the solution to a mixture of the other ingredients to be used in the adhesive. The following additional ingredients are commonly employed in solvent adhesives based on chloroprene polymers: phenolic resins, antioxidants (generally phenolic antioxidants), zinc oxide, magnesium oxide, small amounts (about 1 part per hundred parts of the chloroprene polymer) of water, and additional organic solvents.

Example: Two approximately 17 wt % solutions of the chloroprene polymer (Neoprene AD-30 from DuPont) were prepared by dissolving about 400 g polymer in about 2,000 g of mixed solvent. One was treated at 70°C for 1 hour and one for 2 hours, with 1.3 phr of benzoyl peroxide added in small increments over a 1-minute period. The reaction was terminated by addition to 611 g aliquots of the solution to a masterbatch of adhesive-compounding ingredients prepared by mixing 8 phr (parts per hundred chloroprene polymer) magnesium oxide, 5 phr zinc oxide, 1 phr water, 40 phr tert-butyl phenolic resin (CMK-1634 by Union Carbide), 2 phr fortified hindered phenol (Zalba Special Rubber Antioxidant by DuPont), 100 phr toluene and 10 phr isopropanol.

Canvas-to-canvas bonds were prepared from the adhesives, and the bonds aged 5 days at room temperature. Creep adhesion of the bonds was measured after various periods of time at 80°C under a 2.26 kg load by ASTM D-186 and D-413-39 tests. The results are summarized in the table.

Creep Adhesion—Centimeters Separation

	. . . . . Control		. . . . Process.	
	Untreated	Milled	1 Hour	2 Hours
Adhesive viscosity, cp	1,525	325	175	140
Creep, cm (after min)				
15	4.3	fail*	3.3	2.8
30	7.1	—	5.8	5.3
45	7.4	—	7.6	7.6
60	7.4	—	8.6	8.6
120	7.9	—	11.2	8.9

*After > 12.7 min.

Copolymer of Butadiene, Styrene and Acrylonitrile

Generally rubbery emulsion copolymers of butadiene and styrene are inferior to both natural rubber and unvulcanized block copolymers of butadiene and styrene for the purpose of providing the quality of tack normally required by pressure sensitive adhesives.

G.W. Feeney and G.L. Burroway; U.S. Patent 4,157,319; June 5, 1979; assigned to The Goodyear Tire & Rubber Company however, have formulated an adhesive composition, suitable for a pressure sensitive adhesive, which comprises an admixture of 100 parts by weight of a rubber copolymer of butadiene, styrene and acrylonitrile, if desired, containing preferably about 5 to 10 wt % units derived from styrene, preferably 95 to 90 wt % units derived from 1,3-butadiene, and preferably about 1 to 2 wt % units derived from acrylonitrile with 50 to 150 parts by weight of a tackifier resin selected primarily from rosin esters and olefin/diolefin copolymers containing 5 to 35 wt % aromatic groups. This rubbery copolymer is prepared by water emulsion copolymerizing a mixture of styrene, 1,3-butadiene and acrylonitrile.

Example 1: A butadiene/styrene copolymer was prepared with the following ingredients (all parts by weight): 73 parts 1,3-butadiene, 27 parts styrene, 1 part acrylonitrile, 215 parts water, 5 parts emulsifiers, 0.45 part modifier (mercaptan), and 0.2 part short stop.

For the preparation of the butadiene/styrene copolymer, a reactor was charged with 215 pbw water, 5 pbw emulsifier, 0.35 pbw modifier, 73 pbw 1,3-butadiene, 26 pbw styrene and 1 pbw acrylonitrile. The mixture was adjusted to a temperature of about 30°C and 0.05 part of cumene hydroperoxide free-radical initiator catalyst was added. The copolymerization was allowed to proceed until about 60% of the butadiene, styrene and acrylonitrile monomers had reacted. At this time, the remainders of the styrene and modifier were added. The reaction was allowed to continue for an additional 4 hours at a temperature of 30°C until an overall conversion of the monomer had taken place. The reaction was stopped by the addition of the short stop and the mixture poured onto a drying tray from which the water was evaporated.

The resulting relatively sticky butadiene/styrene copolymer was analyzed to have an ultimate tensile of about 60 psi and an ultimate elongation of 1,200% at 25°C. The tensile strength was determined on an Instron tester with a crosshead speed of about 2 inches per minute. The overall yield of butadiene/styrene copolymer was 85 parts or 85 percent by weight.

Example 2: An adhesive composition was prepared by mixing 70 to 90 pbw toluene, 15 pbw of the butadiene/styrene copolymer prepared according to the method of Example 1 and about 15 pbw of a resin selected from a rosin ester (Staybelite Ester No. 10 from Hercules) or piperylene/2-methyl-2-butene/α-methyl styrene/dicyclopentadiene resin having a softening point of about 75° to 90°C.

Separate mixtures of toluene, copolymer and resin tackifiers are identified below as Experiments A and B. A third experiment, Experiment C, simply substitutes natural rubber of U.S. Pale Crepe No. 1 for the butadiene/styrene copolymer. It is intended as a comparative control.

The actual mixtures were prepared at a temperature of about 25°C with adequate mixing overnight. With the mixture was included 1 part of phenol-type antioxidant for the purpose of aging properties. The resulting mixture was poured onto a polyester film substrate and dried to form an adhesive composition having a thickness of about 1.5 mils. The adhesive composition was then examined for its suitability as a pressure sensitive adhesive by measuring its rolling ball tack, 180° peel strength, 90° peel strength and shear adhesion both in its initial unaged condition and after aging 7 days at about 70°C. Ingredients are shown in the following table.

	. . . . Experiment		
	A	B	C
Ingredients	. . . . . . (parts by weight).		
SBR (Plioflex 1551)	15	—	—
Rubber (Ex. 1)	—	15	—
Natural rubber	—	—	12.5
Diolefin/olefin resin	15	15	15
Antioxidant, phenolic	0.15	0.15	0.125
Toluene	70	90	75

The results of the tests showed the ability to completely substitute the especially prepared butadiene/styrene copolymer for natural rubber for the purpose of preparing these adhesive compositions.

RADIATION-CURED POLYMERS

Containing Quinone Ultraviolet Sensitizers

M.M. Skoultchi and I.J. Davis; U.S. Patent 4,069,123; January 17, 1978; assigned to National Starch & Chemical Corporation describe acrylic-based pressure sensitive adhesives which have been admixed with ultraviolet sensitizers and then crosslinked by ultraviolet irradiation so as to improve their adhesive and cohesive strengths.

These pressure sensitive adhesive compositions comprise a blend of: (1) a pressure sensitive adhesive polymer selected from the class consisting of (a) homopolymers of alkyl esters of acrylic and methacrylic acids, and (b) copolymers of at least one alkyl ester of acrylic and methacrylic acid present in an amount of at least 35% by weight of the copolymer together with at least one other polymerizable comonomer; and (2) from about 0.1 to 10.0% (preferably from 0.5 to 2.0%) as based on the weight of polymer solids, of a quinone compound corresponding to the formulas shown on the following page:

In the above formulas, R is selected from the group consisting of hydrogen, halogen, alkyl radicals, aryl radicals and alkoxy radicals. The copolymerizable comonomer is selected from the group consisting of vinyl ester, vinyl halides, vinylidene halides, ethylenically unsaturated carboxylic acids, nitriles of ethylenically unsaturated monocarboxylic acids, anhydrides of ethylenically unsaturated dicarboxylic acids and C_{1-4} alkyl half esters of maleic and fumaric acids.

The adhesive composition should have a Williams plasticity of up to about 2.5.

The quinone compound is selected from the group consisting of phenanthrenequinone, 1,4-naphthoquinone and 1,2-naphthoquinone.

Example: This example illustrates the preparation of an adhesive composition typical of this formulation. The pressure sensitive adhesive utilized was a 50:50 2-ethylhexyl acrylate:vinyl acetate solution copolymer in an ethyl acetate solvent system. The resin solids content of the lacquer was 40% while its Williams plasticity value was 1.90.

Thereafter, 0.4 part of phenanthrenequinone was dissolved in 100 parts of the abovedescribed pressure sensitive adhesive. Both the modified and unmodified adhesive compositions were cast onto polyethylene terephthalate films and the solvent evaporated to produce adhesive films having a dry thickness of 1.0 mil. The adhesive films were then exposed, for varying periods of time, to a 275-watt General Electric RS sunlamp which was positioned at a 12-inch distance from the film samples. The crosslinked and noncrosslinked adhesives were then evaluated by means of the 180° hold and 180° peel adhesion tests. The results of these evaluations are presented in the following table:

	Ultraviolet Exposure (min)	180° Peel (ounces)	180° Hold (hours)
Control (no sensitizer)	0	27	12.5
Control (no sensitizer)	4	26	12.5
Sensitized adhesive	0	29	12.5
Sensitized adhesive	4	23	48+

The results summarized above clearly show the increased cohesive strength of adhesive copolymers crosslinked in accordance with this process.

The sensitization, crosslinking and testing procedures were identically repeated with the exception that the following polymer compositions were utilized in lieu of the ethylhexyl acrylate-vinyl acetate system: (1) an n-butyl acrylate homopolymer in an ethyl acetate solvent system at a total solids content of 67% by weight; (2) a 65:35 octyl acrylate:vinyl acetate copolymer in an ethyl acetate solvent system at a total solids content of 49% by weight. In each instance, the improved adhesive properties of the crosslinked adhesive were comparable to those described for the 50:50 ethylhexyl acrylate:vinyl acetate system.

Polyoxyalkylene Polymer plus Carbamyloxy Alkyl Acrylate

K.C. Stueben; U.S. Patent 4,111,769; September 5, 1978; assigned to Union Carbide Corporation has formulated a radiation-curable pressure sensitive adhesive composition comprising: (a) from 35 to 80 wt % (preferably 40 to 75 wt %) of a liquid carbamyloxy alkyl acrylate of the formula:

$$CH_2=\overset{\overset{\displaystyle R}{|}}{C}-\overset{\overset{\displaystyle O}{\|}}{C}O-R'-O\overset{\overset{\displaystyle O}{\|}}{C}-NHR''$$

where R and R'' are each, individually, hydrogen or alkyl from 1 to 3 carbon atoms, R' is alkylene of 1 to 3 carbon atoms; and, (b) from 20 to 65 wt % (preferably 25 to 60 wt %) of a polyoxyalkylene homopolymer or copolymer which is characterized by the repeating units:

$$-(-CH_2\underset{\underset{\displaystyle R'''}{|}}{C}HO-)_{\overline{x}}$$

where each R''' is hydrogen or methyl and a sufficient number of the R''' substituents are hydrogen that from 40 to 100% by weight of the oxyalkylene units in the polymer are oxyethylene units and from 60 to 0% by weight are oxypropylene units, and x is a number such that the molecular weight of the polyoxyalkylene homopolymer or copolymer is from 1,700 to 90,000 (preferably from 2,500 to 21,000); and (c) from 0.1 to 10 wt % (preferably 0.5 to 5 wt %) of a photoinitiator.

The preferred polyoxyalkylene polymers are a polyoxyethylene homopolymer or a poly(oxyethylene-oxypropylene) copolymer.

Halogenated Polymerizable Compounds Containing Acrylic Radicals

M. De Poortere, M. Colpaert, P. Dufour and A. Vrancken; U.S. Patent 4,134,814; January 16, 1979; assigned to UCB, SA, Belgium have formulated photopolymerizable adhesives which can be used as lamination adhesives and as contact or pressure sensitive adhesives which have the property of drying very rapidly at room temperature under the effect of visible or ultraviolet radiation, without causing any pollution of the environment.

These halogenated photopolymerizable adhesives comprise:

(a) 20 to 70% by weight of at least one halogenated polymer which (1) has a chain interrupted by at least one oxygen or nitrogen atom (2) contains from 4 to 70% by weight of halogen atoms attached to carbon atoms having the electronic configuration sp^2 (called "active halogen atoms"), (3) has an average molecular weight between 700 and 10,000, and (4) has a glass transition temperature (Tg) between $-80°$ and $20°C$;

(b) 5 to 30% by weight of at least one organic monomer compound containing at least two acrylic and/or methacrylic acid radicals and containing from 0 to 65% by weight of active halogen atoms;

(c) 20 to 70% by weight of at least one organic monomer compound containing an acrylic or methacrylic acid radical and containing from 0 to about 65% by weight of active halogen atoms;

(d) about 1 to 25% by weight of a photoinitiator system comprising (1) 0.5 to 100% by weight of at least one aromatic ketone, (2) 0 to 99.5% by weight of at least one tertiary amine, at least 1 carbon atom of which, in the alpha position with regard to the nitrogen atom, carries at least 1 hydrogen atom, (3) 0 to 90% by weight of an aromatic or nonaromatic alphadione, the sum of (1) + (2) + (3) representing 100% by weight of the photoinitiator system.

The total content of active halogen atoms amounts to 3 to 50% by weight of the total halogenated photopolymerizable adhesive and lamination or contact adhesives based on the halogenated photopolymerizable adhesive.

Example 1:   79 g (0.275 mol) tetrachlorophthalic anhydride, 133 g (0.9 mol) phthalic anhydride, 78 g (1.26 mols) ethylene glycol, 3.05 g (0.025 mol) benzoic acid and 50 ml benzene are placed in a 4-necked 500-ml round-bottomed flask equipped with a stirrer, a nitrogen inlet, a thermometer dipping into the mass and a Dean-Stark water separation column.

The reaction mixture is gradually heated until the mass is liquefied ($100°C$), whereafter stirring is started, while introducing nitrogen into the flask. After heating for 19 hours, the temperature of the mass is $150°C$. The quantity of water distilled off is 11.5 ml and the acidity (alcoholic KOH) is 0.795 meq H^+/g. The intensity of the heating is increased so that, after 1 hour, the temperature reaches $185°C$. The quantity of water distilled off is then 12 ml. The acidity of the solution (alcoholic KOH) is 0.73 meq H^+/g, which corresponds to a conversion rate of 91%. The benzene is distilled off under reduced pressure (temperature of the mass, $150°C$; pressure, 10 mm Hg) while maintaining an inert atmosphere. The flask is emptied while the polyester is still in the molten state and the polyester then allowed to cool. The polyester thus obtained is quite colorless. Analysis: 0.80 meq H^+/g acidity; $20°C$ Tg (differential thermal analysis), 1,700 molecular weight at the top of the peak in GPC; and 14% theoretical active chlorine content.

Example 2: A halogenated polyester is prepared, according to the general mode of operation of Example 1, from 11.2 mol % tetrachlorophthalic anhydride, 27.5 mol % phthalic anhydride, 9.2 mol % adipic acid, 51.2 mol % ethylene glycol, 0.98 mol % benzoic acid and 2.5% by weight benzene. After 10 hours heating, the temperature is 190°C and the acidity is 0.29 meq H$^+$/g, which corresponds to a conversion of 97%. Analysis: 0.30 meq H$^+$/g acidity; 7°C Tg (differential thermal analysis); 3,500 molecular weight at the top of the peak in GPC; 14.6% theoretical active chlorine content.

Example 3: A lamination adhesive is prepared, the composition of which is as follows: 60 pbw polyester of Example 1, 20 pbw diethylene glycol diacrylate, 5 pbw 2-hydroxyethyl acrylate, 5 pbw benzophenone, 10 pbw mixed acrylate β-dimethyl-aminopropionate ester of diethylene glycol.

This composition is applied in a film with a thickness of 4 μ by means of a spiral scraper onto untreated aluminum strips and then polypropylene strips are laminated onto these, using a glass rod. The transparent side is then exposed to a Hanovia 80 W/cm medium-pressure mercury vapor lamp equipped with a semielliptic reflector.

When the adhesive film is dry, the adhesion is such that the films of aluminum and polypropylene cannot be detached without the plastics film giving way. The reactivity of the adhesive, measured by the speed of passage under the lamp starting from which the films of the laminate may be detached without one or other of them giving way, is 2 m/sec. The force necessary for their separation, measured with an Instron apparatus, is higher than 300 g/cm at a speed of 5 cm/min. The adhesive remains just as adherent after 10 successive passages underneath the same lamp. The adhesion of the films of the laminate is preserved after storage for 1 month at 45°C.

Acrylic and Methacrylic Ester Derivatives as Photoinitiators

To prepare a self-adhesive composition of crosslinked copolymers, *G. Guse, E. Lukat, P. Jauchen, W. Lenck and H. Pietsch; U.S. Patent 4,144,157; March 13, 1979; assigned to Beiersdorf AG, Germany* use from 0.01 to 5%, preferably from 0.15 to 2%, by weight, based on the total composition, of a compound taken from the class consisting of acrylic and methacrylic acid-[(2-alkoxy-2-phenyl-2-benzoyl) ethyl] esters and at least one monomer capable of producing a self-adhesive polymer. Such monomers are generally known in the art. Preferably the acrylic and methacrylic ester derivatives (which are used as photoinitiators) contain 1 to 4 carbon atoms in the alkoxy group.

The self-adhesive compositions are produced by mixing the monomers in the appropriate proportions followed by polymerization in the usual manner. The polymerized compositions are then exposed briefly to ultraviolet radiation for crosslinking.

These polymerizable photoinitiators are prepared preferably by esterification of the corresponding α-methylol benzoin alkyl ethers with acrylic or methacrylic chloride at temperatures below 5°C in the presence of organic bases such as pyridine or triethylamine.

The copolymerization of the photoinitiators with the other monomers can be effected according to the known methods of polymerization, such as block-, solution- or emulsion-polymerization. These monomers are the ones known in the art to be capable of yielding self-adhesive polymers. Typically, esters of acrylic or methacrylic acids as well as mono- or dialkyl esters of maleic or fumaric acids are quite suitable. In particular, those esters having alkyl groups of 4 to 12 carbon atoms are particularly preferred.

In addition to the foregoing monomers, up to about 25% by weight based on the total composition of lower acrylic or methacrylic esters, vinyl esters, and vinyl aromatics may also be included. Particularly suitable are methyl or ethyl acrylates or methacrylates, vinyl acetate or propionate, and styrene.

With a Benzoin C_{1-12} Hydrocarbon Alkyl Ether

N.R. Lazear and R.W. Stackman; U.S. Patent 4,150,170; April 17, 1979; assigned to Celanese Corporation have found that improved adhesive strength in pressure sensitive ultraviolet-curable adhesives may be obtained when photoinitiators comprising a combination of benzoin C_{1-12} hydrocarbon alkyl ether and a triplet state sensitizer in initiating amounts are utilized.

The described formulations contain about 0.5 to 20% by weight based on the ultraviolet-curable portion of the system, of a benzoin C_{1-12} hydrocarbon alkyl ether, preferably about 1 to 10% by weight. Examples of such ethers include the ethyl, propyl, butyl, isopropyl, isobutyl, 2-ethylhexyl, amyl, isoamyl ethers of benzoin.

The compositions also contain about 0.001 to 1.0% by weight on the same basis of a triplet state sensitizer or energizer having a triplet state energy of about 40 to 60 kcal/mol. Examples of these materials include pyrene, fluorescein, trans-stilbene, eosin-Y which has the formula:

Generally, the compositions may comprise any of the pressure sensitive compositions previously used in ultraviolet-curable systems. Thus, in most instances, a polymer composition is added to an α,β-ethylenically unsaturated vinyl polymerizable monomer or mixture of monomers which act as diluents. The ratios of these materials will depend on the desired end viscosity, with more monomeric or diluent materials being added to decrease the viscosity. Preferably the pressure sensitive adhesives of the process should have application viscosity in the range of about 20,000 to 200 cp and have a glass transition temperature of about $-10°$ to $-60°C$ or lower, preferably about $-10°$ to $-45°C$.

Containing Poly(Vinyl Alkyl Ether)

Ever-increasing pressures to reduce air pollution and eliminate toxic substances from work environments have created a need for adhesive compositions which are free of volatile solvents. Moreover, elimination of volatile solvents is desirable to reduce energy consumption since heat is required to evaporate volatile solvents when curing the composition. To eliminate volatile solvents, radiation-curable adhesive compositions have been developed which employ reactive solvents that become incorporated into the cured adhesive composition. Multifunctional acrylate compounds, alone or in combination with monofunctional acrylate compounds, have been found to be particularly useful as the reactive solvent portion of these compositions. However, some radiation-curable compositions contain as much as 90 wt % of acrylates, some of which exhibit high toxicity.

K.C. Stueben, R.G. Azrak and M.F. Patrylow; U.S. Patent 4,151,055; April 24, 1979; assigned to Union Carbide Corporation have discovered that compositions containing at least 55 wt % of poly(vinyl alkyl ether) in combination with multifunctional acrylates and, optionally, monoacrylates and photoinitiators, can be radiation cured to a tacky state and display excellent adhesion to a variety of substrates. By the proper choice of components the viscosity of the uncured compositions can be controlled so that the compositions can often be applied as thin films at room temperature or slightly elevated temperature, e.g., $65°C$.

The poly(vinyl alkyl ether) employed is one in which the alkyl portion of the repeating unit contains from 1 to 4 carbon atoms. The poly(vinyl alkyl ether) has a reduced viscosity from about 0.1 to 0.8, most preferably about 0.3, measured at a concentration of 0.1 g of resin per 100 ml of benzene at $20°C$. From a standpoint of producing a composition having good adhesive properties, that is, peel strength, quickstick and shear time, poly(vinyl ethyl ether) is the preferred poly(vinyl alkyl ether). The poly(vinyl alkyl ether) is present in the composition at a concentration of from 55 wt % to 95 wt %.

The multifunctional acrylates used are any of the di-, tri-, tetra-, or pentaacrylates which are known to those skilled in the art of radiation technology as being crosslinking agents in radiation-curable compositions.

Example: A series of radiation-curable pressure sensitive adhesive compositions was prepared by mixing to a uniform composition poly(vinyl ethyl ether), neopentyl glycol diacrylate and benzophenone in the weight percentages indicated in the table below. The compositions were applied at room temperature to a polyethylene terephthalate tape substrate using a drawdown bar at an average film thickness of 1.5 mils. The adhesive coated substrates were cured by exposure in air to ultraviolet light from three 2.2-kilowatt, medium-pressure mercury arc lamps, each 24 inches in length. Total delivered flux from the lamps was 500 watts per square foot. Exposure times for the various formulations are indicated. The irradiated adhesive coated substrates were tested for quickstick, peel strength and shear time. The results, reported in the table below, indicate that the cured compositions generally had an excellent balance of adhesive properties. Values indicated in the table are averages of two samples.

Poly(Vinyl Ethyl Ether)/ Neopentyl Glycol Diacrylate/ Benzophenone (wt %)	Exposure Time (sec)	Peel Strength	Quickstick	Shear Time (hr)
		. . . . . (lb/in)		
78/20/2	4	1.5	—	>60
81/16/3	2	1.7	1.6	>47
81/16/2	4	2.2	2.4	>42
83.5/14.5/2	4	2.2	1.3	>70
84/13/3	2	1.8	1.6	>45
85/13/2	2	2.8	2.5	51
85.5/12.5/2	4	3.2	2.0	>60
86.5/11.5/2	4	2.7	2.3	>90
87/10/3	2	2.5	2.7	>42

ADDITIONAL FORMULATIONS

Polyester Self-Adhesive Films

R. Arpin and P. Tissot; U.S. Patent 4,084,035; April 11, 1978; assigned to Rhone-Poulenc Industries, France have developed self-adhesive films comprising a saturated polyester support layer having, on one face thereof, an adhesive layer which comprises a copolymer of vinyl acetate and a maleate ester of an aliphatic alcohol having 4 to 10 carbon atoms.

Typically, any unstretched, mono- or biaxially stretched or poststretched film made of polyethylene glycol terephthalate can be used as the support layer. The thickness of the films suitably vary from 3.5 to 350 microns. It is, furthermore, possible to render the surface of the film metallic or to print the film.

The other face of the support layer is generally a nonadhesive face and can be covered with a nonstick coating, for example, based on silicone. In this instance, compositions comprising an organopolysiloxane polymer carrying SiH bonds, a catalyst and, optionally, an activator and a hydroxylic methylpolysiloxane polymer are generally used.

The copolymer forming the adhesive layer usually contains 40 to 65% by weight of vinyl acetate and 60 to 35% by weight of a maleate ester. The copolymer has preferably a softening point of from 70° to 110°C and a specific viscosity (measured at 20°C employing an 0.5% by weight solution in cyclohexanone) of from 35 to 65 ml/g. Copolymers of vinyl acetate and 2-ethylhexyl maleate are advantageously employed. The copolymers can be prepared from the monomers in a conventional manner. The adhesive layer can also contain, for example, 1 to 60% by weight of one or more tackifier resins.

Among these resins, colophony and its derivatives, especially its esters such as the methyl ester of dihydroabietyl phthalate, as well as polymers such as polyisobutene or elastomers such as styrene/butadiene copolymers, are generally used.

In the molten state, the adhesive is generally prepared by mixing (hot) the various adjuvants and then adding the copolymer with stirring. Stirring is continued until the mixture is perfectly homogeneous. The molten adhesive can be applied with equipment for extrusion coating, lick coating or transfer-roll coating.

Example: In the example parts and percentages are by weight. A solution containing the following constituents is prepared:

(a) 91 parts of 50/50 vinyl acetate/ethylhexyl maleate copolymer having a softening point of approximately 85°C, a specific viscosity of 50 ml/g in the form of a 0.5% solution in cyclohexanone at 20°C, and a melt viscosity of 150 poises at 200°C;

(b) 9 parts of a dihydroabietyl phthalate, having a softening point (Hercules Drop method) of 60° to 70°C, a saponification number of 129 and an acid number of 10;

(c) 200 parts of acetone; and

(d) 100 parts of toluene.

A polyethylene glycol terephthalate film, 35-cm wide and 23-μ thick is coated with this adhesive mass. After drying, a deposit of 22 g/m^2 is produced.

The linear adhesive power was measured and the following results obtained. Adhesion of the coated face to the coated face: 400 g force necessary for separation; no splitting into layers or detachment of the adhesive layer from the support film was observed. Adhesion of the coated face to the noncoated face: 50 g force for separation; no residue of the adhesive layer on the support film. Adhesion of the coated face to metals (aluminum): 150 g force for separation; there was no residue of the adhesive layer on the metal.

With CH-Acidic Compounds Masking Polyisocyanates

H. Hartmann, W. Druschke, K. Eisentraeger and H. Mueller; U.S. Patent 4,087,392; May 2, 1978; assigned to BASF AG, Germany provide stable pressure sensitive adhesives which are easy to handle and have good adhesiveness and a relatively low viscosity and which can be rapidly spread onto substrates with conventional conveying and coating equipment without the addition of inert solvents and which show no tendency to gel even at temperatures of about 80°C but very rapidly crosslink at temperatures above 100°C.

These pressure sensitive adhesives may be prepared by mixing: (a) from 10 to 90% by weight of a polyol having a molecular weight of from 300 to 8,000 and a hydroxyl number of from 20 to 85, and (b) from 5 to 60% by weight of an amorphous thermoplastic tackifier having a softening point of from 40° to 140°C and a molecular weight of from 200 to 7,500; and reacting the mixture with (c) from 5 to 30% by weight of a polyisocyanate masked with CH-acidic compounds, the percentages being based on the weight of components (a), (b) and (c) together.

The use of masked polyisocyanates as the isocyanate component increases the viscosity of the reaction mixture only significantly when the starting materials are mixed, which means that relatively large amounts of tackifier may be incorporated into the adhesive if desired. Furthermore, metering fluctuations in the preparation of the adhesive reaction mixture are less detrimental to the properties of the product than when conventional nonmasked polyisocyanates are used. Thus the adhesives and the tack and also the properties of the adhesive tapes coated with the adhesives are easy to reproduce on an industrial scale.

Another advantage is that, unlike conventional polyisocyanates, the masked polyisocyanates are not sensitive to moisture, which means that the reaction taking place between the polyol/tackifier mixture and the masked isocyanates is not impaired by atmospheric moisture and moisture in the substrate. Moreover, masked polyisocyanates exhibit virtually no vapor pressure, and therefore, involve no physiological hazards and are environmentally acceptable.

Suitable starting materials which may be used are polyetherols having a hydroxyl number of from 30 to 50; polyesterols with hydroxyl numbers from 35 to 50; and hydroxyl-containing acrylate copolymers having molecular weights of from 1,500 to 8,000 and a pour point below +15°C.

The tackifiers are solid, amorphous, hard-to-brittle, thermoplastic materials having softening points preferably between 75° and 135°C and generally having an average molecular weight preferably from 500 to 1,500, and compatible with the polyol. Of the suitable tackifiers, those are preferred which have an acid number of less than 100 and preferably one from 0 to 70, e.g., terpene resins, terpene phenol resins and ester resins. It is particularly advantageous to use mixtures of different tackifiers (b), for example, mixtures of 2 to 4 different resins. Where a mixture of resins of different softening points is used, for example, a mixture of resins having softening points at about 70°C and resins having softening points at about 120°C, the resulting adhesives may be used over a larger temperature range than is the case with adhesives based on only one resin.

The starting materials (c) are adducts, i.e., reaction products of CH-acidic compounds and polyisocyanates. Examples of masking agents, i.e., CH-acidic compounds, are malonic esters such as diethyl malonate, ethyl acetoacetate, malonic dinitrile, acetyl acetone, methylene disulfone, dibenzoyl methane, dipivalyl methane and dimethyl acetone dicarboxylate.

Preferred masked diisocyanates are those based on isophorone diisocyanate and malonic esters such as, in particular, diethyl malonate, these generally being derived from C_{1-4} alkanols or acetyl acetates of C_{1-4} alkanols such as, in particular, ethyl acetoacetate. The masked polyisocyanates may be prepared by known methods. Examples of catalysts which may be used for the reaction between CH-acidic compounds and polyisocyanates are small amounts of sodium or a concentrated sodium methoxide solution.

Components (a) and (c) are preferably used in proportions such that there are present, per equivalent of hydroxyl group of the polyols, i.e., per hydroxyl group of component (a), from 0.6 to 1.4, most preferably from 0.8 to 0.98 equivalent isocyanate groups of the masked polyisocyanate, i.e., masked isocyanate groups of component (c). This leads to adhesives that may contain free hydroxyl groups, which may be an advantage in many applications.

The adhesives are prepared, for example, in the following manner. The polyol, preferably having a water content of less than 0.1% of its weight, is mixed with the resin or resin mixture in such a manner as to give a homogeneous polyol/resin mixture. To this end, the components are usually heated, with stirring, to temperatures above the softening point of the resin, for example, from 40° to 160°C, until a clear homogeneous mixture is formed. Depending on the nature of the polyol and the nature and amount of tackifier used, the resulting mixture has a viscosity, at room temperature, of from 2,000 to 50,000 millipascal seconds.

Following cooling of the resin/polyol mixture, the masked crosslinking agent is added at a temperature at which the masked polyisocyanate may be readily worked into the mixture, for example, at about 80°C in the case of the diethyl malonate/isophorone diisocyanate adduct. Other conventional auxiliaries such as dyes, fillers, pigments, plasticizers, aging retardants, antioxidants, light stabilizers, reactive diluents, desiccants, wetting agents and, optionally, solvents may be added. It is also possible to add conventional catalysts to accelerate the subsequent reaction of the hydroxyl groups with isocyanate groups. Examples of suitable catalysts for this purpose are dibutyl tin dilaurate and tin(II) octoate and also tertiary amines such as triethylene diamine or mixtures of said materials. These catalysts are advantageously added after the addition of the masked polyisocyanates.

The adhesive may be used in the form of a relatively thick layer. The main field of application is the manufacture of adhesive tapes, adhesive sheeting and adhesive labels, in which case the adhesive is applied as a thin layer to a fabric or web of, for example, paper, cellulose, cellulose acetate, cellophane; polyesters such as polyethylene glycol terephthalate; polyamides such as polycaprolactam, polyvinyl chloride; and polyolefins such as polyethylene and polypropylene; and polystyrene. The layer thickness is generally above 5 microns and advantageously above 25 microns and less than 200 microns, this being equivalent to rates of application of more than 5 grams and advantageously more than 25 but less than 200 grams per square meter.

Following shaping or application of the adhesive mixture, this is allowed to react to completion at temperatures above the dissociation temperature of the masked polyisocyanate. The viscosity of the mixture then increases rapidly and the mixture soon gels and solidifies to form a soft elastic mass having excellent adhesive properties. If, after application, the composition is heated for a brief period to temperatures of 100° to 200°C, depending on the isocyanate adduct used, the composition solidifies within a few minutes and the coated sheeting can be wound up, if necessary, after insertion of a parting sheet, for example, a polyethylene tetrafluoride sheet or wax paper. In the following examples the parts and percentages are by weight.

Example 1: A mixture of 100 parts of a trifunctional polyethylene polypropylene etherol (mean molecular weight 4,500, hydroxyl number 36) and 50 parts of terpene rosin (melting range 117° to 130°C) is heated under reduced pressure to a temperature of 135°C, during which process it is intimately mixed and dehydrated. The mixture is then allowed to cool to 80°C, whereupon 17.7 parts of an adduct of 1 mol of isophorone diisocyanate and 2 mols of diethyl malonate and 0.72 part of tin(II) octoate are added.

A polyester sheet is coated with 30 g/m² of this adhesive composition. Curing is effected for from 1 to 2 minutes at 120°C to give a self-adhesive sheet having the following adhesive values: 500 g peeling strength, more than 24 hours shear strength at 20°C and more than 24 hours shear strength at 50°C, 1,400 g/cm² probe tack.

Example 2: Example 1 is repeated except that 16.5 parts of an adduct of 1 mol isophorone diisocyanate and 2 mols of ethyl acetoacetate are used in place of the masked polyisocyanate mentioned in Example 1. After curing for 2 minutes at 170°C, the following adhesive values are obtained: 300 g peeling strength, more than 24 hours shear strength at 20°C and more than 24 hours shear strength at 50°C, 1,200 g/cm² probe tack.

Anaerobically Cured

M. Douek, G.A. Schmidt, B.M. Malofsky and M. Hauser; U.S. Patent 4,092,374; May 30, 1978; assigned to Avery Products Corporation and Loctite Corporation have provided anaerobic pressure sensitive adhesive compositions which can be applied from or as sheets, tapes and the like to substrates to be bonded by cure upon the exclusion of oxygen. These anaerobic resin systems containing one or more anaerobic resins combined with a thermoplastic polymer system containing one or more high molecular weight thermoplastic polymers, the combination of which, alone or including a tackifier, constitutes a pressure sensitive adhesive system upon evaporation of essentially all of the solvent present. In these compositions there is also included an initiator system which is latent until made active by substantial exclusion of oxygen, preferably in combination with a suitable accelerator.

In one embodiment, if the anaerobic pressure sensitive adhesive system contains free transition metal ions, then at least the peroxy initiator may be encapsulated in microspheres which, upon rupture and upon the exclusion of oxygen, will initiate cure. In another embodiment, a suitable metal accelerator may be encapsulated.

To constitute a suitable pressure sensitive adhesive of this nature, the net composition when free of solvent should have, prior to cure, a static shear strength of at least about 2 minutes at a 250-gram load per 0.25 square inch and a $180°$ peel value of at least about 0.5 pound per inch, preferably at least 1.0 pound per inch, when using standard test methods.

By the term "anaerobic resin system" there is meant one or more anaerobic resins having at least one, preferably two, polymerizable acrylate ester moieties, normally on the ends of the backbone, which will polymerize or cure in the presence of a peroxy initiator and upon the substantial exclusion of oxygen or air, and preferably also in the presence of a suitable accelerator system. Illustrative of these anaerobic resins are polymerizable acrylate esters. Of particular utility as adhesive monomers are polymerizable di- and other polyacrylate esters since, because of their ability to form crosslinked polymers, they have more highly desirable adhesive properties. The preferred anaerobic monomers are triethylene glycol dimethacrylate; the reaction product of hydroxypropyl methacrylate with methylene-bisphenyl-4,4'-diisocyanate, a polymer formed by methacrylate capping of a 1:1 adduct of toluene diisocyanate and hydrogenated 2,2-bis(4-hydroxyphenyl)propane as well as mixtures thereof.

Typical of the thermoplastic polymers which can be used are polyvinyl chloride, polyvinyl ethers, polyvinyl acetates; acrylic based polymers, polyurethanes, polyesters, polyamides, natural and synthetic elastomers and the like, as well as mixtures thereof. The preferred thermoplastic polymers are polyvinyl chloride, polyurethanes, polyesters and acrylic based polymers.

By a "catalyst system" there is meant an acid or base catalyzed system typically containing at least one peroxy initiator, preferably at least one accelerator which preferably is a nitrogen-containing compound and preferably, although not necessarily, at least one stabilizer against free radical polymerization for the anaerobic resin system.

For certain applications, care should be taken in preparing these anaerobic pressure sensitive adhesives to cope with trace transition metal ions which may be present in each constituent of the composition, including the anaerobic resin system and the thermoplastic polymer system, typically picked up from the vessels and systems used in their production.

If necessary, the effect of transition metal ions in the compositions may be controlled by means, such as chelation, known to the art. When accelerated cure is desired, however, transition metal compounds, e.g., copper salts, may advantageously be used as primers or activators external to these pressure sensitive anaerobic compositions. One way to cope with the metal ions is to scavenge them prior to or following their admixture in a mutually compatible solvent prior to casting of the pressure sensitive adhesive layer onto a release liner, which may also require scavenging.

If the constituents of these compositions are properly scavenged, the active metals can be reintroduced, but in another form. Rather than being active in the compositions, their effect may be rendered latent by encapsulating them in microspheres such that they will not be in contact with the peroxy initiator until the microspheres are ruptured upon the application of pressure to react with the peroxy compound to accelerate cure.

An alternative route to prevent premature cure during shelf life or storage is to encapsulate the peroxy compound alone or with its accelerators. Then the active metals or metal ions can be left in the pressure sensitive adhesive composition without fear that premature cure or deactivation will occur.

In the example, both of the following anaerobic resin systems were employed for formulation of anaerobic pressure sensitive adhesive compositions.

Resin 1: Approximately 75% of a reaction product of 2 mols of hydroxypropyl methacrylate with 1 mol of methylene-bis-phenyl-4,4'-diisocyanate and 25% triethylene glycol dimethacrylate.

Resin 2: Approximately 66% of a polymer formed by hydroxypropyl methacrylate capping of a 1:1 adduct of toluene diisocyanate and hydrogenated 2,2-bis(4-hydroxyphenyl)propane, 26% hydroxypropyl methacrylate, 7% acrylic acid and 1% methacryloxypropyltrimethoxysilane.

The following test methods were employed in evaluating the pressure sensitive properties and properties of the cured end-products. In determining the pressure sensitive adhesive properties, the adhesive composition was cast on a suitable support such as paper or Mylar: static shear strength—Federal Test Method Standard No. 147B, Method 20.1 (load 250 g); 180° peel—ASTM D-1000/68, (dynamic, 12 inches per minute); lap shear—ASTM D-1002/64.

Example: To a heated flask equipped with a stirrer and a reflux condenser there were added 1,800 g of toluene, 1,200 g of Resin 1 and 300 g of Resin 2. The mixture was heated with stirring at 70°C until a homogeneous solution was formed. To the stirred solution there was added 300 g of a thermoplastic vinyl chloride copolymer (VAGH-2706, Union Carbide Corp.) and the mixture stirred until it again became homogeneous. To the resultant mixture there was added with stirring 180 g of an aqueous alcoholic solution containing a chelating agent for trace transition metal ions. The solution was held at a temperature between 40° and 50°C and stirred for 3 hours and the chelated transition metal ions removed.

To this solution there were added 70 g of cumene hydroperoxide containing quinone, 37 g of benzoylsulfimide and 37 g of methylene-bis-dimethylaniline to form a catalyst system solution.

The anaerobic pressure sensitive adhesive solution was coated onto the release surface of a backing sheet fabricated from a plastic film and a paper having a silicone release coating. Coating weight after solvent removal was 28 g/m². Another release sheet was applied to protect the anaerobic pressure sensitive adhesive.

A portion of the anaerobic pressure sensitive adhesive was tested for pressure sensitive properties. When applied to a paper support, the 250 g static shear value was 5.5 minutes. The 180° peel test value on a Mylar support was 1.75 lb/in and failure was cohesive. Surface tack was about 2 inches.

Two aluminum alloy plates measuring 1" x 4" x 1/16" were each etched on one end with a mixture of chromic and sulfuric acids to form a roughened surface. To one roughened surface there was applied a 1/2" x 1" layer of the anaerobic pressure sensitive adhesive. The roughened end of the other plate was placed on the anaerobic pressure sensitive adhesive in overlapping relation thereby excluding oxygen and initiating cure. The copper in the aluminum alloy accelerated cure and the anaerobic pressure sensitive adhesive bonded plates were allowed to cure for 24 hours at room temperature.

The bonded plates were tested in an Instron tester and the bond was found to give a lap shear tensile value of 900 psi.

Silicone Composition Requiring No Curing

J.R. Hahn; U.S. Patent 4,117,028; September 26, 1978; assigned to Dow Corning Corporation has developed a composition suitable for use as a pressure sensitive adhesive which consists essentially of (1) from 30 to 70% by weight of a block copolymer consisting essentially of BAB or (ABA)$_n$ types where n is an integer of 2 or greater, A is a polydimethylsiloxane block having at least 15 siloxane units, B is a polystyrenic block having at least 15 styrenic units per block, the percent by weight B being from 5 to 25% by weight of the total weight of A and B; and, (2) from 30 to 70% by weight based on the total weight of (1) and (2) of a benzene-soluble copolymer of SiO_2 units and $Me_3SiO_{0.5}$ units in which the mol ratio of $Me_3SiO_{0.5}$ units to SiO_2 units is from about 0.6:1 to 0.9:1.

A typical example of (1) is a block copolymer of styrene and dimethylpolysiloxane. The ingredient (2) employed herein is well known and can be prepared by the procedure shown in U.S. Patent 2,676,182 which involves the reaction of silica sols prepared from sodium silicate with trimethylchlorosilane or hexamethyldisiloxane and by the method of U.S. Patent 2,857,356 which involves the cohydrolysis of trimethylchlorosilane with tetraethoxysilane.

The pressure sensitive adhesives of this process can be applied to a wide variety of substrates and are characterized by the fact that they adhere to practically any known surface including such difficult substrates as polytetrafluoroethylene or substrates coated with silicone release agents. Thus, these pressure sensitive adhesives can be used to make pressure sensitive adhesive tapes in which the substrate could be metallic, siliceous or organoplastic in nature.

Example: Block copolymer (1) employed in this example contains 20% by weight styrene and 80% by weight dimethylpolysiloxane. It was prepared by polymerizing styrene with dilithiostilbene to give a calculated block size of 10,400 molecular weight. The living polymer was then reacted with hexamethylcyclotrisiloxane and the resulting polymer was coupled with trifluoropropylmethyldichlorosilane to give a $(ABA)_n$ block copolymer having a weight average molecular weight of 298,000.

25 g of this block copolymer were dissolved in 1,1,1-trichloroethane and a 71% solution of (2) in xylene was added in an amount to give 50% by weight of (1) and 50% by weight of (2). A thin film was put on aluminum foil and dried at room temperature to give a film which measured about 0.82 mil. After sitting at room temperature for 3 hours, the tape was found to be as follows: adhesion, 42 oz/in (453 g/cm); hold time, >42 hr; and the adhesive was tacky to the touch.

Addition of Silane

A high-quality rapid-setting adhesive is described in Briggs et al, U.S. Patent 3,890,407. Such an adhesive is used to adhere vinyl or rubber strips to exterior molding or bumpers of automobiles and trucks. While the Briggs et al adhesive is excellent and has many uses, an improvement is required to have these adhesives withstand the extremes of weathering and in particular withstand extended periods of exposure to moisture and high humidity.

J.M. Craven; U.S. Patent 4,118,436; October 3, 1978 has found that the addition of a silane to the Briggs adhesive allows it to withstand the above conditions without reducing the strength of the adhesive bond.

The adhesive composition is of a chlorosulfonated polyethylene, at least one acrylic monomer, an amine aldehyde condensation product, and, optionally, a peroxy catalyst; the improvement used therewith comprises 0.1 to 10 parts by weight per 100 parts by weight of the amine aldehyde condensate, of a silane of the formula

$$Y-N-CH_2-\overset{\overset{\displaystyle R^3}{|}}{\underset{\underset{\displaystyle R^4}{|}}{C}}-CH_2-\overset{\overset{\displaystyle X}{|}}{\underset{\underset{\underset{\underset{\displaystyle R^2}{|}}{O}}{|}}{Si}}-O-R^1$$

where $-R^1$, $-R^2$, $-R^3$ and $-R^4$ are $-H$ or an alkyl group having 1 to 5 carbon atoms; $-X$ is $-O-R$, $-R$, $-OH$ where R is an alkyl group having 1 to 4 carbon atoms; and Y and Z are $-H$ or a monovalent hydrocarbon-containing radical having 1 to 20 carbon atoms.

The adhesive is used as a two-component adhesive. A mixture of the chlorosulfonated polyethylene, acrylic monomer or monomers and the optional peroxy catalyst is applied to one side of a second article to be bonded. The resulting coated sides of the articles are placed in abutting relationship to form an adhesive bond.

Two silanes which are particularly preferred are N-2-aminoethyl-3-aminopropyltrimethoxy silane and 3-aminopropyltriethoxysilane. It is preferred to dissolve the silane in a solvent such as an alcohol or a ketone and then add the resulting solution to the amine aldehyde condensate.

Example: (All parts are by weight.) 100 parts of isopropyl alcohol are mixed with 1.3 parts of a silane $[NH_2(CH_2)_2NH(CH_2)_3Si(OCH_3)_3]$, and 0.63 part of water are added. The mixture is allowed to stand for 16 hours at room temperature after which 100 parts of 808 Accelerator (DuPont) are mixed in.

An adhesive composition is prepared by blending the following constituents:

	Parts by Weight
Hypalon (chlorosulfonated polyethylene)	28.5
Butyl methacrylate monomer	56.2
Methacrylic acid monomer	14.3
1,3-Butylene dimethacrylate monomer	1.0

The above constituents are mixed together and then a sufficient amount of copper naphthenate is added to provide 5 parts of copper per million parts adhesive.

A test is conducted in which a red vinyl strip $\frac{9}{16}$" wide and $\frac{3}{16}$" thick is bonded to aluminum automobile side-molding strip. The vinyl strip is sanded and in one test the above-prepared accelerator solution is applied to the vinyl strip and dried, and in another test the accelerator solution is applied to the aluminum side molding and dried and the adhesive is applied to an opposing part. In both tests the vinyl strip and molding are brought into abutting relationship and a bond is formed. A black vinyl strip ¼" wide and ⅛" thick is similarly bonded to an aluminum automobile side-molding strip.

The peel strength of the adhesive bond at 90°F is measured with a force gauge (Chatillon Model DP-50) after 1-day exposure at room temperature. The bonds are exposed to 100% relative humidity at a temperature of 100°F and the peel strength is measured after 8 and 14 days. After the 14-day exposure the bonds are held at room temperature conditions for 37 days and the peel strength is measured.

The data show that when the accelerator solution which contains a silane is applied to the aluminum molding it forms bonds with superior peel strengths initially and after exposure to high humidity in comparison to the bonds similarly prepared and formed with an accelerator which does not contain silane.

Vinyl Acetate-Ethylene Copolymer Aqueous Emulsion Base

T. Oyamada, C. Tomizawa, M. Domoto and S. Narisawa; U.S. Patent 4,128,518; December 5, 1978; assigned to Sumitomo Chemical Co., Ltd., Japan have found that an emulsion of vinyl acetate-ethylene copolymer useful as a base material for a pressure sensitive adhesive may be prepared by emulsion polymerization of vinyl acetate and ethylene in the presence of a nonionic dispersing agent. It has an ethylene content of 15 to 40% by weight and contains a benzene-insoluble part of less than 30% by weight and has an intrinsic viscosity of the benzene-insoluble part of less than 30% by weight and further has an intrinsic viscosity of the benzene-soluble part of 0.4 to 1.4 dl/g (at 30°C).

In the formulation, a nonionic dispersing agent is used as the dispersing agent in the emulsion polymerization reaction of vinyl acetate and ethylene, various nonionic surfactants are used as an emulsifier, and various nonionic water-soluble high molecular compounds are used as a protective colloid.

The surfactant used may be a polyoxyethylene alkyl or alkylphenol ether, a polyoxyethylene sorbitan fatty acid ester, a polyoxyethylene-polyoxypropylene block copolymer, or the like, and is used in an amount of 1 to 8% by weight based on total weight of the monomers.

The nonionic water-soluble compound used as the protective colloid may include a polyvinyl alcohol, a partially hydrolyzed polyvinyl alcohol having a saponification degree of at least 60 mol % and a degree of polymerization of 300 to 2,700, a cellulose derivative (e.g., methylcellulose, ethylcellulose, hydroxyethylcellulose, carboxymethylcellulose), or the like. The nonionic water-soluble high molecular compound may usually be used in an amount of 0.1 to 1% by weight based on the total weight of the monomers.

The vinyl acetate-ethylene copolymer includes the so-called crosslinked type vinyl acetate-ethylene copolymer emulsion which is obtained by emulsion polymerization of an unsaturated monovinyl monomer having a functional group together with the vinyl acetate monomer and ethylene monomer. The cross-linked type vinyl acetate-ethylene copolymer emulsion is particularly useful for the preparation of a pressure sensitive adhesive which requires a high heat resistance and a low heat sensitivity.

The unsaturated monovinyl monomer having a functional group includes glycidyl-group-containing monomers (e.g., glycidyl acrylate, glycidyl methacrylate), N-methylol- or alkoxy-group-containing monomers such as methylol compounds (e.g., N-methylolacrylamide, N-methylolmethacrylamide) or alkyl ethers thereof (e.g., methyl ether, ethyl ether, or butyl ether of the methylol compounds), carboxyl-group-containing monomers (e.g., acrylic acid, itaconic acid, maleic acid), or the like. The unsaturated monovinyl monomer may be copolymerized in an amount of 0.1 to 5% by weight based on the weight of the vinyl acetate-ethylene copolymer.

The preparation of the vinyl acetate-ethylene copolymer emulsion having an intrinsic viscosity of 0.4 to 1.4 dl/g is preferably carried out by a method using a chain transfer agent. Mercapto acids (e.g., thioglycolic acid, thioglycolic acid esters), aliphatic mercaptans (e.g., octylmercaptan, dodecylmercaptan) may be used as chain transfer agents.

The emulsion polymerization is carried out in the presence of a polymerization catalyst such as a water-soluble radical initiator which is usually used in the conventional emulsion polymerization, for instance, hydrogen peroxide, potassium persulfate, ammonium persulfate, which may be used alone or in the form of a redox catalyst, i.e., in a combination thereof with a reducing agent [e.g., Rongalite (i.e., sodium formaldehyde sulfoxylate), L-ascorbic acid, ferrous sulfate]. The polymerization temperature is preferably in the range of $30°$ to $60°C$. The polymerization pressure is preferably in the range of 30 to 100 kg/cm^2 in order to obtain the desired copolymer having an ethylene content of 15 to 40% by weight.

This vinyl acetate-ethylene copolymer emulsion is incorporated in a pressure sensitive adhesive composition as a base material, optionally, together with other emulsions, such as an acrylic emulsion.

When the vinyl acetate-ethylene copolymer emulsion is blended in an amount of 5 to 35 parts by weight, preferably 10 to 25 parts by weight (as the solid components) with 100 parts by weight (as the solid components) of an acrylic ester copolymer emulsion, the blend shows a highly improved adhesion to a polyolefinic resin base.

Modified Phenol Resin to Increase Cohesion

E. Schunck, H. Schmelzer and R. Sattelmeyer; U.S. Patent 4,131,709; December 26, 1978; assigned to Hoechst AG, Germany have developed an adhesive composition based on the following:

(1) 100 parts by weight of a block copolymer with at least two polymer blocks (A) and at least one polymer block (B), where the polymer blocks (A) comprise nonelastomeric blocks with an average molecular weight of from 2,000 to 125,000 derived from an aromatic hydrocarbon substituted by one of the groups consisting of the members monoalkenyl and monoalkylidene and where not more than 25% of the original double bonds are hydrogenated, and polymer block (B) is an elastomeric block with an average molecular weight of from 10,000 to 250,000 derived from a conjugated diene, where at least 75% of the aliphatic unsaturated bonds are hydrogenated; and,

(2) 10 to 200 parts by weight of at least one cohesion-increasing resin selected from the group consisting of an alkylphenol resin, an arylphenol resin, a phenol-colophony adduct and a rosin-modified phenol resin.

The adhesive compositions may, if desired, additionally comprise:

(3) 10 to 200 parts by weight of a resin or mixture of resins improving the tackiness.

The quantity of resins (2) used is preferably 30 to 70 parts by weight, and the quantity of resins (3) is preferably 20 to 100 parts by weight.

Block (A) generally has an average molecular weight of preferably 5,000 to 50,000 and block (B) generally has an average molecular weight of preferably 30,000 to 150,000. Both blocks may be synthesized from the monomers mentioned above.

Particularly suitable alkyl- or arylphenol resins are the condensation products of p-substituted phenols and aldehydes, preferably formaldehyde, in a molar ratio of 1:0.9-2.5, and most preferably 1:1.2-2.2, prepared using acid or alkaline catalysts. The alkylphenols are preferably those having alkyl groups with 1 to 12 carbon atoms, such as p-n-propylphenol, the p-substituted-n- or tert-butyl- and amylphenols and p-n-nonylphenol. A suitable arylphenol is, for example, p-phenylphenol. These resins may be prepared by conventional processes.

Suitable phenol-colophony adducts include, for example, those prepared from phenol and natural resin acids in a molar ratio of 5:1 to 1:5, preferably 3:1 to 1:3, in the presence of Friedel-Crafts catalysts and having melting points of from 100° to 150°C, preferably 110° to 140°C (determined according to the capillary method), and acid numbers of 30 to 90, preferably 50 to 80.

Suitable rosin-modified phenol resins are, for example, reaction products of phenol-aldehyde resins, preferably phenol-formaldehyde resins with natural resin acids which are esterified with polyhydric alcohols, preferably glycerol, under pressure and with heating.

As the resins (3) for improving tackiness, terpene phenol resins are particularly suitable. Suitable terpene phenol resins may be prepared by condensing α- or β-pinenes (or the industrial mixtures thereof, e.g., turpentine oil) with phenol. The terpene phenol resins primarily improve the tackiness of such mixtures and thus increase the range of applications of the more cohesive block copolymers. By combining the resins (2) with the terpene phenol resins, any desired variation of thermal stability and length of tackiness of the mixtures may be obtained.

As the solvents for the adhesive compositions, conventional solvents may be used, and, if desired, in the form of mixtures, particularly aromatic and aliphatic hydrocarbons, esters and ketones, and preferably a mixture of ethyl acetate, toluene and light petrol in the volume ratio 1:1:1 and having a boiling range from 80° to 100°C. Generally, 200 to 600, preferably 300 to 400 parts by weight of solvent are used to 100 parts by weight of component (1).

These adhesive mixtures are suitable as contact adhesives for use in the shoemaking, building and motor industries, particularly for bonding nonporous materials such as leather, rubber, metals, plastics and textiles, but preferably for bonding polyvinyl chloride.

Containing an Anionic Surface Active Agent

I. Ijichi, S. Kai, H. Teranishi, A. Morioka and K. Seki; U.S. Patent 4,152,309; May 1, 1979; assigned to Nitto Electric Industrial Company, Ltd., Japan have developed a pressure sensitive adhesive which can be used for making adhesive tapes, labels, etc. without the use of organic solvents. Their formulation comprises a liquid diene polymer having an average of 2.0 to 2.5 functional groups capable of reacting with an isocyanate group, an isocyanate compound with 2 or more isocyanate groups, and a tackifier in which an anionic surface active agent is added in an amount of 0.2 to 10 parts by weight per 100 parts by weight of the total weight of the liquid diene polymer and the isocyanate compound.

When the composition, mainly comprised of one or more liquid diene based polymers, one or more isocyanate compounds, and one or more tackifiers, is coated on a support in thin-layer form (about 30 to 50 microns thick) and heated to cause polymerization and crosslinking, particle-like voids are produced on the coating layer, thereby resulting in unevenness, i.e., thick areas and areas having substantially no thickness on the support. Anionic surface active agents are used to overcome these problems and commercially available anionic surface active agents can be used. Parts, in the example below, are by weight.

Example: 100 parts of a liquid diene polymer, polybutadiene containing a —COOH group at each end thereof and having an average molecular weight of 1,500 (Nippon Soda Co., Ltd.'s Nisso PB-C), and 70 parts of a terpene resin (Pennsylvania Industrial Chemical Corp.'s Piccolyte A-115) were mixed in a kneader. In addition, 11.5 parts of diphenylmethane diisocyanate, 0.05 part of cobalt naphthenate, and 0.5 part of one of the anionic, cationic, nonionic, and amphoteric surface active agents shown in the table were added to prepare various compositions. These various compositions were coated on a 150-micron thick plasticized polyvinyl chloride sheet at a thickness of about 30 microns (dry basis), and then heated at 120°C for 2 minutes and wound up in the form of a roll to produce samples.

The properties and characteristics of these samples as pressure sensitive adhesive tapes were evaluated by the following evaluation methods. The results obtained are shown in the table.

Progress of Polymerization Reaction: This test was conducted to evaluate to what extent the coating layer coated on the support undergoes polymerization upon heating, and if the coating layer is satisfactorily converted into a pressure sensitive adhesive layer. First, it was determined if, when pressed with a finger, the adhesive stuck to the finger. Second, it was determined if a sample whose adhesive layer was stuck to a stainless steel plate, could be peeled off without leaving any adhesive on the stainless steel plate:

o polymerization proceeds sufficiently; no adhesive layer remains on a finger or the stainless steel plate

△ adhesive layer somewhat remains on a finger or the stain- less steel plate

x polymerization proceeds insufficiently; the adhesive layer remains on a finger or the stainless steel plate, or stringing of the adhesive layer is caused.

State of Adhesive Layer: The adhesive layer was visually examined to see if the adhesive layer was formed on the support in a uniform thickness or evenly, or if particle-like unevenness (voids caused by the support repelling the adhesive layer) were formed on the adhesive layer:

◇ adhesive layer whose entire surface was smooth, having a uniform thickness

□ 10 or less voids of a diameter of 2 mm or less present per square meter

y more than 10 voids of a diameter of 2 mm or less pres- ent per square meter; or voids of a diameter of more than 2 mm were present.

Adhesive Strength: The sample was slit into 20 mm widths and adhered to a stainless steel plate having a mirror surface, pressed by applying a load of 2 kg with a roll and peeled off at a peeling rate of 300 mm/min with force parallel to the stainless steel plate. The 180° peeling force at this time was measured at a temperature of 20°C with a Tensilon Measure produced by Toyo Measuring Instrument Co., Ltd.

Cohesive Force: This term designates the holding power of the pressure sensitive adhesive tape. The sample was slit to a definite area (in this case, 1 x 2 cm) and stuck to a support (a Bakelite plate). While pressing by applying a load of 1 kg with a roll, a definite load (in this case 300 g) was applied parallel to the support at 20°C for 1 hour, and then the distance the sample moved down along the surface was measured. Of course, the sample whose distance deviated was small had high holding power and was considered to be an excellent pressure sensitive adhesive tape.

Evaluation of Pressure Sensitive Adhesive Tapes

. . . . . .Surface Active Agent.		Polymerization Progress	State of Adhesive Layer	Adhesive Strength (g/20 mm width)	Cohesive Force (mm/hr)
Type	Trade Name				
Anionic					
Sodium salt of fatty acid	Nonsal DK	o	◇/□	480	0.32
Sodium sulfonate	Newlex R	o	◇	420	0.45
Sodium sulfonate	Neocol P	o	◇	390	0.52
Alcohol sulfate	Liponol LL-103	o	◇	450	0.64
Phosphoric acid diester salt	Gaffac RS-410	o	◇	450	0.45
Cationic					
Amines (main components)	Nimine S-204	o	y	230	2.5
Quaternary ammonium	Catinol HB	o/△	y	280	3.6
Nonionic					
Ethers (main components)	Nonion E-220	x	y	120	dropped in 2 min
Alkylphenol	Nonion NS-210	x	y	250	dropped in 5 min
Esters (main components)	Nonion P-10	△/x	y	300	dropped in 25 min
Amphoteric					
Imidazoline derivatives	Ovasolin 40A	△	y	290	150
Amine and aliphatic acid	Anon BF	x	y	110	dropped in 7 min

Samples produced using the anionic surface active agents were good in progress of the polymerization reaction and the state of adhesive layer; furthermore, pressure sensitive adhesive tapes obtained therefrom were similar in adhesive strength and cohesive force to the plasticized polyvinyl chloride adhesive tape on the market. On the other hand, in samples where a cationic, nonionic or amphoteric surface active agent was used, circular voids of a diameter of 1 to 5 mm were produced on the whole surface. The interior of the voids was such that the adhesive did not stick, areas were found where product appearance was poor, and polymerization reaction proceeded insufficiently.

APPLICATIONS

Printable Release Coats for Adhesive Tapes

The solvent release coating developed by *A. Blum and C. Bartell; U.S. Patent 4,070,523; January 24, 1978; assigned to Borden, Inc.* utilizes polyketone resins. Those manufactured by Union Carbide or the Krumbhar Division of Lawter Chemicals (K-1717 resins) may be used. The preferred resin, K-1717B, has the following desirable properties:

- Solubility in a variety of solvents comprising ketones, alcohols, esters, chlorinated hydrocarbons, aromatics, etc., as well as maximum tolerance for aliphatic solvents.

- Wide range of compatibility with various resins and elastomeric film formers.

- Excellent adhesion to difficult surfaces such as hydrophobic plastic films and aluminum foils.

- Clarity and oxidative stability—due to stable structure—combined with low viscosity of solutions.

- Excellent intercoat adhesion which is critical for good ink anchorage.

Example: This example illustrates fresh and aged characteristics of such a printable release coat:

Resin Components

	Parts by Weight
Vinyl acetate/vinyl alcohol resin (T-24-9)	18
Polypropylene/glycerol terephthalate copolymer (Vitel 307)	60
Polyketone resin (K-1717B)	20
Release agent (Odup)	2
Total	100

Solvent Components

Toluene	1,357
MEK	205
Methyl Cellosolve	338
Total	1,900

A 5% solution of the above was applied with a No. 4 Mayer Bar on a 1-mil Mylar film, air dried, and heated under infrared lights for 2 minutes at approximately 200°F. After cooling, 3 strips of a standard cellophane pressure sensitive tape were placed on the coated surface and pressed by rolling twice in both directions with a 4½-pound rubber roller. A similar procedure was utilized to apply 3 strips of the same pressure sensitive tape to a non-release-coated 1-mil Mylar film.

Both overtaped sets were shelf and oven (120°F) aged for 12 days and the force required to separate the 1-inch wide cellophane tape strips from the Mylar backing (T-peel adhesion) was measured on an Instron machine. In addition, the rolling ball tack was measured to check any detackification of pressure sensitive adhesive. The test data, averaged from several readings, are shown in the table.

The data indicate good tack retention for the applied cellophane tape and a rather high level of release efficiency (T-peel values of coated Mylar roughly one-fifth of readings for noncoated substrate).

	. . . . Mylar	
	Release-Coated	Plain
T-peel adhesion, ounce/inch		
12-day shelf	10.0	54
12-day 120°F oven	12.5	58
Rolling ball tack, inches		
12-day shelf	0.5	0.42
12-day 120°F oven	0.6	0.42
Appearance of dried coating	clear	

Drafting Films for Use in Electrostatic Copiers

In the self-adhesive drafting film market it is standard procedure to use sheets consisting of a film support coated on the back with a pressure sensitive adhesive and on the front with a matte coating. A second film having a release layer on its top surface is applied to the back of the first film. A release layer is applied to the second film so that upon its removal the adhesive layer is exposed. These films are used for making corrections in engineering drawings by marking the film's matte surface with the desired material, cutting out the marked portion of the drafting film, removing the release liner and sticking the so-prepared portion of the film on the drawing. In this manner, corrections can be made quickly without requiring erasure of the original drawing. In another use of these films, features which are used frequently in the drawing can be imprinted on the film and the drawing prepared using the feature from the self-adhesive drafting film.

R.H. Hankee and A.T. Akman; U.S. Patent 4,074,000; February 14, 1978; assigned to Xerox Corporation have devised a layered pressure sensitive self-adhesive film for use in preparing and correcting architectural, construction and engineering drawings. The film comprises layers from the top down of a matte coating capable of receiving and bearing graphic information; a layer of an organic resinous material in film form as support for the matte coating; a layer of pressure sensitive adhesive; a release coating; a layer of an organic resinous material in film form as release liner; and, a back coating which is a layer of an organic resinous binder material having dispersed therein particulate silica of particle size ranging from 3 to 20 microns in the longest dimension and a water-insoluble antistatic agent of a type and in an amount sufficient to cause the film to have a surface resistivity in the range of from 10^9 to 10^{14} ohms per square.

The drafting film is schematically represented by Figure 2.1. The top layer, i.e., matte coating, is designated **11**. Sheet materials capable of receiving and holding graphic information generally comprise a film base whose exposed surface carries a synthetic resinous matte film-forming composition. The need for a matte coating on the film base is due to the fact that indicia markings on uncoated plastic surfaces will readily rub, peel or flake off. Poor adhesion of the indicia marking on plastic surfaces is particularly troublesome in the drafting field where plastic films, plastic-coated drafting cloths, etc., are commonly imaged with ink, pencil, crayon or the like. Plastic materials, therefore, require a precoating, generally from a lacquer, of a matte surface in order to accept the marking material directly. Dispersion of particulate material in the resin binder provides a surface having lands and valleys which enhances adhesion of the marking material to the sheet. Any film-forming resinous material such as Mylar may be used as the support material for the matte coating.

Figure 2.1: Drafting Film

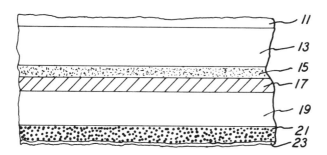

Source: U.S. Patent 4,074,000

The matte lacquer coating is preferably a cellulosic film-forming polymer dissolved in a suitable solvent. In order to provide a matte finish suitable for indicia marking, the coating will include a particulate substance such as an organic material which is incompatible with the film-forming resin or an inorganic pigment such as a finely powdered glass having a particle size ranging from 2 to 6 microns in the particle's longest dimension or amorphous or diatomaceous silica of similar particle size. Titanium dioxide may also be added to impart a white background color and give the matte surface the necessary roughness for ink reception.

The next layer of the film, designated as **15**, is the pressure sensitive adhesive. It is a general practice to coat the pressure sensitive adhesive over the release side of the release liner **19** first and then transfer it to the drafting film during lamination. Since the release coating **17** limits the wetting over its surface, the adhesive coating is applied from a solvent base or hot melt rather than as a water-base coating.

There are two types of adhesive; permanent and removable. In pressure sensitive adhesive drafting films, a removable type of pressure sensitive film, preferably of the repositionable type, is used. Acrylic type adhesives are desirable for use in films which are to be imaged in electrostatographic copiers because of their thermal stability. This is the case because the imaged film is heat treated in the copying process to fuse the toner particles into the sheet. Specific examples of acrylic adhesives which may be used are No. 7598 (Paisley Products), No. RA-96 (Monsanto Polymer and Petrochemicals Co.). Traditionally, solvent based adhesives comprise a solution of rubber and resins or a solution of a polymerized acrylic. Solution acrylics, while being more expensive than the rubber based products, are generally chosen for their excellent elasticity, good ultraviolet and aging stability, and good electrical properties.

The next layer 17 is required as a release coating for the pressure sensitive adhesive. This coating must be formulated to produce a fairly low release level so as to provide easy peel-off. The release level should, however, be high enough so that the sandwich will not come apart in the copier during the imaging process. Silicone resins are preferred for use as the release coating although other materials such as fluorocarbons, e.g., Teflon, may be used. Various silicone resins marketed by the General Electric Company have been found to be suitable.

The next layer of the drafting film is the release liner designated as 19. This layer serves as a support for the silicone release coating and those other layers of the film which are stripped from the backing layer and applied to the document being created. The release liner is formulated of a resinous film-forming material which may be clear or opaque. An opaque backing layer is preferred because it provides an easy method of identifying the matte drafting layer side from the backing. The release liner is typically a Mylar-type polyester or a Mylar film with polypropylene impurities or mineral fillers to cause an opaque appearance.

The final layer of the drafting film is the friction back coating designated as 21. The friction coating is prepared by formulating a solution of a suitable film-forming resin which is adhesive to the surface of the backing layer 19 and dispersing the particulate material therein. Those film-forming resins previously mentioned as being useful in other layers of the film may also be used to form the backing layer and back coating. Typically, the size of the particulate material will range from about 3 to 20 microns in its longest dimension. By adding enough of the particulate material to the resin solution to provide from 0.5 to 15 weight percent of the particulate material based on the dry resin, a film having a somewhat rough back surface is provided. In this manner, individual sheets of drafting film stacked in a pile are encouraged to slide over one another and thereby feed properly in the automatic feed mechanisms with which most electrostatographic copiers are equipped.

An antistatic agent is also added to the friction coating layer to dissipate the static charges from the film which are the result of ionization of air or contact with an electrically excited object. Suitable antistatic agents may be alkyl quaternary ammonium chlorides such as Arquad 18-50, Arquad T, or Armostat 310 (all from Armour Industrial Chemical Co.).

Switch-Proof Label

F.P. Williams; U.S. Patent 4,082,873; April 4, 1978; assigned to Monarch Marking Systems, Inc. describes switch-proof labels useful for marking objects in a manner such that if one were to attempt to transfer the label to another object the label would be destroyed or defaced to such an extent that its transference would be noticeable.

The label comprises a laminate made up of a transparent or translucent outer sheet having an information-containing pattern printed on its inner surface, the printed inner surface having a coating of pressure sensitive adhesive film thereon. The printed pattern has a lesser affinity for the outer sheet than the printed pattern has for the adhesive. The affinity of the adhesive for the surface to which the laminated label is adhered and to the printed pattern is greater than the affinity of the printed pattern for the outer sheet. The free side of the adhesive film of the label is covered by a release sheet.

Once applied to a substrate, if removal of the label is attempted, the label delaminates in a manner such that the outer sheet separates leaving at least a portion of the adhesive layer, having at least a portion of the printed pattern adhering thereto, adhered to the substrate.

The label comprises an external layer of flexible transparent or translucent plastic film, with sufficient transparency or translucency so that a pattern printed on one side is visible through the film. Preferred films include polyester films, and a particularly useful film of this type is the highly oriented polyester known as Mylar film.

On the interior surface of the exterior film there is reverse printed an information-containing pattern. The method of printing can be any printing process useful in printing upon plastic films. The chemical composition of the ink employed to print the pattern is not critical, however, the ink must produce a printed pattern which has greater affinity for (i.e., adhesion to) the adhesive layer than to the inner surface of the outer layer. A particularly useful ink is a flexographic letterpress ink consisting of 10% of a phthalocyanine blue pigment and 90% of 25:75 resin-vehicle mixture, where the resin is a modified phenolic resin and the solvent consists of (by volume) 80% ethyl alcohol, 10% ethylene glycol monoethyl ether and 10% n-propyl alcohol.

In a preferred embodiment, a 1-mil Mylar film was reverse printed with the above ink and the printing dried by 140°F forced air through a slit nozzle one-half inch away from the printed side of the film. The printed layer was then coated on its printed side with a layer of pressure sensitive adhesive.

A pressure sensitive adhesive, useful in conjunction with the ink is a 55% solution of thermosetting acrylic solution polymer in 75% ethyl acetate and 25% toluene (by volume), having a Brookfield viscosity of between 12,000 and 18,000 cp at 25°C. Representative physical data of a 1-mil dry film of this adhesive applied to a Mylar film (cured at 250°F for 2 minutes) are as follows.

Quickstick (rolling ball-incline plane):
Inches of fall per inches of travel = 1.2

180° Peel Adhesion (PSTC-1 test method):
Initial = 56 ounces
Overnight = 76 ounces

20° Hold (½" x ½" adhesive strip, 20 chrome-plated
bar, 200 g weight) = 19 hours

50°C Creep (1" x ½" adhesive strip attached at the
vertical to stainless steel plate, 250 g weight) =
24+ hours

Williams Plastometer (100°C) = 1.73

The use of this adhesive provided a laminated label which delaminated upon removal from the article to which it was affixed. When the film was lifted from the labeled article, the printing remained adhered to the adhesive layer.

Rubbery Cushion with Adhesive Layer for Automotive Use

The product developed by *J.N. Brown; U.S. Patent 4,086,388; April 25, 1978; assigned to Minnesota Mining and Manufacturing Company* is a rubbery cushion which has a pressure sensitive adhesive layer and may be used outdoors. Such cushions may be used as bumper buttons or strips for automobile hoods and gasoline-filler doors. In this process, rubbery cushions of oil-plasticized thermoplastic elastomer compositions are provided with pressure sensitive adhesive layers which are not subject to undue softening. The rubbery cushions can be adhered by their pressure sensitive adhesive layers to substrates and remain firmly in place indefinitely in the face of repeated shearing stresses, even under prolonged exposure to adverse weather conditions. Briefly, there is involved:

a rubbery cushion comprising a thermoplastic elastomer composition including at least 2 parts of a plasticizing oil per 100 parts by weight of the elastomer,

a thin chemically-interfacially-bonded dual-layer film, one layer of which is a polyolefin which is fused to the rubbery cushion, the other layer of which is an oriented polyester polymer, and

a pressure sensitive adhesive layer adhered to the polyester polymer.

The polyolefin layer of the dual-layer film is inseparably bonded both to the elastomer composition and to the polyester layer which acts as a barrier preventing the plasticizing oil from reaching the adhesive in amounts sufficient to have a noticeable softening effect, even after long periods of time. The chemical interfacial bond between the layers of the dual-layer film may be created as disclosed in U.S. Patent 3,188,265. The dual-layer film may be as thin as about 0.02 mm but should not exceed about 0.4 mm so as not to impart undue rigidity at the pressure sensitive adhesive; preferably it is 0.05 to 0.25 mm in thickness.

A preferred polyester polymer of the dual-layer film is biaxially-oriented polyethylene terephthalate film, being readily available commercially at modest cost. Other useful polyester polymers include polybutylene terephthalate. The polyolefin of the dual-layer film may be polyethylene which is low in cost and readily fuses to the thermoplastic elastomer compositions at moderately elevated temperatures.

Preferred thermoplastic elastomers are block copolymers such as Kraton G (Shell Chemical Co.), Telcar (B.F. Goodrich), TPR (Uniroyal) and Somel (DuPont). In all of the block copolymers used, the majority of the end blocks should be crystalline.

Particularly preferred pressure sensitive adhesves are the acrylic copolymers described in U.S. Patent Reissue 24,906. Also preferred are terpene-resin-modified block copolymer adhesives as described in U.S. Patent 3,389,827. Less preferred but nevertheless useful are pressure sensitive adhesives based on mixtures of natural or synthetic rubber and resinous tackifiers although they tend to resinify over a period of time.

The pressure sensitive adhesive layer may incorporate a foam as disclosed in Canadian Patent 747,341. If the pressure sensitive adhesive layer may be exposed to gasoline or other solvents, a preferred foam-containing pressure sensitive adhesive layer is disclosed in U.S. Patent 3,993,833.

Clear Sheet to Protect Signs from Vandalism

It is generally well known that in recent years there has been increased activity by vandals to deface roadway and street signs, particularly with spray-can paints which are very difficult and troublesome to clean off afterwards, so that many such signs must be replaced.

J.H. Bishopp and G. Spector; U.S. Patent 4,090,464; May 23, 1978 has proposed the provision of a vandal guard sheet which is made of clear plastic or vinyl so that the sign is visible therethrough, which can be quickly and easily adhered to the sign so that spray paint is deposited upon the sheet instead of upon the sign. The sheet can thereafter be readily peeled off and replaced by a fresh sheet.

Figure 2.2 is a perspective view of the proposed product showing a defaced guard sheet being removed so as to be replaced by a fresh guard sheet. Referring to the figure, shown on the following page, the reference numeral 10 represents a vandal guard sheet wherein there is a sheet of clear plastic or vinyl-type material which has its entire one side coated with a clear adhesive 11 which is covered by a protective backing paper that can be peeled off so as to expose the adhesive at such time that the guard is intended to be installed upon a front face or side 13 of a roadway sign panel 14. The sheet is made in the same size and shape of the sign so to completely protect the sign face from scrawling 15 (made by vandals upon signs) but which, by employment of the guard sheet, thus defaces the sheet instead of the sign face, so that after the sign information is excessively obliterated, a highway or street sign maintenance crew can quickly and easily simply peel off the defaced sheet and replace it by a fresh sheet, thereby eliminating the difficult or impossible chore of removing such defacing from the sign.

Figure 2.2: Vandal Guard Sheet

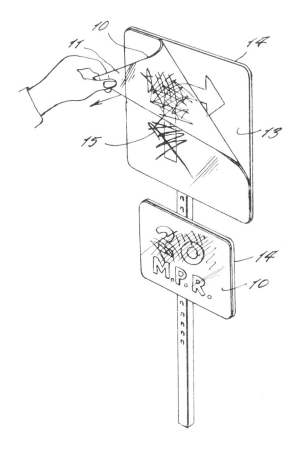

Source: U.S. Patent 4,090,464

Transfer Adhesives

Pressure sensitive transfer adhesives are pressure sensitive adhesives in the form of self-supporting films which are generally provided on release liners, as for example, polyethylene film or a paper coated with a silicone or other release agent from which the films can be readily stripped. The transfer adhesive is commercially supplied in roll form and should be capable of being unwound without any offsetting of adhesive at the edges or elsewhere. After application of the transfer adhesive to the desired substrate, the liner is peeled off to expose the adhesive film, permitting the adherence thereto of another sheet material or application to another substrate, for example, in laminating applications. One use of such transfer adhesive is in forming "sandwich-type" laminated electrical conductors or the like, in which case an adhesive having suitable electrical properties is selected.

Another use for such transfer adhesives is a means whereby a converter may apply an adhesive and liner to a backing, such as a metal foil ribbon (which may, for example, have a label insignia imprinted on the reverse surface). The ultimate user of the laminated product then removes the liner at the time of use to expose the pressure sensitive adhesive, permitting the foil to be adhered to a desired substrate in a dry manner by mere contact and light pressure.

C.R. Freeman and C.I. Sauer; U.S. Patent 4,102,835; July 25, 1978; assigned to Minnesota Mining and Manufacturing Company have discovered that marked and unexpected reinforcement of the physical strength of a conventional pressure sensitive adhesive and proper maintenance of its fourfold balance of properties are obtained by blending minor amounts of certain block copolymers, which can be elastomeric, with the adhesive solids.

In the resulting reinforced pressure sensitive adhesive solids, the major elastomeric component comprises one of the homopolymers or random sequence polymers commonly used in tackifiers of inherently tacky pressure sensitive adhesives, e.g., natural rubber, synthetic polyisoprene, styrene-butadiene rubber, rubbery polyolefins, and the inherently rubbery and tacky acrylate and vinyl ether polymers.

The block copolymers which have been found to provide improved strength are copolymers having 3 or more polymer block structures having a general configuration A–B–A where each A is a thermoplastic polymer block with a glass transition temperature above room temperature, having an average molecular weight between about 5,000 and 125,000; and B is a polymer block of a conjugated diene having an average molecular weight between about 15,000 and 250,000.

Example: The following ingredients were mixed together and then heat-cured in a Mogul mixer at about 240°F for about 20 minutes:

Part A

	Parts by Weight
Natural rubber (masticated crepe)	100
Heat-treated wood rosin	20
Polyterpene resin (115°C, R&B softening point)	40
Zinc rosinate	5
Oil-soluble, heat reactive phenol-aldehyde resin	12
Antioxidant, 4,4'-thio-bis(6-tert-butyl cresol)	2

The mixture was then quenched and diluted to 34% solids by adding heptane. The following ingredients are separately mixed at room temperature and then added and mixed into the heat-cured adhesive formulation:

Part B

	Parts by Weight
Block copolymer of styrene and butadiene 1 polymer block, 70,000 MW butadiene, 2 polymer blocks, 15,000 MW styrene (Shell's Kraton 101)	17
Tackifier (pentaerythritol ester of rosin)	10
Solvent (heptane)	101

The adhesive composition was then knife coated onto a silicone treated paper release liner to a dry thickness of 1.8 mils. A control adhesive was also prepared by using the heptane-diluted heat-cured formulation without addition of the block copolymer. The control was similarly formed into a film on a release liner. The control adhesive could not be wound into a roll and unwound without random transfer of the adhesive from the release liner, whereas the adhesive containing the block copolymer could be unwound without transfer as a self-sustaining integral film.

Adhesive formulations were also prepared by adding Kraton 101 (without the additional pentaerythritol ester of rosin tackifier) to the above conventional rubber-resin adhesive in various amounts. The adhesives were formed into a transfer film by coating on a release liner to a wet thickness of 0.007 inch. The films were dried in an oven to produce dry films having a thickness of 0.002 inch. The physical properties of the films were measured and are summarized as follows:

Kraton 101 (wt %)	Tensile Strength (psi)	Elongation at Rupture (%)	180° Peel* (oz/in)	Permanent Set** (%)
0	30	550–700	49	12.5
7.67	45	475–600	43	3.1
15.3	55	450–550	39	~1
23.0	65	425–500	36	~1

*ASTM Test D-1000; measured by laminating the transfer adhesive liner combination to a 1-mil polyester film, removing the liner, and testing the resulting film-adhesive tape.
**After elongation of 200% for 10 sec; double thickness film used in tests.

Heat-Activated Adhesive Coating

W.R. Lawton; U.S. Patent 4,135,033; January 16, 1979 describes an adhesive coating which has a dry, tack-free surface convertible to a permanently pressure sensitive adhesive layer by the application of heat. The product is useful for labeling, sealing, adhesive-layer transfer, decalcomania, and other packaging applications.

The product is made by forming a thin, dry, nontacky barrier coating on and in the surface layer of any conventional pressure sensitive coating by a simple procedure. The barrier properties of this coating can be destroyed by heat to restore the original properties of the pressure sensitive adhesive layer. The barrier layer is obtained by in situ formation of molecular complexes or compounds within and/or on the surface portion of the adhesive composition.

Two classes of molecular complexes have been found to be particularly useful for the product. The hydrogen bond complexing of hydroxyaromatic compounds with amines has provided an effective method for obtaining the complex barrier coat. Compositions which give solid complexes are readily obtained by reacting stoichiometric amounts of the compounds in methanol. The solid complex will separate and after filtering and drying the dissociation temperature can be determined. The dissociation temperature of the complex relates to the activation temperature of the adhesive. It has been found that the dissociation temperatures can be lowered by including additives which cause a combined heat-solvent effect for dissociation of the complex.

The inclusion of long-chain (8 or more carbons in the chain) hydrocarbon derivatives in urea is an effective means for obtaining the barrier coat. Certain additives will decompose the urea complex at lower than normal activation temperatures for urea inclusion compounds.

Some long-chain normal compounds which are readily complexed by urea are: aliphatic hydrocarbons, aliphatic alcohols, aliphatic fatty acids, aliphatic fatty amides, aliphatic paraffin amines, aliphatic fatty acid esters, aliphatic fatty nitriles, aliphatic ketones, aliphatic ethers, aliphatic halides, aliphatic aldehydes, aliphatic nitro compounds, aliphatic sulfides, aliphatic mercaptans, and unbranched olefins.

Similar results may be obtained by substituting thiourea for urea and using branch-chained hydrocarbon derivatives in place of the normal or straight-chained derivatives listed above.

Example:

Water-Base Adhesive System

Component	Pounds
Amsco 3036, polyethylhexyl acrylate (52%)	259.62
Carboxymethylcellulose (CMC 7L, 10% in water)	13.50
Ammonia (28°Be)	6.75
Methanol	18.91
Water	31.97

CMC 7L (10% in water) to adjust viscosity to 3,000 cp

Hydrogen Bonded Molecular Complex Systems

Hydroxyaromatic	Amine
Tetrabromobisphenol A	Dibenzylamine
Tetrabromobisphenol A	Hexamethyleneamine
Hydroquinone	Hexamethyleneamine
Pyrogallol	Hexamethyleneamine

The adhesive coating is prepared by incorporating 4 parts of amine in 96 parts adhesive base, applying this mixture onto conventional base paper stock such as Kromekote with a 6-mil Bird applicator and allowing to dry at room temperature to give a conventional tacky pressure sensitive adhesive.

A 7% solution of the hydroxyaromatic compound is prepared using methanol as the solvent and the adhesive coating is dipped into this solution to give a wash coat. Upon evaporation of the methanol, the surface of the adhesive layer is dry and essentially tack-free due to the in situ formation of the hydrogen bonded molecular complex between the amine and the aromatic hydroxy compound.

When these nontacky potentially adhesive coated papers are heated for 10 seconds on a hot arch at a temperature of about 135° to 143°C, adhesiveness is similar to the properties of the original adhesive coating without wash-coat treatment. The adhesive tack is still present two weeks after heat activation.

Addition of 4 parts per 100 parts of the adhesive formulation of a third coreactant comprising any one of the following: benzil, hydroquinone-monomethyl-ether, p-hydroxy-acetophenone, 4-methoxybenzil, tert-butyl hydroquinone, 1,5-diphenylpentene-3-dione, or glycollic acid, and wash-coating with this solution results in dry, tack-free coatings which can be activated to full adhesiveness by heating for 10 seconds at 110°C. Without the use of these additives, full adhesiveness is not obtained under the same activation conditions.

A table containing an extensive list of molecular complexes of phenols and amines is contained in the patent.

Backing of Polypropylene Having Specific Characteristics

Pressure sensitive adhesive tapes used for sealing covers of cartons must have good strength in the transverse direction and such tapes having hard polypropylene films have been used. However, such tapes are difficult to tear with finger pressure and must have special cutters and even then they are difficult to cut in the transverse direction.

Y. Miki, H. Nishizawa and Y. Suzuki; U.S. Patent 4,137,362; January 30, 1979; assigned to Nitto Electric Industrial Co., Ltd., Japan have found that pressure sensitive adhesive tapes having good strength in the transverse direction which can be easily cut with the fingers and which do not break during unwinding can be obtained by using as backing a film comprising polypropylene as a main component.

The backing is produced by stretching a long sheet of polypropylene as the main component in a substantially transverse direction to the lengthwise direction of the tape to be produced, the backing having the properties that the shock edge tearing resistance is below 200 g, the elongation at break is above 200% in the lengthwise direction, the tensile strength is above 170 kg/cm^2 in the lengthwise direction and above 800 kg/cm^2 in the transverse direction, and impact strength is above 15 kg-cm/mm^2 in the lengthwise direction and above 40 kg-cm/mm^2 in the transverse direction, and the backing has a thickness of about 10 to 300 μ.

Example: 80 parts by weight of isotactic polypropylene pellets (intrinsic viscosity 2.0) and 20 parts by weight of polyethylene pellets having a low density (specific gravity 0.921, melt index 32 by ASTM D 1238) were blended in a blending device. The mixture was extruded in a melted condition from a T-die using a conventional extruder to produce a sheet having a thickness of 480 μ.

The resulting sheet was stretched 7.0 times in the transverse direction at 130° to 170°C using a tenter and then the sheet was subjected to heat treatment to an annealing of 5 to 8% to produce a backing film. Both surfaces of the resulting backing film were treated with corona discharging to a wet tension of 40 dynes.

An adhesive consisting of 100 parts by weight natural rubber, 50 pbw polyterpene resin, and 440 pbw toluene; a primer of 20 pbw chlorobutadiene rubber, 15 pbw phenol-formaldehyde resin and 150 pbw toluene; and, as a releasing agent, a toluene solution containing 1% of the reaction product of N-octadecylisocyanate and soluble starch were applied to the tape to produce a pressure sensitive adhesive tape having a width of 1 m. After rolling on a paper core, the pressure sensitive tape was cut to make rolls of a pressure sensitive adhesive tape having a width of 50 mm.

The tape was evaluated and found to be cuttable in the transverse direction by only the fingers of both hands. The tape did not break during unwinding and adhered well to the carton it was used to seal. It also showed very adequate strength when tested by a drop-impact and a drum-rolling test (JIS Z 0209).

To Attach Emblems or Trim to Auto Bodies

In manufacturing industrial products such as automobiles, it is known to adhere rigid solid objects such as emblems, the manufacturer's name, body trim, and so forth to the automobile body by means of a relatively thick sheet of resilient foamed plastic such as polyurethane, for example, 30 mils in thickness, coated on each face with a pressure sensitive adhesive. This provides a cushioning effect to take care of the wide variations in the uniformity of the emblems and further provides a cushion so that when the emblem is struck a blow it does not pop off as would be the case with a thin adhesive layer absent the foam layer.

The formulation devised by *F.T. Sanderson and R.E. Zdanowski; U.S. Patent 4,145,465; March 20, 1979; assigned to Rohm and Haas Company,* by providing a permanently pressure sensitive adhesive composition as a thick, backing-free sheet with the resilience of a soft rubber and certain critical characteristics as to the physical nature of the adhesive, makes it possible to completely eliminate the foam sheet and yet have a thick cushioning material which will permanently adhere the rigid body to the automobile or other surface.

It is essential that certain critical properties be present in the adhesive which would not necessarily be required for a pressure sensitive adhesive used in the normal thickness of 0.5 to 2 mils, or in other terms, in an amount less than about 1 lb/50 ft^2 of surface.

These properties include a relatively massive thickness of from 8 to 125 mils, preferably 10 to 70 mils and more preferably below about 40 mils, a Tg of below 0°C, preferably below -15°C, a tensile strength of from 15 to 60 psi, and elongation at break of from 500 to 3,000%. It is preferred that the adhesive sheet have an elastic recovery of from 50 to 90%, a Williams' plasticity of from 2 to 5, a maximum creep distance of 0.1, more preferably 0.05 inch, and substantially 100% elastic recovery when an inch square thereof is compressed in a direction perpendicular to the surface of the sheet of adhesive under a weight of 4 to 25 lb for 0.1 to 20 min.

The preferred manner of obtaining a product having these characteristics is by the use of crosslinking mechanisms to crosslink polymers in the adhesive. Another useful means is the inclusion of a very finely divided inorganic bodying agent. Combinations of the two are particularly useful.

The pressure sensitive adhesive is preferably an addition polymer, particularly one comprising acrylic monomers. One of the monomers utilized in a substantial proportion to prepare the preferred pressure sensitive adhesive is a tackifying or "soft" monomer which may be represented by the following formula: $CH_2{=}CR{-}COOR^1$ where R is H or alkyl having 1 to 4 carbon atoms, and R^1 is the straight-chain or branched-chain radical of a primary or secondary alkanol, alkoxyalkanol or alkylthiaalkanol and having up to about 14 carbon atoms.

As is apparent, an important property of the polymer is the Tg thereof, and consequently the selection of monomers and proportions thereof depends upon their influence on the Tg. The Tg of the polymer must be below 0°C (i.e., it must give a tacky coating) and is preferably below -15°C. Examples of the Tg of homopolymers and the inherent Tg thereof which permits such calculations are as follows:

Homopolymer of	Tg, °C
n-octyl acrylate	-80
n-decyl methacrylate	-60
2-ethylhexyl acrylate	-70
octyl methacrylate	-20
n-tetradecyl methacrylate	-9
methyl acrylate	9
n-tetradecyl acrylate	20
methyl methacrylate	105
acrylic acid	106

These or other monomers are blended to give the desired Tg of the copolymer. For example, a large proportion (e.g., 90% by weight) of a combination of 3 parts of acrylic acid or methacrylic acid and 87 parts of methyl acrylate (all "hard" monomers) with a small proportion (e.g., 10%) of octyl acrylate (a "soft" monomer) provides a copolymer having the desired Tg. Most of the esters of acrylic acid or methacrylic acid having a low Tg are well known in the pressure sensitive adhesive art as tackifying monomers.

Emulsion copolymers are prepared by well-known methods, using a dispersing agent, a free-radical polymerization initiator and/or an accelerator.

The adhesive is preferably subject to latent crosslinking (crosslinking subsequent to polymerization). There is thus included within the copolymer up to 20% by weight of such functional, polar, or reactive monomer, preferably an unsaturated carboxylic acid, half esters and half amides of α-unsaturated dicarboxylic acids, and salts thereof with ammonia, an alkali metal such as sodium, potassium or lithium, or with a volatile water-soluble amine such as dimethylamine or triethylamine, in order to provide the crosslinking functionality.

Among useful bodying agents are the very finely divided channel carbon blacks, zinc oxide, calcium carbonate, calcium silicate, amorphous hydrated silica, amorphous hydrated alumina, the clays, and magnesium carbonate. Particularly preferred is amorphous or colloidal silica. The particle size of such inorganic bodying agents should be less than 25 microns, preferably less than 5 microns, and still more preferably less than 3 microns.

Example 1:   A glass bottle is charged with 100 g of distilled water and a Teflon-covered magnetic stirrer bar. The container is placed on a magnetic stirring plate, and stirring speed adjusted to give a vortex reaching nearly to the bottom of the liquid. 22 g zinc oxide is sifted into the water and stirred until a well-mixed slurrry is obtained. At this point, 40 ml of 28% ammonium hydroxide are added, followed by 42 g of glycine, and stirring continued until a clear solution is obtained.

A second container is charged with 100 g of an aqueous dispersion containing 55 weight percent of a tacky acrylic emulsion copolymer prepared in a conventional manner of a 98 to 2 weight ratio mixture of butyl acrylate and methacrylic acid. The emulsion is stirred using a magnetic stirrer assembly, and zinc ammonium glycinate solution, prepared as described above, is added in an amount to give about 0.25 equivalent of zinc per carboxyl. After mixing for several minutes, an aqueous solution of 7% ammonium hydroxide is added to the emulsion until the pH is raised to 9.5 as indicated by a Leeds and Northrup Model 7400-A-2 pH meter.

100 g of the adhesive emulsion is charged to a stirred reactor; stirred slowly while adding 1.1 g of Gantrez M-155 (polyvinyl methyl ether from General Aniline & Film Corp) for 15 minutes before adding 1.0 g of ethylene glycol; and continuing to stir slowly; finally adding 3.3 g of Cab-O-Sil M-5 (colloidal

silica from Cabot Corp.) and stirring until a smooth, creamy mastic mixture is obtained.

Example 2: Charge 100 g of the adhesive from Example 1 and 1.1 g Gantrez M-155 to the reactor as in Example 1. While stirring slowly, add 1.0 g of a poly-acrylic acid thickener. Continue to stir while adding 3.3 g Cab-O-Sil M-5 and stir until a smooth creamy mixture is obtained. Add 2 g of ammonium hydroxide and stir for about 5 additional minutes.

A film of the finished mastic is cast on polyethylene terephthalate film by use of a Gardner knife set to a 30-mil gap. After several hours of air drying, the sample is baked in an air-circulating oven at 190°F for 15 minutes. Following storage at 78°F and 50% relative humidity for about 36 hours, a strip 1" x 8" is cut for testing. The strip is applied to a red oxide pigmented epoxy primed steel panel and a 4.5-pound roller is slowly passed over the laminate twice to insure proper bond formation. After 20 minutes of laminate aging, adhesion of the sample is measured by peeling the sample at a 180° angle at 10 inches per minute using an Instron Tester Model F/TM. The peel adhesion is 9.4 pounds per inch.

The following adhesives, modified to have the degree of crosslinking and proper body by the use of bodying agents, as in Example 1, to provide the critical physical properties of the adhesive-free film, are useful in this process.

Example 3: An emulsion polymer of 600 g of 2-ethylhexyl acrylate, 1.4 g of triethylene glycol dimethacrylate, and 150 g of dimethylaminoethyl methacrylate, prepared by sequential addition of monomers.

Example 4: A solution polymer in ethyl acetate and toluene of 35 parts of vinyl acetate, 60 parts of ethylhexyl acrylate, and 5 parts of hydroxyethyl methacrylate to which polymer is added 0.12 equivalent of tetrabutyltitanate per equivalent of active hydrogen.

Mixing a Polysiloxane with the Adhesive

Because of the increased movement towards the so-called "do-it-yourself" type of home improvements and decorating, many articles such as wall coverings, counter coverings, tile appliques, shelving covers and edgings, etc. are now sold with pressure sensitive adhesive coatings thereon. One of the problems with such pressure sensitive adhesive coated articles is that, in use, they are extremely difficult to position and, if necessary, reposition, even immediately after the application to the substrate.

M.M. Sackoff, J.R. Smith and B.E. Walling; U.S. Patent 4,151,319; April 24, 1979; assigned to United Merchants and Manufacturers, Inc. have discovered an improved method for the production of pressure sensitive adhesive coated laminates wherein a first sheet having a release surface thereon is coated with a pressure sensitive adhesive to form a first laminate, and the first laminate is then subjected to conditions suitable to dry or cure the pressure sensitive adhesive, and the pressure sensitive adhesive side of the thus-treated first laminate is married to the inner surface of a facing layer having inner and outer surfaces. The method involves the intimate mixing with the pressure sensitive adhesive prior to coating onto the release surface of a means for decreasing the "zero-minute peel value" of the facing layer-pressure sensitive adhesive layer combination. More particularly, by carefully selecting the material which is intimately mixed with the pressure sensitive adhesive a pressure sensitive adhesive coated laminate is obtained which is easily positioned and adhered to substrates and which can be removed and, if necessary, repositioned with ease during the application time.

Specifically, the material used to decrease the zero-minute peel value is a polysiloxane and must be capable of being intimately mixed and dispersed throughout the pressure sensitive adhesive. By virtue of the intimate mixing of the polysiloxane material with the pressure sensitive adhesive, the need for a separate and precisely controlled coating step wherein the polysiloxane is applied to the pressure sensitive adhesive is obviated. The number of process steps required during actual assembly of the pressure sensitive adhesive laminate is thus decreased without any sacrifice in the utility of the end product obtained.

The product produced according to this process after removal from the release surface, i.e., when it is about to be applied to the substrate, possesses a relatively low-peel value, i.e., a low degree of affinity for the substrate. As a consequence, if, when applied to the substrate, the particular sheet is not straight or in the desired position, it is easily removed for repositioning. Moreover, if two areas of the pressure sensitive adhesive coated side of the sheet should touch one another, they are easily pulled apart without disturbing the uniformity of the coating of the pressure sensitive adhesive or stretching or wrinkling of the facing layer.

Example: A release sheet was prepared from kraft paper (42 lb per ream) by coating it with a conventional silicone release finish (e.g., Dow Corning Syloff 23 with DC 23A catalyst), the amount of coating being about 0.5 lb per ream. The paper was coated on the machine finished side. The coating was carried out using an 80-line quadrangular gravure coating roll.

The solids content of the silicone release material in Tolusol 50 (a 1:1 mix of toluene and heptane) was 3.8%. After coating, the coated paper was subjected to drying and curing conditions in an oven at about 350° to 400°F for about 1 minute.

The kraft paper used in the usual commercial embodiments generally carries printed information, e.g., application instructions, suggested use, measuring scales to facilitate cutting, etc., on the nonrelease side.

A dimethylsiloxane-oxyalkylene block copolymer containing oxyethylene and oxypropylene units and having a viscosity of about 2,250 cp (No. 4 spindle at 30 rpm) was dissolved in water with a propeller-type stirrer to form a solution having a ratio of polysiloxane to water of about 1 to 3, respectively, by weight. This solution was slowly added to a butyl acrylate-vinyl acetate copolymeric pressure sensitive adhesive having a solids content of about 50% by weight, a viscosity of about 480 cp (No. 3 spindle at 60 rpm) and a plasticity of about 1.8 mm contained in a 5,000-gallon tank fitted with a double-blade mixer. The concentration of polysiloxane in the pressure sensitive adhesive was 0.2% by weight based on the solids content of the adhesive. The contents of the tank were then mixed by the double-blade mixer at 60 to 65 rpm for 16 hours and the resulting mixture coated onto the treated release layer by reverse roll coating. Alternatively, a 55-line quadrangular gravure roll can be used.

Thereafter, the coated release sheet was dried at about 240°F for about 5 to 10 seconds to yield a dry deposit of polysiloxane modified pressure sensitive adhesive of from about 1 to 2 mm/cm^2.

Finally, a polyvinyl chloride film having a thickness of about 3.5 mils was married to the thus-coated release sheet by passing the combination through a nip at a pressure of about 30 psi. The final product was rolled into a suitable package for subsequent cutting and packaging. Sample pieces of the laminate produced were tested for zero-minute peel, face-to-face peel, ease of application, etc. and were found to possess overall properties superior to those of a laminate produced in similar fashion but with the absence of the polysiloxane.

Adhesives for Rubber

RUBBER TO TIRE CORD ADHESION

Metal Deactivators as Adhesive Promoters

G.K. Cowell and D.J. Cherry; U.S. Patent 4,077,948; March 7, 1978; assigned to Ciba-Geigy Corporation have found that unusually good bonds are produced between vulcanizable elastomeric compositions and metals where there is added to the vulcanizable elastomeric composition a compound with metal deactivating activity, and the compositions are then vulcanized while in contact with the metal. Compounds of particular suitability within this broad definition are found to have at least two nitrogen atoms directly bonded to each other, the remaining valence requirements of each nitrogen atom being fulfilled through bonding to hydrogen or to an organic residue. Hydrazine derivatives are thus especially suitable compounds.

They have also found that a particularly useful adhesion promoter system is obtained by combining the metal deactivators with an adhesive system produced from the reaction of a methylene acceptor and a methylene donor reactable therewith. The methylene donor contains at least one trivalent nitrogen connected to at least one CH_2 radical and is capable of generating methylene groups in the presence of the methylene acceptor. It is suggested that the methylene donor in the presence of heat yields methylene either as formaldehyde or as methylene radical, which is reactable with the methylene acceptor to produce a resinous matrix in the rubber while at the same time promoting desirable adhesion between metal and rubber surfaces (see U.S. Patent 3,517,722).

Examples of methylene donors are dimethylolmelamine, trimethylolmelamine, partially etherified trimethylolmelamine, and fully etherified hexamethylolmelamine. The most well-known type of methylene acceptor for these systems is resorcinol.

The metal deactivating compound, used in an amount of from 0.05 to 10 weight percent based on the weight of the elastomer, has the formula:

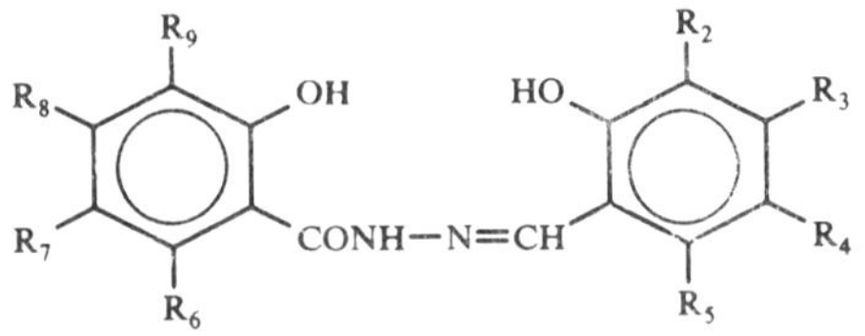

where R_2, R_3, R_4, R_5, R_6, R_7, R_8 and R_9 are each independently hydrogen, lower alkyl, lower alkoxy or a phenyl group.

Examples of accelerators are mercaptobenzothiazole or N-cyclohexylbenzothiazole-2-sulfenamide.

Example: Bonding of Styrene-Butadiene Copolymer (SBR) to Brass Plated Steel Wire — (a) Preparation of SBR Stock: 150 parts of a SBR Master-Batch No. 1605 (Ashland Chemical) containing 100.00 parts SBR and 50.00 parts carbon black FEF (N-550) were placed in a size B Banbury Mixer, initially heated to 21.1°C and having a rotor speed of 116 rpm. After mixing for 0.5 minute, 2.82 parts of zinc oxide and 0.71 part of Age-Rite Resin D (2,2,4-trimethyl-1,2-dihydroquinoline polymers as antioxidant) were added and mixed for another minute. Stearic acid, 0.71 part, was then added and mixed for an additional 2.5 minutes, then swept down and the mixing continued for 2 more minutes for a total mixing time of 6 minutes, at which time the entire mix was dumped at a temperature of 173.6°C.

(b) Sample Preparation: The stock obtained after Banbury mixing was cut into smaller samples, ranging in size from 50 to 100 g. These were subsequently milled on a two-roll mill at 82.2°C for 7 minutes during which time 1.24 parts of sulfur, 0.64 part of Santocure NS, 0.14 part of tetramethylthiuram monosulfide and either 1.00 or 1.60 parts of the additive compound were added. After 7 minutes of milling, the stock was sheeted.

The sheeted stock was cut into 0.5" x 8.0" strips and stacked to form two strips each weighing approximately 16 g. The strips were freshened (lightly washed) with n-hexane prior to being placed in the curing mold.

The mold was preheated in a hydraulic compression press maintained at 148.9°C for a minimum of 3 minutes. The mold was removed from the press and loaded according to the following sequence: brass support plate (8" x 0.5" x 0.062"), rubber strip, brass plated steel wire samples, rubber strip, brass support plate and top plate of mold. The sample was cured for 24 minutes at 300°F under a ram force of 10,432.63 kp. After the 24 minutes curing time, the mold was removed from the press and the rubber sample containing both the support plates and wires were removed from the mold and allowed to cool to ambient temperature (about 22°C). A description of this sample preparation procedure is found in ASTM Test D-2229-68.

(c) Test Method: After standing at ambient temperature for 18 to 24 hours, the wires were pulled from the rubber stock using an Instron Testing machine according to ASTM Test D-2229-68, but including the modifications described by A.E. Hicks et al in *Rubber Chemistry and Technology* (1972).

(d) Test Results: The test results reported below were obtained according to the procedures described above. Pull-Out Force in the table is the average force required to pull the wire from 1 cm of rubber. The blank value is the average force required to pull the wire from a formulation containing all of the ingredients except the additive compound. All formulations with the same stock designation number were prepared from the same rubber stock.

Adhesion of SBR to Brass Plated Steel Wire

Additive Compound	Concentration (phr)	Stock Designation	Pull-Out Force (kp/cm)
	1.00	1	12.06
	1.60	1	15.2
Blank	—	1	5.92
	1.00	2	8.56
Blank	—	2	5.98

Two-Dip Adhesive System

H.J. Langer and W.J. McKillip; U.S. Patent 4,078,115; March 7, 1978; assigned to Ashland Oil, Inc. have provided a two-dip adhesive system which is especially adapted for bonding steel wire to rubber in the manufacture of a variety of reinforced vulcanizates. In the first dip an organic solution of a polymer in the form of an addition polymerization product consisting essentially of a vinyl aminimide/4-vinylpyridine copolymer is applied to wire and thereupon heat cured. Following the application of the first coating, the substrate is coated with a conventional RFL (resorcinol-formaldehyde condensate-latex) composition and heat cured.

In a further embodiment, an adhesive system for bonding a polyester filamentary element to rubber is provided. The application of the respective coatings is the same as described above except that the first coating is that of an aqueous solution of an alkylated vinyl aminimide/4-vinylpyridine copolymer and a polyepoxide.

Example 1: Resin A — Into a suitable reaction vessel equipped with stirrer, thermometer, reflux condenser and gas inlet tube were charged 270 parts by weight dimethyl-(2-hydroxypropyl)amine methacrylamide (DHA), 105 parts of 4-vinylpyridine, 8.45 parts diacetone acrylamide, 384 parts isopropanol, and 8.45 parts azobisisobutyronitrile (AIBN). After deaerating the reactor contents by evacuation to incipient boiling and flushing with nitrogen, the copolymerization was conducted with stirring for 5 hours at 70°C. The viscous polymer solution was stripped on a rotary evaporator to provide a yellow solid which was ground to a coarse powder.

Resin B — In a manner similar to that employed in preparing Resin A, 22.4 parts DHA were copolymerized with 8.4 parts of 4-vinylpyridine in 31 parts of isopropanol in the presence of 0.3 part AIBN. After 4 hours of polymerization at 70°C the resultant copolymer solution was yellow and very viscous. The crude polymer was purified by dissolving with methanol and reprecipitated with acetone. The yield of purified polymer was 50%. The inherent viscosity of the polymer was 0.33 at 0.5 g of polymer in 100 ml of methanol. Gel permeation analysis of the copolymer showed a narrow molecular weight distribution and a weight average chain length of 630 A.

Example 2: This example serves primarily to illustrate the effectiveness of Resins A and B for bonding steel wire to rubber in a two-dip operation wherein the wire is coated in the initial dip with the indicated copolymers. The type of wire used for the evaluation was a national standard single strand brass plated (0.16 inch diameter) steel wire.

In the first dip, the wire was passed through a 15% solids solution of the copolymer in methanol. In each instance before dipping, the wire was cleaned with perchloroethylene and dried for about 30 minutes at room temperature. Following dipping, the copolymer treated wire was cured at 445°F for 80 seconds. Next, the polymer coated wire was passed through a standard RFL formulation, details of which are set forth in the table below. Following the second dip as described, the coated wire was then cured under the same conditions as previously noted.

Static adhesion evaluation of the coated wire was thereupon determined in accordance with the standard test (ASTM 2229-68). A vulcanizable rubber compound of the following recipe was employed in this test:

	Parts by Weight
Natural rubber No. 1 smoked sheet	35
cis-1,4-polybutadiene rubber	20.0

(continued)

	Parts by Weight
SBR rubber	67.5
FEF carbon black	55.0
Zinc oxide	10.0
Stearic acid	1.0
Pine tar oil	7.5
Antioxidant BLE (Naugatuck Chem. Co.)	0.5
Sulfur	3.0
Resin bonding (Arofene 7209, Ashland Chem. Co.)	1.67
Benzothiazol disulfide	1.2
DPG (diphenylguanidine)	0.2

The adhesive values noted for their respective test runs together with the composition of the RFL dip solution are outlined in the following table. The average adhesion value specified was determined from eight separate samples of the coated wire.

	.... Second Dip					Average Adhesion Value
	RF**	HCHO	Latex***	NaOH	Water	
First Dip*	................(%)................					(lb)
Resin A	9.56	4.08	52.41	2.14	31.8	44.8
Resin B	9.56	4.08	52.41	2.14	31.8	46

  *15% resin-85% methanol
  **Resorcinol/HCHO fusible resin (Arofene 779, Ashland Chem. Co.)
  ***Vinylpyridine latex (Gen-Tac, General Tire and Rubber Co.)

Adding Metal Compound to Adhesive

R.J. Kelly, R. Miller and D. Adams; U.S. Patent Reissue 29,699; July 11, 1978; assigned to Uniroyal, Inc. have found that the addition of a selected metal-containing compound to an adhesive used to adhere polyester articles such as cord, film, etc., to a natural or synthetic rubber base results in the adhesion between the polyester article and the rubber base being substantially increased over the adhesion obtained when the selected metal-containing compound is not added to the adhesive.

The natural and synthetic rubbers used may be compounded with the usual conventional ingredients including sulfur or sulfur-yielding vulcanizing agents, organic or inorganic accelerators of sulfur vulcanization, and if desired, reinforcing fillers or pigments such as carbon black, silica, etc., as well as any such appropriate ingredients as antioxidants, antiozonants, processing aids, extender oils, tackifiers, lubricants, reclaim or other extenders, etc., as appropriate to the particular rubber used and the particular article being manufactured. The composition may contain more than one rubber, such as a blend of natural rubber and cis-polybutadiene. Preferred rubbers to be used are natural rubber, polybutadiene, poly(butadiene-styrene) or blends thereof.

Physically, the polyester may take the form of filaments, yarns, fabrics, tapes, fibrillated tapes, and films, or other shaped or molded forms presenting a high ratio of surface to volume.

The two-dip adhesive system used is described in U.S. Patent 3,307,966. The first dip comprises the metal compound as described in admixture with a liquid carrier containing from about 0.05 to 15.0% by weight of solids of (a) a polyepoxide having an average of at least two epoxy groups in each molecule, a melting point below 150°C, an average molecular weight below 3,000 and an epoxide equivalent below 2,500; and (b) an aromatic polyisocyanate.

Metals which may be used in the compound added to the adhesive are selected from the group consisting of sodium, potassium, rubidium, cesium, strontium, silver, cadmium, barium, cerium, uranium, titanium, vanadium, chromium, tin, antimony, manganese, iron, cobalt, nickel, copper, zinc, lead, bismuth, calcium and zirconium. Especially preferred metal-containing compounds are the zinc, copper, and cadmium compounds.

Example: This example illustrates the application of the adhesive to bond polyester tire cord to rubber using a two-dip system.

Twisted 3440 denier, polyester (polyethylene terephthalate) tire cord consisting of 3-ply cord (in which each ply has 9.2 Z turns in the singles and 9.2 S turns in the plying operation) is run through a first dip having the following composition: 22.0 g phenol blocked MDI (Hylene MP, DuPont); 1.5 g dioctyl sodium sulfosuccinate (Aerosol O.T.); 9.00 g epoxide (Epon 812); 1,200 g water; and 0.00023 mol zinc acetate per 1,000 g coating composition.

The total solids on the cord after this first-dip treatment is 0.5 to 1.5% (dry weight, based upon the weight of the cord). The cord is then cured for 40 seconds at 445°F while applying 2.0% stretch. The cord is then run through a second dip (RFL dip) having the following composition: 100 g vinylpyridine latex [40% solids consisting of 70% butadiene-15% vinylpyridine-15% styrene terpolymer (Pyratex)] ; 2.5 g ammonium hydroxide; 9.0 g resorcinol-formaldehyde condensation product (Koppers Penacolite 2170); 4.5 g 37% formalin.

This composition is diluted with water to 20% solids. The cord is then dried and cured at 390°F for 90 seconds. The percent solids of RFL picked up on the cord is 2.5 to 4.5% dry weight based upon the weight of the cord.

Adhesion results are obtained using the test method which is described in U.S. Patent 3,549,481. The testing showed that if zinc acetate is added (as in the first dip above) to the adhesive within the prescribed range (i.e., 0.00001 to 0.010 mol metal compound per 1,000 g coating composition) a substantial increase in adhesion is obtained.

Addition of Epoxytrialkoxysilane to Silicone Elastomers

In spite of the advances made in obtaining bonding of elastomeric materials to substrates, problems still remain. The adhesion of the previous acetoxy-containing room temperature vulcanizable silicone elastomers to glass and aluminum surfaces, particularly on long exposure to moisture, has not been satisfactory.

J.R. Hilliard; U.S. Patent 4,115,356; September 19, 1978; assigned to Dow Corning Corporation has found that the addition of epoxytrialkoxysilanes to room temperature vulcanizable silicone elastomer compositions containing organotriacetoxysilane improves the adhesion of the cured elastomer to glass and aluminum surfaces. The improved adhesion is particularly evident if the adhesion is determined after substantial exposure of the bonded surfaces to water.

The formulation which he arrived at is one which is storage stable in the absence of moisture and which vulcanizes at room temperature on exposure to moisture. It comprises a mixture of

(A) 100 parts by weight of polydiorganosiloxane consisting essentially of repeating units of the formula R_2SiO having a viscosity of at least 0.10 Pa·s at 25°C in which each R represents a monovalent hydrocarbon or halogenated monovalent hydrocarbon radical of from 1 to 18 carbon atoms, the siloxane having an average of from 1.5 to 2 silicon-bonded hydroxyl radicals per molecule; and,

(B) from 0.5 to 15 parts of an organotriacetoxysilane of the formula

$$R'Si(O\overset{\displaystyle O}{\overset{\displaystyle \|}{C}}CH_3)_3$$

in which R' represents a monovalent radical selected from the group consisting of lower alkyl, fluorinated lower alkyl, alkenyl, and aryl radicals, there being at least one organotriacetoxysilane molecule per hydroxyl radical of polyorganosiloxane (A); and

(C) from 0.1 to 2.5 parts of a compound of the formula

$$H_2C\overset{\displaystyle O}{\overset{\displaystyle /\,\backslash}{}}CH(R'')Si(OR''')_3$$

in which R'' is a radical attached to silicon by a carbon-to-silicon bond and selected from the group consisting of divalent hydrocarbon radicals of from 1 to 10 carbon atoms and divalent radicals of less than 10 carbon atoms which are composed of carbon, hydrogen, and oxygen atoms, the last being in the form of ether linkages, not as an atom of a ring; and R''' is selected from the group consisting of aliphatic hydrocarbon radicals of less than 5 carbon atoms.

Silicone elastomer compositions with silicon-bonded acetoxy functionality are well known in the art and are available commercially. The preferred epoxytrialkoxysilane is glycidoxypropyltrimethoxysilane.

Example: A mixture of the following ingredients was made in the absence of moisture (all parts are parts by weight): 100 parts polydimethylsiloxane fluid with a viscosity of approximately 1.25 x 10^{-2} m^2/sec at 25°C endblocked with approximately 20% trimethylsiloxy radicals and 80% hydroxyl radicals; 10 parts fumed silica with a surface area of 200 m^2/g; 5.6 parts of a mixture of 50 wt % methyltriacetoxysilane and 50 wt % ethyltriacetoxysilane with 0.1 part of dibutyltin diacetate based on 100 parts of silane mixture.

100 g each of the above mixture were then mixed with varying amounts of glycidoxypropyltrimethoxysilane as shown in the table and packaged without exposure to moisture. Samples for measuring adhesive properties were prepared by cleaning the substrate with trichloroethylene, then with acetone. A 1.6 mm thick layer of the mixture was applied to the substrate using a drawdown technique and shims of the proper thickness. An aluminum screen was imbedded in the surface of the mixture and another 1.6 mm thick layer was applied over the screen immediately. The sample was then cured for 7 days at 25°C and 50% RH.

A strip 25.4 mm wide was cut the length of the cured sample through the cured composition to the substrate. The end of the flexible cured composition was cut loose from the substrate at the substrate surface for a distance sufficient to attach a tensile testing machine to the end when the flexible cured composition and reinforcing layer was bent back 180°. The cured composition layer was then separated from the substrate at a rate of 51 mm/min. During the pull, the flexible layer was cut down to the substrate surface at least 3 times to assure the maximum strain at the interface between the cured composition and the substrate. The average of the peak loads after undercutting to the substrate surface was reported as the peel strength. The appearance of the peeled surface was observed. The adhesive values obtained are shown in the table.

Similar samples were prepared, cured and then immersed in boiling water for the indicated time to determine if the bond would withstand the effects of water immersion. After the water immersion the samples were cooled and tested as described above, noting particularly if the separation of the cured composition from the substrate was a cohesive or adhesive failure. An adhesive failure is failure between the cured composition and substrate; a cohesive failure is failure within the cured composition, leaving an amount of cured composition adhered to the substrate.

	. . . . . Unprimed Glass		. . . . Unprimed Aluminum			
	. . . Initial	. . 1-Hour Boil . .	. . . Initial . . .	. . 2-Hour Boil . .		
X		Failure*		Failure		Failure		Failure
(g)	kN/m	(%)	kN/m	(%)	kN/m	(%)	kN/m	(%)
0	1.6	100	0.35	0	0.43	0	0.17	0
0.5	1.4	100	2.26	0	1.5	55	0.44	0
1.0	1.6	100	0.37	0	1.7	100	0.37	0
2.0	2.5	100	4.5	100	4.7	100	3.1	100
3.0	. . . . incomplete cure							

Note: X = concentration of glycidoxypropyltrimethoxysilane.
Kilonewton per meter (kN/m) = 5.72 pounds per inch (ppi).

*Percent of cohesive-type failure.

Containing Resorcinol-Aldehyde and Catechol-Aldehyde Resins

The commonly used cord-to-rubber adhesives are water dispersions of latex and resorcinol-formaldehyde resins with other ingredients such as carbon black and blocked polyisocyanates added for specific end results.

Catechol-aldehyde resins have not previously been identified with superior adhesive properties. However, *R.S. Bhakuni and R.F. Laske; U.S. Patent 4,132,693; January 2, 1979; assigned to The Goodyear Tire & Rubber Company* have discovered that they impart certain very desirable properties to reinforced rubber articles, namely improved aged adhesion and lower fabric stiffness.

The process of manufacturing a typical resorcinol-formaldehyde latex adhesive is generally comprised of the following steps:

- Dissolve resorcinol in water in the mixing tank.
- Add a suitable aldehyde such as formaldehyde in solution form to the above solution.
- Add sodium hydroxide (or another suitable base such as ammonium hydroxide or potassium hydroxide) slowly to the mixing tank.
- After the resin is formed add the latex.
- Age for approximately 12 to 24 hours.

Generally, the mol ratio of resorcinol to formaldehyde (or other aldehyde) is from 1:10 to 3:1, and a ratio between 1:3 and 1.4:1 is preferred. When sodium hydroxide is used as the base catalyzing the resinification reaction, the mol ratio of sodium hydroxide to resorcinol is generally from 0.05:1 to 0.4:1.

The process is usually carried out at atmospheric pressure and room temperature although a cooling means is sometimes employed to remove the heat of the resinification reaction.

The latex is selected for the properties which it imparts to the fabric and to the final product. Vinylpyridine/diene terpolymers and copolymers (such as styrene/1,3-butadiene/α-vinylpyridine terpolymer and 1,3-butadiene/α-vinylpyridine copolymer) are often used for their excellent adhesive properties. Other latexes which are used are: natural rubber, neoprene, acrylonitrile/butadiene copolymer, polybutadiene, ethylene/propylene/diene terpolymer, styrene/butadiene copolymer (SBR) and chlorosulfonated polyethylene. Often the latex is chosen to be compatible with the rubber which will later be applied to the reinforcing cord, for example, SBR latex for SBR rubber. The dry weight ratio of latex solids to resin ranges from 14:1 to 1:9 and is preferably from 12:1 to 1:1.

This formulation has modified the normal process for manufacturing adhesive dip as explained above. A catechol-aldehyde resin is made by reacting catechol and aldehyde in the presence of a base in water at a temperature of from 40° to

95°C. The mol ratio of catechol to aldehyde is in the same range as the resorcinol-aldehyde ratio stated previously. The mol ratio of sodium hydroxide (or gram equivalent of another base) to catechol is from 0.1 to 1.0 and preferably from 0.4 to 0.8. A sufficient quantity of the resin is normally made for use in several batches of cord dip adhesive. This operation is called preforming.

The preformed catechol-aldehyde resin is added as an ingredient to the adhesive between the third and fourth steps of the process given previously. The amount of resorcinol-aldehyde resin made is reduced by the corresponding amount of catechol-aldehyde resin with which it is replaced. The total amount of resin remains roughly the same as in the older recipes. The weight ratio of resorcinol-aldehyde resin to catechol-aldehyde resin is preferably from 40:60 to 70:30.

This formulation, without diminishing fabric properties obtained using the old resorcinol-formaldehyde latex dips, has increased adhesion of the fabric to rubber in aged samples and decreased fabric stiffness. An increase in aged adhesion or higher retention of adhesion upon aging of dipped fabric is of great commercial importance when dipped fabric is shipped around the world and kept in inventory for extended periods. Decreased fabric stiffness is desirable in tire manufacturing for reduced defective tires due to blister and blows.

These cord dip adhesives can be utilized with reinforcing cords of rayon, nylon, polyester, and aramid polymers.

Incorporation of Microcrystalline Wax in Formulation

Adhesive-coated tire cord, other fabrics, and single end cords which are subsequently, after storage and/or shipment, to be adhered to rubber layers, must be protected from the adverse effects of exposure to the products of combustion present in fuel oil- or kerosene-fired dryers and curing ovens as well as to atmospheric oxidation. It has been found that such fabrics and cords, after exposure to the products of combustion present in fuel oil- or kerosene-fired dryers and curing ovens, demonstrate little or no adhesion to rubber layers in the construction of tires and other end products. In the face of natural gas and propane shortages, many textile companies are forced to use oil or kerosene as the fuel in their treating ovens.

Conventional adhesive dip formulations for application to fabric made from polyamide, aramid, polyester or rayon cord, which is subsequently to be adhered to shaped rubber articles prior to the vulcanization thereof, comprise a dip of a resorcinol-formaldehyde condensate and a latex of a vinylpyridine-butadiene-styrene terpolymer or of a butadiene-styrene copolymer. *R.E. Hartz; U.S. Patent 4,137,358; January 30, 1979; assigned to Uniroyal, Inc.* found that the effects of the products of combustion present in fuel oil- or kerosene-fired dryers and curing ovens, as referred to above, and of atmospheric oxidation can be overcome by the incorporation of from 3 to 10 weight percent, based on the dry weight of rubber in the resin-rubber dip, of a microcrystalline wax or a

Fischer-Tropsch wax having a melting point of 160° to 215°F and a molecular weight of up to 1,500, or a mixture of such waxes.

The production of microcrystalline wax from tank bottoms has been described in U.S. Patent 2,443,840. These waxes have a melting point ranging from 170° to 200°F. They contain essentially long chain normal and branched paraffins with some naphthenic-type compounds. Also suitable for use in this formulation is the plastic microcrystalline wax of somewhat higher melting point which is derived from various lube oil stocks and petrolatums. Another suitable hydrocarbon wax is the Fischer-Tropsch wax. Such waxes are prepared commercially by the gasification reaction of carbon monoxide and hydrogen with a catalyst.

These wax additives can be purchased in the form of a solids emulsion or can be prepared in the form of a slurry by mixing various proportions of the additive and water with an appropriate dispersing system. A suitable procedure for preparing a dispersion or slurry is to mix 49 parts of the wax, 99 parts of distilled water, and 2 parts of a dispersing aid, e.g., dioctyl sodium sulfosuccinate, and stirring until a uniform dispersion is formed (all parts are by weight).

The properties which suitable waxes should have, and which are possessed by the microcrystalline and Fischer-Tropsch waxes, are: nonoxidizable; melting point of 160° to 215°F and suitable viscosity so as to be able to flow and bloom to the surface of the adhesive for protection; compatibility so as to be able to migrate into the rubber stock during cure in order to maintain adhesion of the rubber to the cord; and compatibility with resorcinol-formaldehyde-latex (RFL) dip so as not to change the stability of the dip.

In a typical procedure for the application of the adhesive, a tire cord is run through a first dip having the following composition (described in U.S. Patent 3,307,966): 22.0 g phenol blocked MDI (Hylene MP, DuPont), 1.5 g dioctyl sodium sulfosuccinate (Aerosol O.T.), 9.0 g epoxide (Epon 812), 1,200 g water.

The total solids on the cord after this first dip treatment is 0.5 to 1.5% (dry weight based upon the weight of the cord). The cord is then cured for 40 seconds at 445°F while applying 2.0% stretch.

The cord is then run through a second dip (RFL dip) having the following composition: 100 g vinylpyridine latex [40% solids consisting of 70% butadiene, 15% vinylpyridine, 15% styrene terpolymer (Pyratex)]; 2.5 g ammonium hydroxide; 9.0 g resorcinol-formaldehyde condensation product (Koppers' Penacolite 2170); 4.5 g formalin (37%) diluted with water to 20% solids; and, microcrystalline or Fischer-Tropsch wax—5 parts wax solids to 100 parts dry rubber solids.

Modified resorcinol-formaldehyde condensation products can also be substituted, e.g., those in which another dihydric phenol is used or in which a third reactive component is included.

The cord is then dried and cured with oil or kerosene as fuel at 390°F for 90 seconds. The percent solids of RFL picked up on the cord is 2.5 to 4.5% dry weight based upon the weight of the cord.

Example: This example illustrates the application of the wax additives to protect polyester tire cord from the adverse effects of fuel oil or kerosene as the fuel for the drying and curing oven, prior to adhesion of such fabrics to rubber. Twisted 3440 denier polyester (polyethylene terephthalate) tire cord consisting of 3-ply cord (in which each ply has 9.2 Z turns in the singles and 9.2 S turns in the plying operation) was put through the dip and heating procedure previously set forth.

The table summarizes the comparison of initial adhesion of the polyethylene terephthalate tire cord treated with oil or kerosene as fuel with and without additive.

Additive	Amount*	Initial Appearance/ Pounds Pull**
None	0	3.3/25.0
N,N'-ethylene bis-stearamide	5	3.5/24.0
Microcrystalline emulsion***	3	4.8/30.0
Microcrystalline emulsion***	6	5.0/31.0
Fischer-Tropsch dispersion†	3	4.8/30.5
Fischer-Tropsch dispersion†	5	5.0/31.0

*Grams wax solids per 100 grams dry rubber solids (latex).
**Measured according to U.S. Patent 3,549,481.
***Mobilcer Q with 165°F melting point.
†H-1-N-3 (Moore and Munger) with melting point of 170°F.

Adhesive Particularly Useful for High-Modulus Yarns

An adhesive coating composition described by *P. Bourrain and P. Giroud; U.S. Patent 4,148,963; April 10, 1979; assigned to Rhone-Poulenc-Textile, France* makes it possible to increase the adhesion of heat-stable, high-modulus yarns, based on polyesters and yarns of the modal type as well as glass fibers, towards compounded rubbers based on natural rubber and synthetic rubber. This composition consists of the following:

	Parts by Weight
Resin Solution A	
Softened water	547–8,954
Sodium hydroxide	1–2
Precondensed resorcinol/formaldehyde resin	
(Penacolite B₁A)	48–57
40% strength formaldehyde	14.75–17.5
Latex Solution B	
Latex based on vinylpyridine (Gen-Tac)	805–875
Polychloroprene latex	46–92
28% strength ammonium hydroxide	80

(continued)

	Parts by Weight
Latex Solution C	
Softened water	200
28% strength ammonium hydroxide	15
Dicarboxylated butadiene-styrene resin	
(Pliolite 4121)	350

In the composition the proportion by weight of solids relative to the weight of the bath is 6 to 30%, the proportion by weight of dry resin relative to the total weight of solids is 8 to 10%, the ratio of dry latex to the weight of dry resin is 8:1 to 11:1, the proportion by weight of dry chlorinated latex relative to the total weight of dry latex is 5 to 10%, the proportion by weight of dry latex relative to the total amount of solids is 86 to 87% and the final pH is 9.6±0.3.

By high modulus there is to be understood here a modulus of elasticity or Young's modulus greater, on average, than 250 grams per 1.1 denier (dtex), this modulus being measured in accordance with standard specification ASTM D-2101 (Part 25, 1968).

The coating is more particularly applicable to the high modulus yarns which furthermore have a high tenacity and/or a high breaking energy, that is to say of at least 50 g/dtex, and preferably of at least 100 g/dtex. The high modulus yarns are obtained by spinning polymers of inorganic origin such as boron or of organic origin, or are obtained by carbonization or graphitization of filaments of polymers based on acrylonitrile. As filaments of organic origin, there may be mentioned those produced from polymers which generally contain aromatic nuclei to which rigid radicals are fixed.

Examples of these polymers are the aromatic polyamides, more particularly those having a para-structure of the type of poly-para-phenylene terephthalamide or poly-para-benzamide; the arylaliphatic polyamides or copolyamides of the poly-hexamethylene terephthalamide type or copolyamides which contain aromatic diacids or aliphatic saturated, unsaturated and/or cyclic diacids and aromatic or aliphatic diamines; the copolyamides derived more particularly from terephthalic acid, adipic acid and tetramethylenediamines or para-phenylene-diamines (French Patent 2,272,118; December 19, 1975); the polyoxadiazoles such as polyarylene-1,3,4-oxadiazole; the copolyoxadiazoles and the polyvinyl alcohol-based polymers having a high tenacity and high modulus.

The resin solution A is prepared by introducing the starting products, in the stated order, while stirring at ambient temperature. This solution undergoes a ripening process for a few minutes, while stirring at ambient temperature.

The latex solutions B and C are separately prepared in an identical manner, while stirring and at ambient temperature. The latex solution C is poured, while stirring at ambient temperature into the latex solution B, and the resin solution A is then poured into the mixture, while stirring at ambient temperature.

The adhesive-coating composition is preferably left standing for about 24 hours so as to give, after the subsequent treatments, adhesive-coated yarns which have optimum mechanical and physical characteristics. The final pH of the compositions is generally 9.6 ± 0.3.

The adhesive-coating bath thus obtained has a viscosity which is stable over a period of time, and also has good mechanical stability when the filaments are passed over draining rollers; this means that no coagulation of the adhesive-coating bath takes place on the draining devices.

The adhesive-coating treatment by means of this composition is carried out on various types of yarns which have in general been sized and coated with an adhesive primer beforehand in accordance with any process known to those skilled in the art.

After passing through the adhesive-coating bath, which is kept at ambient temperature, the yarns are drained to render the layer of coating uniform and are dried at a temperature which depends on the speed of forward travel of the yarn (but at least $100°C$), using any appropriate device such as an oven, etc. after which they are again heat-treated at a high temperature, generally above $130°C$ in order to crosslink the resin. The cords treated in this way are generally wound up or doubled directly after the crosslinking treatment.

These adhesive-coating compositions make it possible to obtain yarns having great flexibility and good adhesion, while the tensometric properties (breaking load, tenacity and elongation at break) remain unchanged. The yarns adhesive coated in this way possess both good tack and good slip.

Example: An example of the adhesive composition is made by stirring the following ingredients at ambient temperatures:

	Parts by Weight
Solution A	
Softened water	956.5
Sodium hydroxide	1
Penacolite B_1A	57
40% strength formaldehyde	15.5

A latex solution B, which is left to ripen for 3 to 4 minutes, and a latex solution C, are prepared separately, while stirring and at ambient temperature; these solutions contain, respectively:

	Parts by Weight
Solution B	
Latex based on vinylpyridine	805
Polychloroprene	92
28% strength ammonium hydroxide	80

(continued)

	Parts by Weight
Solution C	
Softened water	200
28% strength ammonium hydroxide	15
Pliolite 4121 (dicarboxylated butadiene- styrene resin)	350

Latex solution C is poured into solution B while stirring and solution A is then poured into the mixture thus obtained, while stirring at ambient temperature. The coating solution thus prepared has the following composition:

Solids content of the composition	25%
Dry weight of resin/solids content	9.8%
Dry weight of latex/solids content	86%
Dry weight of latex/dry weight of resin	8.7/1
Dry weight of chlorinated latex/total of latexes	10%

The bath is left to stand for 24 hours.

The adhesive-coating composition thus obtained (pH 9.6) has a stable viscosity and good mechanical stability.

Adhesive-Coated Glass Fibers

The process developed by *P. Bourrain and P. Giroud; U.S. Patent 4,157,420; June 5, 1979; assigned to Rhone-Poulenc-Textile, France* is one for coating glass fibers in such a way that the adhesion between them and rubbers based on natural rubber is increased.

The adhesive composition consists of a resin solution A produced from:

	Parts by Weight
Softened water	433
Resorcinol	36.6
Sodium hydroxide	1–3.50
40% strength formaldehyde	21.3–50.0

and a latex solution B composed of:

	Parts by Weight
Softened water	0–292
60% strength natural latex	213–428
40% strength latex based on vinylpyridine	640–960

in which the total weight of solids relative to the bath varies between about 30 and 35%; the formaldehyde/resorcinol molar ratio varies between 0.85 and 2; the dry weight of latex relative to the resorcinol/formaldehyde resin varies between about 8.5/1 and 11/1; the final pH is 9.6±0.2; and, the proportion by weight of dry resorcinol/formaldehyde resin relative to the amount of all the products varies between 10 and 12%.

The fibers passing through the bath are subjected to vibratory movements communicated directly to the fibers.

Preferably, there is used, jointly with the resin solution A, a latex solution C composed of:

	Parts by Weight
Water	0–292
60% strength natural latex	213–428
28% strength ammonium hydroxide	2.6–10.5
Latex based on vinylpyridine	640–960

By latex based on vinylpyridine there is to be understood a styrene/butadiene/vinylpyridine terpolymer commercially available as Gen-Tac (General Tire and Rubber Co.) or Ugitex V.P. (Rhone-Poulenc).

Solution A, the various components of which are introduced in the precise order indicated above, is stirred at $20 \pm 2^\circ C$ for a few hours to ripen the resin. The final pH, after ripening, must be 9.0 ± 0.5. The latex solutions B and C, into which the components are also introduced in the abovementioned order, are also prepared while stirring.

Solution A is then poured into solution B or C, while stirring at ambient temperature. The final compositions have a pH of 9.6 ± 0.2. The actual adhesive-coated compositions are left to stand for 24 hours so as to obtain, after the subsequent treatment, an optimum result in respect of the mechanical and physical characteristics of the adhesive-coated glass fibers.

Glass fibers which have been adhesive-coated in this way are useful for manufacture of tires, conveyor belts, hoses, etc.

TIRE SEALANTS

Butyl Rubber-Based Composition Cured Using Special Method

J.V. Van Ornum; U.S. Patent 4,068,027; January 10, 1978; assigned to Rocket Research Corporation has found that an effective sealant composition may be formulated using a carbon reinforced curable butyl rubber matrix and certain modifiers to achieve the necessary mechanical strength, thermal stability, and sealing capabilities required of a commercially acceptable self-healing tire puncture sealant. This sealant composition comprises a combination of partially crosslinked (i.e., partially cured) high and low molecular weight butyl rubbers, a tackifier, and a carbon reinforcer. The weight ratio of high molecular weight to low molecular weight butyl rubber may vary from 20/80 to 60/40. The tackifier constitutes about 55 to 70 weight percent of the composition and the carbon reinforcer constitutes up to about 17 weight percent of the composition, the balance being the crosslinked rubber constituents.

To aid in maintaining sufficient tackiness and thermal stability at elevated temperatures, a thermoplastic and elastomeric partially-hydrogenated block copolymer may be included up to about 10 weight percent of the composition, the block copolymer having a general configuration of A–(B–A)$_{1-5}$ where prior to hydrogenation each A is a monovinyl arene polymer block and each B is a conjugated diene polymer block.

Such a sealant composition may be cured by means of a quinoid curing agent and a free radical crosslinking activator. The method of formulating butyl rubber based sealant compositions using these curing materials is based on the discovery that a dramatic reduction in the gel time can be achieved by adding the activator to the other components in two portions, rather than all at once. The first activator portion is added and after a short period of time, the product of this first reaction is an ungelled, uncured, base component which has an appreciable pot life and which gels almost immediately when mixed with the remainder of the activator. Larger first activator portions result in shorter gel times but also in shorter pot lives for the base components, this latter feature providing a practical upper limit to the size of the first portion. For spray applications, the base component and the second activator portion are mixed just prior to spraying.

Using this method, an adequate thickness of the above sealant composition can be applied to a tire in a single spraying operation.

Carbon-Reinforced, Partially Crosslinked Butyl Rubber

The self-healing tire puncture sealant composition developed by *J.V. Van Ornum and P.L. Stang; U.S. Patent 4,113,799; September 12, 1978; assigned to Rocket Research Corp.* is adapted for application to the internal surface of a rubber tire under temperatures varying from -20° to 270°F.

The sealant composition comprises a combination of partially crosslinked (i.e., partially cured) high and low molecular weight butyl rubbers, a tackifier, and a carbon reinforcer. The weight ratio of high molecular weight to low molecular weight butyl rubber may vary from 20/80 to 60/40. The tackifier constitutes about 55 to 70 weight percent of the composition and the carbon reinforcer constitutes up to about 17 weight percent of the composition, the balance being the crosslinked rubber constituents.

To aid in maintaining sufficient tackiness and thermal stability at elevated temperatures, a thermoplastic and elastomeric partially-hydrogenated block copolymer may be included up to about 10 weight percent of the composition, the block copolymer having a general configuration of A–(B–A)$_{1-5}$ where prior to hydrogenation each A is a monovinyl arene polymer block and each B is a conjugated diene polymer block.

The sealant composition may be applied by a variety of means. For purposes of tire sealing, the sealant composition may be formulated as a sprayable composition that cures in situ or as a composition that is first cured in sheet form and then applied. For other purposes, the sealant composition may be extruded or brushed onto a substrate. A suitable solvent such as toluene may be employed in the preparation of the sealant composition. The weight percentages specified, however, are on a solvent-free basis, unless otherwise noted. The preferred weight range for the high molecular weight butyl rubber is from 100,000 to 400,000. The preferred molecular weight range for low molecular weight butyl rubber is from 10,000 to 30,000.

Crosslinking of the butyl rubber constituents may be effected by one of the known sulfur or quinoid systems. A quinoid curing system is preferable to vulcanization for the tire sealing applications where the sealant must be capable of lasting years in a harsh environment. Quinoid cures depend on crosslinking through the nitroso groups of aromatic nitroso compounds. In the quinoid curing system, p-quinone dioxime (G-M-F) and p,p-dibenzoylquinone dioxime are preferred as the curing agents.

The curing agent/crosslinking activator combination which has been found to result in the shortest gel time is a p-quinone dioxime/benzoyl peroxide combination. The preferred concentration of p-quinone dioxime is 2 to 4% by weight of butyl rubber. The preferred concentration of benzoyl peroxide is 7 to 10% by weight of butyl rubber.

Preferred tackifiers are fluid monoolefin polymers of moderate viscosity such as those consisting of essentially butylene (1-butene, 2-butene and isobutylene) with the balance being isoparaffins having average molecular weights in the range of 500 to 5,000, terpene polymer resins such as polymerization products of β-pinene, and low molecular weight styrene polymer resins such as polymerization products of α-methylstyrene.

The reinforcing agent provides tensile strength to the sealant. It may be any one or more of a large number of well-known substances provided that one of these substances must be finely divided carbon. Carbon black provides reaction sites for the curing process, and preferably comprises at least 1% of the solids by weight. The substance comprising the remainder of the reinforcing agent may either be carbon black or some other suitable substance selected on the basis of the desired color of the sealant. The reinforcing agent should be present in an amount not exceeding 17% of the solids by weight.

The block copolymer constituent, prior to hydrogenation, is composed of A blocks of monovinyl arene polymers including styrene, α-methylstyrene, ring alkylated styrenes, and the like, as well as mixtures thereof; and B blocks of conjugated diene polymers having 4 to 10 carbon atoms per monomer molecule, including butadiene and isoprene. The A blocks make up the end groups and typically comprise about one-third of the copolymer by weight; and the B blocks make up the mid groups and the balance of the copolymer. The copolymer is partially hydrogenated so that the conjugated diene block segments are substantially fully saturated. The monovinyl arene polymer block segments are not appreciably saturated. Hydrogenation in this fashion enhances the utility of the block copolymer as an oxidation and high temperature-degradation resistant constituent of the sealant composition. The average molecular weight of the copolymer is in the range of 60,000 to 400,000. Block copolymers of this type are described in U.S. Patent 3,595,942.

Sealant compositions were made by mixing the ingredients in the weight proportions shown in the table below. All the sealants were satisfactory except "H," which was too soft.

Sealant Compositions

Ingredient	A	B	C	D	E	F	G	H
	. . . . (parts by weight).							
High MW butyl rubber*	9.5	9.5	10.03	14.25	9.0	14.5	10.4	4.8
Low MW butyl rubber**	14.5	14.5	15.04	9.5	13.0	9.5	6.9	19.2
Tackifier***	61.75	61.75	65.16	62.0	63.75	61.75	60.0	61.75
Carbon black	9.5†	9.5†	4.76††	9.5†	9.5†	9.5†	17.2†††	9.5†
Block copolymer§	4.75	4.75	5.01	4.75	4.75	4.75	5.0	4.75
p-Quinonedioxime	2.5	2.5	2.5	2.5	2.5	2.5	4.0	2.5
Benzoyl peroxide§§	7.9	7.0	7.9	10.0	7.9	7.0	10.0	7.9

*Copolymer consisting of 98.5% isobutylene and 1.5% isoprene by weight, having average molecular weight between 10,000 and 300,000.

**Copolymer consisting of 96% isobutylene and 4% isoprene by weight, having average molecular weight between 10,000 and 30,000.

***Polymer consisting of 98% isobutylene and 2% isoparaffin, having average molecular weight between 500 and 5,000.

†Furnace black having a surface area of 235 m^2/g, arithmetic mean particle diameter of 17 mμ, and a pH of 9.0.

††Furnace black having surface area of 96 m^2/g, arithmetic mean particle diameter of 29 mμ, and a pH of 8.0.

†††Carbon black composition consisting by weight of 2 parts of high-abrasion furnace black, 1 part semireinforcing furnace black, and 1 part medium thermal black.

§Block copolymer having configuration A–(B–A)$_{1-5}$, A representing a polystyrene block and B representing a hydrogenated polyisoprene block, the isoprene making up about two-thirds of the compound by weight, and the average molecular weight being between 70,000 and 150,000.

§§In parts by weight per 100 parts of total butyl rubber.

Butyl Rubber plus Saturated Hydrocarbon Polymer

The composition described by *K. Kageyama and M. Iwakura; U.S. Patent 4,116,895; September 26, 1978; assigned to The Yokohama Rubber Co., Ltd., Japan* for use as a puncture sealant for tubeless tires comprises a butyl rubber emulsion, at least one additional rubber component selected from the group consisting of diene-type unsaturated hydrocarbon polymer emulsions and a natural rubber latex, at least one saturated hydrocarbon polymer emulsion, a crosslinking agent for the rubbers and a crosslinking activator.

The butyl rubber emulsion employed preferably includes those which are prepared by emulsifying isobutylene-isoprene copolymer or partially halogenated (e.g., chlorinated or brominated) isobutylene-isoprene copolymer with any surface active agent. The total solids content of the butyl rubber emulsion is preferably 60% by weight or more.

The types of diene-type rubber emulsion employed in the sealant composition are those which are compatible with the butyl rubber emulsion, and which can be cured (or crosslinked) with the crosslinking agents and crosslinking activators to form a three-dimensional structure. Examples of such diene-type rubbers are polyisoprene, polybutadiene, styrene-butadiene copolymers, acrylonitrile-butadiene copolymer, ethylene-propylene-diene terpolymer, and their derivatives.

The diene-type rubber emulsion and/or the natural rubber latex can be blended with the butyl rubber emulsion in any blending ratio. However, in view of the sealing property of the sealant after it is applied to the tire as the sealant layer, the amount of the diene-type rubber emulsion and the natural rubber latex, i.e., the additional rubber component, is preferably within the range of 5 to 150 parts by weight, more preferably 20 to 100 parts by weight, in terms of the solids content of the emulsion and latex, based upon 100 parts by weight of the butyl rubber.

The saturated hydrocarbon polymer emulsion employed in this sealant composition suitably includes those which contain a polymer having a number average molecular weight of 500 to 100,000 and which are derived from at least one monoolefin having 4 through 6 carbon atoms. Examples of such polymers are: polybutene obtained from the polymerization of isobutene; polyisobutylene obtained from the polymerization of isobutylene; polypentenes obtained from the polymerization of one or more pentenes; polyhexenes obtained from the polymerization of one or more hexenes; and polyolefins obtained from the co-polymerization of monoolefins having 4 through 6 carbon atoms. These polyolefins are liquid or semisolid at room temperature and give adhesion properties to the sealant composition.

The curing of this sealant composition is preferably carried out under moderate conditions, for example, at a temperature within the range of room temperature to approximately 70°C, in order not to decrease the working properties of the manufacturing step and the quality of the products. For this reason, a so-called quinoid curing agent such as p-quinone dioxime or p-quinone dioxime dibenzoate is preferably used as a crosslinking agent, and organic or inorganic peroxides are also used as a crosslinking activator together with the crosslinking agent. Examples of suitable organic peroxides are: benzoyl peroxide, lauroyl peroxide; 2,4-dichlorobenzoyl peroxide; tert-butyl peroxybenzoate; bis-p-monomethoxy benzoyl peroxide; bis-p-nitrobenzoyl peroxide; 2,5-dimethyl-2,5-bis-benzoyl peroxy-hexene; cumene hydroperoxide; and tert-butyl hydroperoxide. Hydrogen peroxide is also used. Examples of the suitable inorganic peroxide are manganese peroxide, lead peroxide, and the other metal peroxides which do not react with water.

In the quinoid curing system, the amount of the quinone dioximes employed in the sealant composition is preferably within the range of from 3 to 10 parts by weight based on 100 parts by weight of the solid rubber component. Further, the peroxides are usually used in an amount of from 0.5 to 1.5 equivalent weight to 1 equivalent weight of the quinone dioximes.

The sealant composition can further include, if desired, inorganic fillers (pigments) such as, for example, silicic anhydride (silica), silicic acid, clay, talc, mica, calcium carbonate, alumina, titanium white, and carbon black and organic fillers such as microcrystal of cellulose. The blend of these fillers ensures the increase of the solid content of the emulsion, the prevention of cold flow and flow during and immediately after the coating and the improvement of the heat resistance of the crosslinked sealant layer. The amount of the fillers, if employed, is preferably 50 parts by weight or less based on 100 parts by weight (solids content) of the emulsion.

The sealant composition is generally applied to materials to be sealed (i.e., the inside surface of the tire, etc.) by a conventional spraying or coating technique. For the purpose of obtaining better use of this sealant composition, it is preferable that the first blend comprising the butyl rubber emulsion and the crosslinking agent and, if used, the diene-type rubber emulsion and/or the natural rubber latex, and the second blend comprising the unsaturated hydrocarbon polymer emulsion and the crosslinking activator, are separately prepared, and that the two blend systems are mixed with each other immediately before the sealant composition is used.

After the combined sealant composition is applied to the predetermined portion by spraying or coating, it is allowed to stand at an ambient temperature or is heated to a temperature of up to 60° through 70°C to thereby vaporize the water content in the sealant composition and simultaneously cause the curing or crosslinking reaction. Since the sealant layer thus formed has a high elongation property and an outstanding self-sealing property, it exhibits an excellent effect as a puncture-sealing material for tubeless pneumatic tires and seals the tires against puncture.

TACKIFIERS

For Use in Sulfur-Curable Alpha-Olefin Polymers

Tackifiers are employed in elastomeric compositions to produce wider practical ranges of elastomer properties, such as stiffness and tack. Tackifiers are used to improve adhesion, and sometimes cohesion, without necessarily stiffening or softening the elastomer composition.

The tackifiers described by *T.M. Galkiewicz, K.C. Petersen and J.L. Sullivan; U.S. Patents 4,073,776 and 4,073,826; both dated February 14, 1978; both assigned to Schenectady Chemicals, Inc.* are particularly useful in elastomer products which require green tack in their construction. These elastomers include hydrocarbon or hydrocarbon-acrylonitrile, or halohydrocarbon elastomers.

Normally-solid elastomers which are alpha-olefin (e.g., monoolefin or diolefin) hydrocarbon polymers which are sulfur-curable are becoming increasingly important for producing a variety of useful products. For broader application, however, such elastomers must be adapted to have excellent tack properties.

For various applications it is desirable that elastomers have good tack properties prior to curing, in that for bonding purposes it is desirable that the tack property is sufficient to allow the elastomer to bond, on contact, with a force which is sufficiently high to oppose delaminating forces appearing during the fabrication of elastomer products, but low enough to permit clean separation prior to cure if the need arises.

This formulation is directed to tackifiers utilized as tire stock tackifiers, cement tackifiers and tackifiers for any rubber application requiring green tack. Examples of rubber products which require good green tack in their construction include tires, belts and hoses.

The tackifiers used in this formulation are the reaction products of alkyl phenol novolac resins with oxirane-bearing materials. In particular, the alkyl phenol novolacs which are reacted with the oxirane-bearing materials are formed with alkyl phenols, where the alkyl group on the alkyl phenol contains 4 to 16 carbon atoms. These tackifiers are of particular utility in improving the tack of normally-solid, alpha-olefin hydrocarbon polymers which are sulfur-curable. These sulfur-curable alpha-olefin hydrocarbon polymers are referred to in the prior art as elastomers or rubbers and include both the natural and synthetic rubbers.

These tackifiers are the reaction products AB. A represents the residue of an alkyl phenol novolac. B represents the residue of an oxirane-bearing material. The reaction product of the reaction between an alkyl phenol novolac resin and an oxirane-bearing material can be referred to simply as AB.

The alkyl phenol novolac resin is formed by the well-known condensation of phenols with aldehydes. The alkyl phenol used is a phenol which is substituted by a hydrocarbon radical (e.g., alkyl) of 4 to 16 carbon atoms. Conveniently, the alkyl substituent is in the para-position of the phenol moiety. Commercial alkyl phenols can contain minor quantities of disubstituted alkyl phenols as well as ortho-monosubstituted phenols. Para-alkyl phenols, commercially available and containing such disubstituted alkyl phenols and monosubstituted phenols may be used to prepare the tackifiers. Preferably, the alkyl substituent on the phenol contains 8 to 12 carbon atoms. The alkyl substituent on the phenol can be a straight-chain alkyl containing 4 to 16 carbon atoms or a branched-chain alkyl of 4 to 16 carbon atoms. Exemplary novolacs are novolacs based on p-tert-butylphenol, p-tert-octylphenol, p-nonylphenol, p-dodecylphenol and p-hexadecylphenol.

The aldehyde used to condense the alkyl phenol can conveniently be formaldehyde. However, formaldehyde can be replaced as a reactant totally or in part by acetaldehyde, acetone, propionaldehyde, isobutyraldehyde, butyraldehyde, or benzaldehyde as defined by the formula for the alkyl phenol novolac:

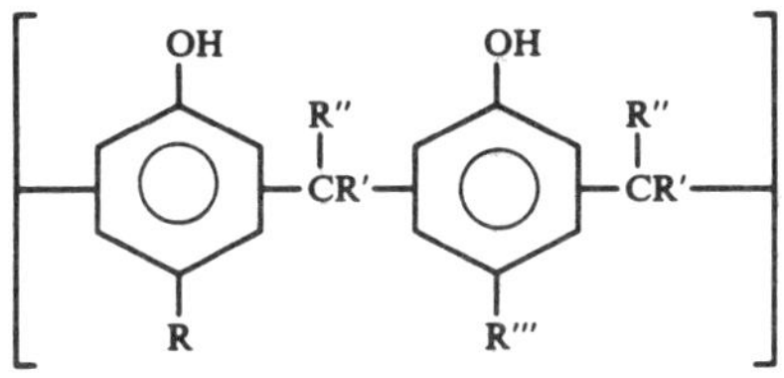

where each of R and R''' is the same or different and is an alkyl of 4 to 16 carbon atoms; and where R' and R'' are the same or different and can be a hydrogen, a lower alkyl of 1 to 4 carbon atoms, aromatic (e.g., phenyl) or heterocyclic (e.g., furyl), and if R' is an alkyl group of more than 1 carbon atom or phenyl or heterocyclic then R'' must be hydrogen.

The condensation of the alkyl phenol with an aldehyde can be undertaken by any convenient method. In a preferred procedure, prior to the addition of the aldehyde to the alkyl phenol, the alkyl phenol, the acid catalyst and the solvent are brought to reflux temperatures, with subsequent addition of the aldehyde. While any solvent inert to the reaction conditions can be employed, preferably a solvent is employed which has a boiling point of 100°C or more, such as xylene or toluene.

Any oxirane reactant can be allowed to react with the alkyl phenol novolac to form AB. The epoxy equivalent weight of the reactant oxirane-bearing materials is between 58 and 4,000. Parenthetically, the term "oxirane" refers to materials containing epoxy groups. Thus, generally, oxirane-bearing materials based on mono-, di-, polyglycidyl ethers, glycidyl esters, glycidyl hydantoins, aliphatic oxiranes, cycloaliphatic oxiranes and aromatic oxiranes can be employed. Useful bisphenol A-derived epoxy resins are manufactured by Shell Chemical Company under the trade name Epon resins. Epon resins in the series Epon 828 to 1009, as well as the glycerine-derived Epon 812, are epoxy resins which may be used.

The alkyl phenol novolac and the oxirane-bearing material are mixed and distilled between temperature ranges of 130° to 250°C under pressures ranging from atmospheric to less than 1 inch of mercury.

Optimum results are obtained when 8 to 12% by weight of oxirane-bearing material is reacted with alkyl phenol novolacs. Generally, with epoxy resin levels within the 5 to 20 weight percent range based on the phenol load, the preferred formaldehyde mol ratios for resins based, for instance, on octyl-, nonyl-, or dodecylphenol would be 0.60 to 1.2 (F/P). The important consideration is that the final product has the proper compatibility with the elastomer compound.

The oxirane material-alkyl phenol novolac reaction products are excellent tackifiers for various elastomers. By elastomers is meant normally solid, alpha-olefin hydrocarbon, halohydrocarbon and hydrocarbon acrylonitrile polymers containing sulfur-curable unsaturation. Thus, the term elastomers includes natural and synthetic rubbers. By elastomers is meant styrene-butadiene copolymer, natural rubber, ethylene-propylene, nonconjugated polyene, e.g., nonconjugated diene, butyl, chlorobutyl rubbers, e.g., cis-isoprene polymer, polybutadiene, polychloroprene, butadiene-acrylonitrile copolymer. By halohydrocarbon polymers is meant halogenated hydrocarbon polymers. These include chlorobutyl rubbers as well as polychloroprene. By hydrocarbon-acrylonitrile elastomers is meant copolymers of acrylonitrile and one or more olefins, e.g., butadiene-acrylonitrile copolymers and elastomeric acrylonitrile-butadiene-styrene terpolymers.

Example: A tackifier was prepared in the following manner: 1,258 g of dodecylphenol, 150 g of xylene and 14 g of toluene sulfonic acid were added to a three-necked five-liter flask equipped with an agitator, a thermometer and a condenser. The temperature was raised to 100°C and then 430 g of 37% uninhibited formaldehyde was added over approximately 20 minutes. The reaction was maintained at reflux temperature for 4 hours after the addition of formaldehyde. Thereafter, 850 g of xylene and 500 g of water were added to the reaction mixture, followed by the addition of 7 g of 50% sodium hydroxide solution which brought the pH of the reaction mixture to about 4 to 5.

The mixture was agitated for 30 minutes at 100°C. The water layer was then allowed to separate and was decanted. The resin was then washed at 100°C for 30 minutes with 1,000 g of water. The water layer was decanted and the resin was washed again with 1,000 g of water in a similar manner. The water layer was again decanted and the resin distilled to 180°C under 25 inches of vacuum.

At distillation temperature of about 180°C, 15 g of Epon 829 (liquid epoxy resin based on bisphenol-A with an epoxy equivalent weight of 193-203 and viscosity of 30,000 to 70,000 cp) were added to the distillation vessel and distillation was continued to 220°C under 25 inches of vacuum. The resin was held at 220°C for 1 hour and then poured into a pan to cool. The yield was approximately 1,327 g with a ring and ball melting point of 97°C.

The resin produced in accordance with this example was tested in the elastomer below. All the ingredients except the resin were milled on a two-roll rubber mill to give a master batch. The specific conditions employed were as follows: all materials were milled in at 175°F over a period of 4 to 6 minutes. Generally, some variations from these conditions are acceptable, particularly 3 to 10 minutes at 100° to 250°F, depending on equipment, etc.

The tackifier resin was milled into the rubber stock as the last ingredient and the stock was sheeted off at a thickness of about one-quarter inch. As referred to in the following composition, Synpol 1708 is a 37½% oil-modified styrene-butadiene elastomer; Circosol 4240 is a naphthenic oil; Neozone D is N-phenyl-β-naphthylamine; and Santocure is N-cyclohexyl-2-benzothiazolesulfenamide.

	Parts by Weight
Synpol 1708	68.75
Smoked sheet	50
HAF black	65
Circosol 4240	16.5
Zinc oxide	3.0
Stearic acid	1.0
Neozone D	1.0
Santocure	1.2
Diphenyl guanidine	0.3
Sulfur	2.0
Resin	3.0

The elastomer containing the tackifier made in the example showed good tack values, even after 96 hours.

Phenolic Tackifiers Which Enhance Adhesion to Reinforcing Elements

During the fabrication of rubber articles, it is common for unvulcanized calendered sheets to be laminated to obtain the desired structural configuration. This technique of building up layers of uncured elastomer is used extensively in the tire building industry, but also finds utility in the manufacture of other rubber articles such as mechanical goods and hoses. In order that the unvulcanized composites have the necessary mechanical stability toward handling and storage, the elastomeric materials must have sufficient tack so that the desired configuration is retained through the vulcanization step. This tack is the ability of unvulcanized elastomer to adhere to itself or to another elastomer which also has tack. This adhesive property, known as "building tack" plays an important role in the production of rubber goods. In the manufacture of tires, building tack holds the inner liner, beads, plys, sidewalls and tread together prior to vulcanization.

Tackifiers are often added to synthetic rubbers and blends of synthetic and natural rubbers to aid in tack retention before vulcanization, to reduce compound viscosity, and to function as plasticizers. Phenolics are often advantageously used as tackifiers, but tend to decrease the adhesion of the rubber to wire or reinforcing fibers.

E.J. Weaver and J.N. Mitchell; U.S. Patents 4,146,512 and 4,146,513; both dated March 27, 1979; both assigned to Ashland Oil, Inc. have formulated phenolic tackifiers which do not adversely affect the adhesion of rubber to the reinforcing element. These compounds are prepared by the reaction of the corresponding alkylphenol-formaldehyde resin containing terminal methylol groups with di-2-hydroxyalkylamine or morpholine, or by the reaction of an alkylphenol-formaldehyde novolac with di-2-hydroxyalkylamine or morpholine in the presence of formaldehyde, or by the reaction of an alkylphenol with formaldehyde and morpholine or di-2-hydroxyalkylamine. These compounds can also be prepared by the reaction of the corresponding alkylphenol-formaldehyde product with N-methylol morpholine.

Tackifiers made by these processes when the amine group is morpholine can be represented by the formula:

where R is alkyl; R^1 is

R^2 is R^1 or a mixture of R^1 with H or $-CHR^3OH$, or both, provided that when R^2 is a mixture, R^1 is the predominant constituent of the mixture; R^3 is H or lower alkyl of 1 to 4 carbon atoms; and where m plus n is at least 1.

The phenolic component for the compounds of this process is a difunctional hydroxybenzene, having an alkyl substituent in the para position which contains 1 to 14, preferably 4 to 12, carbon atoms. Phenols with additional substituents in the meta position are operable, but are not preferred.

The alkylation procedure used to prepare the alkylphenols is usually carried out under acid catalysis with equimolar amounts of phenol and the alkylating agent, but normally a portion (2 to 5%) of the alkylation takes place in the ortho position without deleterious effect on the performance of the tackifier. The presence of a small amount of dialkylphenol, obtained by use of greater than stoichiometric amount of alkylating agent, has been found in some cases to lead to higher adhesion and tack values than expected. Typical alkylating agents are diisobutylene (octylphenol), tripropylene (nonylphenol), and tetrapropylene (dodecylphenol). The amines used to prepare these tackifier resins are morpholine and amines which can be dehydrated to morpholine such as diethanolamine and di-2-hydroxypropylamine.

The molar ratio of alkylphenol/formaldehyde/amine can vary widely. At one end of the spectrum would be 2,6-diaminomethylene-4-alkylphenol with a molar ratio of 1/2/2. At the other extreme would be a ratio of alkylphenol/formaldehyde/amine of 20/28/2. The ratio selected depends, among other factors, on (a) the physical properties of the resin which are desired for easiest handling of the resin and the tackifier rubber; (b) the melting point desired; and (c) the cost limitations. Thus, although fluid compositions are effective, it has been found that preferred tackifiers have softening points between 85° and 135°C, and most preferably between 95° and 120°C.

A typical rubber composition employing a tackifying compound of the type that was used here includes brass coated steel, polyester or polyaromatic amide reinforcing fibers and the vulcanization product of uncured natural or synthetic rubber or their mixtures, extender oil, carbon black, curative agents and from 2 to 5 parts per hundred parts of rubber of a tackifying compound of the type employed according to this formulation.

Examples: The table shows the preparation of various resins. Examples 1 through 4 are standard types of tackifiers commercially available. Examples 5 through 7 have been modified with polyamines, giving resins which have not shown significant improvement in adhesion, and are deleterious to tack retention. Examples 8 through 14 show the preparation of the described compounds based on morpholine and diethanolamine. The procedures, A–E, are as follows.

Procedure A: The phenolic component and formaldehyde (50% aqueous) and acid catalyst were charged simultaneously to a reaction vessel and reacted at 100°C. After the prescribed reaction time, the product was vacuum dehydrated to the desired softening point.

Procedure B: The phenolic component, acid catalyst and azeotroping solvent were heated at least to the boiling point of the solvent, and aqueous formaldehyde was added incrementally while continually azeotropically removing water. Solvent was removed by vacuum distillation when condensation was complete.

Procedure C: The phenolic component in an azeotroping solvent was heated until the phenolic component dissolved. The amine component was added slowly so that the exotherm of dissolution did not cause the temperature to exceed 130°C. In those cases (Examples 8 through 11, 13, 14) where the amine was morpholine, 2 to 10% of triethylamine (TEA) catalyst, based on morpholine and phenol, was used. The formaldehyde (50% aqueous) was then added either as one charge (Examples 1, 3, 8, 10, 13, 14) or intermittantly (Examples 2, 5, 7, 11, 12, 15). When added as one charge, the reaction was run at full reflux and the solvent and water were removed at the completion of the reaction by vacuum distillation. When added incrementally, the water was continually removed azeotropically over the time of formaldehyde addition.

Procedure D: Resin of Example 2 (octylphenol novolac) was dissolved in sufficient toluene to make an 80% solids solution, heated to 110°C, and the secondary amine was added in one charge (TEA catalyst was added when morpholine was used). The temperature was raised to 120°C, and formaldehyde was added incrementally while continually azeotroping the water. The solvent was then removed by vacuum distillation.

Procedure E: Dimethyloloctylphenol was prepared by the base (93% lithium hydroxide and 4% TEA) catalyzed condensation of octylphenol (1 mol) with 50% aqueous formaldehyde (2 mols). The dimethyloloctylphenol product was dissolved in toluene/benzene at 65% solids. Morpholine was added in one charge and the reaction mass was refluxed to completion and vacuum dehydrated to the desired softening point and methylol content of less than 1%. This value for methylol content includes methylol groups and benzylic ether bridges.

Preparation and Evaluation of Various Resins

Ex. No.	Proce- dure	R*	Overall Mol Ratio Phenolic/ CH$_2$O/ Amine	Amine	Time for CH$_2$O Addition (min)	Catalyst Level TEA (%)	Softening Point** (°C)
1	A	octyl***	1/0.9/0	—	—	—	85–105
2	B	octyl	1/0.95/0	—	240	—	110–130
3	B	dodecyl†	1/1.25/0	—	300	—	95–115
4	. . . . Resin of Ex. 1 plus 25% pentaerythritol ester of rosin acids						
5	C	tert-butyl	1/1.15/0.25	TETA††	180	—	113.5
6	C	nonyl	1/1.35/0.25	TETA	180	—	106
7	C	tert-butyl†††	1/2.3/0.5	EDA§	180	—	99
8	C	H	1/3.5/1	morpholine	—	5	96
9	C	tert-butyl	1/3.2/1	morpholine	—	10	paste
10	C	octyl§§	1/2.3/1.1	morpholine	—	9	liquid
11	D	octyl	1/1.2/0.23	morpholine	20	3	104
12	D	octyl	1/1.2/0.23	DEA§§§	25	—	151
13	E	octyl	1/2/0.66	morpholine	—	4	93
14	E	octyl§§	1/1.7/0.67	morpholine	—	4	82
15	C	octyl	1/2/2	morpholine	100	4	liquid

```
          *R is the para alkyl group on the phenol.
         **Ring and Ball softening point, ASTM E 28-67.
        ***From the alkylation of phenol with diisobutylene.
          †From the alkylation of phenol with tetrapropylene.
         ††Triethylenetetramine.
        †††From the alkylation of phenol with tripropylene.
           §Ethylenediamine.
          §§The phenolic was prepared by alkylation of phenol with diisobutylene
             at a molar ratio of olefin/phenol of 1.5/1; therefore, some of the
             active positions have been substituted with alkyl groups.
         §§§Diethanolamine.
```

Testing of the various rubber samples showed conclusively that those containing the described tackifiers (Examples 8 through 14) developed the same degree of tack as conventional, commercial phenolic tackifiers, but showed dramatic increases in static adhesion compared to the conventional phenolic tackifiers.

OTHER FORMULATIONS

One-Coat, Single-Package, Storage-Stable Adhesives

P.J. Jazenski and L.G. Manino; U.S. Patent 4,119,587; October 10, 1978; assigned to Lord Corporation have found that compositions containing as essential ingredients at least one halogenated polymer, at least one aromatic nitroso compound, at least one lead salt of certain organic and inorganic acids, and (optionally) at least one maleimide compound are unexpectedly effective as adhesive materials for bonding a variety of elastomers, including vulcanizable and vulcanized natural and synthetic rubber compositions to themselves or other solid substrates. The compositions are characterized by the unexpected ability to provide strong rubber-to-substrate bonds with good-to-excellent environmental resistance without the necessity of first priming the substrate surface. They provide excellent adhesion for both unvulcanized and vulcanized elastomer compositions without requiring treatment such as chlorination of the rubber surface. In addition to affording one-coat adhesive systems characterized by excellent primary adhesion and environmental resistance, they exhibit excellent shelf-life stability, resistance to sweep, prebake resistance, good layover characteristics, and are effective over a broad spectrum of bonding temperatures, e.g., from about $90°$ to over $180°C$.

Generally, the aromatic nitroso compound will be present in an amount in the range preferably from about 10 to 100 parts by weight per 100 parts by weight of halogen-containing polyolefin. The amount of lead salt is most preferably from about 35 to 75 parts by weight per 100 parts by weight of halogen-containing polyolefin.

The halogen-containing polymers suitable for use can be described as halogen-containing natural and synthetic polyolefinic elastomers. The preferred nitroso materials are the dinitroso aromatic compounds, especially the dinitrosobenzenes and dinitrosonaphthalenes, such as the m- or p-dinitrosobenzenes and the m- or p-dinitrosonaphthalenes.

Particularly preferred lead salts include dibasic lead phthalate, monohydrous tribasic lead maleate, tetrabasic lead fumarate, and dibasic lead phosphite; and include mixtures thereof.

Preferred maleimide compounds include the N,N'-linked bismaleimides which are either joined directly at the nitrogen atoms without any intervening structure or in which the nitrogen atoms are joined to and separated by an intervening divalent radical such as alkylene, cycloalkylene, oxydimethylene, phenylene (all three isomers), 2,6-dimethylene-4-alkylphenol, or sulfonyl. m-Phenylenebismaleimide is a presently preferred compound and is available as HVA-2 from DuPont.

These adhesive compositions have been found to be particularly suitable for bonding a wide variety of elastomeric materials, including both vulcanized and vulcanizable elastomeric materials, to themselves or to other substrates, particularly inorganic substrates. Elastomers which can be bonded include, without limitation, natural rubber, polychloroprene rubber, styrene-butadiene rubber, nitrile rubber, ethylene-propylene copolymer rubber, ethylene/propylene/diene terpolymer rubber, butyl rubber, polyurethane rubber, parel-type elastomers, and the like.

Substrates other than the elastomers per se which can be effectively bonded include fabrics such as fiber glass, polyamides, polyesters, aramids (e.g., Kevlar) and the like; and metals and their alloys such as steel, stainless steel, lead, aluminum, copper, brass, bronze, Monel metals, nickel, zinc, and the like, including treated metals such as phosphatized steel, galvanized steel, and the like; glass; ceramics; etc.

Example: Adhesive compositions are prepared by KD mill mixing the following ingredients:

	. . . .Composition, pbw		
	A	B	C
Chlorosulfonated polyethylene	15	15	15
Chlorinated natural rubber	6	6	6
p-Dinitrosobenzene	18	18	18
Dibasic lead phthalate	0	10	10
N,N'-m-phenylene-bismaleimide	0	0	4
Carbon black	5	5	5
Silica	2	2	2
Toluene/chlorotoluene/1,1,1-trichloroethane	.to 22% total solids content .		

The adhesive compositions are employed to bond unvulcanized natural rubber stock to grit-blasted, degreased, cold-rolled steel. The assemblies are cured at $154°C$ for 5 minutes with prebakes of 0 and 10 minutes. Boiling water and peel adhesion (ASTM D-429 method B, modified to $45°$) tests are made with the following results. In the data reported in the example, failure is expressed in terms of percent of failure in the rubber body, e.g., 95 R means that 95% of the failure occurred in the rubber body, with the remaining failure being between the adhesive composition and the metal (CM).

As can be seen from the data, adhesives B and C, which are representative of the herein-described compositions, are substantially superior to adhesive A, especially in regard to environmental resistance.

Adhesive	Prebake Minutes	Boiling Water, 2 Hours	. . . .Peel Adhesion. . . .	
			Pounds per Inch	Failure
A	0	100 CM	40	100 R
B	0	27 R	50	95 R
C	0	82 R	66	100 R
A	10	100 CM	44	100 R
B	10	62 R	43	100 R
C	10	80 R	38	100 R

Organopolysiloxane Resin as Adhesion Promoter for Silicone Rubbers

Many adhesives and methods have been suggested for adhering silicone rubbers to substrates such as metals, glass and ceramics. Primers of alkoxysilanes or acetoxysilanes have been used. However, rust has occurred on metal substrates under the applied silicone rubber where the adhesion has not been perfect.

R. Mikami and K. Mine; U.S. Patent 4,122,127; October 24, 1978; assigned to Toray Silicone Company, Ltd., Japan have formulated a primer which will overcome this problem. The primer comprises a mixture of (A) an organopolysiloxane resin which has the following general formula:

$$[R'_n X_m SiO_{(4-n-m)/2}]_p$$

in which formula R^1 is a monovalent hydrocarbon group containing 1 to 8 carbon atoms, n has an average value of 0.90 to 1.80, X is a hydroxyl group or an alkoxy group, m is such to give a minimum value of 0.01 weight percent of X based on the weight of (A), the sum of m + n does not exceed 3.0, p has a value greater than 1; and (B) an organopolysiloxane represented by the formula:

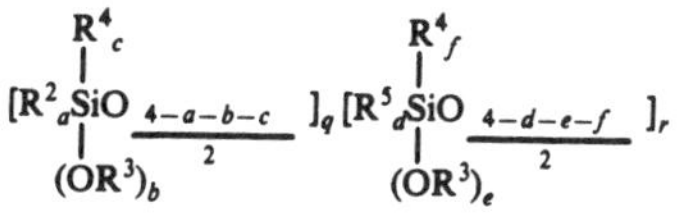

$$[R^2{}_a\underset{\underset{(OR^3)_b}{|}}{\overset{\overset{R^4{}_c}{|}}{Si}}O_{\frac{4-a-b-c}{2}}]_q [R^5{}_d\underset{\underset{(OR^3)_e}{|}}{\overset{\overset{R^4{}_f}{|}}{Si}}O_{\frac{4-d-e-f}{2}}]_r$$

where R^2 is a monovalent organic group containing at least one epoxy group, R^3 and R^4 are independently hydrogens or monovalent organic groups, R^5 is a monovalent hydrocarbon group containing at least one unsaturated group, a, b, c, d, e and f each have the value of 0 to 3, the sum of a, b, and c is less than 4 and the sum of d, e, and f is less than 4, q is 0 or a positive integer and r has a value of 2 or more.

Example: A toluene solution that contained 60 wt % of phenylmethylpoly-siloxane resin containing an average of 1.29 phenyl groups and methyl groups per silicon atom (the ratio of phenyl groups to methyl groups is 0.43) and also having 4.0 wt % of hydroxyl groups was designated (C).

An organopolysiloxane resin solution prepared by mixing 5 parts of γ-amino-ethylaminopropyltrimethoxysilane, 15 parts of methyltrimethoxysilane, and 0.5 part of dibutyltin acetate in 100 parts of a xylene solution of 50 wt % organopolysiloxane resins which contained an average of 1.50 phenyl groups and methyl groups per silicon atom (the ratio of phenyl groups to methyl groups is 0.57) and also contained 0.5 wt % of hydroxyl groups was designated (D).

By adding 1 or 2 of the organopolysiloxanes (shown below) as indicated in the following table, the primer compounds were prepared.

Organopolysiloxane 1:

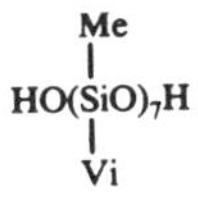

$$HO\underset{\underset{Vi}{|}}{\overset{\overset{Me}{|}}{(SiO)_7}}H$$

Organopolysiloxane 2:

$$\underset{}{\overset{\overset{O}{/\backslash}}{CH_2}}-CHCH_2OCH_2CH_2CH_2\underset{\underset{OMe}{|}}{\overset{\overset{OMe}{|}}{Si}}O(\underset{\underset{Vi}{|}}{\overset{\overset{Me}{|}}{Si}}O)_7\underset{\underset{OMe}{|}}{\overset{\overset{OMe}{|}}{Si}}CH_2CH_2CH_2OCH_2CH\overset{\overset{O}{/\backslash}}{-}CH_2$$

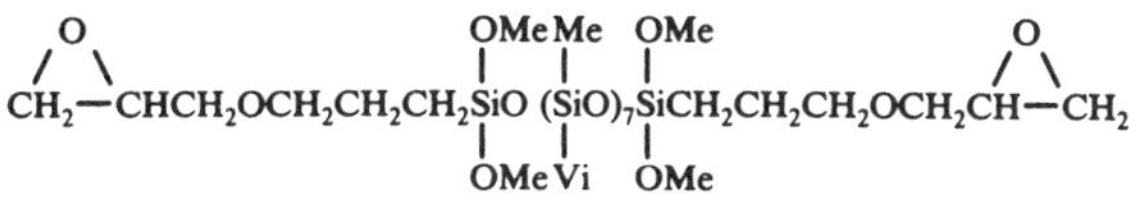

Vi = vinyl; Me = methyl

Cold pressed steel plates were treated with these primers. The primer using organopolysiloxane resin (C) solution was cured at $150°C$ for 30 minutes, and a film was formed. The primer using (D) was dried at room temperature for 1 hour and the film was formed. Using one part of vulcanizing agent RC-4 (Toray Silicone Co.) to 100 parts of silicone rubber, press-molding on the steel plates was carried out. The measurement of the adhesive force was conducted according to JIS K 6103. The results are shown in the table.

Organopolysiloxane Resin Solution		Organopolysiloxane		
(C)	(D)	1	2	Adhesive Force
. . . . (parts).				(kg/cm^2)
100	—	—	2	8
—	100	1	—	8-10
—	100	3	—	8-10
—	100	—	1	13-14*
—	100	—	3	12-13*

*100% cohesive rupture

All of the runs showed excellent adhesion between the steel plates and the silicone rubbers. The five types of primers were coated on steel plates. After curing, they were left at room temperature for 20 days. Formation of rust was not observed in any of them. Using these steel plates which were treated with primers and left for 20 days, press-molding was conducted and the adhesive force was measured. In all cases, there were no negative effects on the adhesive force caused by the time lapse.

In another patent *assigned to the Toray Silicone Company, Ltd., K. Mine and M. Yokoyama; U.S. Patent 4,157,357; June 5, 1979* have described a formulation for a curable organopolysiloxane which gives improved adhesion to various substrates without the use of a primer.

The composition consists of a mixture of:

(a) an organopolysiloxane containing at least 2 hydrogen atoms bonded to silicon atoms and having the general formula $R_aSiO_{(4-a)/2}$ where a has an average value of 1 to 3 and R is hydrogen, hydroxy radicals or unsubstituted or substituted monovalent hydrocarbon radicals which do not contain aliphatic unsaturation;

(b) an organopolysiloxane with at least one $A(R'O)_2Si—$ group and at least two lower alkenyl radicals bonded to silicon atoms, in each molecule, where A is a monovalent hydrocarbon radical containing an epoxy group, R' is a lower alkyl radical containing 1 to 6 carbon atoms, the sum of the number of hydrogen atoms bonded to silicon atoms in component (a) and the number of lower alkenyl radicals bonded to silicon atoms in component (b) is five or greater where there is present 0.1 to 10 mols of lower alkenyl groups bonded to silicon in (b) for each mol of hydrogen atoms bonded to silicon atoms in (a); and

(c) 0.1 to 40 parts by weight per 1 million parts by weight of components (a) and (b) of an addition reaction catalyst for the addition reaction of the hydrogen atoms bonded to silicon atoms with the lower alkenyl groups bonded to silicon atoms, such as platinum.

In the formula $A(R'O)_2Si-$, A may be selected from the group consisting of: β-epoxyethyl, α-chloro-β-epoxyethyl, γ-glycidylpropyl, β-(3,4-epoxycyclohexyl)-ethyl, and γ-(3,4-epoxycyclohexyl)propyl.

Example 1: 100 parts of a copolymer composed of 90 mol % dimethylsiloxane units, 10 mol % hydrogen methylsiloxane units and having hydrogen dimethylsilyl groups as the end groups and with a viscosity of 100 cs and 50 parts of a compound with the formula:

$$
\begin{array}{c}
\underset{\text{Me}}{|} \quad \underset{\text{Me}}{|} \\
\underset{\text{O}}{|} \quad \text{Me} \quad \underset{\text{O}}{|} \\
\overset{O}{\diagup\!\diagdown} \qquad \underset{|}{O} \quad \underset{|}{Me} \quad \underset{|}{O} \qquad \overset{O}{\diagup\!\diagdown} \\
CH_2\text{---}CHCH_2OCH_2CH_2CH_2SiO(SiO)_4SiCH_2CH_2CH_2OCH_2CH\text{---}CH_2 \\
\underset{|}{O} \quad \underset{|}{Vi} \quad \underset{|}{O} \\
\underset{\text{Me}}{|} \quad \underset{\text{Me}}{|}
\end{array}
$$

and an ethanol solution of chloroplatinic acid (10 ppm of platinum relative to the total amount of the polysiloxanes) were mixed thoroughly. This mixture was injected between an aluminum plate and a glass plate. These plates were placed in an oven and the mixture hardened by heating at 150°C for 30 minutes. After cooling to room temperature, the two plates were pulled apart. At this time, the silicon elastomer layer was broken, i.e., a so-called cohesive rupture occurred.

Example 2: 100 parts of dimethylpolysiloxane with hydrogen dimethylsilyl end groups and a viscosity of 5,000 cs, 2 parts of $Si(OSiMe_2H)_4$, 4.0 parts of the compound with the formula:

$$
\begin{array}{c}
\underset{\text{Et}}{|} \qquad \qquad \underset{\text{Et}}{|} \\
\underset{\text{O}}{|} \quad \text{Me} \quad \text{Me} \quad \underset{\text{O}}{|} \\
\overset{O}{\diagup\!\diagdown} \qquad \underset{|}{O} \quad \underset{|}{} \quad \underset{|}{} \quad \underset{|}{O} \qquad \overset{O}{\diagup\!\diagdown} \\
CH_2\text{---}CHCH_2OCH_2CH_2CH_2SiO(SiO)_7(SiO)_3\text{---}SiCH_2CH_2CH_2OCH\text{---}CH_2 \\
\underset{|}{O} \quad \underset{|}{Me} \quad \underset{|}{Vi} \quad \underset{|}{O} \\
\underset{\text{Et}}{|} \qquad \qquad \underset{\text{Et}}{|}
\end{array}
$$

and 30 parts of powdered silica which had been surface treated with trimethyl-silyl groups to give hydrophobicity (specific surface area, 200 m^2/g) were mixed well. 20 ppm of platinum dispersed on carbon powder relative to the total amount of the copolymers were then added and the mixture was mixed thoroughly. This mixture was then placed between a polyimide film and a copper plate. This was placed in an oven and hardened by heating at 200°C for 20 minutes. After cooling to room temperature, the polyimide film was pulled in a 180° peel direction. The polyimide film tore before the silicon elastomer layer was ruptured.

Improving Adhesion of Rubbery Ethylene Copolymers

Copolymers of ethylene with at least another α-alkene or ethylene copolymers containing a second α-alkene are popular synthetic rubbery polymers because of their excellent resistance to ozonization, weather and aging. Such resistance renders these polymers attractive for various purposes.

However, conventional ethylene-α-alkene and ethylene-α-alkene-polyene elastomers are deficient with respect to tack properties and properties of adhesion. Tack and adhesion connote the adhesion of unvulcanized rubber to itself, that is, connote the adhesion of two surfaces of two strips of unvulcanized rubber of identical composition when those two pieces are contacted with each other under moderate pressure. Good tack properties of elastomers are essential in the manufacture of various articles, in particular multipartite or multilayered articles such as driving belts, conveyor belts and car tires.

H.J.G. Paulen; U.S. Patent 4,131,634; December 26, 1978; assigned to Stamicarbon, BV, Netherlands describes ethylene copolymers which are characterized by satisfactory tack values by virtue of the nature of their composition. They also are produced in a process which does not require exposing them to light.

The composition of the tacky rubbery mixture is based on a rubbery copolymer of ethylene, at least one other α-alkene, which may be propylene, and one or more polyenes. The mixture must have an elongation at rupture of at least 800% and a tensile strength of between 1.3 and 50 kg/cm². The process consists of mixing:

(1) 20 to 75 parts by weight of a rubbery copolymer containing 60 to 80% by weight of ethylene, 20 to 40% by weight of at least one α-alkene and 0 to 20% by weight of at least one polyene and having in the unvulcanized state an elongation at rupture of over 800% and a tensile strength of between 10 and 100 kg/cm², with

(2) 80 to 25 parts by weight of at least one polymer composed of a conjugated diene selected from the group consisting of styrene-butadiene copolymers, polybutadiene and polyisoprene, and

(3) a tackifier resin together with optionally conventional fillers and additives.

If polyisoprene is used as the conjugated diene, at least 90% of it should be in the cis form.

Preferable tackifiers are alkylphenol and formaldehyde reaction products having a molecular weight of from 200 to 3,000, in which the alkyl group contains from 8 to 12 carbon atoms, such as the Amberols.

Example: A mixture was prepared of a copolymer consisting of ethene, propene, and ethylidene norbornene (EPDM) and a styrene-butadiene copolymer (SBR 1500). Per 100 dl of polymer, the following substances were added:

	Parts
Zinc oxide	5
Stearic acid	1
FEF black	50
Aromatic oil	40
Zinc dibutyldithiocarbamate	2
2-mercaptobenzothiazole	0.5
Tetramethylthiuram disulfide	0.5
Sulfur	1.5
Amberol S T 140 F*	5

*A tackifier resin formed by polycondensation of alkyl phenol and formaldehyde.

The tensile strength and the elongation at rupture were determined on both the mixture and the mixture compounded with additives. The compounded mixture was also tested for tack, which was determined by means of a tackmeter. The results of the testing are given in the table below.

	. . . . Mixture.			
	A	B	C	D
EPDM*	100	50	40	0
SBR 1500	0	50	60	100
Mixture				
Tensile strength, kg/cm^2	27.5	>11	7	1.9
Elongation at rupture, %	2,350	>2,365	2,160	390
Compounded				
Tensile strength, kg/cm^2	>2.2	>2.4	>2.7	1.2
Elongation at rupture, %	>2,365	>2,365	>2,365	260
Tack value, g/5 mm	400	3,700	2,200	300

*The EPDM used [ML (1 + 4) 125°C Mooney value 59] consisted of 62% by weight of ethene, 28% by weight of propene, and 10% by weight of ethylidene norbornene.

Wood, Corrugated Board and Paper Adhesives

WOOD ADHESIVES

Flame-Resistant Adhesive

J.A.L.F. Chollet; U.S. Patent 4,066,463; January 3, 1978 has developed a flame-resistant composition which has an inorganic and an organic component. The inorganic component comprises a combination of three constituents, viz: (a) 20-90 wt %, by weight of the total composition, of a concentrated aqueous solution of an alkali metal silicate; it is preferred to use a $38°$-$40°$ Bé potassium silicate solution; (b) 5-25 wt %, by weight of the total composition, of a clay, preferably powdered kaolin; and (c) 2-7 wt %, by weight of the total composition, of deflocculated short asbestos fibers, i.e., subjected prior to use to a mechanical deflocculating treatment to loosen and separate the fibers from one another.

The organic component consists of a 30-70% (preferably about 50%) aqueous solution of a carboxymethylcellulose, of a starch ether or of a dextrin or a mixture thereof. The organic component comprises 0.2-2 wt % (dry weight) by weight of the total composition. The adhesive may be used for bonding wood to wood, metal to wood, metal to fiberboard, wood to plastic, etc.

The process for preparing the adhesive comprises (a) charging the silicate solution into a container and rotating it in a given direction; (b) intimately dispersing the aqueous organic component solution in the stirred silicate while rotating the organic component solution in the direction opposite to that in which the silicate was rotated; (c) on completion of the dispersion, adding the clay with continued stirring and (d) thereafter dispersing the deflocculated asbestos in the stirred mixture. The inorganic filler, when used, is added after step (b) and prior to addition of the clay.

168

Example: The two following adhesive compositions are prepared according to the abovedescribed method:

	Composition 1 (wt %)	Composition 2 (wt %)
Sodium silicate	85	76
Kaolin	7	16
Deflocculated asbestos fibers	4	4
Solvitose CL*	2	2
Methyl siliconate	2	2
	100	100

*Starch ether, 50% aqueous solution.

Both compositions exhibit outstanding tack and, after setting are classified "flame-resistant" and "fireproof" according to standard specifications.

The tensile strength is found to be 11-15 kg/cm^2 at normal temperature when black sheet iron is bonded to concrete; after 7 days at 40°C and 95% relative humidity, the tensile strength is 11-18 kg/cm^2.

Agglomerated wood-fiberboards were bonded to concrete with the above compositions; the agglomerate breaks at 8-9 kg/cm^2 at normal temperature but the bonding plane remains intact. Plywood bonded with the composition broke at 12-14 kg/cm^2; the bonding plane remained intact. Compositions exhibiting analogous properties were obtained by replacing the starch ester with carboxymethylcellulose of dextrin.

Amylaceous Extender for Plywood Adhesive

In most plywood adhesive formulations, the synthetic resin is combined with various fillers and extenders, which not only reduce the overall cost of the adhesive but also perform various functions in the adhesive system. Fillers are added primarily to reduce the cost of the plywood adhesive and to add particulate matter to the adhesive. This particulate matter assists in reducing penetration of the resins into the wood and fills small cavities between veneers, thus preventing starved joints.

Extenders are usually amylaceous containing materials, such as cereal flours and starches. In phenol-formaldehyde glues, these extenders provide a source of starch, which becomes gelatinized under the highly alkaline conditions present in phenolic resin systems. The gelatinized starch provides the glue with increased viscosity and tack. These properties allow for more control of the glue during application, provide improved glue lines and limit penetration of the resin into the wood plies.

With urea-formaldehyde glues, most amylaceous materials are not satisfactory. Since urea-formaldehyde resins are neutral or slightly acidic and do not contain alkali, the starch does not become gelatinized.

Wheat flours are the only principal starch-based extenders that are employed today in urea-formaldehyde plywood glues. These products are unique in that they can provide viscosity in certain adhesive systems without the starch portion of the flour becoming gelatinized.

R.J. Alexander and R.K. Krueger; U.S. Patent 4,070,314; January 24, 1978; assigned to Krause Milling Company have found that certain high fiber-containing by-products from the grain milling and processing industries can function as extenders in producing urea-formaldehyde plywood glues or adhesives, comparable to those made with wheat flours.

These materials include cereal grain milling by-products, such as corn bran, sorghum bran, oat hulls, barley hulls, rice hulls, rice bran and wheat bran, oilseed processing by-products such as soybean hulls and cottonseed hulls, and cereal grain processing by-products, such as spent brewer's grains, expelled or extracted corn germ and malt husks.

An unexpectedly high percentage of a fibrous material yielding high glue viscosities can be combined with those products which produce low viscosity glues to provide a mixture imparting a satisfactory viscosity to urea-formaldehyde adhesives.

This negative synergistic action affords the economic advantage of allowing the use of large percentages of the less expensive fibrous materials to produce plywood glues having properties comparable to those glues formulated with the more expensive wheat flours.

Example: Fibrous By-Products — Various finely ground fibrous cereal-derived by-products were tested on glue extenders in comparison to wheat flour. The important property of glue viscosity as measured by a Brookfield viscometer at room temperature was used for the comparisons.

These materials were compared in the following glue formula, and the initial glue viscosity results are shown on the following page. In practical interior grade plywood production, the useful glue viscosity range is about 1,000 to 5,000 centipoises.

Material	Parts by Weight
Water	228
Catalyst	21
Filler	25
Extender	175
Hobart mixed for 5 min	
Urea-formaldehyde resin	420
Hobart mixed for 10 min	

Extender	Viscosity at 20 rpm, cp
Wheat flour	2,150
Rice hulls	910
Sorghum hulls	1,240
Oat hulls	1,330
Corn bran (dry milled)	3,400
Oat feed	4,000
Rice bran	5,000
Barley hulls	10,800
Extracted corn germ	16,000
Wheat bran	20,000
Corn bran (wet milled)	21,500
Spent brewer's grains	35,000
Malt husks	62,000
Soybean hulls	97,000

Premixed Catalyzed Vinyl Acetate Polymer

The object of the formulation provided by *M.C. Woo; U.S. Patent 4,085,074; April 18, 1978; assigned to National Casein of New Jersey* was to provide an adhesive emulsion comprising a polymerized crosslinkable vinyl acetate resin, a crosslinking catalyst and a crosslinking inhibitor, which adhesive provides a cured water-resistant Type II bond and possesses increased shelf life and can therefore be used over a relatively long period of time for direct application to porous and semiporous substrates for gluing in appropriate pressing equipment.

The foregoing objects are obtained by providing an aqueous resin emulsion containing vinyl acetate polymerized with alkyl esters of acrylic acid or methacrylic acid, such as methyl acrylate, ethyl acrylate, butyl acrylate, methyl methacrylate and ethyl methacrylate and N-alkylol derivatives of amides of alpha,beta-unsaturated carboxylic acids, such as N-methylol acrylamide, a protective hydrocolloid emulsifier such as polyvinyl alcohol or cellulosic dispersing agents, e.g., hydroxyethylcellulose, carboxymethylcellulose, etc., a crosslinking catalyst in the form of an acidic metal salt such as aluminum chloride, aluminum nitrate, chromic chloride and chromic nitrate, and a crosslinking inhibitor in the form of an amine such as ammonia, the alkyl and alkanol amines, e.g., mono-, di- and triethylamine and mono-, di- and triethanolamine.

The preferred adhesive composition includes a resin emulsion, comprising a monomer mixture of 90-92% vinyl acetate, 4-6% ethyl acrylate and 2-5% N-methylol acrylamide, polymerized with the use of a redox or free radical type catalyst system in an aqueous medium containing about 5% polyvinyl alcohol. The solids content of the resultant emulsion is 50% by weight.

The resin emulsion is white, has a pH of 4.5-5.5, a viscosity of 3,000-5,000 cp (25°C) and a SG of 1.09. To this resin emulsion is added an aqueous solution of $AlCl_3$, specifically 1.5 to 12.0 pbw of the $AlCl_3$ solution to 100 pbw of the resin emulsion, the solution containing 0.21 g-eq $AlCl_3$/100 g solution, thus providing 0.0032-0.025 g-eq of $AlCl_3$/100 g of resin emulsion, and triethanolamine (TEOA) in a mol ratio of TEOA/$AlCl_3$ up to 0.7-1.7 maximum. The following table lists illustrative compositions and their properties.

	Examples							
	1	2	3	4	5	6	7	8
RE (resin emulsion), g	100	100	100	100	100	100	100	100
Triethanolamine (TEOA) 85 wt %, g	0.5	1.5	1.85	2	5.5	3.67	5	7.4
AlCl$_3$*, aqueous solution, g	2	4	5	8	10	10	12	12
Mol ratio TEOA/AlCl$_3$	0.68	1.0	1.0	0.68	1.5	1.0	1.1	1.67
Initial pH	3.7	3.7	3.65	3.45	3.8	3.4	3.5	3.70
Initial viscosity, cp (25°C)	1,175	1,050	1,050	550	900	800	600	550
Shelf or pot life, days	186	180	111	207	171+	178+	190+	30+
Overnight cured Dry, psi (% WF)	658 (100)	660 (100)	668 (100)	581 (100)	620 (100)	539 (100)	582 (100)	520 (100)
One-half hour boil, psi (% WF)	466 (33)	316 (0)	299 (0)	307 (0)	254 (0)	234 (0)	276 (0)	48 (0)
48-hour soak, psi (% WF)	534 (35)	452 (25)	385 (2)	486 (14)	326 (0)	339 (22)	400 (14)	348 (0)
Two-week cured Dry, psi (% WF)	645 (100)	604 (100)	681 (100)	535 (100)	636 (100)	548 (100)	574 (100)	528 (100)
2-cycle boil, psi (% WF)	384 (1)	311 (0)	288 (0)	214 (0)	194 (0)	198 (0)	256 (0)	Del
48-hour soak, psi (% WF)	530 (75)	440 (1)	407 (2)	480 (25)	336 (0)	378 (15)	431 (32)	310 (0)

*about 28 wt % AlCl$_3$ (about 0.21 g eq AlCl$_3$/100 g solution).

Note: WF = wood failure.
        Del = delaminated.

To obtain an adhesive composition which will possess a shelf life which is increased over that of the resin emulsion plus catalyst alone and also possess a cured water resistance bond of at least 50 psi in a one-half hour boil test, the minimum $TEOA/AlCl_3$ mol ratio is about 0.1 to 0.2.

The procedures used in testing the adhesive compositions were as follows. The test samples were prepared in accordance with ASTM D-906-64 and applied to the birch test panels at a glue spread rate of 40 lb/1,000 ft^2 single glue line, assembly time of 1 minute, a press temperature of 121°C, a press time of three minutes and a pressure of 150 psi.

The one-half hour boil test was carried out in accordance with ASTM D-1151-72 using a preconditioning of one day and immersion in boiling water for one-half hour and determining the average strength after exposure. The 2-cycle boil test was carried out in accordance with Commercial Standard CS35-61 (U.S. Dept of Commerce).

The forty-eight hour soak test is carried out in accordance with the procedure of ASTM D-1151-72, wherein the specimen is immersed in water at a temperature of 23°C for 48 hours and then subjected to the shear strength test.

The adhesive composition is directly applicable to semiporous and porous substrates, such as hardwoods, softwoods, hardboard, particle board, high-pressure laminates and similar materials in appropriate pressing equipment in open assembly time up to about 5 minutes and closed assembly time up to about 20 minutes. The adhesive is primarily applicable to hot press (pressures of 25-200 psi) or high-frequency cure although it will produce an effective Type II water-resistant bond when pressed at room temperature (25°C) for about 40 minutes.

Extender of Powdered Tree Foliage

In the wood bonding industry, the conventional adhesive extender used is wheat flour. Wheat flour prices have tripled in recent years and supply is getting difficult. Fillers used are usually the powder of corn cobs, tree barks or walnut shells. Their supply is inconsistent and their prices have also increased greatly. The search for a substitute extender and filler material has become one of the most important tasks to the plywood and laminating industry.

S. Chow; U.S. Patent 4,082,903; April 4, 1978; assigned to Canadian Patents and Development Limited, Canada has found that ground tree foliage is a very satisfactory and economical extender or combined extender-filler in wood adhesive compositions. The foliage can be dried and ground to a powder, or can be wet ground to a mulch. The starting foliage need not be dried out or "dead" but can be in green condition. The foliage can be obtained from conifers and other evergreens, or from deciduous trees. Suitable evergreens include pine, spruce, fir, hemlock, cedar and redwood, while suitable deciduous trees include poplar, birch, maple, elm and basswood. The ground foliage has been found to be an active extender, i.e., it is not just a diluent or filler, but contributes some adhesive or bonding capacity to the composition under adhesive curing conditions. For wood bonding the usual adhesive types are phenolic resins and amine- or amide-aldehyde resins. Particular resin glues in widespread use are phenolformaldehyde condensation products (novolacs) and urea-formaldehyde condensation products (intermediate-stage). Other suitable resin glues include polyvinyl acetate, urea-melamine-formaldehyde and phenol-resorcinol-formaldehyde glues. The polyvinyl acetate resin glues are usually in aqueous emulsion form. The foliage powder is very suitable for use in dry powder adhesives.

The foliage powder is mixed into the adhesive similarly to other extenders and fillers. Any technique that gives uniform mixtures for spreading before curing is suitable. The amounts of foliage incorporated can vary widely depending on the type of adhesive and bond strengths required. For plywood adhesives from 1 to 65% by weight of the total solids in the composition would be acceptable with the preferred range being about 2 to 15% foliage. For lower strength adhesives such as waferboard and particle board type, the amount of foliage can range up to 95% by weight of the composition solids.

The strength will generally decrease with increasing amounts of foliage but even at levels of 80 to 95% foliage, bond strengths of the order of 60 to 70 psi can be achieved (such strengths are acceptable for certain uses, e.g., interior-use particle board or furniture panels).

The particle size of the foliage powder may range from –325 mesh up to about –40 mesh, or diameters up to about 0.4 mm or more. The particle size is not critical but should be fine enough to permit uniform adhesive mixtures to be formed.

Phenol-Formaldehyde Resin for Waferboard Production

R.C. Vasishth and P. Chandramouli; U.S. Patent 4,086,125; April 25, 1978; assigned to Cor Tech Research Ltd., Canada describe the use of the acid-catalyzed, phenol-formaldehyde resins described in U.S. Patent 4,017,437 (see page 127, Satriana, *Adhesives Technology Annual.* Noyes Data Corporation, Park Ridge, NJ 1978) in the production of panelboard, such as waferboard, particle board and plywood.

Example: A 1.8:1 mol ratio phenol-formaldehyde resin (about 75% NV solids) cooked by the procedure of Example 1 of U.S. Patent 4,017,437 was emulsified as described in Example 2 of the same patent by dispersion in a 1% aqueous solution of Natrosol HXR 250 (hydroxy ethyl cellulose) to give a stable emulsion of about 39 to 43% NV solids.

Samples of the emulsion were catalyzed with various amounts of p-toluene sulfonic acid (as a 50% aqueous solution) and sprayed onto Aspen wafers of 4 and 6% moisture content at a resin level of 2.5% based on ultimate board weight.

In comparative tests, the resin first was catalyzed with 2% p-toluene sulfonic acid (used as a 50% aqueous solution) and was emulsified in analogous manner to that described above at least 3 hours after catalyst addition.

The resin-coated wafers in each case were laid into a mat and hot pressed at 210°C to a thickness of $5/16$ inch at a density of 40 lb/ft^3 for the required cycle time, cooled and tested. The results are reproduced in the following table:

Moisture Content (%)	Press Cycle (min)	Catalyst Level (% of resin)	Post or Pre Catalysis	Internal Bond* Strength (psi)
6	3.5	5	Post	Delaminated
6	5	5	Post	Delaminated
6	3.5	8	Post	Delaminated**
6	3.5	12	Post	31**
6	3.5	20	Post	30**
4	3.5	5	Post	55
6	3.5	2	Pre	62***
6	3	2	Pre	55***

*An internal bond strength exceeding 50 psi is considered good.
**Darkened spots were evident over the board surface.
***Boards very light in color.

The results of the above table show that the emulsions that are made from resins to which acid is added prior to emulsification act as faster curing adhesives, require considerably less acid and yield a better-looking product.

Solid Particulate Adhesives Obtained by Spray Drying

There are many disadvantages to the use of water-soluble or dispersible base-catalyzed phenolic adhesives used for making waferboard, hardboard, etc., for example, instability on storage, expense of transportation, and detriment to the bonded product during the heating necessary to remove the water.

A. Berchem, K.K. Sudan and E.M. Gres; U.S. Patent 4,098,770; July 4, 1978; assigned to Reichhold Chemicals Limited, Canada have found that a range of phenol-aldehyde resins, hitherto known only as aqueous solutions or dispersions, can be obtained as a particulate solid by spray drying such an aqueous suspension or solution. It has been found that the capability of such resin solutions or dispersions being processable to a particulate solid by spray drying is markedly enhanced by the incorporation therein of a nonphenolic polyhydroxy compound.

In a preferred embodiment this process provides a method of preparing a phenol-formaldehyde resin in solid particulate form comprising reacting together, in an aqueous medium, phenol and formaldehyde, in a ratio of 1.0 mol of phenol to between 1.8 and 3.5 mols of formaldehyde in the presence of between 0.1 and 1.0 mol of a basic catalyst; continuing this reaction until a desired viscosity is reached; modifying the reaction system by adding thereto between 0 and 30%, based on the weight of the phenol originally taken, of a nonphenolic polyhydroxy compound; spray-drying the thus-modified liquid resin in a pressure nozzle dryer; and chilling the spray-dried particulate solid for storage.

The basic catalyst can be an alkali hydroxide, or carbonate, or an alkaline earth oxide or hydroxide. The preferred catalyst is sodium hydroxide, but barium hydroxide, calcium hydroxide and calcium oxide all produce usable resins.

The nonphenolic polyhydroxy compound can be chosen from glycols, such as ethylene glycol, propylene glycol, butylene glycol, and diethylene glycol; polyhydroxy ethers, such as the Voranols, and other polymeric hydroxy materials, for example the Carbowaxes; and polyhydroxy compounds such as glycerol, sorbitol, cane sugar, etc.

Internal bond strength (IB) or tensile strength perpendicular to the surface is determined by testing preconditioned 2 x 2 inch specimens by ASTM procedure D 1037.

The specimens are bonded to the loading blocks which are engaged in heads of the testing machine. The specimens are subjected to stress to failure with the load passing through the center of the specimens in a continuous fashion at a uniform rate.

Internal bond strength is calculated from the maximum load at failure per square inch. The location of the line of failure is also reported.

Viscosity was measured by means of a Gardner-Holdt Viscosity Comparator, to stated letter viscosity.

Example 1:   Resin A is prepared from 1 mol phenol, 2 mols formaldehyde (44% solution), 0.65 mol sodium hydroxide (50% solution) and enough water to make up 45% NV.

Phenol, formaldehyde, water and $\frac{1}{3}$ amount of base required were heated under reflux with agitation to 60°C and held at 60°C with cooling for 30 minutes. The balance of sodium hydroxide was added and the mixture held at 80°C to viscosity D. The resin was cooled rapidly for storage.

Example 2: Resin B is prepared by spray drying the above liquid Resin A at 45% solids to give a free flowing powder with no flow or cure but infinite solubility in water.

Example 3: Hardboard was prepared from both resins A and B. Resin B was reconstituted with water to 45% solids and then diluted further to 8% solids. The dilute resin solution was sprayed into a wood furnish slurry buffered at pH 4.5 at 2% resin level. The slurry was drained and the mat pressed at 350°F for 6 minutes to form hardboard. Hardboard was also prepared with original liquid resin A. Physical testing: A, IB 34 psi and B, IB 38 psi.

Use of Lignin Derivatives

K.G. Forss and A.G.M. Fuhrmann; U.S. Patent 4,105,606; August 8, 1978; *assigned to Keskuslaboratorio-Centrallaboratorium AB, Finland* have formulated an adhesive to be used in the manufacture of plywood, fiberboard, particle board, and similar wood products which comprises the combination of a phenol-formaldehyde resin and lignin derivatives, such as lignosulfonates or alkali salts of alkali lignins.

It has been unexpectedly discovered that the molecular weight of the lignin derivatives is of critical significance as far as the adhesive characteristics of the product are concerned; thus at least 65% and preferably over 70% by weight of the lignosulfonates should have molecular weights over that of glucagon, and at least 40%, and preferably over 45% by weight of the alkali lignins should have molecular weights in excess of that of glucagon as determined by gel chromatography, or to put it in another way, at least 55% and preferably 65% by weight of the lignosulfonates shall have molecular weights in excess of 5,000, and at least 35% and preferably over 50% by weight of the alkali lignins shall have molecular weights in excess of 5,000 as determined by gel chromatography calibrated by calibration substances with known molecular weights.

Lignosulfonates and alkali lignins with a molecular weight distribution required by the process, may be obtained, from spent sulfite liquor or black liquor, for example in a manner described in U.S. Patent 3,825,526, or by any kind of precipitation, or by ultrafiltration, or by any other method.

Example: Sodium lignosulfonates, of which 67% (w/w) had molecular weights exceeding 5,000, were used in the manufacture of the adhesive. A water solution containing 50% (w/w) of these lignosulfonates had a viscosity of >80,000 cp (Brookfield RVT viscosimeter, 50 rpm, 23°C). A water solution containing 10% of these lignosulfonates had a pH of 8.2.

160 grams of these lignosulfonates were dissolved in 240 grams of water. The solution was mixed with 600 grams of a resole-type phenol-formaldehyde resin, having a solid content of 40%. 10 grams of paraformaldehyde was added and the adhesive was mixed for 60 minutes. The viscosity of the adhesive was 216 cp at 23°C.

Five panels of plywood were made; five test pieces were taken from each of them for the determination of shear stress and five for the determination of shear stress after boiling. The results of the determinations are given in the following table (average values of 25 test pieces as well as confidence limits in 95% probability).

The panels fully met the requirements of the Finnish standard for exterior grade plywood, and have properties corresponding to those obtained when using commercial phenol-formaldehyde resin.

	Dry	**After Boiling**
Shear stress	35.1±1.7 kp/cm^2	22.2±1.2 kp/cm^2
Wood failure	94%	91%

Rapid Hardening at Room Temperature

A. Kameyama, Y. Matsunaga and H. Saito; U.S. Patent 4,107,119; August 15, 1978; assigned to Nippon Kasei Chemical Co., Ltd., Japan have produced an adhesive composition which is rapidly hardened at room temperature and has excellent water resistance.

The adhesive composition is composed of (a) a resinous composition produced by mixing an aromatic primary amine with an aqueous solution or emulsion containing a polymer or copolymer of acrylamide and (b) formaldehyde or a derivative thereof as main components.

When the adhesive composition of the process is used as an adhesive agent for plywoods, combined substrates, furniture etc., it is possible to add a filler such as clay, talc, kaoline, asbestos, woody powder, wheat starch, walnut shell powder, coconut shell powder etc.

In the application of the adhesive composition, it is necessary to coat the adhesive composition on the bonding surface in haste because it is rapidly hardened at room temperature. In order to prevent the gelation of the adhesive composition before the coating, it is possible to coat the resinous composition (a) on one surface and to coat formaldehyde or a derivative thereof or a condensate type resin containing a large amount of formaldehyde on the other surface being bonded.

The process may be illustrated by certain examples, wherein the terms of part and percent mean part by weight and percent by weight unless otherwise specified.

Example 1: In a reactor equipped with a stirrer, a thermometer, a reflux condenser and a nitrogen gas inlet, 159.7 parts by weight of deionized water, 2.0 parts by weight of a nonionic surfactant of polypropyleneglycol polyethyleneglycol ether (Pronon-208), 2.0 parts by weight of a nonionic surfactant of polyoxyethylenenonylphenol ether (Nikkol NP-18), 0.2 part by weight of potassium persulfate and 8.0 parts by weight of vinyl acetate were charged.

The reaction was carried out at 70°C for 30 minutes in a nitrogen atmosphere. An aqueous solution of 60.9 parts by weight of vinyl acetate and 14.2 parts by weight of acrylamide in 30 parts by weight of deionized water, was added dropwise to the reaction mixture at 70°C for 1 hour.

After the addition, the reaction was continued at 70°C for 5.5 hours and the reaction mixture was cooled to obtain an aqueous emulsion of a resin.

100 parts by weight of the resin was admixed with 0.072 mol of each of the following amines and then, a diluted hydrochloric acid was added to adjust pH as shown.

When 10 parts by weight of the resinous composition was admixed with 2 parts by weight of 37 weight % formalin at room temperature ($23°C$), the gelation times were as follows.

Type of Amines	pH	Gelation Time (sec)
Aniline	4.0	230
m-Toluidine	4.0	170
2,4-Tolylenediamine	4.0	195
o-Aminophenol	4.2	780
m-Aminophenol	5.0	160
p-Aminophenol	4.7	290
m-Phenylenediamine	4.8	360
2,3-Xylidine	3.7	250
p-Anisidine	4.7	190
4,4'-Methylenedianiline	4.8	160

Example 2: 22.9 parts by weight of m-phenylene diamine was dissolved in 100 parts by weight of 30 weight percent aqueous solution of polyacrylamide (degree of polymerization 530) and dilute hydrochloric acid was added to adjust the pH to 4.5 whereby a resinous composition was obtianed.

When 10 parts by weight of the resinous composition was admixed with 3 parts by weight of 37 weight percent formalin, the gelation time was 170 seconds.

The adhesive composition was coated on one surface of a lauan plywood having a thickness of 4 mm, a width of 25 mm and a length of 80 mm and the other lauan plywood was superposed on the surface to be parallel to the direction of the grain of wood. The superposed plywoods were compressed under 10 kg/cm^2 at room temperature ($23°C$) for 10 minutes.

The tensile shearing strength in normal state and the strength under cyclic boil test (according to Japanese Agricultural Standard) were measured after 24 hours from the release of the pressure. The adhesion strength (normal state) was 31.1 kg/cm^2, and the strength under cyclic boil test was 18.1 kg/cm^2.

In performing the cyclic boil test the specimen was submerged in boiling water for 4 hours and then dried at a temperature of $60°±3°C$ for 20 hours. It was again submerged in boiling water for 4 hours and then kept in the water until its temperature went down to room temperature.

Addition of Powdered Green Tea

Thermosetting phenolic resin adhesives may be prepared by reacting phenols, such as phenol or cresol, with aldehydes, such as formaldehyde or acetaldehyde, in a strongly alkaline medium. The phenolic adhesives are generally not compatible with ordinary acidic hardeners, and therefore require a higher curing temperature and a longer curing time in comparison with other thermosetting resin adhesives made from urea resins or melamine-urea co-condensed resins.

Such higher curing temperatures and longer curing periods for phenolic adhesives inevitably lead to lower production yields of adhered products such as plywood. As a result, the commercial applications of phenolic resin adhesives have been significantly limited despite their excellent properties such as good bonding strength and water resistance.

N. Nakamura, Y. Saeki and S. Nemoto; U.S. Patent 4,109,057; August 22, 1978; assigned to Sumitomo Durez Company, Ltd., Japan have found that the addition of powdered green tea to a thermosetting phenolic resin adhesive results in a significant improvement in the curing time of the resin as well as improved physical properties of the bonded products derived therefrom.

The formulation further provides a method for reducing the cost of phenolic adhesive compositions since the powdered green tea employed may be obtained as the inexpensive waste produced during the refining of crude tea in green tea production and which is usually discarded thereafter.

The phenolic resins employed are thermosetting condensation products, i.e., resoles, obtained by reacting one mol of a phenol and one to three mols of an aldehyde in the presence of an alkaline catalyst, such as sodium hydroxide. Examples of the phenols which may be used to prepare these phenolic resins are phenol and substituted phenols such as cresol, and the like, as well as mixtures thereof. Suitable aldehydes have 1 to 8 carbon atoms and include formaldehyde, formalin, paraformaldehyde, acetaldehyde, propionaldehyde, and the like, and mixtures thereof.

The powdered green tea suitable for use may be derived from a variety of sources, which include the powdered waste produced during the refining of crude tea, as well as powdered tea prepared by pulverizing green tea commercially available for tea brewing.

During the manufacture of green tea, the tea leaves are normally pulverized, in order to destroy the cellular fiber, and subsequently dried. The product, known as crude tea, is classified by particle size into several grades. The finest grades or sizes have the appearance of a powder and have little commercial value. Preferably, the powdered green tea will have an average particle size of less than about 50 mesh.

Example: Phenol and formaldehyde, in a 1:2 molar ratio, respectively, were charged to a reactor and reacted under reflux conditions at a pH of about 11 in the presence of sodium hydroxide as catalyst until the viscosity at 25°C reached about 1.5 poises, thereby producing a phenolic resin with the following properties.

Cure time	20 min/135°C
Nonvolatile content	41%
Compatibility with water	≥49 times at 25°C
Free formaldehyde	0.7%
pH	10.8 at 25°C, glass electrode
Viscosity	1.6 poises/25°C

100 parts of this phenolic resin were added at ambient temperature to 1-30 parts of powdered green tea, 90% of which was 50 mesh size or finer, prepared by pulverizing commercially available green tea. The curing times of these adhesives are shown in the following table, which also contains green-tea-free adhesive data as a point of reference.

Phenolic resin, parts	100	100	100	100	100	100
Green tea, parts	1	5	10	20	30	0
Cure time (min/135°)	19	17	15	12	9	20

Four plywood samples were prepared with these adhesives under the following conditions. A control sample was also prepared with an adhesive not containing green tea.

	. . . . . .Composition.				
	1	**2**	**3**	**4**	**Control Example**
Phenolic resin, parts	100	100	100	100	100
Green tea, parts	1	5	10	20	0
Walnut shell, parts	10	8	5	0	10

. . . . **Experimental Conditions**. . . . .

Composition:	Lauan veneer (thickness 2 mm, moisture content 7% or less) 3 ply
Spread:	36±1 g/30 x 30 cm
Cold press:	20°C, 10 kg/cm^2, 30 minutes
Hot press:	135±1°C, 10 kg/cm^2, 150 seconds*

*Hot press time per 1 mm thickness of single veneer is 25 seconds.

Bonding strengths of the plywood obtained are as follows (all figures are averages of ten measurements):

	. . . . .Composition				
	1	**2**	**3**	**4**	**Control Example**
Shear strength (kg/cm^2)*	2.5	5.7	12.1	11.9	0
Wood failure (%)*	0	15	48	52	0

*Boiled for 72 hours according to JIS, special class.

It is thus apparent that the addition of green tea to a thermosetting phenolic resin adhesive significantly shortens the curing time and improves the bonding strength in comparison with the control examples which do not contain green tea.

Bark-Extended Adhesives

A resinous adhesive system has now been discovered by *S. Hartman; U.S. Patent 4,113,919; September 12, 1978; assigned to Champion International Corporation* which has components that are wholly derived from trees, and which is formed in situ during the manufacture of adhesively bonded wood products utilizing such adhesive system.

A tree-derived, resinous adhesive composition can be provided by admixing a finely divided, alkali-treated wood bark, tannin and an aldehyde. The resulting admixture is subjected to condensation conditions to form a reactive bark-extended tannin-aldehyde resinous adhesive. Thus, in the case of the formation of plywood, the alkali-treated bark, tannin and aldehyde admixture may be applied to the surface of a plurality of wood plies, which are thereafter pressed under conventional temperature and pressure conditions in order to form a laminate. Under such conditions, the bark, tannin and aldehyde are caused to condense, in situ, and form a bark-extended tannin-aldehyde resinous adhesive.

Wood bark of various different species of trees may be utilized; however, the bark of coniferous trees is preferred. Thus, for example, bark from various pines may be employed and also the barks of western hemlock, Douglas fir, oak, western white fir, Sitka spruce, coast redwood, eastern hemlock, and the like.

The bark prior to use is reduced to a powder, having a particle size of 60-200 mesh and preferably 100 mesh. The term "tannin," as employed in describing this process, is intended to mean the plant tannins which may be obtained, for example, from the bark of trees by crushing the bark into small pieces and then washing and boiling the bark with water in order to extract the tannin. After the separation of insoluble matter, a resulting thick, reddish brown, viscous extract is evaporated, leaving crude tannin as a hard cake. The cake may then be purified by extracting with an alcohol-ether mixture.

Wattle tannin is a preferred tannin for use in the formula. Wattle tannin is extracted from the bark of the *Acacia mollissima* tree with water. The conventional method of preparing wattle is to disintegrate the wood and bark, and extract the bark and/or wood with water. The solution of wattle tannin and water is evaporated to 85% concentration and the concentrated wattle tannin is spray dried. Wattle tannin is the commercial catechol tannin or flavotannin product. The alkali used to treat the bark is preferably sodium or potassium hydroxide.

Suitable amounts of paraformaldehyde added to the weight of the bark-tannin composite are between 3 and 8%. The amount of finely divided bark of particles of 80 to 200 mesh, preferred size being 100 mesh, utilized is in the range of between about 10 and about 50, preferably between about 15 and 30% by weight of the tannin-aldehyde resin.

Example 1: 350 g of Ponderosa pine bark (100 mesh) is admixed with 200 g (50%) sodium hydroxide. An exothermic reaction develops. Enough water is slowly added to begin the leaching process; the exotherm continues for several minutes. As the exotherm diminishes, more water is added to a total of 750 g. The pH of this alkali bark slurry is adjusted to between 5.5 and 7, preferred pH being about 6.1.

1,000 g of above pH-adjusted alkali bark is added under stirring to 1,000 grams of wattle tannin solution (50%). To the resultant homogeneous blend there is added 100 g paraformaldehyde (catalyst). The ratio of paraformaldehyde to tannin (50%) solution amounts to 1:10. The paraformaldehyde is used at a preferred range of 5% of the tannin solution (50%). The broad range is from 3 to 10%.

The resulting admixture is used as the adhesive precursor in preparing plywood panels. The adhesive precursor is spread on both sides of the core of a three-ply panel at a rate of 60 to 85 lb/1,000 ft^2. The panel is placed in a press held at a temperature of 280°F (range 250°-280°F) while at a pressure of 180 psi (range 160-180) for a period of 6 minutes (range 4-7 minutes).

Example 2: 500 parts of a pH-adjusted alkali bark mixture (Example 1) was added to 1,000 parts of 50% solution of wattle tannin. To this was added under agitation, 50 parts wheat flour and 50 parts paraformaldehyde. Plywood panels were prepared under the same conditions as Example 1. Shear specimens were run on the prepared plywood samples. The results are given below:

| | . . . . . Press Conditions. . . . . . | | | . . . . Shear Strength . . . . | |
Adhesive	Temperature (°F)	psi	Time (min)	V/P, psi	% Wood Failure
Example 1	280	180	6	190	80
Example 2	280	180	6	220	85

V = Vacuum
P = Pressure

A number of processes for the use of bark as an extender for phenolic resin adhesives have been patented. However, they are all for use in bonding wood from western trees, such as Douglas fir and are not effective in veneers using southern pine, which are more difficult to bond.

In another patent by *S. Hartman and M. Ozkan; U.S. Patent 4,144,205; Mar. 13, 1979; assigned to Champion International Corporation* there is described a high quality phenolic resin adhesive for use in the production of plywood and the like, in which bark dust from a variety of trees such as the fir, pine, cedar, and hemlock, may be used as an extender, which has the capacity to resist almost unlimited exposure to moisture, heat, and weather, permits a press time on the order of seven minutes, and can be used to adhere veneers of the southern pine species as well as that of the Douglas fir species.

It has been found that a phenol-aldehyde resin having a low molecular weight, soluble in ethanol is required, that the bark particles must have a small particle size to make them reactive with formaldehyde (–100 mesh), and that the greater the concentration of the alkali used, the greater the bark-formaldehyde reactivity. Further, if the bark is treated with a surfactant prior to its alkali treatment, the formaldehyde reactivity of the southern pine bark increases, whereas the formaldehyde reactivity for western bark increases with time.

The formulas for two control phenolic resin adhesives are as follows.

	Formula 1 (parts by weight)	Formula 2 (parts by weight)
Water	500	665
Resin	1,231.8	300
Corn cob	450	350
Wheat flour	250	200
. . . . Mix 5 Minutes		
Caustic	170	130
. . . . Mix 20 Minutes.		
Resin	2,168	2,360
. . . . Mix 3-5 Minutes		
Total mix	4,700	4,000
Total resin	1,403	2,660
% Total solids	46.5	44.1
% Resin solids	29.8	28.5

As is evident from the above, two resin additions are required in the preparation of the control formulas. These resins are characterized as low molecular weight resins by gel permeation chromatograph (GPC) analysis. Both are soluble in all proportions in ethanol.

It has been discovered that the control phenol-aldehyde resin adhesive may be extended in two ways. First, a percentage of the phenol-aldehyde resin may be replaced by the same percentage of alkali-treated bark. Second, all of the corn cob used in the adhesive may be replaced by an equal amount of treated alkali bark while also reducing the percentage of phenol-aldehyde resin. It can be appreciated that such substitution results in a more economical adhesive.

Example: For the purpose of replacing 10 and 20% respectively of the phenol-aldehyde resin with an alkali-treated bark, a premix was prepared comprising bark, sodium hydroxide (50%) and water in a ratio of 1 to 1 to 2. The premix was prepared by adding a 50% sodium hydroxide solution to the bark, mixing the components, then adding the required amount of water.

By varying the percent phenol-aldehyde resin in the first resin addition in the control formulas, any desired viscosity of the subject adhesive may be obtained. By way of illustration, a series of adhesive mixes was prepared using one formula in which the percent of phenol-aldehyde resin in the first addition was varied. These formulations were prepared as follows, where amounts given are in parts by weight.

	45% of Resin in 1st Addition	50% of Resin in 1st Addition	65% of Resin in 1st Addition	80% of Resin in 1st Addition
Bark	73	73	73	73
Caustic (50%)*	60	60	60	60
H_2O	146	146	146	146
H_2O	146	146	146	146
Resin	316.34	351.6	457.5	562.4
Bark	44	44	44	44
Glue X	65	65	65	65
Caustic	13	13	13	13
Resin	386.65	351.6	245.6	140.7
Viscosity, cp**	9,000	8,000	4,200	2,500

*While sodium hydroxide has been used as the caustic in all of the examples,
  it is clear that any alkali metal hydroxide may be used.
**Viscosities determined at 26°C.

Adhesive mixes prepared as above, i.e., where 10 and 20% of the total phenolic-aldehyde resin solids portion of the control formulas were replaced by equal percentages of treated bark, were then used in preparing plywood (southern pine) panels under the following conditions: assembly time, 5 to 30 minutes; prepress, 172 psi, 3½ minutes; press time, 300°F, 200 psi, 7 minutes; and glue spread, 80 to 85 pounds per thousand square feet double glue line.

Shear samples of each panel were then subjected to standard boil and vacuum per pressure shear tests, and the shear values and percent wood failure recorded. The results when compared with the control showed that the replacement of 10 and 20% of the total phenolic resins solids portion with the alkali-treated bark in accordance with this process yielded an equally high quality adhesive. It should be noted that it is not necessary to add hot water or hot alkali to obtain the desired results.

Use of Lignosulfonate to Replace Some Phenol

The process developed by *G.G. Allan; U.S. Patent 4,127,544; November 28, 1978; assigned to Weyerhaeuser Company* is one for treating a lignosulfonate derived from an ammonia base waste sulfite liquor such that the modified ammonium lignosulfonate may be readily substituted for a substantial portion of the phenol ingredient in producing a useful phenol-formaldehyde adhesive. The improved process activates the ammonium lignosulfonate so that it reacts readily with an aldehyde and forms a copolymer of the lignosulfonate-phenol-aldehyde components. The process ensures that incorporation of ammonium lignosulfonate into the resin does not incorporate undesirable water solubility into the final product.

The improved process comprises heating a dried ammonium lignosulfonate residue derived from an ammonia base waste sulfite liquor that has been dissolved in a polar solvent such as phenol, resorcinol, aniline or dimethylformamide to a temperature of 150° to 300°C while under autogenous pressure. The reaction is continued until the resulting ammonium lignosulfonate-phenolic component of the reaction mass is insoluble in water but fully soluble in dilute aqueous sodium hydroxide solution. The reaction mass may, after washing with water to remove unreacted soluble sugars, then be condensed with an active aldehyde such as formaldehyde or furfural to yield a thermosetting copolymer useful as an adhesive that is soluble in dilute aqueous alkali hydroxide but insoluble in water and completely insoluble in water upon curing.

The ammonium lignosulfonate ingredient used in the described process is the residue recovered from drying to a powder the spent sulfite liquor from the ammonia-based sulfite digestion of cellulose-containing materials. The residue comprises principally ammonium lignosulfonate along with a quantity of wood sugars and their derivatives. It is generally necessary to remove these sugars, which may constitute up to 30% by weight of the liquor solids, from the ammonium lignosulfonates because they do not react when the sulfite liquor is treated with an aldehyde.

Allowing the sugars to be incorporated into an ammonium lignosulfonate-phenol-formaldehyde resin imparts a water solubility characteristic to the resin rendering it unsuitable for many purposes. Therefore, it is a major advantage of this process that treating the waste sulfite liquor with phenol at a relatively high temperature results in a phenol-ammonium lignosulfonate product that is insoluble in water while the unreacted sugars and their derivatives retain their water-soluble characteristics. A simple washing with water removes these sugars preventing their undesirable impact upon the final ammonium lignosulfonate-phenol-aldehyde condensation product.

Example: A mixture of oven-dry ammonium lignosulfonate (75 g, Crown Zellerbach's Orzan A) and phenol (150 g) was heated at 200°C under autogenous pressure (75 psi) for 3 hours. The black oil obtained (185 g) did not impart any coloration to water, was slightly soluble in ethanol and methylethylketone and completely soluble in dimethylformamide and 10% aqueous NaOH. The product (1.5 g) dissolved in 1.18% aqueous sodium hydroxide (8.5 ml) and treated with 37% formaldehyde solution gave a tough resinous material after heating for 4 hours on a steam bath.

Curable Amino Resin Binder System for Fiberboard

R.D. Sundie, W.R. Michael, and H.E. Ulmer; U.S. Patent 4,139,507; Feb. 13, 1979; assigned to Allied Chemical Corporation have formulated a one-component, stable, curable amino resin binder system comprising urea, formaldehyde, an oxime compound soluble in the binder system, a buffering agent and optionally a water-miscible C_1 to C_5 alcohol compound and/or melamine.

These binder systems are stable for as long as about one and a half months. They can be easily shipped as a one-component system for commercial use without the necessity of adding a catalyst to initiate curing.

The binder system comprises: (a) 25-50 weight percent urea; (b) 20-40 weight percent formaldehyde; and (c) 20-40 weight percent water, and has a pH of 6.5 to 7.5.

Examples 1 through 13: Various samples of amino resin binder solutions as shown in the table below, are prepared wherein various oxime compounds are used to improve the stability of the binder solutions.

Example	Oxime	Oxime	Component I*	Urea Solution**	Alcohol	Alcohol (wt %)
		. . . . (wt %)				
1	Acetaldoxime	4.65	51.94	41.54	ethanol	1.87
2	Hexanaldoxime	4.65	51.94	41.54	butanol	1.87
3	Formaldoxime	4.65	51.94	41.54	ethylene glycol	1.87
4	Crotonaldoxime	6.52	51.94	41.54	–	–
5	Hexenaldoxime	6.52	51.94	41.54	–	–
6	Benzaldoxime	4.65	51.94	41.54	–	–
7	Methyl ethyl ketoxime	6.52	51.94	41.54	–	–
8	Diethyl ketoxime	6.52	51.94	41.54	–	–
9	Glyoxime	6.52	51.94	41.54	–	–
10	Methyl glyoxime	6.52	51.94	41.54	–	–
11	Methyl ethyl glyoxime	6.52	51.94	41.54	–	–
12	Cyclohexanoxime	6.52	51.94	41.54	–	–
13	α-Hydroxy isobutyl aldehyde oxime	6.52	51.94	41.54	–	–

*Component I = 84 wt % UF Concentrate 85 (Allied Chemical), 5 wt % melamine, 10.7 wt % water, 0.3 wt % ammonium acetate.
**Solution of 50 wt % water and 50 wt % urea.

Examples 14 through 17: Four fiberboards were prepared by spraying the amino resin binder solutions, as set forth in the table below, onto mechanically reduced wood fibers until the particles become impregnated with the resin. The spray was regulated so that the resin solids content of the fiber was 8% of the oven dry (O.D.) weight of the fiber.

The impregnated fibers were formed into a mat. The mats for boards 14 and 15 were compressed for 1 minute between press platens heated to about 145°C using high-frequency heating. The mats for boards 16 and 17 were compressed for 5 minutes between press platens heated to about 145°C; high-frequency heating was not used on these three boards.

	. . . . Examples			
	14	15	16	17
Resin				
Component I, wt %*	51.92	51.92	51.92	51.92
Urea soln, wt %**	41.54	41.54	41.54	41.54
Oxime, wt %	4.67***	4.67†	4.67***	4.67†
Methanol, wt %	1.87	1.87	1.87	1.87
Catalyst	No	No	No	No

(continued)

	. . . . Examples.			
	14	15	16	17
Board				
Thickness, inch	0.471	0.472	0.482	0.476
Density, pcf	45.6	45.9	47.2	47.1
Modular rupture, psi	3,200	3,700	3,300	3,500
Internal bond, psi	56	97	62	53
After 24 hr immersion in water, %				
Water absorption	32.5	32.1	23.9	21.4
Thickness swell,	8.44	8.13	5.33	4.86
Linear expansion	0.40	0.60	0.21	0.38

*Component I is comprised of 84 wt % UF con. 85, 5 wt % melamine,
    10.7 wt % water and 0.3 wt % ammonium acetate.
**A solution containing 50 wt % urea and 50 wt % water.
***Acetoxime.
†Methyl ethyl ketoxime.

The data of the above table shows that fiberboards with acceptable physical properties can be produced with the amino resin binder systems containing an oxime compound.

For Joining Untreated Wood to Wood Treated with Fire Retardant

Many building codes now require that wood components utilized in construction, particularly doors, be treated with fire-retardant chemicals. Such fire doors, as they are known in the trade, generally comprise a fire-resistant core, a wood framing made of stiles and rails around the core, which stiles and rails have been treated with the fire-retardant chemicals and an untreated or fire-retardant-treated wood skin which is glued to the stiles and rails.

The primary object of the process developed by *C.N. Bye; U.S. Patent 4,141,745; February 27, 1979; assigned to National Casein of New Jersey* is to provide a relatively inexpensive, commercially practicable proteinaceous adhesive with improved bond strength and/or increased assembly time for gluing the aforementioned wood skins to the fire-retardant-treated door frame stiles and rails.

This object is achieved with an adhesive composition essentially of two major components—one a dry mix of the protein with solubilizing agents, fillers and preservatives added and the other, water, to which tetrahydrofurfuryl alcohol has been added, the dry protein mix being added to the aqueous tetrahydrofurfuryl alcohol mixture in proportions and for a time necessary to produce a homogenous adhesive mixture for cold press or hot press application to the surfaces of various substrates to be joined.

The preferred protein is casein although mixtures of casein, compatible soluble blood and soy protein may also be used as the protein source. The cold press application is preferred because of easier adaptation to assembly line production, but the adhesive also performs well in hot press applications.

Casein glue is made up of acid-precipitated casein in a fine granular state, 30 mesh and finer, mixed with suitable solubilizing agents. This glue is supplied dry and mixed at the time of use with water. The casein solubilizing agents generally consist of lime (calcium hydroxide) plus sodium salts, including any convenient sodium salt of an acid whose calcium salt is relatively insoluble, provided that it is not hygroscopic and will not react with the lime or casein as long as the mixture is kept dry.

It was found that an adhesive composition with good dry shear bond strength and/or increased assembly time comprises about 13.0 to 22.9% protein, preferably 13.0 to 18.7%, and about 1.0 to 20.0% tetrahydrofurfuryl alcohol, preferably 5.0 to 15.0%, by weight based on the wet mix.

Normal conditions for joining the substrates with the adhesive composition were 25 minutes cold press time and 1-3 days cure time prior to machining. At the lower protein content of about 13.0%, it was found advantageous to increase press time to one hour and allow up to seven days cure-out before machining.

Containing Sulfur

One of the disadvantages of the chemically reactive synthetic resin adhesives used in the plywood industry is that such adhesives have relatively long "curing" times. This disadvantage is especially troublesome because the press-curing in the manufacture of plywood is recognized as a natural bottleneck in a well-designed plant for continuous plywood manufacture. Curing temperatures from $80°-140°C$ are used, depending on the type and application of the plywood, for periods of 5-10 minutes under pressures chosen to bring about compaction, i.e., all plies being brought into intimate contact, and, above that, 4% compression.

An adhesive has been formulated by *H. Shapiro and R.N. Sanders; U.S. Patent 4,152,320; May 1, 1979; assigned to Ethyl Corporation* which has many of the advantages of the chemically reactive thermosetting resin adhesives, but which has a shorter curing time and greater resistance to bond strength erosion by moisture contact and which can be applied using conventional technology.

This cost-effective, fire-retardant, wood-binding adhesive composition contains as binder 30 to 60 weight percent of sulfur and 70 to 40 weight percent of a synthetic thermosetting resin. The resin is selected from urea resins, melamine resins, phenol resins, resorcinol resins and mixtures of these.

The adhesive also contains as the fire-retardant component 1 to 5 weight percent based on the total adhesive weight of a mixture of dicyclopentadiene, styrene and a member of the group consisting of maleic acid, alkyl mono- and diesters of maleic, fumaric and phthalic acid having from 4 to 8 carbon atoms in each alkyl group; brominated styrene; and triphenyl phosphate or methyl substituted triphenyl phosphate having from 1 to 3 methyl groups attached to the phenyl ring.

The process for bonding at least two wood layers using this adhesive includes the following steps: (a) contacting the adherent surfaces to be bonded, at least one of which has the adhesive applied to at least a portion thereof, under sufficient pressure to form a joint; (b) heating the joint to a temperature of about 112° to 150°C for a time sufficient to cure the synthetic resin adhesive binder and to melt the sulfur; and (c) cooling the joint to solidify the sulfur, whereby an effective bond is formed.

The sulfur which forms a part of the mixed adhesive composition of this formulation may be utilized in either liquid, e.g., molten, or solid form, preferably in a finely divided or powdered state. Elemental sulfur of commercial purity in either amorphous form or any crystalline solid form is preferred because of its ease of handling and ability to mix with the resin prior to adding the catalyst. From about 30 to 60% sulfur may be used in the total amount of adhesive binder. Preferably, 40 to 50 weight percent is employed.

Example: A mixed adhesive was prepared by melting in a flat-bottomed flask 100 parts of sulfur to which was added 3 parts of styrene and 3 parts of maleic acid. After addition, the mixture was stirred and then quick-cooled to provide a crystalline powder after grinding. Then 62.4 parts of a commercially available Weldwood Plastic Resin Glue, a phenol-formaldehyde resin, was combined with 37.4 parts of water and a thin paste was made. The sulfur powder containing styrene and maleic acid was then added to the paste along with 4.2 parts of dicyclopentadiene.

After thorough mixing in not more than 2 minutes the pasty mass was spread on 3" x 5" x 1/8" pine test plaques and assembled into a 4-ply stack. The assembled stack was placed in a press and loaded to 667 psig over a 10-minute period with the temperature of the press platen faces being held at 270°F (132.5°C). The pressure eased off over a period of 10 minutes to 400 psig at which the pressure was maintained for another 10 minutes.

During the pressing, sulfur was observed to melt and flow out between the plies. Finally, the pressure was released and the plywood sample cooled to ambient conditions. Penknife-type testing revealed fracturing in the wood phase. Therefore, strength of the ply bonding was judged equivalent to conventional plywood.

CORRUGATED BOARD ADHESIVES

For Increasing Production Rates of Heavyweight Board

The use of water-resistant corrugated paperboard for packaging has increased tremendously in the last several years. Accompanying this increased usage has been the demand for stronger and more specialized products to withstand more rigorous conditions. This demand has been met by the corrugated board industry by furnishing multiwall and heavyweight corrugated paperboard.

However, when such products are manufactured by the corrugators, the production rate is much lower than when single-faced or double-faced paperboard is being produced. Thus, the equipment is operating much less efficiently when such products are being made.

In order to provide for an adhesive which allows the production of heavyweight corrugated board at greater rates, *C.B. Musselman and E.M. Bovier; U.S. Patent 4,116,740; September 26, 1978; assigned to Anheuser-Busch, Incorporated* have formulated a water-resistant adhesive compound using a waxy starch as a component of a high solids carrier in a starch-based corrugating adhesive formulation.

The waxy variety starch can be derived from waxy maize, waxy sorghum, or any species yielding a high (95-100%) amylopectin content starch, and must be modified to make it less viscous so that the cooked solids portion can be increased without affecting the viscosity of the finished adhesive. Such chemical treatments include acid modification, oxidation, enzyme treatment, chemical derivatization, e.g., starch ethers or starch esters, and combinations thereof. Also, included are thermal modifications, such as dextrinizing and autoclaving. The waxy starch is modified to a fluidity of 25 or greater. The modified waxy starch has the following physical and chemical characteristics: inherent viscosity, 0.2 to 2 dl/g at 77°F and 0.5% concentration; solubility, 0.1 to 100% at 77°F; gelatinization temperature, 75° to 180°F; and alkaline fluidity, 25 to 90 ml measured at 25°C.

The percent of cooked or carrier starch solids in the final mix is between about 3 and 12% of the final adhesive mix and preferably is above 4%. Of this, 25 to 100% is gelatinized waxy starch solids. The amount of raw starch solids in the final mix is between about 10 and 35% of the final adhesive mix.

Example: This is an example of a one-tank mix using a ketone-formaldehyde resin (APC 6010 by Anheuser-Busch, Incorporated). In a mixer are placed 250 gallons of water at 120°F, 300 lb of 85 fluidity thin-boiling waxy maize starch, 150 lb of pearl cornstarch, and 68 lb of a 50% solution of sodium hydroxide. This mixture is heated to 145°F and agitated for 20 minutes. This treatment causes the starch to gelatinize.

To this mixture is added 7 lb of borax pentahydrate and 300 gallons of cold water. This is thoroughly mixed into the gelatinized starch mixture and 1,250 pounds of pearl cornstarch are added. The product is thoroughly mixed until smooth. To this is added 14 lb of borax pentahydrate and 60 lb of ketone-formaldehyde resin (APC 6010). These additives are thoroughly mixed into the starch. This starch adhesive when employed as a corrugating adhesive gives excellent adhesion and good water resistance.

In addition, the corrugator, which has a speed with normal water-resistant adhesives and heavy board of 280 ft/min, gives a rate of 350 ft/min with this adhesive using the waxy maize starch.

For Bonding Paraffin-Modified Corrugated Board

The trend in the packaging of fruits, vegetables, poultry, and fish is toward the use of the paraffin-modified corrugated boxes because of their good moisture resistance and low cost. However, paraffin-modified corrugated board is difficult to bond with current hot-melt adhesives and the boxes thus formed are not generally suitable for use at elevated temperatures ($>90°F$). Wooden boxes are too costly and the uncoated corrugated board has poor moisture resistance.

Paraffin-treated corrugated board is available in three types: (1) paraffin-coated, (2) paraffin-impregnated, and (3) paraffin-impregnated and coated. The degree of difficulty in bonding these modified boards with a hot-melt adhesive increases in the order listed.

Amorphous polypropylene provides a relatively good bond on paraffin-modified corrugated board if the bonds are aged for about 2 minutes. The long open time of unmodified amorphous polypropylene is also a problem on modern box forming machines. For example, amorphous polypropylene is quite tacky immediately after solidifying from the melt and it remains tacky for at least about 6 minutes. It does not become virtually nontacky until it has aged for about 3 hours. Therefore, the major deficiencies of amorphous polypropylene in the bonding of paraffin-modified corrugated board include the long open time and long tack time of the polymer as well as the weakness of the bonds at both elevated and low temperatures.

To overcome these deficiencies, *M.F. Meyer, Jr., and R.L. McConnell; U.S. Patent 4,120,916; October 17, 1978; assigned to Eastman Kodak Company* have found that the addition of about 10% of a low-viscosity crystalline polypropylene reduces the open time of amorphous polypropylene to a very low level; the addition of about 10% of a low-viscosity polyethylene such as Epolene C-16 imparts good bond strength at both low ($20°F$) and high temperatures ($120°F$). When all three polymer components are present, the blend has a low (desirable) open time (<3 seconds), will provide rapid bonding on paraffin-modified corrugated board over a broad temperature range ($50°-115°F$), and the bonds obtained are strong over a very broad temperature range ($20°-120°F$).

The low-density polyethylenes used are well-known in the art and can be prepared by the high-pressure polymerization of ethylene in the presence of free radical catalysts and have crystalline melting points (Tm) of about $95°$ to $115°C$. These polyethylenes have a melt viscosity at $190°C$ of about 100 to 30,000 cp and acid numbers of 0 to 60. These polyethylenes have a density of about 0.904 to 0.940 g/cm^3, most preferably about 0.906 g/cm^3. These polyethylenes can be unmodified polyethylene or polyethylene which has been degraded. Also, these polyethylenes can be degraded polyethylenes which have been oxidized or reacted with maleic anhydride or acrylic acid, for example. These polyethylenes can be used in amounts of about 3 to 20% by weight, preferably 5 to 12% by weight.

Suitable commercially available amorphous polypropylenes are the Eastman Eastobond M-5 type polyolefins. They are present in the adhesive formulation in an amount preferably of from 76 to 90% by weight.

The crystalline, hexane-insoluble polypropylenes or propylene-containing co-polymers may be prepared by the method given in U.S. Patent 3,679,775, and can be used in amounts of from 3 to 20% by weight of the adhesive composition, preferably about 10%.

The adhesives should contain from 0.25-1.0% by weight of antioxidant such as Irganox 1010 in combination with Cyanox 1212 (American Cyanamid). The adhesive compositions are prepared by blending together the components in the melt at a temperature of about $180°$ to $230°C$ until a homogeneous blend is obtained, approximately 2 hours. Various methods of blending materials of this type are known in the art and any method that produces a homogeneous blend is satisfactory.

These components blend easily in the melt and a heated vessel equipped with a stirrer is all that is required. For example, a Cowles stirrer provides an effective mixing means for preparing these hot-melt adhesive compositions.

Example: 199 g amorphous polypropylene (melt viscosity = 3,000 cp at $190°C$, glass transition temperature = $-17°C$), 25 g crystalline thermally degraded polypropylene (melt viscosity 2,100 cp at $190°C$, Tm = $155°C$), 25 g polyethylene (melt viscosity 3,075 cp at $190°C$, Tm = $102°C$, crystallization temperature is $86°C$, and an acid number of 4), and 0.25 g Irganox 1010 and 0.87 g Cyanox 1212 antioxidants are blended in a resin pot at $200°C$ with stirring under nitrogen for one hour. This blend has a melt viscosity of 2,290 cp at $190°C$.

A thin film (1 to 2 mils thick) of molten polymer at $190°C$ is spread on a Teflon plate at room temperature. The time the film remains tacky to touch is defined as open time on Teflon. Films of this blend are not tacky to touch after one second.

Test specimens of paraffin-impregnated and coated corrugated board (2" x 5") are folded $180°$ in the middle of the specimens perpendicular to the direction of the flutes. This configuration provides strong "spring back" forces on the bond immediately after the compression force is released. Bonds are made with this blend at temperatures of $73°$ and $115°F$. The melt temperature of the blend is $190°C$ and an open time of $\leqslant 0.5$ second is used. A compression time of only one second is required to make a bond at $73°F$ and only three seconds are required to form a bond at $115°F$.

Paraffin-impregnated and coated corrugated board specimens (2 x 6.5") are bonded with this blend at a melt temperature of $190°C$, open time of 0.75 second, compression time of 1.0 second and compression force of 40 psig using an adhesive bead width of 70 mils along the entire length of the specimen (6.5").

Before the bonding step, the test specimens are cut so that half of the test specimens have the flutes parallel to the long axis and half have the flutes crosswise to the long axis. Thus in each individual bond comprised of two pieces of paraffin-modified board, the flutes are parallel to the long axis in one piece and crosswise to the long axis in the opposite piece. The bonds are aged 24 hours and equilibrated at $30°$, $73°$, and at $120°F$ for 4 hours before testing by slowly pulling them apart by hand. These bonds provide good bond strength with partial fiber tear at $30°F$, are very strong and provide fiber tearing bonds at $73°F$ and provide good bond strength with a substantial amount of fiber tear at $120°F$.

Based on a Vinyl Alcohol Polymer

J.N. Coker; U.S. Patent 4,131,581; December 26, 1978; assigned to E.I. Du Pont de Nemours and Company has formulated an adhesive composition consisting essentially of:

(1) 2 to 35% by weight of a polymer selected from the group consisting of polyvinyl alcohol and ethylene/vinyl alcohol copolymers containing at least 50 mol percent vinyl alcohol;

(2) 10 to 80% by weight of at least one solvent for the abovementioned polymer. This solvent is crystalline at $40°C$ and is selected from the group consisting of urea, solid alkyl substituted ureas containing a total of up to 9 carbon atoms, thiourea, biuret, ϵ-caprolactam, solid aliphatic amides containing up to 6 carbon atoms, solid polyhydric compounds and ammonium carboxylate salts.

(3) 5 to 80% by weight of at least one viscosity reducing diluent selected from the group consisting of water, liquid polyhydric compounds, liquid alkyl substituted ureas having up to 9 carbon atoms in the alkyl group, liquid aliphatic amides containing up to 6 carbon atoms, and dimethyl sulfoxide; and

(4) 0 to 60% by weight extender selected from the group consisting of starch, dextrin, clay, silica, carbon black, talc, calcium carbonate, barium sulfate and vinyl polymer latices.

It is necessary that when the viscosity-reducing diluent is present in an amount of above 45% by weight the adhesive composition is a suspension and the polymer is polyvinyl alcohol which has been subjected to a heat treatment at a temperature of $70°$ to $190°C$ dispersed in a liquid solvent comprising 42 to 100% by weight of methanol, 0 to 13% by weight of water, and 0 to 45% by weight of a solvent chosen from the group consisting of 2 to 5 carbon monohydric alcohols, esters, ketones, ethers, hydrocarbons and chlorohydrocarbons in which the PVA is insoluble. This reduces the cold water solubles content of the polyvinyl alcohol while maintaining its solubility in $80°$ to $100°C$ water.

These adhesives containing at least 50 mol percent vinyl alcohol can be applied by usual, coating, spraying or hot melt techniques and are useful for bonding many sorts of cellulosic materials. Some examples of their use are as follows.

Example 1: A mixture of 400 g of urea and 400 g of water was placed in a Brabender mixer and was agitated. As soon as most of the urea dissolved, 200 g of polyvinyl alcohol (Elvanol 90-50, degree of hydrolysis 99.0-99.8%, 4% aqueous viscosity 12-14 cp at 20°C, from Dupont) was added as rapidly as possible, producing a thin slurry. Heat was applied. As the temperature of the solution increased progressive thickening of the mixture was observed, its viscosity reaching a maximum at 40° to 50°C. As heating was continued, the mixture thinned out, clarified and became quite fluid at a temperature of 80° to 90°C (viscosity of 2,000 cp); this operation required 15 to 20 minutes. The mixture was placed in a polyethylene bag and chilled to below 5°C to induce gelation. The gelled product was chopped and stored in a moisture-tight container to prevent loss of water.

Example 2: Using a Cowles dissolver, 1,500 g of urea and 200 g of water were heated to 125° to 135°C with agitation. To the molten mass was added 500 g of polyvinyl alcohol (Elvanol 90-50) at as fast a rate as possible. Agitation was continued until the mixture was fully melted and had clarified. At this point it was poured onto Teflon sheeting where it immediately crystallized. The resulting slab was broken up and pulverized. The finely divided product was either used in this form or was compression molded into cartridges for application by hot melt gun. The product was stored in a moisture-tight container.

Example 3: A 6" roll coater assembly equipped with a heated and jacketed pan was used for applying the melted adhesive to the applicator roll. This equipment was provided with a dual feed of kraft paper rolls such that a laminated structure could be produced. The experiment was carried out by first melting the adhesive composition of Example 1 in a separate jacketed reservoir and holding it at a temperature of about 85° to 95°C. Once the paper feed had been started, operating at 20 ft/min, the melted adhesive was pumped into the heated applicator pan at the rate at which it was applied to the paper. In this manner, a structure in which two sheets of kraft paper were bonded with 1 to 2 mil adhesive was produced. The adhesive could be applied smoothly without gaps and no undesirable adhesive buildup occurred on the applicator roll as the run proceeded. At the completion of the experiment the equipment was cleaned up easily by hot water washing. The bonded structure exhibited excellent high humidity and ambient temperature water resistance. It was readily repulpable in hot water.

Example 4: In this experiment a 25" corrugator was used with the applicator roll heated to 85°C. Approximately 1 pound of the blend of Example 1 was melted and was placed on this roll where it was quickly spread into a continuous coating on the roll by a doctor blade assembly. A single-faced corrugated structure was passed over the roll at about 100 ft/min such that the tips of the flutes were wetted by the adhesive. The second facing was then applied and the resulting structure passed through several pressure rolls before being cut about 10 to 12 feet down the line. The adhesive exhibited excellent rheology on the machine as long as its water content was maintained. Adequate "green strength" had developed in the structure by the time it reached the collector station and fiber tear was observed after the structure had been allowed to age 3 to 5 minutes.

The corrugated structure exhibited little or no undesirable "ribbing" (in direct contrast to structures prepared with conventional corrugating adhesives). Equipment cleanup at the end of the trial was relatively simple due to the ease of removing adhesive buildup from the applicator roll and also because of the ready solubility of the adhesive in hot water. Examination of the resulting corrugated structure indicated excellent performance for the adhesive. The dry pin adhesion (corrugated flute to liner bone) proved to be superior to starch (e.g., 85-100 lb vs 50-70 lb) at about one-half to two-thirds the loading (e.g., 1.5 lb vs 2.0-3.0 lb per 1,000 ft^2 board).

The corrugated board bonded with the adhesive of this example exhibited excellent ambient temperature water resistance in the ply separation test, fiber tear being observed after 10 days' immersion. In contrast starch bonded structures delaminated after 1-2 minutes' immersion. The polyvinyl alcohol-urea bonded structure was readily repulpable in hot water.

Addition of Silica to Improve Thermal Stability

If glycolic acid polymers are to be used for hot melt adhesives in the corrugated paper manufacturing industry, they must meet certain specifications for "creep resistance" (resistance to loss of stability at elevated temperatures) which they have not yet been able to do.

R. Bacskai; U.S. Patent 4,156,676; May 29, 1979; assigned to Chevron Research Company has found that, if minor amounts of silica are incorporated into hot melt adhesive compositions comprising a glycolic acid polymer, the creep resistance of the adhesive compositions is significantly increased. In fact, adhesive compositions which failed for use in the manufacture of corrugated paper have been improved by the incorporation of a minor amount of silica to the point where they easily pass the creep resistance requirements of that industry. The amount of silica necessary need only be from 0.1 to 10% by weight.

Examples 1 through 4: Four silica-modified glycolic acid copolymers were prepared by mixing 1, 2, 4, and 5% by weight of silica and a previously prepared glycolic acid copolymer. The silica, Cab-O-Sil M-5, was obtained from the Cabot Corporation. It is reported to have a nominal particle size from 0.007 to 0.014 micron and a surface area of 200 to 400 m^2/g.

The glycolic acid copolymer was prepared by mixing 371 grams of glycolic acid (4.875 mols) 16.75 grams of diglycolic acid (0.125 mol), 7 grams of ethylene glycol (0.0115 mol), and 2.1 grams of trimethylolethane (0.0175 mol) in a 500 ml flask equipped with a short distillation head, a nitrogen capillary, and an oil bath. The mixture was heated to 218°C for one hour at atmospheric pressure, two hours at 150 mm and 2 hours at 0.3 mm mercury. 271.12 g of product were removed from the flask. 265.80 g of the recovered product was ground into fine particles, and heated at 150°C for 96 hours at 0.3 mm mercury. 221.74 g of polymer were recovered. The polymer had a molecular weight of about 13,000.

The four silica-modified copolymers were prepared by mixing about 40 grams of copolymer with from 0.4 g (1%) to 2.0 g (5%) of silica in a Brabender mixer. Mixing was carried out at about 220°C for 2.5 minutes at a mixer speed of 180 rpm. Each of the four copolymers passed an accepted creep resistance test.

For comparative purposes, a sample of the glycolic acid copolymer which was not modified by silica and samples of the copolymer modified by the incorporation of 1% talc, 1% graphite, 1% calcium carbonate, 5% talc, 5% graphite, and 5% calcium carbonate were also tested. None of these compositions passed the creep resistance test.

Example 5: In this example, silica-modified glycolic acid copolymer was prepared in situ, by mixing 74.2 g of glycolic acid (0.975 mol), 3.35 g of diglycolic acid (0.025 mol), 1.4 g of ethylene glycol (0.023 mol), 0.42 g (0.0035 mol) of trimethylolethane, and 0.50 g of Cab-O-Sil M-5 in a 100 ml flask equipped with a short distillation head, a nitrogen capillary, and an oil bath. The mixture was heated at 218°C for one hour at atmospheric pressure, one hour at 150 mm, and one hour at 0.3 mm mercury. The recovered product was ground and heated at 150°C for 96 hours at 0.3 mm mercury. The resulting polymer had a molecular weight of about 34,000. The polymer, comprising 1.8% by weight silica, easily passed an accepted creep resistance test.

Examples 6 and 7: Two silica-modified glycolic acid homopolymers were prepared by mixing 1% by weight of silica and a previously prepared glycolic acid homopolymer. The silica was Cab-O-Sil M-5.

The glycolic acid homopolymer was prepared by heating 304.2 g of glycolic acid at a temperature of 218°C for one hour at atmospheric pressure, one hour at 150 mm, and one hour at 0.3 mm mercury. The product was ground and heated at 150°C for 96 hours at 0.3 mm mercury.

The two silica-modified homopolymers were prepared by mixing each of two 40 g aliquots of the homopolymer with 0.4 g of silica. Both samples easily passed an accepted creep resistance test.

Starch Carrier Composition

Corrugated paperboard consists of sheets of flat and corrugated paper, bonded together with adhesive. It is commonly made by (1) passing a sheet of paper, which is referred to as the medium, between fluted rolls, usually heated, to form corrugations; (2) applying an adhesive to the tips of the corrugations, known in the art as flutes, on one side of the medium; (3) bringing a flat sheet, which is referred to as the liner, in contact with the adhesive-coated flutes, and (4) bonding the two sheets by the application of heat and pressure.

In the interest of production efficiency, it is desirable to operate the corrugator at the fastest possible rate, usually from about 300 to 700 ft/min depending primarily on board weight. Because the production rate is limited by the speed of formation of the adhesive bond, corrugating adhesives must be capable of rapid increases in viscosity so that the bond may be made quickly. Starch paste is the preferred adhesive in the industry, although adhesives based on other inexpensive materials, such as flour, are also used.

The adhesive paste is generally formulated in two portions. The primary mix, or carrier, is an extremely viscous solution of gelled (dissolved) starch. The secondary mix contains raw, i.e., ungelled, starch plus additives to expedite the bond formation and to give desired special properties to the paste. When the two portions of the adhesive are combined, the carrier holds the raw starch particles in suspension and imparts sufficient viscosity to the mixture to permit its application by conventional techniques.

The adhesive bond is established by heating the paste in situ so that gelation occurs when it is sandwiched between the flute tips and the liner. At the gel temperature, the raw starch component dissolves and absorbs water, causing a rapid increase in the viscosity of the adhesive. Thus, the carrier starch and raw starch act as co-binding agents in joining the medium to the liner.

Ordinary, or domestic, starch pastes do not produce water-resistant bonds. In order to attain water resistance, modifiers such as thermosetting resins are added to the finished starch pastes. Although acid-cured urea-formaldehyde, melamine-formaldehyde, melamine-urea-formaldehyde resins provide the best water resistance, they are not commonly used because they require higher temperatures to gel. In starch paste technology, sodium hydroxide, or a similar alkaline material is used to reduce the gel temperatures of the pastes to about $140°$-$155°F$ so that they will set rapidly during bond formation. When acidic conditions are employed, the gel temperatures of the pastes are about $155°$-$165°F$.

The carrier portion of the typical adhesive is generally prepared by cooking a mixture of starch, water, and sodium hydroxide. Sodium hydroxide, which is known in the art as caustic, reduces the gel temperature of the starch in proportion to the concentration of alkali in the solution. Thus, the caustic which is added to the primary mix reduces the gel temperature of the carrier starch and, to a lesser degree, that of the raw starch in the finished paste. However, excessive caustic can degrade the starch and the cellulose in the paperboard, discoloring and weakening the finished product. This degradation may also induce an unstable viscosity in the starch paste. Finally, caustic is difficult to handle safely.

Carriers containing caustic are particularly unsuitable for acid-curing resin systems. Since these pastes develop maximum water-resistance in an acidic environment, the caustic-derived carrier must be neutralized before the resins are added. When the caustic is neutralized, however, its beneficial effect on the gel temperature of the raw starch is lost, and the high gel temperatures of these pastes make reductions in corrugator speeds necessary. Furthermore, if corrugating plants make boards with both alkaline and acidic pastes, the entire adhesive-mixing apparatus has to be thoroughly cleaned during changeover to prevent glue bond failure during the transition, and, in some cases, clogging of the glue system.

These hindrances to productivity sparked the development of the alkaline water-resistant pastes. The alkaline pastes are not as water resistant as the acidic pastes, but when the liner boards themselves have only a limited resistance to water, the alkaline adhesives produce adequate bonds. However, these alkaline adhesives are incompatible with the acidic resin systems which are incorporated in rigid-when-wet corrugated boxes and have become a weak link in the box structure.

A. Sadle and T.J. Pratt; U.S. Patent 4,157,318; June 5, 1979; assigned to International Paper Company have, therefore, developed a formulation which uses urea as the gelatinizing agent for the carrier starch. They have found that the quantity of urea required to gel the carrier at or near room temperature will also reduce the gel temperature of the raw starch in most corrugating adhesives to a value in the optimal temperature range.

In accordance with their formulation, the carrier portion of the adhesive is composed of 1 part starch, about 3-5 parts water, and 0.5 to 4, preferably 1 to 3, or most preferably 2 to 2.5 parts urea. First, the starch is mixed with some or all of the water to form a stirrable slurry; about 2 parts water are usually sufficient for this purpose. The urea is then added, along with any remaining water, and the solution stirred to insure uniform distribution of the urea. Gelatinization will take place almost immediately. The gelled carrier is then diluted, if necessary, with enough additional water to make a pourable solution, and added to a secondary mix containing about 3 to 8 and preferably about 4 to 6 parts raw starch as well as enough water to give an overall starch concentration of about 15 to 35% by weight.

The urea-based carrier here described is compatible with all common adhesive additives, including urea-formaldehyde, melamine-formaldehyde, melamine-urea-formaldehyde, resorcinol-formaldehyde, phenol-formaldehyde, and ketone-aldehyde resins, polyvinyl acetate, and copolymers of vinyl acetate and ethylene. It reduces the gel temperature of a paste independent of its pH. Using this carrier, water-resistant corrugated board may be produced at about the same rate as domestic corrugated board. In plants which produce both varieties of board, this carrier may be incorporated in both paste formulations, reducing waste and eliminating the time and cost of cleaning the adhesive system during changeover.

The adhesive produced by this process has a gel temperature within the range of about 140°-155°F. It has a very short texture, which is advantageous for economical application levels. If a stringier texture is desired, it may be produced in accordance with prior art methods, for example, by briefly cooking the primary mix or by adding borax to the secondary mix. Since urea does not degrade the carrier starch as does caustic, paste made with this carrier is more viscous than a caustic paste containing the same amount of starch.

Example 1: 450 lb of 50% aqueous urea solution is added to a slurry containing 225 lb of water and 100 lb of pearl starch. This mixture, the carrier, is agitated until clear, about 15 minutes. The gelled carrier is diluted with 500 lb of water and then added to the secondary mix, which was prepared by mixing 1,125 lb of water, 500 lb of pearl starch, and 16 lb of borax.

Example 2: A standard domestic paste with a caustic-derived carrier is made for comparison with the paste in Example 1. The carrier is made by adding 16 lb of caustic dissolved in 42 lb of water to a slurry made by mixing 415 lb of water with 105 lb of pearl starch. The carrier is agitated for about 15 minutes and then diluted with 250 lb of water. The diluted carrier is then added to a secondary mix which was prepared by mixing 1,500 lb of water, 500 lb of pearl starch, and 16 lb of borax.

Table 1 compares the properties of the urea-based adhesive discussed in Example 1 with those of the caustic-derived adhesive discussed in Example 2. It will be seen that the only significant difference is in the pH of the two adhesives, the urea-based paste being considerably less alkaline.

Table 1

Property	Urea	Caustic
Starch solids (%)	18.1	18.7
Viscosity (Stein-Hall sec)	35	40
Gel temperature (°F)	142	145
pH	9.0	12.5

The two pastes were run on the same corrugator at commercially acceptable speeds using identical board components. Comparative board properties are listed in Table 2.

Table 2

Property	Urea	Caustic
Short column crush (lb/in)	36.2	39
Flat crush (psi)	37.2	55
Pin adhesion, dry (lb/5 in²)	67.4	68

PAPER ADHESIVES

Glue Stick

W.A. Pletcher and R. Wong; U.S. Patent 4,066,600; January 3, 1978; assigned to Minnesota Mining and Manufacturing Company supply a formulation for friction-activatable adhesive compositions containing a linear polyester, a tackifier and, optionally, a plasticizer therefor. These compositions can be used as adhesive sticks by rubbing the end of a stick thereof (conveniently the stick is about 1 centimeter square in cross section and about 8 centimeters in length and is wrapped with paper, foil, or other similar material) rapidly in a back and forth or circular motion on a surface to be bonded, and then pressing a second surface against the sticky coating thus formed. Thus, the adhesive is applied by scribbling as with a child's crayon.

A second surface is pressed against the first within a short time after the rubbing operation is completed (usually within 3 minutes thereof) and a firm bond is formed within a few minutes thereafter. Various materials, such as paper, fabrics, leather, light wood, foils, and the like can be conveniently and durably bonded in this way. The compositions can be easily removed from clothing, furniture, etc.

These solvent-free adhesive compositions comprise a mixture of:

(1) 25 to 95 parts by weight of a thermoplastic linear polyester having crystallizable ester units of the formula:

$$-\overset{O}{\overset{\|}{C}}R_1\overset{O}{\overset{\|}{C}}-OR_2O-$$

where R_1 consists of divalent radicals remaining after removal of the carboxyl groups from one or more saturated aliphatic dicarboxylic acids and/or aromatic dicarboxylic acids, R_1 containing from 2 to 8 carbon atoms when it is an aliphatic radical and 6 to 13 carbon atoms when it is an aromatic radical, R_2 consists of divalent radicals remaining after removal of the hydroxyl groups from one or more saturated aliphatic diols containing from 2 to 12 carbon atoms, the polyester having a DTA melting temperature of from about $40°$-$75°C$ and an inherent viscosity of at least 0.2 dl/g at $25°C$, as measured in a 0.5 g/dl solution of polymer in chloroform at $25°C$,

(2) 5 to 50 parts by weight of a tackifier for the polyester having the ability to lower its melting temperature and to raise its glass transition temperature,

(3) 0 to 25 parts by weight of a plasticizer for the polyester having the ability to lower its melting temperature and glass transition temperature and to reduce its cohesive strength.

The sum of (1), (2) and (3) is 100 parts by weight resulting in a mixture with the following properties:

(a) a softening temperature of at least $40°C$,

(b) a write-on factor of at least about 0.06,

(c) an open time from about $1/3$ to 10 min,

(d) peel adhesion in the open state of not less than 50 g/cm, and

(e) peel adhesion in the closed (solid, nonbondable) state of not less than about 100 g/cm.

Example 1: Detailed Polyester Preparation — A 3-neck flask is fitted with a mechanical stirrer, a Dean-Stark trap-condenser, a thermometer, and a gas inlet for maintaining an inert atmosphere within the flask. The following are charged in the flask: 58.4 parts sebacic acid, 41.6 parts 1,4-cyclohexanedimethanol, and 0.1 part Irganox 1010 (antioxidant).

Inert gas is introduced into the flask and the contents of the flask are brought to 170°C by means of a heating oil bath. The mixture is stirred and held at this temperature for about 3 hours. During this time, water resulting from the condensation is collected in the trap. The temperature of the mixture is maintained at 170°C, while the pressure is then reduced from about 5 to 0.25 mm Hg. These conditions are maintained for about one-half hour to remove additional volatile material. About 0.1 part of tetrabutyltitanate catalyst is then added while maintaining the inert atmosphere. The temperature and pressure of the mixture are brought to 180°C and 0.15 mm Hg and these conditions are maintained for approximately 1.5 hours.

The polymer solidifies to a tough, flexible, colorless, opaque material having a melt viscosity of 1,200 cp at 175°C, an inherent viscosity of 0.27 dl/g, a DTA melt temperature (T_m) of about +46°C, a glass transition temperature (T_g) of –48°C and an acid number (or value) of 13.

Example 2: Detailed Adhesive Preparation — To 82.6 parts of the polyester of Example 1, at 180°C, and under an inert atmosphere, is added 13.2 parts of Isoterp 95 tackifier with mixing for approximately 5 minutes. 4.2 parts of dioctyl phthalate and 0.8 part of titanium dioxide pigment are then added with mixing for about 10-15 minutes and the composition is poured from the reaction vessel and allowed to solidify at room temperature.

This compatible material is tough and flexible and has a melt viscosity of 1,000 cp at 175°C, an inherent viscosity of 0.24 dl/g, a glass transition temperature (T_g) of –38°C and an acid number of 10.4. This adhesive has a softening temperature of 42°C, a write-on factor of 0.1112, an open time of 6 minutes, a peel adhesion in the open state of 250 g/cm and in the closed state of 245 g/cm.

A different formulation for a solid adhesive "glue stick" is described by *N. Araki, Y. Yotsuyanagi, S. Nagasawa and T. Okitsu; U.S. Patent 4,073,756; February 14, 1978; assigned to Konishi Co., Ltd., Japan* and comprises:

(a) an N-fatty acid acylated amino acid or a salt thereof;

(b) a compound having in its molecule a polyoxyalkylene structure containing at least four oxyalkylene units;

(c) a water-soluble or water-dispersible polymer having adhesive properties; and

(d) water or a mixture of water and one or more organic solvent(s) and/or one or more plasticizer(s).

Examples of useful neutralized fatty acid acylated amino acids are monosodium and disodium N-stearoylglutamates. Most preferred examples of the N-fatty acid acylated amino acids are those consisting of a saturated fatty acid moiety having 14 to 18 carbon atoms, e.g., myristic acid, palmitic acid, stearic acid, etc., and an amino acid moiety, i.e., aminodicarboxylic acid, e.g., aspartic acid, glutamic acid, etc., such as N-myristoylaspartic acid, N-palmitoylaspartic acid, N-stearoyl-aspartic acid, N-myristoylglutamic acid, N-palmitoylglutamic acid, N-stearoyl-glutamic acid, and the alkali metal salts of these acids.

Of the gelling agents used, free N-fatty acid acylated amino acids are most pre-ferred, and mono-alkali-metal salts thereof such as the monosodium, mono-potassium, or monolithium salt.

The proportion of the N-fatty acid acylated amino acid(s) and/or the salt(s) thereof used as the gelling agent(s) in these adhesive compositions is about 2 to 30% by weight, preferably about 4 to 20% by weight, of the total amount of the adhesive composition.

A water-soluble or water-dispersible polymer having adhesive properties is used as the adhesive component. Various known natural, semisynthetic, or synthetic polymers may be used as such polymers. Polymers having a mean molecular weight of about 5,000 to 2,000,000 preferably 10,000 to 1,500,000 are gener-ally used in this process.

Typical examples of the synthetic polymer are copolymers or homopolymers having a carboxyl group or a carboxylic anhydride group, where the proportion of the carboxyl and/or carboxylic anhydride group containing monomer is pref-erably 10 to 100 mol %, more preferably 30 to 100 mol %, based on the total mols of monomer(s) used, e.g., a copolymer of vinyl methyl ether and maleic anhydride, polyacrylic acid, a copolymer of ethylene and acrylic acid, polyvinyl acetate, polyvinyl methyl ether, polyvinyl alcohol, etc. Examples of the semi-synthetic polymers used in this formulation are cellulose derivatives such as methyl cellulose, ethyl cellulose, carboxymethyl cellulose, hydroxymethyl cellu-lose, etc., and starch derivatives such as carboxymethyl starch, etc. Examples of natural polymers used are starch and dextrin. These polymers are required to have adhesive properties and to be water-soluble or water-dispersible.

The proportion of the adhesive component or polymer blended in the adhesive composition of this process is generally about 5 to 50% by weight, preferably 15 to 35% by weight, of the total amount of the adhesive composition.

In this process, water or a mixture of water and an organic solvent is used as the solvent. The solid adhesive composition has a pH value of about 2 to 7, prefer-ably about 2 to 5, which can be easily handled without being accompanied by the problems in conventional solid adhesives, is excellent in shape stability at use, can be easily and uniformly spread over the adherent surfaces of papers, etc., and shows excellent adhesive properties.

Example: Into a 200 ml 4-necked flask equipped with a stirrer, a reflux condenser, and a thermometer were charged 57.0 g of water and 10.0 g of Pluronic L-64 and, after further adding to the mixture 25.0 g of a 1:1 molar copolymer of vinyl methyl ether and maleic anhydride having a mean molecular weight of 250,000 and 8.0 g or N-stearoylglutamic acid with stirring, the resultant mixture was heated to 80°C. The mixture was further stirred for one hour at 80°C to provide a homogeneous starch-like mixed solution. The product was poured into a lipstick-type container having a capacity of 5 ml and then allowed to cool to room temperature, whereby the product gelled after 5 minutes. In addition, the pH of the adhesive obtained was 3.6.

For Temporary Attachment of Coupons to Cartons

It is common practice to package, within cartons for food products, coupons of various sorts, and this is usually done for promotional purposes. It is desirable to assemble the coupon and the carton at the point of manufacture of the latter, because this saves the step of subsequent insertion of the coupon, which would normally have to be performed by the food packer. To enable assembly at the point of carton production, it is necessary that the coupon be adhered to the carton, and this must be done with a very specialized adhesive.

More particularly, the adhesive must be sufficiently tacky in the wet state to maintain the coupon and the carton at the time of initial assembly, and to withstand relatively high machine speeds; reducing production rates to accommodate coupon attachment is, of course, undesirable.

The adhesive must also exhibit sufficient strength in the dry state so that the coupon does not become displaced or detached during subsequent setting-up operations for the carton, and during insertion of the contained product. On the other hand, the material must possess such release properties as will permit the consumer to remove the coupon without damage to it or the carton. Finally, after removal the adhesive should exhibit low levels of tack, so that the consumer can collect a number of coupons in a stack; adhesion under such circumstances would be a source of considerable annoyance.

D.C. Wiesman; U.S. Patent 4,071,491; January 31, 1978; assigned to American Can Company has formulated an adhesive which satisfies these requirements and which comprises an aqueous dispersion of isobutylene rubber, aliphatic hydrocarbon wax having a melting point of at least about 130°F, and a lower polyolefin. The dispersion contains, on a weight basis, about 45 to 55 parts of the rubber, and 5.2 to 2.8 parts of the wax, and about 4.7 to 2.5 parts of the polyolefin, sufficient water being included to provide 100 parts of the formulation. The dispersion will have a viscosity of about 2,500 to 6,000 centipoises.

In preferred embodiments, the formulation will also contain about 0.14 to 0.5 part by weight of a thickening agent, and normally, it will contain about 35 to 45 parts by weight of water. A particularly preferred formulation includes about 49 parts of the rubber, 4.0 parts of the wax, 3.6 parts of the olefin, and 0.225

part of thickener, the viscosity of the dispersion being in the range of about 3,000 to 4,000 centipoises. The rubber utilized will most advantageously be a copolymer of about 97 weight percent isobutylene and about 3 weight percent of isoprene, the polyolefin will most advantageously be polyethylene, and the wax will most advantageously be a paraffin.

Finally, the rubber will desirably have an average particle size of about 0.3 micron, with about 95% of the particles thereof falling within the range of 0.1 to 0.8 micron.

Example: In this example all parts are by weight. An adhesive formulation was prepared by admixing 78.5 parts of Exxon Corporation's Butyl Latex 100; 1.5 parts of Borden's Polyco 296 W as a thickener; 10 parts of wax dispersion, Velvetol 77-07 by Quaker Chemical Corp.; 10 parts of a polyethylene called Velvetol 77-18, also from Quaker; and about 0.2 part Diamond Shamrock's defoamer Foamaster VF.

When four dots of this adhesive were used to attach a parchment paper coupon to various kinds of coated and uncoated paperboard stock, machine trials showed that it resulted in sufficient adhesion in the wet state so that carton-making operations could be carried on at full speed without the coupon becoming detached. In all instances release was good, with no damage to either the coupon or the substrate. The coupon surface was nonblocking, and substantially free from tacky residue.

Delayed-Tack Adhesives

Delayed-tack adhesive compositions, as known in the art, are nontacky at room temperatures, become adhesive or tacky by application of heat, and remain adhesive or tacky for an interval after heating is discontinued. Such compositions have found widespread use as adhesive films and as coatings on substrates, such as paper, plastic sheet and metal foil, in the production of, for example, labels and tapes.

Recently, a requirement for delayed-tack adhesives which work at very low temperatures has become more widespread. A typical application of this would be a label coated with a delayed-tack adhesive which is required to adhere well to a package at very low storage temperatures. At such temperatures, e.g., -20° to -40°C, previously proposed activated delayed-tack adhesive films, although they do not recrystallize, do become very rigid and consequently the adhesive quality, or tack, of the layer is much reduced.

N.K. Henderson and E. Thomson; U.S. Patent 4,091,162; May 23, 1978; assigned to Smith & McLaurin Limited, Great Britain have developed a delayed-tack adhesive composition comprising particles which have a soft and tacky polymer core surrounded by a hard and nontacky polymer shell in admixture with a solid modifier, such as a solid plasticizer.

The delayed-tack adhesive composition can be prepared by a method comprising forming, in a liquid, a dispersion of particles of the soft and tacky polymer, forming around each of the particles a shell of the hard and nontacky polymer, providing in the dispersion particles of the plasticizer and removing the liquid at a temperature below the melting point of the plasticizer.

The hard and nontacky polymer shell makes the particles resistant to blocking when incorporated in the unactivated adhesive composition. On activation the plasticizer dissolves the shell and releases the core polymer, whereby the advantages which are derived from the use of a soft and tacky polymer are obtained.

A typical delayed-tack composition in dispersion is in the form of a latex containing polymeric particles, the centers or cores of which consist predominantly of molecules of a polymer which is soft and tacky. An example of such a soft, tacky polymer is poly(2-ethylhexyl acrylate). The outer layers or shells of the particles consist predominantly of molecules of a polymer which is hard and nonblocking. An example of such a polymer is polystyrene.

An example of a polymerization reaction found to give a suitable core-shell polymer dispersion is given below.

	Parts
Initial Charge	
Water	200
Sodium bicarbonate	0.65
Lankropol K.M.A. (a sodium dialkyl sulfosuccinamate in 60% aqueous solution)	5.25
Monomer 1	
2-Ethylhexyl acrylate	89
Styrene	5
Monomer 2	
2-Ethylhexyl acrylate	25
Styrene	25
Acrylic acid	2
Monomer 3	
2-Ethylhexyl acrylate	2
Styrene	50
Acrylic acid	2
Initiator/Emulsifier Solution 1	
Water	50
Lankropol K.M.A.	23.2
Potassium persulfate	0.4
Initiator/Emulsifier Solution 2	
Water	30
Potassium persulfate	0.2
Ethylan H.A. (nonylphenol polyglycol ether)	2.0

The production procedure is as follows. The initial charge is heated to 85°C while stirring. 10% of the initiator/emulsifier solution 1 is added. Monomer 1 is added dropwise over a period of 2 hours. The rate is controlled so that the monomer reacts instantaneously, i.e., the reaction rate is controlled by the rate of monomer addition. The remainder of the initiator/emulsifier solution 1 is added simultaneously with monomer 1 at a rate such that the total amount is added over a period of 3¼ hours, i.e., the total time for the addition of monomers 1 and 2. The temperature is lowered to 70°-75°C and monomer 2 is added over a period of 1¼ hours, again so that the monomer reacts immediately. After the addition of monomer 2 is complete, the temperature is increased to 80°-85°C for thirty minutes. The temperature is lowered to 70° to 75°C. Monomer 3 and initiator/emulsifier solution 2 are added over a period of 1¼ hours, such that monomer 3 reacts immediately. After all monomer 3 has been added, the temperature is increased to 80°-85°C and the reaction continued for one hour.

The reaction product is a stable polymer latex of 43% solids content which will be called "polymer latex A." In this, the polymer prepared from monomer 1 forms particle cores which have around them outermost shells of polymers prepared from monomer 3. It is possible that an intermediate layer of polymer which had been prepared from monomer 2 may be present in some of the particles.

Example: The following delayed-tack formulation was provided:

	Parts
Water	41.2
Dicyclohexylphthalate (modifier)	40.2
Polypale Ester 10	9.6
Polymer latex A	38.0

The first three components were ground in a ball mill or pebble mill with a dispersing agent until a suitable particle size was obtained and then the fourth component was added. The resulting aqueous dispersion was coated on normal label base paper in a manner known in the art and dried at a temperature below the melting point of the dicyclohexylphthalate. In this way the adhesive was not activated. The dry coating weight was 25 g/m^2.

The paper with the delayed-tack adhesive composition coating was compared with a conventional delayed-tack label paper specially formulated for low temperature performance.

The conventional deep-freeze labels gave poor to fair adhesion with no fiber-tear, whereas the adhesive labels prepared by the method of this Example gave 20 to 100% fiber-tear.

Pressure-Sensitive Adhesive Tapes and Labels

Block copolymers have been formulated in the past to produce a number of types of adhesive compositions. The basic patent in this field, U.S. Patent 3,239,478 shows combinations of block copolymers with tackifying resins and paraffinic extending oils to produce a wide spectrum of adhesives. It is particularly advantageous to employ extending oils in the adhesive composition since extending oils are less expensive than the block copolymers, and in addition, typically improve the tack of the adhesive.

However, a standard adhesive comprising a block copolymer, tackifying resin, and paraffinic extending oil does not have adequate adhesion to porous substrates such as paper and cardboard.

G.T. Coker, Jr.; U.S. Patent 4,097,434; June 27, 1978; assigned to Shell Oil Co. found that a particular extending oil having a relatively low saturates content can be employed in adhesive compositions and can be very useful in bonding porous substrates.

The adhesive comprises:

(a) 100 parts by weight of a block copolymer having at least two monoalkenyl arene polymer blocks A and at least one elastomeric conjugated diene block B, the blocks A comprising 8 to 55% by weight of the copolymer;

(b) about 50 to 200 parts by weight of a tackifying resin compatible with block B; and

(c) about 10 to 100 parts by weight of a plasticizer having a saturates content of less than about 15% by weight, this plasticizer being a rubber compounding oil.

The saturates content of the specific plasticizer employed herein is very critical. At a plasticizer saturates level of less than 15%, the holding power increases dramatically to levels that are commercially acceptable. This finding is particularly surprising in view of the prevalent prior art practice of employing only those plasticizers having a saturates content of over about 50%.

The block copolymers employed in the formulation are thermoplastic elastomers and have at least two monoalkenyl arene polymer end blocks A and at least one elastomeric conjugated diene polymer midblock B. The number of blocks in the block copolymer is not of special importance and the macromolecular configuration may be linear, graft or radial depending upon the method by which the block copolymer is formed.

Typical block copolymers of the most simple configuration would have the structure polystyrene-polyisoprene-polystyrene and polystyrene-polybutadiene-polystyrene. A typical radial polymer would comprise one in which the diene block has three or more branches, the tip of each branch being connected to a polystyrene block.

Other useful monoalkenyl arenes from which the thermoplastic (nonelastomeric) blocks may be formed include alpha-methylstyrene, tert-butylstyrene and other ring alkylated styrenes as well as mixtures of the same. The conjugated diene monomer preferably has 4 to 5 carbon atoms, such as butadiene and isoprene. A much preferred conjugated diene is isoprene.

The block copolymer by itself is not tacky or sticky. Therefore, it is necessary to add a tackifying resin that is compatible with the elastomeric conjugated diene block. A much preferred tackifying resin is a diene-olefin copolymer of piperylene and 2-methyl-2-butene having a softening point of about 95°C, available commercially as Wingtack 95.

The amount of tackifying resin employed varies from about 50 to 200 parts per hundred rubber (phr), preferably, between about 50 and 150 phr.

At saturates levels above 15% by weight, the adhesion of the composition to porous substrates is not commercially acceptable for masking tapes. A standard masking tape by which to base commercial acceptability has a crosslinked-natural rubber adhesive having a holding power of at least 35 minutes.

The saturates content of the plasticizer is determined by clay-gel analysis where the procedure utilized is according to ASTM test D-2007. The clay-gel analysis also determines the percent by weight of asphaltenes, polar compounds, and aromatics. Preferably, the aromatics content of the plasticizer should be above about 55% and the combined polar compound plus aromatics content, above about 85%.

The plasticizers employed are rubber compounding oils. Rubber compounding or extending oils are well known in the art, and are typically lube oil extracts. For this formulation, these oils must have a saturates content of less than about 15% by weight and should, preferably, have a specific gravity of about 0.9 to 1.1 and a viscosity at 212°F of about 80 to 1,000 SSU. Commercially available rubber compounding oils of the foregoing description include Dutrex oils, particularly numbers 739, 896, 898, and 957 (Shell).

The amount of rubber compounding oil employed is also critical to the adhesive characteristics of the adhesive. The amount of oil varies from about 10 to about 100 phr, preferably about 40 to 80 phr.

The adhesives may be prepared by either blending block copolymer, oil and tackifying resin in a solvent, such as toluene, and removing the solvent by a steam stripping operation or they may be prepared by merely mixing the components at an elevated temperature, e.g., at about 150°C. In addition, if desired, the adhesive compositions may be cured, for example, by known irradiation techniques.

A preferred use of the formulation is in the preparation of pressure-sensitive adhesive tapes by a method such as that disclosed in U.S. Patent 3,676,202 or in the manufacture of labels. The pressure-sensitive adhesive tape comprises a flexible backing sheet and a layer of the pressure-sensitive adhesive composition coated on one major surface of the backing sheet. The backing sheet may be a plastic film, paper or any other suitable material and the tape may include various other layers or coatings, such as primers, release coatings and the like, which are used in the manufacture of pressure-sensitive adhesive tapes. One unique property of this adhesive composition is its excellent adhesion to porous, as well as nonporous substrates. Typical porous substrates include cardboard, paper, foamed polyurethane, and foamed-open cell polystyrene.

Containing Triethanolamine Plasticizer to Prevent Paper Curl

Water remoistenable adhesive-coated papers are well known in the art. They are commonly referred to as gummed labels or tapes. These gummed products are commonly obtained by preparing a solution of a water-soluble gum such as animal glue or dextrin, applying a thin film thereof to paper, and evaporating the solvent to give a dry, nontacky, potentially adhesive material on paper or other substrates.

Practical applications for adhesive-coated papers and products manufactured in this manner are limited by the tendency of the gummed sheets or tapes to curl or roll-up. Not only does this become a problem in the application of the adhesive product to a desired substrate, but printers cannot properly run these adhesive-coated papers in their presses. The curl effect is due to the differing response of the adhesive coating and the paper backing on exposure to changing humidity conditions.

W.R. Lawton; U.S. Patent 4,113,900; September 12, 1978 have found that the objectionable characteristics observed with flat gummed products prepared by known methods can be overcome by a very simple process of manufacture. A finely ground gummed adhesive is dispersed in an organic solution containing an aliphatic or aromatic hydrocarbon solvent; a material selected from the group consisting of polyfunctional amines, alcohol amines, or polyhydric aliphatic alcohols; and sufficient secondary solvent such as an alcohol to cause the polyols, polyamines, or polyalcohols to form a homogeneous solution with the hydrocarbon solvent.

This dispersion, after equilibrium has been reached, is coated onto paper or other suitable substrates by the usual solvent coating methods such as by reverse roll or by wire-wound rod coaters and dried at temperatures ranging from ambient atmospheric temperatures to over 149°C. The resultant product is a sheet with a potentially adhesive coating of fine particles tightly bound to the substrate and also partially bound to one another. This adhesive-coated product remains essentially flat when exposed to atmospheric humidity conditions varying from less than 20% to essentially 100% relative humidity. The activation time for obtaining tack is excellent and the adhesive properties are essentially equivalent to those obtained with the unmodified adhesives applied to the same coating weight.

Example: 540 parts toluene, 162 parts methanol, 60 parts triethanolamine and 400 parts animal glue (150 mesh). The toluene, methanol, and triethanolamine were placed in a tank and the mixture highly agitated for 3 hours after addition of the powdered glue.

This coating solution was added to the coating pan and stirred continuously during the coating operations. A 60" wide Black-Clawson reverse roll coater was used to apply the coating at a running speed which was varied from 100 to 500 ft/min and a coating weighing 6 to 8 lb/3,000 ft^2 (dry) was applied to a 35 lb bleached sulfite base paper. Drying temperatures in the 50 foot oven were varied from 121° to 204°C. A satisfactory adhesive-coated product was obtained during all operating conditions. Paper from this run was printed both by offset and letterpress without any problems due to curl or dusting of the coated product.

Adhesives Usable in Products Which May Be Repulped

There is currently a tendency towards use of hot melt adhesives in the manufacture of paper reinforcing tapes, books, sanitary napkins and diapers. In order to be fully satisfactory as a paper adhesive, a hot melt adhesive must be capable of supply in a handleable form for convenient application as a hot melt, and must provide bonds of satisfactory strength, flexibility and heat resistance. Desirably also, the composition of the adhesive will be such that when paper materials bonded with the adhesive are subjected to a repulping operation in order to reclaim the paper, the composition will not adversely affect the repulping of the paper.

I.E. Fakla and R. Grote; U.S. Patent 4,129,539; December 12, 1978; assigned to USM Corporation have found that an adhesive composition which may be applied as a hot melt to provide good adhesive bonds to cellulosic materials, and which can be rendered adhesively ineffective in copious quantities of water, for example in a repulping operation involving mastication with cold water of pH 6 to 9, may comprise polymeric binder material which is insoluble in cold water and which has free carboxyl groups, and sufficient aliphatic amino alcohol reactive in the presence of water with the carboxyl groups to promote solubility or dispersibility of the polymeric material in cold water.

The preferred amino alcohols are secondary or tertiary amino compounds in the molecules of which there are not less than two lower aliphatic residues attached to the nitrogen atom, each including a hydroxyl group. Examples of these materials are diisopropanolamine, diethanolamine, triethanolamine, or N-methyl diethanolamine, and mixtures thereof. From amino alcohols having two hydroxyl groups per molecule, diisopropanolamine and diethanolamine are preferred. The amino alcohol may be used in quantities of about 2 to 10, more preferably about 3 to 6, parts by weight per 100 parts by weight of the composition.

A preferred polymeric binder material is a vinyl acetate-crotonic acid copolymer having from about 2 to 10% crotonic acid, a most preferred material having about 2% by weight carboxylic acid groups. Pentaerythritol ester of colophonium having an acid number in the range of 196 to 212 may also be used, in which case the solubilizing effect required is somewhat less because such polymers tend to be more soluble in cold water. These materials preferably have melt viscosity values in the vicinity of 3,000 cp at 175°C and a softening point of 105°C (ring and ball).

In a preferred group of adhesive compositions intended to have resistance to high humidity and good flexibility of bonds, it is advantageous to include a substantial proportion of the vinyl acetate-crotonic acid copolymer in the composition. In a composition intended for bonding nonwoven disposables, e.g., sanitary napkins and diapers, 2 to 10 parts by weight amino alcohol, a mixture of 0 to 20 parts by weight hydroxypropyl cellulose of 10 to 10,000 cp (2 % by weight solution in water) and 30 to 55 parts by weight vinyl acetate-crotonic acid copolymer are used, and as placticizer a mixture of 5 to 55 parts by weight polyethylene ether glycol having a MW of 10,000 to 20,000 and 5 to 10 parts by weight triethylene glycol. These compositions exhibit excellent flexibility and can be applied as extremely thin beads from the melt at high speed.

In a preferred group of adhesive compositions intended for making reinforced paper tapes for packaging, 2 to 10 parts by weight of the amino alcohol per hundred parts by weight of the composition, and 10 to 65 parts by weight pentaerythritol ester of colophonium per 100 parts by weight of the composition are used, together with an ethylene-vinyl acetate copolymer (containing 28 to 40% by weight vinyl acetate and having a melt index of 20 to 300) or an ethylene-ethyl acrylate copolymer (containing 18% by weight ethyl acrylate and of a ML of 5 to 20); preferably these copolymers are used in admixture in a ratio weight of 1:1 to provide from 10 to 30 parts by weight of the composition.

In these compositions a plasticizer mixture of 10 to 30 parts by weight polyethylene ether glycol having a molecular weight of 10,000 to 20,000 and 10 to 20 parts by weight dibutyl phthalate or butyl benzyl phthalate per 100 parts by weight of the composition may be used. These compositions exhibit high adhesion and cohesion, a low viscosity on the order of less than 6,000 cp at 175°C (measured on a Rotovisco), solubility in cold water (below 20°C) under repulping conditions, low weight loss when held at 175°C over two hours, and a comparatively short open time. The illustrative compositions were made by mixing materials according to the formulations given in this table.

| | . . . . Illustrative Composition . . . . | | | | | |
Component	First	Second	Third	Fourth	Fifth	Sixth
Colophonium ester of pentaerythritol	42	43	–	–	15	16
Vinyl acetate-crotonic acid copolymer	–	–	40	42	41	41
Hydroxypropyl-cellulose	–	–	–	5	–	–
Polyethylene ether glycol 20,000 MW	16	17	40	42	25	25
Ethylene-vinyl acetate copolymer	12	12.5	–	–	–	–
Ethylene-ethyl acrylate copolymer	12	12.5	–	–	–	–
Diisopropanolamine	–	–	–	5	–	–
Triethanolamine	6	5	5	–	4	3
Venetian turpentine	–	10	–	–	–	–
Dibutyl phthalate	11.5	–	–	–	–	–
Calcium carbonate	–	–	10	–	10	10
Triethylene glycol	–	–	5	6	5	5
Antioxidant	0.50	0.50	0.50	0.50	0.50	0.50

The ethylene-vinyl acetate copolymer used was Elvax 250 by DuPont. The ethylene-ethyl acrylate copolymer used was Copolymer DPDB 6169 by Union Carbide. The colophonium ester of pentaerythritol used in the first illustrative composition was Resin B106 by Hercules. The colophonium ester of pentaerythritol was Pentalyn 255 by Hercules. The hydroxypropyl cellulose used was Klecel by Hercules. The vinyl acetate-crotonic acid copolymer used was Mowilith CE5 by Hoechst.

The first and second illustrative compositions were found particularly suitable for production of adhesive paper tapes by a process in which the adhesive is melted in a trough of a reverse roller coater, applied to paper, and then a loose reinforcing web of nylon or polyester applied into the adhesive layer.

The third and fourth illustrative compositions were found particularly suitable for use in bonding paper to paper, and paper to nonwoven materials for non-woven disposables for example napkins and diapers. The composition exhibited good thermal resistance (less than 4% weight loss when held at 175°C for 2 hr) under application conditions using an extrusion gun fed from a bulk applicator, and an ability to draw thin beads at high application speeds up to 75 m/min relative speed between the workpiece and the applicator nozzle. The bonds had excellent flexibility, and good adhesion and cohesion. The compositions had resistance to high humidity and yet good dispersibility on immersion in cold water.

The fifth and sixth illustrative compositions were found particularly suitable for bookbinding using a machine having roller applicators for applying adhesive to the spines of books at about 175°C, and an adhesive reservoir maintained 20°C below application temperature. These compositions were found to have good adhesion and cohesion, a sufficiently short open time to allow production of more than 8,000 books per hour, and excellent thermal resistance under application conditions (less than 4% weight loss over 2 hours at 175°C).

Suitable for Bookbinding or Making Paper Bags

Recently, hot-melt adhesives have been developed as non-solvent-type adhesives which permit high-speed processing and which have excellent adhesive properties suitable for mass production. Hot-melt adhesives are mainly used for processing paper goods which can be collected after use and reclaimed. However prior hot-melt adhesives as are commonly used have serious defects from the viewpoint of the economics of collecting and reclaiming the used paper because they do not dissolve or disperse in water, and, therefore, are difficult to reclaim or they deteriorate to lower the commercial value of the reclaimed paper.

Therefore, *M. Sumi and J.-i. Suenaga; U.S. Patent 4,140,668; assigned to Unitika Ltd., Japan* have formulated water-soluble or -dispersible adhesives which have softening points of 100° to 150°C, which melt at about 180°C, which have a melt viscosity of 1,000 to 8,000 cp at 180°C, which do not evaporate when they are heated, and which have excellent heat stability.

These adhesive compositions comprise polyvinyl alcohol having about a 30 to 60 mol percent residual acetate group content and are produced by dissolving polyvinyl acetate having a degree of polymerization of about 60 to 200 in methanol, the amount of the methanol ranging from about the stoichiometric amount necessary to obtain a polyvinyl alcohol having the desired degree of hydrolysis to about 2 times as much as the abovedefined stoichiometric amount, and carrying out alkaline hydrolysis and, if desired, a suitable amount of a plasticizer and/or an ethylene-vinyl acetate copolymer can be added thereto.

These hot-melt adhesive compositions can be used for bookbinding, for sealing corrugated paper, cartons, etc.; they do not block the screens or adhere to a metal wire screen roll in paper manufacturing equipment, nor spot in reclaimed paper goods, and thus effectively economize paper resources.

The plasticizer preferred for use in the adhesive is a polyethylene glycol having an average molecular weight of 400 to 600. It is used in an amount of from 5 to 35 weight percent, based on the polyvinyl alcohol (PVA). The ethylene-vinyl acetate copolymer, used in an amount of 15 to 35% based on PVA, should have a melt index of about 400. Also advantageously used is about 1 weight percent of phosphorous acid based on the PVA.

The adhesive compositions made as described easily dissolve in water at 20° to 30°C. They soften at 120° to 140°C, melt at about 160°C and have a melt viscosity of 1,000 to 6,000 cp at 180°C.

Hot Melt Bookbinding Adhesive

In the binding of paperbacked books, the adhesive is applied from a wheel applicator to the back of signatures which have been collected, trimmed, and notched. Then the paperback is attached before the adhesive has solidified.

In the binding of hardbound books, a polyvinyl acetate latex primer is applied to the trimmed and notched signatures before application of the adhesive. The purpose of the primer is to provide improved page pull and page flex values for the bound book. The primers are brittle at 20°F. Many hardbound books are put through a rounding process in which the back of the books is given a semiround shape. Therefore, it would be an advance in the state of the art to provide a hot-melt adhesive system which will withstand the rounding process for hardbound books and provide good page pull, page flex, and low-temperature flexibility for both paperback and hardbound books.

M.F. Meyer, Jr. and R.L. McConnell; U.S. Patent 4,140,733; February 20, 1979; assigned to Eastman Kodak Company have found that a multicomponent blend comprised of a maleated polyethylene, wherein all or part of the polyethylene component is maleated, an ethylene/vinyl acetate copolymer (or ethylene-alkyl acrylate copolymer), and a tackifying resin provides good page pull and page flex values on either paperbacked or hardbacked books. These compositions have a melt viscosity low enough to be used in a wheel applicator and they are flexible as well as nonbrittle at temperatures substantially below 20°F.

It was found that the addition of about 15% of a tackifier such as Eastman Resin H-130 and about 10% of an ethylene/vinyl acetate copolymer containing about 15 to 30% vinyl acetate to selected maleated polyethylenes provide an adhesive having good flexibility at temperatures below 20°F and melt viscosities below 10,000 cp. Books bound with this blend have good page pull (≥4.0 lb/in) and good page flex (≥300) values on either primed or unprimed books.

The low density polyethylenes are well known in the art and can be prepared by the high-pressure polymerization of ethylene in the presence of free radical catalysts and have crystalline melting points (Tm) of about 95° to 115°C. These polyethylenes have a melt viscosity at 190°C of about 100 to 30,000 cp and acid numbers of 0 to 60. They have a density of about 0.904 to 0.940 g/cm³, most preferably about 0.906 g/cm³. These polyethylenes can be unmodified polyethylene or polyethylene which has been degraded. Also, these polyethylenes can be degraded polyethylenes which have been oxidized or reacted with maleic anhydride or acrylic acid, for example. These polyethylenes can be used in amounts of about 45 to 85% by weight, preferably about 55 to 75% by weight.

The ethylene/vinyl acetate copolymers used in the process may contain from about 10 to 35 weight percent vinyl acetate but the preferred range is from about 18 to 30 weight percent. The melt index of these ethylene copolymers is preferably in the range of about 100 (melt viscosity 82,000 cp at 190°C) to about 200 (melt viscosity 40,000 cp at 190°C).

Example 1: 611.1 grams of low density polyethylene having a melt viscosity at 190°C of 3,075 cp, a saponification number of 4.0 and Tm of 102°C (Epolene C-16); 373.7 grams of low density polyethylene having a melt index of 220, melt viscosity of 41,000 cp at 190°C and a Tm of 100°C (Epolene C-13); 224.2 grams of a hydrocarbon tackifier having a ring and ball softening point of 130°C (Eastman Resin H-130); 149.5 grams of ethylene/vinyl acetate copolymer having a melt index at 190°C of 169 and a vinyl acetate content of 28% (Elvax 220); and 4.7 grams of Cyanox 1212 and 1.4 grams of Irganox 1010 antioxidants are blended in a 5-liter round-bottom flask at 200°C under nitrogen with stirring for three hours. This blend has a melt viscosity of 5,900 cp at 190°C in the Tinius Olsen Melt Indexer.

A 24-mil compression-molded film of this blend conditioned at 20°F for 8 hours does not break or crack when manually flexed. This composition is used to bind books in a Sulby Minabinda at a melt temperature of 350°F. Books bound on this binder give page pull values of 5.3 lb/in on a Collins Page Pull Tester at a crosshead speed of 4 in/min and page flex values of greater than 300 at 60 cpm with a 2 pound load in a Collins Page Flex Tester. These books did not break in the binding when the books were opened on a flat surface at a temperature of 20°F.

This example shows that a combination of the polyethylene, tackifier, and ethylene/vinyl acetate copolymer, as prepared above, provides the low melt viscosity needed for use on a wheel applicator while maintaining good adhesive properties and good page pull and page flex values and good low-temperature properties.

Example 2: The procedure of Example 1 is followed except that the books to be bound are first primed with a polyvinyl acetate-based emulsion and dried (about 3 mil layer) before the hot-melt adhesive (about 18 mil layer) is applied to the book. The primed and adhesively bonded books provide page pull values of 5.3 lb/in and page flex values of greater than 300. The polyethylene-based hot-melt adhesive forms a strong bond to the primer and does not separate from the primer when the books are opened.

SPECIAL PROCESSES

For Producing Urea-Formaldehyde Adhesives

P.M. Puig; U.S. Patent 4,090,999; May 23, 1978 has developed a process for making urea-formaldehyde adhesives with a standard solids content of 65% by directly reacting the ingredients and eliminating the necessity of a final distillation of the adhesive. The process consists of the following steps:

  (a) catalytically oxidizing methanol with air;

  (b) condensing the gases emanating from the catalytic oxidation of stage (a) whereby condensable products therein are condensed in the form of a solution;

  (c) separating the solution containing the condensed products obtained in stage (b) from the uncondensed gases remaining after the condensation stage;

  (d) cooling the uncondensed gases from stage (c) and washing the cooled gases in a washing column with a solution of cold polymerized formaldehyde which takes up the methanol and formaldehyde present in the uncondensed gases;

  (e) distilling the solution obtained in stage (c) in order to recover separately a concentrated aqueous formaldehyde solution and methanol;

  (f) distilling the solution obtained in stage (d) in order to recover the methanol taken up into the solution and to isolate the formaldehyde in the form of a solution thereof;

  (g) combining a part of the formaldehyde solution obtained in stage (f) with the condensed, distilled formaldehyde solution obtained in stage (e); and

  (h) mixing the concentrated aqueous solution of formaldehyde obtained in stage (e) with urea under conditions of predetermined pH and temperature in order to obtain urea-formaldehyde adhesives or resins.

Manufacture of Paper Envelopes with Microencapsulated Glue

In the manufacture of envelopes from an envelope sheet, two kinds of glue are normally used. For the bottom and side portions which are "permanently" sealed together, a back gum containing from about 60 to 70% of solids is used. On the envelope flap, a remoistenable seal gum is applied. In the process of manufacture normally used, the first step is to apply the remoistenable seal gum to the portion of the envelope sheet, which will later become the lid of the envelope.

The coated envelope sheet is then dried. Each envelope sheet is then scored in the places where folds are desired and back gum is applied to the portion of the envelope sheet where the bottom flap and the side flaps will be sealed. The back gum is also applied in liquid suspension or solution. The bottom and sides are then immediately folded up to finish the envelope.

One of the disadvantages of this process is that, once the back gum is applied, the bottom and side flaps must be immediately folded up and sealed. For reasons of storage and handling, it would be desirable to be able to apply the back gum without immediately folding the bottom and side flaps. At the present time, it is not possible to collate envelope sheets on which both the seal gum and the back gum have been applied.

It would also be advantageous to apply the seal gum in the form of micro-capsules to the seal flap as this would permit the end user to seal the envelope without moisture. This feature would be of particular advantage to large mailers using inserting machines.

The process described by *H.R. Lillibridge; U.S. Patent 4,134,322; January 16, 1979; assigned to Champion International Corporation* provides an improvement in the process for manufacturing envelopes whereby the back gum is micro-encapsulated. A suspension of the microcapsules is applied to the appropriate area on the envelope sheet prior to the drying of the seal gum. Both adhesives are dried at the same time and the envelope sheet can be collated without first having to fold the bottom and sides. When it is desired to finish the manufacturing process, the bottom and sides are folded and appropriate pressure is applied. The microcapsules on the bottom are ruptured thereby spreading the gum and causing adhesion.

The encapsulating material for the back gum should be of a hydrophobic nature; i.e., it should be water-insoluble. The encapsulating material can be a thermoplastic resin containing nonionizable groups, examples of which are polyvinyl chloride, polystyrene, polyvinyl acetate, vinyl chloride-vinylidene chloride co-polymers, cellulose acetate and ethylcellulose.

The critical feature in choosing the encapsulating material, in addition to water insolubility, is the rupture point of the capsule. It is important that the capsule shall be able to contain the back gum up to the point of pressure employed in the folding and sealing step.

It is also important that the melting point of the encapsulating material be sufficiently high so that it will not melt during ordinary storage conditions; thus, the melting point of the encapsulating material should be about 50°C. The rupture point for the capsule should be below about 50 psi; i.e., between about 5 to 50 psi. The microcapsules have been more fully described in U.S. Patents 3,875,074 and 3,886,084.

The process comprises the steps of (1) coating a remoistenable seal gum to the lid portion of an envelope sheet, (2) coating a microencapsulated back gum to the side portion or bottom portion of the envelope sheet, the back gum being applied to the portion of the envelope sheet where contact between the side and bottom portion is to occur, and the steps (1) and (2) performed substantially simultaneously, (3) drying the envelope sheet, (4) scoring the envelope sheet along lines defining the configuration of the envelope, (5) folding the envelope along the scored lines forming the bottom and sides of the envelope, and (6) applying sufficient pressure to the area coated with the microencapsulated back gum so as to cause rupture of the microcapsules and adhesion of the bottom and side portions of the envelope.

Adhesives for Textiles and Plastics

ADHESIVES FOR TEXTILES

Tape for Use on Cloth Diapers

G.F. Bateson, F.W. Brown and S.M. Heilmann; U.S. Patent 4,074,004; Feb. 14, 1978; assigned to Minnesota Mining and Manufacturing Company have provided a normally tacky and pressure-sensitive adhesive-coated sheet material which readily bonds to the surface of fabric and which remains firmly bonded even after the fabric subsequently is saturated with moisture. Tape products of the process thus have a particular utility in the formation of closures for cloth diapers. Such tapes have also proved extremely useful in preparing labels for shirt collars or other clothing which will be subjected to numerous launderings and in forming starting tabs for disposable wet wipe products.

The formulation is based on the unexpected discovery that copolymers of certain acrylate monomers with one or more of certain conjugated vinyl monomers yield remarkable and unexpected benefits when incorporated in a tape product. There is also a specific relationship between (1) the combined percent of the conjugated vinyl monomer residues in the copolymer and (2) the inherent viscosity of the copolymer.

Adhesives of the process bond readily to irregularly surfaced fibrous substrates such as paper, cloth or nonwoven fabrics; surprisingly, they also remain firmly adhered even when the substrate is thereafter saturated with moisture and subjected to repeated flexing. Pressure-sensitive adhesive tape products made with conventional adhesives are incapable of even approaching the performance, in wet environments, of tapes made in this way.

The adhesives are copolymers formed from 100 parts by weight of homopolymerizable ethylenically unsaturated monomers which consist essentially of

(1) from about 88 to 99 parts by weight of monomers consisting of

(a) 85-100 weight percent acrylic acid ester of at least
one nontertiary alkyl alcohol the molecules of the
alcohol containing 1 -14 carbon atoms, the average
being about 5-12 carbon atoms, and

(b) correspondingly 15-0 weight percent of at least one
monomer selected from the group consisting of vinyl
acetate, styrene, vinyl ethers and alkyl methacrylates, and

(2) correspondingly from about 12 to about 1 part by
weight of conjugated vinyl monomers selected from the
classes listed below in the amounts shown, the monomers
being selected in types and amounts to impart tacky and
pressure-sensitive adhesive properties to the copolymer:

(a) 0-10 parts by weight of at least one carboxylic acid
selected from the group consisting of acrylic acid,
methacrylic acid and itaconic acid,

(b) 0-7 parts by weight of at least one amide containing 3-4 carbon atoms, and

(c) 0-10 parts by weight of at least one nitrile containing 3-4 carbon atoms.

The degree of polymerization is such that y (the inherent viscosity in tetrahydrofuran) is less than $(14 - x/5)$, where x = the weight percent of monomers falling
under (2) above. Stated more specifically, these polymers possess a combination
of properties such that they fall in Area I on the accompanying Figure 5.1. Areas
II and III of the drawing depict successively more preferred adhesives.

Example: An adhesive was prepared in the following manner. To a 1-quart
(about 1-liter) round glass bottle are charged 212.5 g solvent blend containing
49.867% each of toluene and ethyl acetate and 0.265% 2,2'-azobis(isobutyro-
nitrile); 18.8 g isooctyl acrylate containing 1% tertiary dodecyl mercaptan;
157.92 g isooctyl acrylate; and 11.28 g acrylic acid.

The bottle was purged thoroughly with nitrogen, capped, and placed in a rotating
water bath, tumbled for 21 hours at a bath temperature of 53°C and then for an
additional 3 hours at a temperature of 60°C. The bottle was then removed from
the bath. The resulting solution polymer was found to have the following properties: solids content, 46.7%; viscosity at 25°C, 850 cp as measured by a Brookfield viscometer using a #3 spindle at 12 rpm. The inherent viscosity in tetrahydrofuran was 0.73 dl/g at a concentration of 0.1744 g/100 ml.

Figure 5.1: Formulation of Moisture Resistant Adhesive Tapes

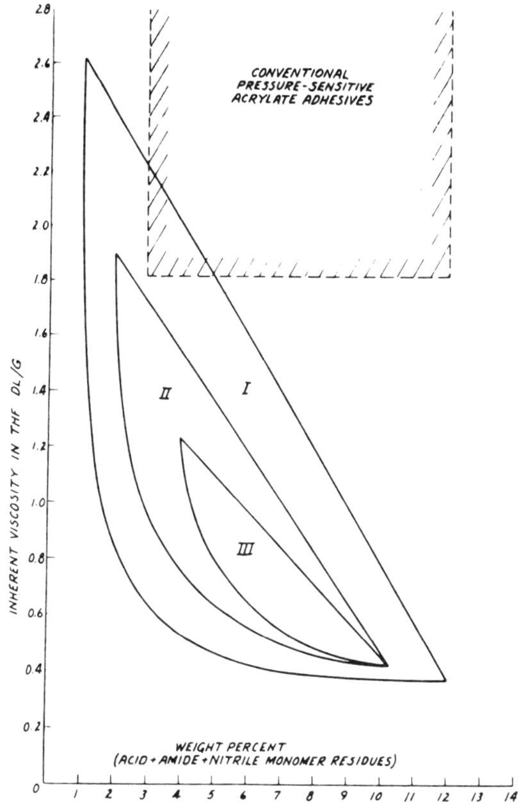

Source: U.S. Patent 4,074,004

Hot Melt Adhesives for Bonding Fabrics

H.S. Starbuck, W.R.Jones and J.E. Mahn; U.S. Patent 4,078,113; March 7, 1978; assigned to General Fabric Fusing, Inc. describe a method for producing transparent or pigmented films from resin particles of a particular class of laurolactam-containing polyamide copolymers and/or terpolymers. Such films are suitable as coatings and/or hot melt adhesives for fabrics and other materials. The method comprises drying the resin while in the solid state until it is substantially water free, extruding the dried resin particles while heating to a temperature sufficient to volatilize and drive off unreacted monomer contaminants, but below the degradation temperature of the resin, expelling the molten resin from the extruder onto heated, sheeting dye means to form a film therefrom, dropping the film onto a substrate backing material, and running the film and backing through a roller to achieve a uniform and controlled thickness.

Such hot melt adhesives find application for such apparel uses as labels, emblems, appliques, shirt fronts, patches, flies, collars and cuffs on such delicate and other fabrics, such as linen, hemp, jute, cotton, nylon 6,6, acrylic, nylon 6, spandex, wool, rayon acetate and others. The resulting bonds may be dry cleaned or laundered at normal temperatures without damage, due to the excellent chemical and heat resistance of the resins used.

The transparent film is made from resin particles consisting essentially of a mixed polyamide condensation product of laurolactam monomer and a co-condensible monomer member selected from the group consisting of caprolactam, hexamethylenediamine adipate, hexamethylenediamine sebacate and mixtures thereof. The laurolactam is present in amounts of from 80 to 20% by weight and the co-condensible monomer member in amounts of 20 to 80% by weight.

Modified Polyamide Hot Melt Adhesive Powders

The process described by *J. Hefele; U.S. Patent 4,080,347; March 21, 1978; assigned to Kufner Textilwerke KG, Germany* for making modified polyamide hot melt adhesive consists of adding an agent for lowering the melting point and/or one for lowering the fusion viscosity of the powdered polyamide.

Preferably copolyamide powder with an initial melting point of preferably from $100°$ to $135°C$, is tempered together with powdered added agent with an initial melting point of preferably from $55°$ to $95°C$, at a temperature above $55°C$ but to a maximum of $3°C$ above the initial melting point of the optimum molecularly uniform distribution.

The tempered mixture which comprises various granule sizes may suitably comprise from 40 to 97% by weight of copolyamide powder and from 3 to 60% by weight of powdered added agent.

During the tempering process, the original granular additive diffuses under the given temperature conditions into the polyamide granules, and at the end of the tempering process, which may last from a few hours to some days, a largely uniform end product is obtained in powdered or very easily powderable form, in which each granule is of a fairly similar nature, and in which each polyamide granule originally present is molecularly penetrated by the additive.

Suitable powdered melting point and fusion viscosity depressing agents for tempering with polyamides or copolyamides are, for example, o,p-toluenesulfonamide (a commercially available mixture of o- and p-toluenesulfonamides with an initial melting point of about $95°C$), o,p-toluenesulfonic cyclohexylamide, p-toluenesulfonic cyclohexylamide, and caprolactam. Known polyamide plasticizers may constitute up to 60% by weight of the mixture, and are of particular use where the main consideration is a lowering of the melting point and viscosity.

The mixture may also contain other agents which also reduce the melting point and fusion viscosity, such as dicyclohexylphthalate and fatty acids such as palmitic or stearic acid and their mixtures, such as stearin or acid waxes, like oxidized polyethylene or oxidized montan waxes, in an amount of about 5 to 10% by weight.

The modified hot melting adhesive powders prepared by this method may be used for coating flexible flat sheets by the powder print-on and spray processes. Flat sheets coated in this manner and mostly cut to size are heat-sealed to other flat sheets, likewise mostly cut to size, using pressure and heat.

This process, carried out in fixing presses and also known as "fixing," or "ironing" or "heat-sealing coating," is suitable, for example, for coated fabric and knitting liners, lining materials, fleeces, and foam material, for producing a bond with top materials of articles of clothing, with hat and shirt collar top materials, with natural and synthetic leather and furs of leather and fur clothing and shoes, and for ironing under upholstery covering, floor covering, curtains, and carpets. Fabric, knitting and fleece liners with a coating of a very low melting point material have proved particularly suitable for fixing natural and synthetic furs, because by using liners coated by the powder print-on process for their fixing, any damage to the temperature-sensitive fur hairs is prevented, without loss of their cleaning resistance in the usual cleaning agents such as perchloroethylene.

Example: A very evenly distributed mixture was prepared comprising: 300 parts of 6/6, 6/12-copolyamide powder, 30 parts of caprolactam powder, and 280 parts of o,p-toluenesulfonamide powder.

The two finely gound additives, namely caprolactam and o,p-toluenesulfonamide, had an average granule size of under 20 μ, and the copolyamide powder (MP 116°C) had the following characteristics:

	Percent
Monomer proportions	
Lauric lactam	30
Caprolactam	40
Hexamethylenediammonium	
adipate	30
Particle size distribution	
Above 250 μ	0.5
180–250 μ	9.0
125–180 μ	29.0
90–125 μ	38.5
60–90 μ	14.0
Below 60 μ	9.0

The mixture was tempered at rest in a plastic bag for about 80 hours at 68°C, and then allowed to cool. After about three days, the tempered mixture assumed a crystalline condition and disintegrated to powder by light manual rubbing. The tempered product was sieved through a 300 μ brush sieve. About 1% of sintered residue remained on the sieve. A sieve analysis of the sieved powder (MP 66°C) gave the following particle size distribution.

	Percent
Above 250 μ	5.0
180–250 μ	34.5
125–180 μ	33.0
90–125 μ	22.0
60–90 μ	5.0
Below 60 μ	0.5

The powder was intensively mixed with 0.2% of magnesium stearate and was applied by the powder print-on process in points corresponding to an 11 mesh grid on a rayon-cotton liner of approximately 80 g/m^2 weight. The applied quantity was approximately 25 g/m^2.

The coated liner was used for heat-sealing furs and leather. The temperature acting on the adhesive points during the distribution process was about 85°C. The bond was resistant to cleaning in perchloroethylene, and resistant to washing at a wash temperature of 30°C.

Heat Curable Multilayer Composite Adhesive Sheet

A composite sheet curable by a free radical reaction at relatively low temperatures within relatively short periods of time was prepared by *Y. Hori, H. Takahashi, M. Sunakawa, I. Ijichi and K. Kamei; U.S. Patent 4,091,157; May 23, 1978; assigned to Nitto Electric Industrial Co., Ltd., Japan.* The composite sheet is useful as an adhesive sheet or a surface layer-forming material of a sheet material such as a decorative sheet.

They found that when a layer-forming unsaturated compound material containing a normally solid or liquid free radical reactive unsaturated compound (to be referred to as an unsaturated compound) and a layer-forming initiator material containing a normally solid or liquid free radical initiator (to be referred to as an initiator) are formed into separate layers at a temperature which does not cause a loss of the radical reaction initiating ability of the initiator material, the resulting heat-curable multilayer composite sheet has a long shelf life and is curable within short periods of time.

Advantageously, the unsaturated compound used is an acrylic unsaturated polymer expressed by the following formula:

$$CH_2=\underset{\underset{X}{|}\ \underset{O}{\|}}{C}-C-O-Y-O-\underset{\underset{O}{\|}\ \underset{X}{|}}{C}-C=CH_2$$

wherein X is a hydrogen atom or methyl group, and Y is a residue of a member selected from the group consisting of polyesters, polyurethanes, epoxy compounds and polyglycols, most preferably having from about 100 to about 300 carbon atoms, which contain a (meth)acrylic acid group at both ends of the molecule and which have a molecular weight of not more than about 5,000, preferably about 1,000 to about 5,000. Such acrylic unsaturated polymers are commercially available.

Examples of the initiators suitably used are those which decompose at relatively low temperatures (e.g., about 60° to about 150°C) to generate free radicals, such as, (1) peroxide-type free radical initiators such as dialkyl peroxides such as di-tert-butyl peroxide, tert-butyl cumyl peroxide or dicumyl peroxide; diacyl peroxides such as benzoyl peroxide or acetyl peroxide; hydroperoxides such as tert-butyl hydroperoxide or cumene hydroperoxide; ketone-type peroxides such as methyl ethyl ketone hydroperoxide; peracids such as peroxyacetic acid or peroxybenzoic acid; peroxy carbonate; or peroxy oxalate; (2) azo-type free radical initiators such as azobisisobutyronitrile or azobis-tert-butyronitrile and (3) inorganic free radical initiators such as potassium persulfate or ammonium persulfate.

Example 1: 150 parts of an acrylic unsaturated polymer (Polyester Acrylate ND-1, Toa Gosei Kagaku KK) and 100 parts of a nonreactive carrier [Biron-200, Toyo Boseki, KK, which is a linear saturated polyester resin which contains hydroxyl groups or carboxyl groups at both ends of the molecule and has a specific gravity of 1.255 (25°C), a molecular weight of 15,000 to 20,000, a glass transition point of 67°C and a melting point of 180° to 200°C] were dissolved in methyl ethyl ketone to form a 60% solution. The solution was coated on a strippable paper (obtained by treating glassine paper with a silicone resin), and dried at 120°C for 3 minutes to form an unsaturated compound material layer having a thickness of 80 microns. This unsaturated compound material layer was tacky.

Separately, 50 parts of benzoyl peroxide and 100 parts of the abovementioned polyester resin were dissolved in methyl ethyl ketone to form a 20% solution. The solution was impregnated into a nonwoven polyester fabric and dried at 50°C for 5 minutes to form an initiator material layer with an initiator pickup of 7 g/m^2 (an initiator-impregnated nonwoven fabric).

The initiator material layer was contacted with the unsaturated compound material layer on the strippable paper, and bonded by means of a metal roll and a rubber roll at a pressure of about 1 kg/cm^2 and at room temperature for about 1 second to form a composite sheet.

Example 2: 150 g of a polyester acrylate and 100 g of a nonreactive carrier were dissolved in 200 g of methyl ethyl ketone by stirring. The resultant solution was coated on a strippable paper (prepared by treating glassine paper with a silicone resin) and dried at 130°C for 3 minutes to form an unsaturated compound material layer having a thickness of 70 microns.

Separately, 10 g of benzoyl peroxide was dissolved in 200 g of toluene, and the solution was impregnated into a pulp decorative paper (150 microns thick) having a wood grain pattern thereon, and allowed to dry in air to form an initiator material layer (initiator-impregnated sheet) with an initiator pickup of 5 g/m^2.

The unsaturated compound material layer was laminated to both surfaces of the initiator-impregnated sheet using a hot metal roll whose surface temperature was maintained at 60°C to form a composite sheet. In more detail, two sheets of the unsaturated compound material layer were laminated to both surfaces of the initiator layer in a "sandwich" shape. In this laminating, the roll pressure was 1 kg/cm^2 and the period of pressing was about 1 second.

The resultant sheet had good flexibility at a temperature of 10°C or more, and when stored in the dark at 50°C at a relative humidity of 60 to 70%, it retained its initial properties even after 7 months.

After storage for 4 months at a relative humidity of 60 to 70%, the strippable paper was removed from the composite sheet, the composite sheet superimposed on a plywood veneer sheet and the assembly hot-pressed at a pressure of 5 kg/m^2 and a temperature of 130°C for 3 minutes using a hot press. In spite of the fact that the hot-press conditions were mild, the coating of the resulting decorative sheet was not found to be deteriorated in any manner even after 1 hour in boiling water when evaluated in accordance with JIS K 6902. The resulting decorative sheet also withstood the impact of a steel ball (28 g) which dropped onto it from a height of 500 mm, and thus exhibited superior impact resistance. The pencil hardness thereof (determined by JIS K 5400) was 3H. The shear bond strength between the coating (i.e., the cured composite sheet) and the plywood veneer sheet was 150 kg/cm^2 at room temperature, which represents a firm bonding sufficient for decorative sheets.

Partially Crystalline Copolyesters for Fusion of Textiles

Fusion adhesives in powder form are preferred in textile lamination. The powders are sintered onto the lining materials, usually woven fabrics, by means of special applicator systems, and then cemented to the facing materials with the application of heat and pressure. Adhesives which are resistant to dry cleaning agents are needed as the adhesives.

A polyester adhesive fusion coating mass which was crystalline, had a melting point less than 130°C, resisted the action of dry cleaning solvents, and which had a substantial melt viscosity, say in the range of 100–20,000 poises was needed.

K.G. Sturm and K. Brüning; U.S. Patent 4,094,721; June 13, 1978; assigned to Dynamit Nobel AG, Germany provided these characteristics with a linear saturated crystalline polyester of an acid moiety and a moiety of a dihydric alcohol, at least 40% of the acid moiety being terephthalic acid moieties, and containing moieties of 1,4-butanediol and 1,6-hexanediol, the ratio of 1,4-butanediol moieties to 1,6-hexandediol moieties being 1:90–90:10.

It was found that linear saturated crystalline polyesters derived from terephthalic acid and a mixture of 1,4-butanediol and 1,6-hexanediol have good crystallinity notwithstanding melting points of 40° to 130°C. Such polyesters, moreover, have a glass transition temperature of -10° to +30°C, a maximum logarithmic damping decrement from 0.6 to >1.3 and a reduced viscosity measured on a 1 wt.-% solution in a 60-40 mixture of phenol and 1,1,2,2-tetrachloroethane of

0.5 to 1.5. The difference between the glass transition temperature and the melting temperature of such polyesters is equal to or less than 100°C. Such polyesters have good viscosities in the melt and resist the action of dry cleaning solvents. They can be employed as fusion coating masses in various forms, either in the form of heated molten masses or in the form of solids. As solids, they can be employed in the form of strips or in the form of powders. They are highly useful as solvent-free adhesives and satisfy the requirements imposed by the textile and shoe industries.

The terephthalic acid moieties can be provided by the acid chloride or a monoalkyl or dialkyl ester. Part of the terephthalic acid moiety may be replaced by an aliphatic dicarboxylic acid such as adipic, azelaic or sebacic acid, or by isophthalic acid.

Example: A copolyester in whose preparation terephthalic acid and isophthalic acid were used in a molar ratio of 85:15 and butanediol-1,4 and hexanediol-1,6 were used in a molar ratio of 50:50, and which had a reduced viscosity of 0.95, a glass transition temperature Tg of 29°C, MP of 110°C and a maximum logarithmic damping decrement of 0.79, was ground in suitable grinding apparatus to a powder of a fineness of 60 to 200 μm. The powder was applied to webs of fabric by means of an engraved roller, and was sintered thereon to form drops of fusion adhesive.

The fabric treated in this manner can be bonded to other fabrics in ironing machines in periods of approximately 10 to 15 seconds, at temperatures between 140° and 160°C, without the loss of the textile character of the laminate. The laminate is resistant to the cleaning agent perchloroethylene which is used in chemical cleaning, and to machine washing at 60°C.

Electron-Beam Curing of Adhesives for Flocking

The process developed by *S.V. Nablo and A.D. Fussa; U.S. Patent 4,100,311; July 11, 1978; assigned to Energy Sciences Inc.* is one for curing texturing material such as flock and its supporting adhesive secured to a heat-sensitive substrate that inherently limits the degree of thermal curing that may be employed and consequently the speed of curing.

It comprises applying an electron-curable adhesive layer to a heat-sensitive substrate, flocking fiber material upon the adhesive layer, passing the assembly of substrate and adhesively secured flocking material past a predetermined region, directing electron beam energy at the predetermined region upon the flocking material and through it onto the adhesive layer, adjusting the electron beam to produce an electron dose of the order of 2 megarads ± 50%, of energy of the order of 150 keV ± 30%, and with a line speed of passing the predetermined region preferably of the order of about 20 to 80 meters per minute, in order to cure the flocking adhesive without affecting the heat-sensitive substrate. At the 2 megarad treatment level, less than 5 cal/g of energy are delivered to the adhesive. Assuming an adhesive specific heat of 0.3, temperature elevations of only 10° to 15°C are expected during the curing process, with much lower figures for the underlying web which receives almost no energy directly, and typically has a thermal capacity much greater than that of the adhesive film.

Example: An 8 mil heat-sensitive sheet vinyl substrate, knife-coated with a few mils thickness (about 4 mils) of Dow XD 7530.01 acrylic epoxy adhesive, upon which nylon flock fibers were electrostatically flocked in conventional fashion with 6 Denier X 1.25 mm (50 mil) fibers, was found to be successfully adhesive-and-flock cured with line pass rates of 60 meters per minute, with an Energy Sciences' Electrocurtain TM model CB 150. The apparatus was adjusted to produce a dose of 2 megarads with an electron energy of 150 keV. Parallel studies were done with the same system at line speeds of 30 meters per minute and a treatment level of 3 megarads. Flocking material thickness on the order of two to fifty times the adhesive layer thickness are thus curable and have been demonstrated at flock lengths of up to 4.5 mm.

Similar results were also obtained with Hughson urethane adhesive RD-2275-58 on the upper surface of a 5 mil thick vinyl wall covering, with pressure sensitive adhesive and release paper already applied to the rear or lower surface. In both of the above cited examples, the abrasion resistance and solvent resistance of the flock textured surface were at least factors of 3-5 better than that realized using conventional thermally cured epoxy or acrylic-latex emulsion adhesives.

Successful similar adhesive-flocking curing has been effected with the same apparatus and adjustments on temperature-sensitive styrene substrates, including styrene sheets up to 100 mils in thickness. Wood and paper substrates, and a variety of sheet and foam polyethylene and polyurethane as well as wallpaper-like paper-foil laminates, have also been employed to carry the electron-beam-cured adhesive-flocking coating, as have other types of electron-curable adhesives, such as acrylic-latexes and both aromatic and aliphatic urethanes, epoxy esters and other types of flocking materials such as rayon and polyesters, all with dose, energy and line speed adjustments within the before-stated ranges.

Hot Melt Linear Block Copolyether-Polyester Compounds

Hot melt adhesives are well known in the art. For example, certain kinds of polyethylene, polyethylene-vinyl acetate, polyvinyl chloride and polyamide are conveniently applied to textiles in a molten state so as to form an adhesive bond upon cooling. Among these known adhesives, polyolefin and polyvinyl chloride have a poor bonding strength to textiles, especially to textiles composed of polyester fibers. They easily lose their adhesive strength at elevated temperatures.

Polyamides, which are the most practical hot melt adhesive for textiles at present, exhibit good adhesive strength and good resistance to dry cleaning agents, but the properties of the textiles bonded therewith are not necessarily excellent and furthermore their adhesive strength is often lost after washing with a detergent and hot water or after steaming.

*C. Tanaka, Y. Yoko and M. Masanobu; U.S. Patent 4,130,603; Dec. 19, 1978;
assigned to Toray Industries, Inc., Japan* have synthesized a thermoplastic
linear copolyether-polyester, consisting essentially of a dicarboxylic acid unit
derived from 40 to 60 mol % of terephthalic acid and 60 to 40 mol % of isophthalic acid, a glycol unit derived from 1,4-butylene glycol and a polyether
unit derived from polytetramethylene glycol having a molecular weight of 600
to 1,300 wherein the polyether unit constitutes 10 to 33% by weight on the
basis of the block copolyether-copolyester. The block copolyether-copolyester has a relative viscosity of 1.3 to 1.7 and melting point of 95° to 145°C,
and is an excellent hot melt adhesive for textiles, giving an adhesive joint with a
greatly improved softness, liveliness and resistance to dry-cleaning and hot-water
washing.

Example: Into a glass flask having a stainless steel stirrer with helical ribbon
type screw, 78.6 parts of dimethyl terephthalate, 96.0 parts of dimethyl isophthalate, 120 parts of 1,4-butanediol and 41.7 parts of poly(tetramethylene
oxide) glycol having a molecular weight of about 1,000 were placed together
with 0.08 parts of tetrabutyl titanate. The mixture was heated with stirring at
210°C for 2 hours while distilling off methanol from the reaction system. The
recovered methanol was 52.0 parts which corresponded to 90% of the theoretical weight. The reaction temperature was then raised to 250°C and the pressure
on the system was reduced to 0.2 mm Hg for a period of 60 minutes. Polymerization was continued for 80 minutes under these conditions.

The relative viscosity of the product in o-chlorophenol at 25°C was 1.43 and the
polymer showed a melting point of 120°C.

The polymer resin was crushed into powder under a liquid nitrogen atmosphere
and the resulting powder having a particle size of 60 to 180 μ was applied onto
some fabrics and steam-ironed at 150°C under a pressure of 300 g/cm^2 for 13
seconds. The peel strength was measured. The results were as follows:

Fabric	Applied Amount g/m^2	Peel Strength After Ironing g/in.
Cotton	30	1,750
Cotton	15	1,100
Polyester	30	inseparable
Polyester	15	2,200
Wool	30	1,500
Wool	15	1,250
Polyester/cotton	30	2,700
Polyester/cotton	15	1,850

Bonding Hydrophobically Treated Substrates

A process by *G. Hoss and E. de Jong; U.S. Patent 4,141,869; February 27,
1979; assigned to Plate Bonn GmbH, Germany* relates to a heat-sealing thermoplastic adhesive of a thermoplastic polymer which is resistant to washing and

dry cleaning, and contains from 0.05 to about 5% by weight, based on the weight of the polymer, of at least one hydrophobic silicone compound.

These heat-sealing thermoplastic adhesives develop outstanding adhesion in the bonding of substrates (front fixing) which have been hydrophobically treated, for example with silicones, such as raincoat fabrics, etc. The adhesion which they develop remains intact, even after repeated dry cleaning and after washing.

Suitable thermoplastic polymers for the process are any polymers of the type which are already used for the heat-sealing of textiles. Preferred thermoplastic polymers are, for example, copolyamides of polyamide-forming amide salts containing at least 10 carbon atoms, polyamide-forming ω-amino acids and/or polyamide-forming lactams; crystallizable copolyesters based on terephthalic acid/diol; and thermoplastic polyurethanes.

The polymers may contain plasticizers. This applies in particular to copolyamides. Benzene and toluene sulfonic acid ethylamides are particularly preferred. The plasticizer is preferably used in a quantity of up to about 25% by weight, based on the total quantity of polymer and plasticizer.

The preferred hydrophobic silicone compounds are the compounds normally used for the hydrophobic finishing of textiles, preferably the so-called silicone oils.

The quantity in which the hydrophobic silicone compounds are used is preferably between about 0.5 and about 3% by weight, based on the weight of the polymer. In general, excellent results are obtained with about 1 to 2% by weight of hydrophobic silicone compounds.

Example 1: 100 parts by weight of a standard commercial-type thermoplastic adhesive for the washing-resistant and dry-cleaning-resistant bonding of textile substrates based on copolyamides 6, 66 and 12 (Platamid H 105, a product of Plate Bonn GmbH: melting range 115°-125°C, melt index 4 g/10 min at 160°C according to DIN 53 735), in the form of a powder with a grain size range from 0 to 200 μ, were coated in an intensive mixer with 1 part by weight of a standard commercial-type silicone oil (Hydrofugeant 20 218). By means of a powder spot-printing machine, the powder was then spot-printed through an 11-mesh screen onto an interlining of the type normally used in the clothing industry in a coating weight of 20-21 g/m² and sintered in a heating tunnel. The interlining had a weight per square meter of approximately 110 g/m² and consisted of 100% cotton in the warp direction and 100% rayon in the weft direction.

The coated interlining was then ironed onto a standard, heavily siliconized raincoat material of 65% Diolen polyester and 35% cotton with a weight of 165 g/m² (Ninoflex 14 916) under a pressure of 350 g/cm² and with a temperature/time program in an electrical press heated on one side, of the type commonly used in the clothing industry.

5 cm wide strips were cut from the composite material thus produced and subjected to a separation test in a tension tester. The bond strength values obtained are shown in the table. The composite material was also dry cleaned 5 times, after which bond strength was again measured.

Example 2: In the thermoplastic adhesive described in Example 1, the quantity of silicone oil added was increased to 1.5 parts by weight. The bond strength values obtained with the thermoplastic adhesive thus modified are shown in the table.

Example 3: 100 parts by weight of the copolyamide powder used in Example 1 were mixed with 1 part by weight of a standard commercial-grade silicone "Silikonoel ZG 318" and used for bonding under the same conditions as in Example 1. The bond strengths obtained are shown in the table.

Example 4: By way of comparison, the copolyamide powder used in Examples 1, 2 and 3 was applied to the interlining without any addition of silicones and ironed onto the raincoat material. The bond strengths obtained are shown in the table.

Example	Copolyamide (g/m^2)	Additive	Conditions of the Ironing Press*	Bond Strength** ...(kp/5 cm) ... Normal	After Dry Cleaning 5 Times
1	20-21	1% of Hydro-fugeant 20 218	150/15	1.9	1.8
			150/18	2.5	1.8
			150/13+5 D	3.1	2.5
			160/15	2.5	2.1
			160/18	1.9	2.3
			160/13+5 D	3.2	3.0
2	21-22	1.5% of Hydro-fugeant 20 218	150/15	1.5	1.6
			150/18	1.6	1.7
			150/13+5 D	3.2	2.4
			160/15	2.3	1.7
			160/18	2.1	1.9
			160/13+5 D	3.0	3.0
3	19-20	1% of Silikonoel ZG 318	150/15	2.2	1.7
			150/18	2.6	2.1
			150/13+5 D	3.0	2.7
			160/15	2.2	1.7
			160/18	2.2	2.2
			160/13+5 D	2.9	2.4
4	19-20	None	150/15	0.7	nm
			150/18	0.7	nm
			150/13+5 D	1.1	nm
			160/15	0.7	nm
			160/18	0.7	nm
			160/13+5 D	1.6	nm

*150 = platen temperature of press in degrees C.
 15 = pressing time in seconds.
 13+5D = pressing time of 13 seconds followed by 5 seconds steaming.
**Bond strength values are averages of 10 tests. nm = not measured; bond strengths too low.

"Breathing" Lining Material Fused to a Garment by Hot Pressing

The lining material of the process developed by *K. Tischer and W. Föttinger; U.S. Patent 4,148,958; April 10, 1979; assigned to Firma Carl Freudenberg, Germany* consists of the combination of a random fiber nonwoven fabric serving as an interfacing, with a nonwoven fabric of oriented, especially longitudinally oriented, staple fibers serving as lining, and on the basis of its structure it has good physiological properties hitherto unattained in lining materials.

The known desirable characteristics of random fiber nonwoven material as interfacing are combined in accordance with the process with the properties of oriented fiber nonwoven material favorable to moisture transport, the oriented fiber nonwoven replacing the previously customary lining materials. By the bonding together of the two nonwovens, the specific weight of the oriented fiber nonwoven material can be kept exceedingly low. For example, while a conventional lining has a specific weight of about 65 to 130 g/m^2, it is possible in the case of this laminated material to reduce the oriented fiber content to a fraction of this weight.

An adhesive composition is provided in a known manner on the back of the random fiber nonwoven serving as the interfacing. Polyamides, polyethylenes, polyvinyl chlorides, polyvinyl acetates and mixtures thereof are suitable as binding compositions. The adhesive composition can be applied either by the powder sprinkling method or by means of printing machines.

Example: The following example will serve to explain the process. A random fiber sliver having a specific weight of 22 g/m^2 is covered with a fiber sliver made from longitudinally oriented fibers having a specific weight of 10 g/m^2. The two slivers are together impregnated with a commercial acrylate (e.g., Acronal 35 D of BASF) by the foam impregnation method. The specific weight of the binding agent amounts to 11 g/m^2.

For the random fiber fabric a fiber mixture of the following composition is used: 50% of highly crimped nylon, 3.3 dtex/51, number of crimps per cm approximately 20; 25% of polyester fibers 3.3 detx/60; and 25% of viscose fibers 1.7 dtex/40.

The fiber mixture of the longitudinal sliver is composed as follows: 60% of nylon fibers 1.7 dtex/40, number of crimps per cm 5 to 10; 20% of polyester fibers 1.7 dtex/40; and 20% of viscose fibers 1.3 dtex/40.

The fabric consolidated by foam impregnation is imprinted with a pattern on the side of the longitudinal sliver by means of a rotary printing machine. The printing paste consists of a conventional pigment binder, a dye, and a commercial acrylate thickener. The deposit after drying amounts to approximately 8 g/m^2.

The random fiber side of the material is provided with a commercial ternary copolyamide having a melting point of 125°C by means of a powder dotting machine. The deposit amounts to 15 g/m^2.

The laminate material obtained in the manner described above can be used in any desired manner for the lining and stiffening of articles of clothing, only a single working procedure being required. The composite lining can be joined to a men's jacket fabric by placing the random fleece face against the underside of the jacket fabric and pressing with a Hoffman press.

Powder Adhesives Suitable for Fusible Interlinings

A fusible interlining is a fabric which has been coated on one side with a discontinuous pattern of fusible adhesive. When the interlining is bonded to a base fabric in a garment, it provides body and shape to the garment without impairing the ability of the fabric to breathe. Fusible interlinings are used in the manufacture of men's and women's suits, in shirt collars and cuffs, and in the waistbands of trousers.

Certain copolyesters are potentially useful for fusible interlining applications which do have adequate bond strength on rainwear fabrics and which do not tend to strike through on dark fabrics, but powdering of these copolyesters by cryogenic grinding techniques is relatively expensive.

R.L. McConnell and D.A. Weemes; U.S. Patent 4,155,952; May 22, 1979; assigned to Eastman Kodak Company found that melt blending of small amounts of low-viscosity polyethylenes with these polyesters or copolyesters results in substantially improved grinding rates and, therefore, substantially decreases the cost of manufacturing powder. These melt blends were found to provide nontacky and nonblocking blends, readily reduced to powder by cryogenic grinding techniques. The yields of powder were also found to be substantially higher than those obtained when grinding the nonblended polyester or copolyester.

This formulation is directed to semicrystalline polyester/low viscosity polyethylene melt blends, which comprise an intimate melt blend of a semicrystalline polyester having an inherent viscosity ranging from about 0.4 to 1.2, a melting point of about 80° to 175°C and an apparent heat of fusion (ΔH_f) of $\leqslant 10$ calories per gram, and a low molecular weight polyethylene having a melt viscosity ranging from about 50 to 30,000 cp at 150°C, a density at 25°C of about 0.90 to 0.980, and an acid number of about 0-80. The low molecular weight polyethylene is present in the blend in concentrations ranging from about 3 to about 30% by weight, with preferred concentrations being about 5 to 15 weight percent.

The apparent heat of fusion (ΔH_f) of polymers is the amount of heat absorbed when crystallizable polymers are melted. ΔH_f values are readily obtained using thermal analysis instruments, such as the Perkin-Elmer DSC-2 Differential Scanning Calorimeter or the Du Pont Model 990 Thermal Analyzer with different scanning calorimeter cell.

Some examples of preferred polyester polymers are those derived from about 70 mol percent terephthalic acid, 30 mol percent adipic acid, 73 mol percent ethylene glycol, and 27 mol percent 1,4-butanediol.

Another preferred polyester polymer may be derived from 10 mol percent iso-phthalic acid, 90 mol percent terephthalic acid, and 100 mol percent 1,6-hexanediol having a melting point of about 140°C and an apparent heat of fusion of about 8 calories per gram; 20 mol percent isophthalic acid, 80 mol percent terephthalic acid and 100 mol percent 1,6-hexanediol having a melt-ing point of about 125°C and an apparent heat of fusion of about 5 calories per gram; and 35 mol percent isophthalic acid, 65 mol percent terephthalic acid and 100 mol percent 1,6-hexanediol having a melting point of about 140°C and an apparent heat of fusion of about 2 calories per gram.

Preferred polyester polymers may further be derived from 40 mol percent iso-phthalic acid, 60 mol percent terephthalic acid and 100 mol percent 1,4-butane-diol having a melting point of about 140°C; 50 mol percent isophthalic acid, 50 mol percent terephthalic acid and 100 mol percent 1,4-butanediol having a melt-ing point of about 129°C and an apparent heat of fusion of about 2 calories per gram; and 60 mol percent isophthalic acid, 40 mol percent terephthalic acid and 100 mol percent 1,4-butanediol having a melting point of about 140°C.

Preferred polyester polymers may still further be derived from 60 mol percent isophthalic acid, 40 mol percent terephthalic acid, and 100 mol percent 1,4-cyclohexanedimethanol.

The process of making the adhesive involves melt blending particulate or pel-leted material selected from a semicrystalline polyester having an inherent viscosity ranging from about 0.4 to 1.2, a melting point of about 80°-175°C and an apparent heat of fusion of ≤10, and a low molecular weight polyethylene having a melt viscosity ranging from about 50 to 30,000 cp at 150°C, a melting point of about 90° to 135°C, a density at 25°C of about 0.90 to 0.980, and an acid number of about 0-80. The low molecular weight polyethylene present in the melt blend is in concentrations ranging from about 3 to about 30% by weight. The blended materials are melt extruded at a temperature of about 150°-250°C into a cooling medium to form a predetermined extruded shape, which is chopped into pellets or otherwise granulated. Then the pellets or granulated materials are ground under cryogenic temperature conditions into a powder.

The cooling medium is generally water having a temperature of about 5° to about 50°C.

Process for Patterned Deposition of Thermoplastic Adhesive

The process developed by *J. Hefele; U.S. Patent 4,139,613; Feb. 13, 1979; assigned to Kufner Textilwerke KG, Germany* relates to the patterned deposition of powdered thermoplastics adhesive material on the outer surface of a textile or other porous-flexible surface form or substrate, in particular by means of an engraved component provided with a pattern of depressions arranged in an outer face thereof and in which depressions the adhesive powder is insertion raked, the surface form to be coated being heated and positioned on the outer face of the engraved component and then taken off from it together with the adhesive powder disposed in the depressions.

After being taken off the adhesive powder lies lightly clustered in the form of small heaps in a pattern corresponding to the arrangement of the depressions on that side of the surface form to be coated which had been brought into contact with the engraved component. These small heaps of powder can then through further temperature working be sintered together to form a sealed bead.

The process provides for the patterned deposition of powdered thermoplastic adhesive materials on the outer surface of a textile, by means of an engraved component provided with a pattern of depressions arranged in an outer face, in which depressions the adhesive powder is insertion raked. The surface to be coated is then heated and positioned on the outer face of the engraved component and then taken off from it together with the adhesive powder in the depressions, wherein there is a first insertion raked in the pattern of depressions in the engraved component an adhesive powder material and then on this powder a further adhesive powder material is insertion raked in the depressions, so that both the powder layers superposed one on the other are taken up by the surface form positioned on the engraved component.

Thus in accordance with the process there is first insertion raked in the pattern of depressions an adhesive powder material with specific characteristics and then on this powder there is insertion raked a further adhesive powder material, preferably with characteristics differing from those of the first insertion raked powder, so that both the powder layers superposed the one on the other in the depressions are taken up by the surface form positioned on the engraved component.

The adhesive powders insertion raked in a separate manner one after the other in the depressions in the engraved component may have differing physical or chemical properties. The process is suitable not only for coating with two different sorts of adhesive powder, but serves also for the deposition of two patterned coatings with like physical or chemical properties. The surface material used in this process may be a material selected from the group consisting of woven textile material, knitted textile material, weft inserted knitted textile material, fleece, nonwoven fabric, synthetic foamed material and synthetic leather.

A preferred mode of use is that which relates to the production of stiffening inserts for clothing articles in which, for the affixing of the insert onto the outer face part of the clothing article, an adhesive in the form of a patterned adhesive undercoating with a higher melting viscosity, or higher melting point is used and having superposed thereon is a patterned upper coating with a lower melting point. With such inserts there is formed by the fixing of the undercoating which takes place on pressing, due to its higher melting point, a barrier for the upper coating so that the latter does not sink into the insert material but binds itself to the outer face material.

The process is further utilizable in the production of netting material made of thermoplastic additives with differing adhesive characteristics on both sides of the netting. The production of such adhesive netting can be carried out through the use of an engraved component in which the patterned depressions in its outer face are formed by mutually intersecting grooves. Into these depressions there is first dispersed an adhesive powder with a higher melting viscosity and then onto that there is dispersed a further adhesive powder with a lower melting viscosity.

These two superposed powder coatings can be drawn off the engraved component by an intermediate carrier formed by a kraft paper, which may be a one-sided siliconized kraft paper, positioned on the engraved component and then become sintered. After cooling, the drawn off adhesive network may be readily stripped off the silicon paper and be applied directly through insertion and pressing between two porous-flexible surface forms, e.g., textiles or leather. When cut up it can together with the silicon paper be readily transferred to other surface forms through quick pressing over of the adhesive free side of the intermediate carrier. Thus the adhesive coating with the higher melting viscosity becomes directly superposed on the surface form.

In this way the adhesive coating with the lower melting viscosity, which lay under the silicon paper, is then placed over the coating with the higher melting viscosity. The latter forms on later pressing with an overlying material, or for example by pressing on of a surrounding edge band, the previously mentioned barrier for the overlying coating. In this way there can be achieved high adhesion strength even with light machine pressing. The adhesive netting produced by this process can also with advantage be used as adhesive backup for gold lace.

Detailed drawings of the apparatus for applying these powdered hot melt adhesives to textiles, furs, or leather are provided in the patent, together with description of the workings of the apparatus.

ADHESIVES FOR PLASTICS

For Bonding Vinyl Coverings

Vinyl resins have been used to coat or cover a wide variety of materials. Most materials such as acrylonitrile-butadiene-styrene (hereinafter referred to as ABS) may be coated with a vinyl resin without the use of the adhesive materials simply by dipping or spraying the material being coated with the resin coating. However, vinyl resin coverings particularly on materials like ABS are likely to be nonpermanent, in that a good bond between the materials is not formed. Within a short time after coating, peeling and cracking, etc., of the polymeric surface results. Further, if the coating or covering is to be used on surfaces which are subject to frictional contact, e.g., rubbing, etc., the covering or coating slips.

C.H. Wollen; U.S. Patent 4,074,014; Feb. 14, 1978; assigned to Leon Chemical & Plastics, Division of U.S. Industries, Inc. has found that strong and permanent adhesives useful for bonding vinyl resin coverings to surfaces can be made by employing mixtures of phenolic resins and methacrylate resins having a MW greater than about 100,000, preferably about 400,000. The methacrylate resin is most preferably a methyl methacrylate.

If the adhesive is used with vinyl resin coverings, contains plasticizer, and the plasticizer is no more than limited in compatibility with the resins, superior bonding results, particularly on polymeric surfaces, showing a synergistic combination.

The adhesive preferably comprises from about 1 to about 95% (all percent are by weight resin solids) phenolic resin and from about 5 to about 99% methacrylate resin, more preferably from about 50 to about 80% phenolic resin and from about 20 to about 50% methacrylate resin, and most preferably from about 60 to about 70% phenolic resin and from about 30 to about 40% methacrylate resin.

Example 1:   Adhesives are formulated by intermixing varying amounts of a phenolic resin solution, DS-9-2816 (15% solids, Jamestown Finishes, Inc) and a methacrylate resin, having a MW of about 400,000 Acryloid, 55D42 (8% solids, Rohm and Haas Company). The adhesive is spray or dip coated on an ABS molded surface which is subsequently covered by injection molding the following material over the ABS substrate.

	Parts
Polyvinyl chloride (QYNL)	100
Diisodecyl phthalate	91
Epoxy phthalate	8
$CaCO_3$	20
Stearic acid	¼
Ba-Cd-Zn chelator stabilizer	2½

The composite is then tested as to the adherence of the vinyl coating to the ABS before and after aging. The adhesive strength of the resulting bond is given in pounds per inch width. The PVC covering has a Durometer rating of 70 and thus itself tears at around 50 pounds per inch of width. The following table summarizes the results of these tests.

Parts 55D42	Parts DS-9-2816	Adhesive Strength* lb/inch width
10	1	47–52
7	1	51–53
4	1	51–57
1	1	50–53

(continued)

Parts 55D42	Parts DS-9-2816	Adhesive Strength* lb/inch width
1	4	49–57
1	7	47–52
1	10	46–50

*In all cases, the vinyl coating tore before adhesive failed.

Example 2: ABS Cyclolac T tensile bars coated with only methacrylate resin (55D42) and coated with a 1:2, methacrylate (55D42) to phenolic resin (DS-9-2816), adhesive mixture were strained to 0.8% over a 15½ inch diameter wooden drum. Various plasticizers contained in vinyl coverings are painted onto the adhesive-covered surface of the ABS until the surface crazed. The following table sets forth the results of these tests.

Plasticizer	55D42	Adhesive Mixture, 1 part 55D42:2 parts DS-9-2816
Tricresyl phosphate	½ hour	1 hour
Octyl decyl phthalate	4 hours	6 hours
Epoxy tallate	>1 week	>2 weeks
Liquid Ba-Cd-Zn stabilizer	>1 week	>2 weeks
Diisodecyl phthalate	16 hours	>2 weeks
8:91:2½ epoxy tallate, diisodecyl phthalate and Ba-Cd-Zn chelator stabilizer	24 hours	>2 weeks

As seen in this example, the phosphate plasticizer, which is highly compatible with the adhesive resin mixture acts to penetrate the resin and attack the ABS substrate rapidly as compared to the other limited incompatibility plasticizers showing the synergistic combination of resins and preferred plasticizers.

Two-Package Organopolysiloxane Systems

J.R. Schulz; U.S. Patent 4,087,585; May 2, 1978; assigned to Dow Corning Corporation has found that certain hydroxylated, vinyl-containing polysiloxane and epoxy-containing silanes can be used to provide silicone compositions which when cured in contact with substrates exhibit adhesion properties. Compositions obtained by mixing vinyl-containing polyorganosiloxane, organosilicon compound having silicon-bonded hydrogen atoms, a platinum catalyst, a hydroxyl-containing polysiloxane having at least one silicon-bonded hydroxyl radical and at least one silicon-bonded vinyl radical and a silane having an epoxy organo group and an alkoxy group, show adhesion to substrates when cured in contact with the substrates.

These compositions can be, for example, elastomer-forming compositions and compositions which produce gels. An adhesion additive can be prepared by mixing the hydroxyl-containing polysiloxane and the epoxy silane under essentially anhydrous conditions at a temperature below 50°C. The resulting adhesion additive can then be mixed with vinyl-containing polyorganosiloxane and organosilicon compound having silicon-bonded hydrogen atoms to provide compositions. The resulting compositions have extended pot life and when applied to substrates and heat cured show improved adhesion properties.

The compositions of a preferred elastomer comprise a product obtained by mixing (A) a polydimethylsiloxane endblocked with dimethylvinylsiloxy units or methylphenylvinylsiloxy units in which the average ratio of organo radicals per silicon atom is in a range of 2.0025 to 2.02; (B) a polymer having at least 3 silicon-bonded hydrogen atoms per molecule consisting of trimethylsiloxy units, dimethylsiloxane units and methylhydrogensiloxane units where the average number of siloxane units per molecule is less than 50; (C) a reaction product of chloroplatinic acid and a vinylsiloxane having at least two dimethylvinylsiloxy units per molecule and any additional siloxane units being dimethylsiloxane; (D) a polysiloxane having an average of about two silicon-bonded hydroxyl radicals per molecule, 3 to 15 siloxane units selected from dimethylsiloxane unit and methylvinylsiloxane unit, and an average of at least one silicon-bonded vinyl radical per molecule; (E) a silane of the formula

$$CH_2\overset{\displaystyle O}{\overset{\displaystyle /\ \backslash}{CH}}CH_2OCH_2CH_2CH_2Si(OCH_3)_3$$

and an extending filler (F). In this composition (A) is present in an amount of 100 parts by weight, (B) is present in an amount of 0.5 to 10 parts by weight, (C) is present in an amount sufficient to provide from 5 to 50 parts by weight platinum per one million parts by weight of (A), the combined weight of (D) and (E) is present in an amount of 0.25 to 2 parts by weight per 100 parts by weight of the composition exclusive of the weight of (D) and (E) where the weight ratio of (E) to (D) is from 0.4 to 1.5, (F) is present in an amount of 20 to 150 parts by weight and the ratio of silicon-bonded hydrogen atoms in (B) to the silicon-bonded vinyl radicals in (A) is from 1.2 to 4.

The most preferred compositions of this embodiment are those having lower viscosities and where the average ratio of organo radicals per silicon atom in (A) is in a range of 2.0065 to 2.02; the extending filler is finely divided quartz; and the pigment is carbon black, which improves the flame retardant property of the cured products. These compositions can contain polymethylvinylcyclosiloxane in amounts of 0.01 to 0.5 parts by weight. These compositions when heat cured show improved adhesion to substrates such as glass, aluminum, copper, stainless steel and many organic plastics such as polyesters.

In another process for improving the adhesion of organopolysiloxane compositions to substrate without the use of a primer, *K. Mine and M. Yokoyama; U.S. Patent 4,082,726; April 4, 1978; assigned to Toray Silicone Company, Ltd., Japan* describe the following mixture:

> (1) an organopolysiloxane containing at least two lower alkenyl groups bonded to silicon atoms which are in the same molecule;

(2) an organopolysiloxane containing at least two hydrogen atoms bonded to silicon atoms which are in the same molecule, in which the total number of lower alkenyl groups and hydrogen atoms bonded to silicon in the mixture is at least 5;

(3) a catalyst; and

(4) an organosilicon compound which has at least one $Q(R''O)_2Si-$ group and at least one lower alkenyl group or hydrogen atom bonded to silicon atoms which are in the same molecule, wherein Q is a monovalent hydrocarbon radical containing at least one epoxy group and R'' is a lower alkyl group.

(1) may be a dimethylvinylsilyl endblocked polydimethylsiloxane, and (2) a trimethylsilyl endblocked methylhydrogenpolysiloxane and (1) and (2) are present in the ratio of 1.0 silicon-bonded lower alkenyl group in (1) to 1.0 silicon-bonded hydrogen group in (2). (3) is chloroplatinic acid and is present as 10 ppm of platinum in the form of the chloroplatinic acid in ethanol. (4) is present as 1.0 weight percent based on the total weight of the components and may have any of the following average formulas, for example:

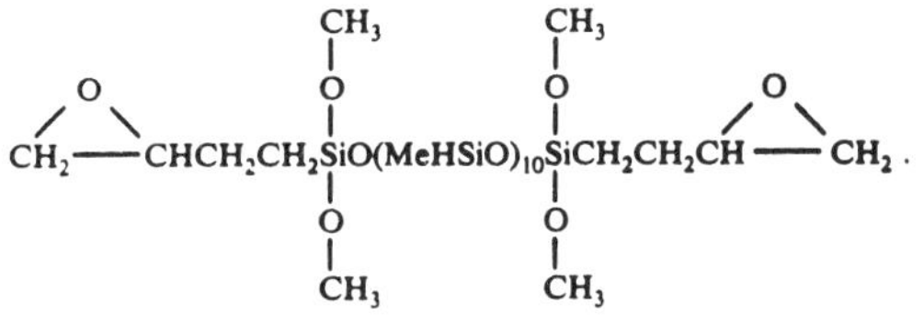

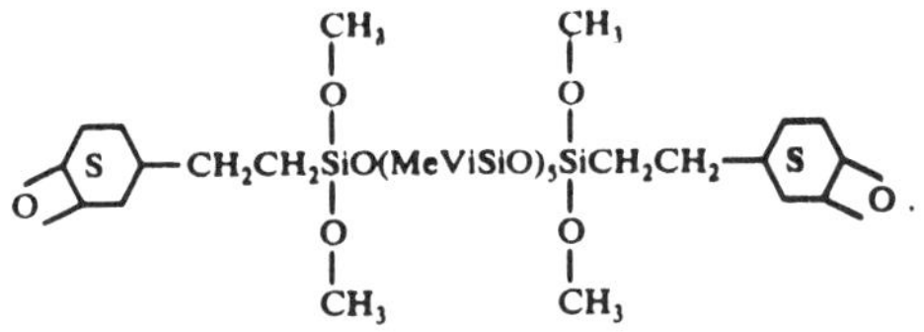

or

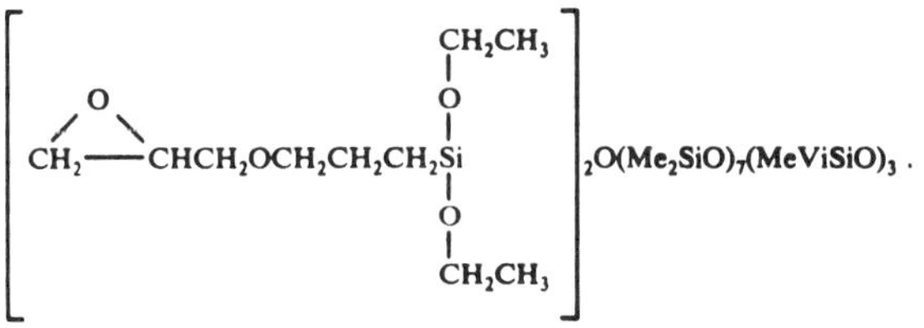

(In the formulas, Me = methyl, Et = ethyl, Vi = vinyl.)

The curable organopolysiloxane composition is prepared by mixing all four in-gredients (1)(2)(3)(4). If all four ingredients are mixed together, the cure reaction initiates at room temperature and therefore it is preferred not to so mix the four ingredients all at the same time unless the curable organopolysiloxane is to be used right away.

In order to avoid premature curing, the components (1)(2)(3) and (4) can be separated into two mixtures which are then mixed together just before use. The cured organopolysiloxane adheres very strongly to metal, glass, ceramics, stones, concrete, wood, paper, fibers, plastics, rubbers, etc.

Example: 65 parts of a copolymer of dimethylsiloxane and methylphenyl-siloxane blocked at both ends with methylphenylvinylsilyl groups and having a viscosity of 9,000 cs (mol ratio of dimethylsiloxane units to methylphenyl-siloxane units was 90:10) and 35 parts of a copolymer of SiO_2 units, $Me_3SiO_{0.5}$ units and $Me_2ViSiO_{0.5}$ units (containing 2.5% of vinyl group) were mixed. 4 parts of $Si(OSiMe_2H)_4$ and 1.1 part of the compound represented by the follow-ing formula

and 20 parts of platinum per million parts of the total amount of both the aforementioned copolymers in the form of platinum powder dispersed on car-bon powder were added to 94.9 parts of the mixture thus formed and were vigorously mixed and poured between a polyimide film and a copper plate. This composite was placed in an oven, and was cured by heating at 200°C for 20 min-utes. When the composite thus prepared was cooled to room temperature, one end of the polyimide film was bent in a 180° reverse direction and was drawn. The polyimide film was broken before the silicone elastomer layer was broken.

As a comparison example, when the mixture to which only the compound repre-sented by the above formula was not added, was subjected to an adhesion test under the same conditions, the silicone elastomer layer was then separated from the polyimide surface. Adhesive failure had taken place.

Aqueous Vinyl Ester Emulsions plus Allyl Carbamate

In the formulation of *M.K. Lindemann; U.S. Patent 4,115,306; September 19, 1978; assigned to Chas. S. Tanner Co.* aqueous adhesive emulsions based on copolymers of vinyl esters such as vinyl acetate, especially with ethylene, are provided with improved solvent insolubility and adhesion characteristics in the presence of hydroxy functional protective colloid (polyvinyl alcohol) which is present during the copolymerization, by the inclusion in the copolymer of a monoethylenic carbamate, especially allyl carbamate.

Example: The following mixture is charged to a 17 gallon pressure autoclave equipped with agitator and temperature controls: 15,510 grams of water and 1,373 grams of a combination of two polyvinyl alcohols.

The above solution is premixed in a vessel equipped with jacket heating and agitation. The water is charged to the vessel at room temperature and the poly-vinyl alcohols are added as dry powders with agitation. The mixture is heated to 180°F and held at this temperature for 2 hours, and the solution is allowed to cool and is then filtered through 4 ply cheesecloth. The resulting solution is then added to the pressure autoclave. The following materials are then added in sequence to the autoclave:

Material	Grams
1 Allyl carbamate	156
2 Ferrous sulfate	
(1% aqueous solution)	4
3 Phosphoric acid	10
4 Vinyl acetate	26,000

The autoclave is then purged to remove oxygen by pressurizing to 20 psig with nitrogen and then evacuating to 2 psig. This nitrogen purge procedure is repeated once with nitrogen and then twice with ethylene. After the last purge, the auto-clave is pressurized with ethylene to 600 psig and this pressure of ethylene is maintained until about 95% of the vinyl acetate has been consumed in the polymerization reaction.

The polymerization reaction is initiated and maintained by keeping the contents of the autoclave at a temperature of $57° \pm 2°C$ during the reaction period and by appropriate additions of an oxidant solution and a reductant solution. More particularly, the reaction mixture is maintained in a reducing state by addition of a reductant solution constituted by 300 grams of sodium formaldehyde sul-foxylate in 2,700 grams of water and the exothermic polymerization reaction is controlled by addition of an oxidant solution constituted by 300 grams of a 30% aqueous solution of hydrogen peroxide diluted with 2,940 grams of water.

The reaction mixture is sampled hourly for pH and total solids content, and whenever the batch solids exceeds 58%, it is diluted with water to 56%. When the unreacted vinyl acetate content drops to 5% of its original amount, the re-maining oxidant solution (modified by the addition of 10 grams of tert-butyl hy-droperoxide) is added to permit the reaction to be completed. This is achieved as follows.

When the proportion of unreacted vinyl acetate has dropped to 2% of its original amount, the autoclave is vented to reduce the pressure to 130 psig, and the con-tents are then transferred to a preevacuated pressure vessel where the temperature is maintained until the vinyl acetate content is less than 0.3% of its original amount. The total solids content is then adjusted with water to 55-57% and 20% aqueous sodium benzoate is added to adjust the pH to 4-4.5. An emulsion having the following properties is obtained.

Solids content = 55.5%
Brookfield viscosity (60 rpm–#4 spindle) = 2,500 cp
Insolubility in trichloroethylene (12 hr air dry) = 67.2%
Intrinsic viscosity (dimethyl formamide) = 1.39
Tg (differential thermal calorimetry) 5°C

In the absence of the allyl carbamate, the air dry insolubility would be about 40%, so the 67.2% insolubility obtained represents a marked improvement. The result is an adhesive emulsion exhibiting superior adhesion, particularly to polypropylene surfaces.

Adhesive for Protective Vinyl Film on Gym Mat

In order to produce an acceptable impact absorbing article, for example, a gym mat, a foam material must have the proper shock absorbing and safety features and be soft enough to permit the mat to be rolled for shipping and storage purposes. Polyvinyl chloride foams are usually provided with a flexible vinyl protective finish of from 10-25 mils thick.

A polyethylene crosslinked closed cell flexible foam meets the basic requirement for an impact absorbing material and offers a number of important advantages (in addition to cost considerations) over a PVC closed cell foam, e.g. less shrinkage after coating (1% vs 15% for PVC foam), lighter in weight (approximately 30% less), low temperature flexibility, ease of heat sealing for bonding purposes. While either foam without such a protective film meets the requirements of a gym mat, such a foam is rather weak and will be readily destroyed in use.

To date, the major drawback to the use of gym mat applications of a polyolefinic-containing foam has been the inability to obtain adhesion of a protective vinyl film to the polyolefinic foam.

An adhesive primer composition has been found by *R.J. Guglielmo, Sr.; U.S. Patent 4,129,676; Dec. 12, 1978* to be effective in achieving bonding between the foam and the vinyl protective film. The composition is coated onto the polyolefinic-containing foam prior to application of the vinyl film and is comprised of a chlorinated hydrocarbon solvent in which there are added a material selected from the group consisting of a chlorinated paraffin, an acrylic type ester, and a vinyl copolymer.

The chlorinated paraffin is a liquid resin made by controlled chlorination of paraffins to a chlorination level of 40-80%, such as the chlorowaxes, with a number designation which indicates the degree of chlorination. Acrylic resins are readily available commercially, e.g., Acryloid A-10 from Rohm and Haas. DuPont makes vinyl copolymers such as Elvax 40.

Chlorinated hydrocarbon solvents are well-known. 1,1,1-trichloroethane has been found to be particularly useful since it may be readily handled with standard ventilation requirements.

Example 1: A polyethylene foam was coated by spraying a primer comprised of the following components:

Ingredient	% by Weight
1,1,1-trichloroethane	90
Cyclohexanone resin	8
PVC copolymer resin	2
	100

After 10 minutes, the foam is dry enough to permit application of the vinyl coating which is effected by painting a solution formulation of polyvinyl coating. After 24 hours, the foam is rolled and packed for shipping in the usual manner.

Example 2: The operational steps of Example 1 are followed, except a primer comprised of the following components is used:

Ingredient	% by Weight
1,1,1-trichloroethane	90
PVC copolymer resin	6
Chlorowax No. 70	4
	100

Example 3: A primer having unlimited or universal application is comprised of the following components:

Ingredient	% by Weight
1,1,1-trichloroethane	97.375
Chlorowax No. 70	0.875
Acryloid B-66	0.875
Elvax 150	0.875
	100.000

Containing a Substituted Silane

The adhesive formulated by *P.R. Lakshmanan; U.S. Patent 4,133,789; January 9, 1979; assigned to Gulf Oil Corporation* is useful in bonding a low-energy plastic surface to metal such as, for example, in construction, automotive applications, fabricating, packaging, electrical insulation, and radiation shielding, etc.

It provides an improved tensile lap shear strength when compared to the composition without silane and contains the following components: (1) an ethylene-vinyl ester copolymer; (2) a tackifier selected from the following group: (a) a rosin, (b) a rosin ester, (c) a terpene resin, (d) a terpene-phenolic resin and (e) a rosin-modified phenolic resin; (3) a silane having the following structural formula:

$$R(CH_2)_n[NH(CH_2)_3]_mSiX_3$$

wherein R is selected from the group consisting of amino, mercapto, glycidoxy-propyl, epoxycyclohexyl, methacryloxy and vinyl; n is an integer from 0 to 6; m, when R is amino, is an integer from 0 to 1; m, when R is selected from the group consisting of mercapto, glycidoxypropyl, epoxycyclohexyl, methacryloxy and vinyl, is the integer 0; and X is selected from the group consisting of Cl, OCH_3, OC_2H_5 and $OC_2H_4OCH_3$; and (4) a wax.

Ethylene-vinyl ester copolymers that can be used include vinyl esters having from 2 to 4 carbon atoms. Examples of suitable ethylene-vinyl ester copolymers are ethylene-vinyl acetate, ethylene-vinyl formate, ethylene-vinyl propionate, ethylene-vinyl butyrate and mixtures thereof. The copolymer can contain preferably about 12 to about 25 weight percent of a vinyl ester, and can have a melt index as measured by ASTM 1238 52T of preferably 2.0 to about 100. These copolymers can be prepared by any method generally employed in the art.

Suitable tackifiers can be selected from the group consisting of (a) a rosin, (b) a rosin ester, (c) a terpene resin, (d) a terpene-phenolic resin and (e) a rosin-modified phenolic resin. Tackifiers for use in the formulation have a softening point preferably in the range of about 65° to about 135°C.

Especially preferred silanes for use in this formulation are gamma-amino-propyl-triethoxysilane: $NH_2(CH_2)_3Si(OC_2H_5)_3$; N-beta-(aminoethyl)-gamma-aminopro-pyltrimethoxysilane: $NH_2(CH_2)_2NH(CH_2)_3Si(OCH_3)_3$; and gamma-mercaptopro-pyltrimethoxysilane: $HS(CH_2)_3Si(OCH_3)_3$.

The fourth component of the adhesive composition is a wax. Any suitable wax, natural or synthetic, can be employed. The preferred waxes include petroleum-derived waxes such as, for example, microcrystalline waxes and paraffin waxes; intermediate waxes which are hybrid materials derived from heavy waxy distillates and having physical and functional properties intermediate to those of fully refined paraffin waxes and microcrystalline waxes; and polyethylene.

For purposes of this process a low-energy plastic surface is defined as one which has a critical surface tension (γc) of between about 24 to about 37 dynes/cm (mN/m), e.g., polyethylene ranges from about 25.5 to about 36 dynes/cm, and polypropylene ranges from about 24 to about 34 dynes/cm.

Any metal can be used but those which are preferred are those which have industrial uses such as automotive, electrical and radiation applications. Among these metals are included, for example, lead, copper, bronze, steel, stainless steel and aluminum, and metal blends and/or alloys containing one or more of these enumerated metals.

In general, the adhesive composition can have the composition on a weight percentage basis as set forth in the table below.

Adhesive Composition

	Broad Range, (wt %)	Preferred Range, (wt %)
Ethylene-vinyl ester copolymer	20–55	25–52
Tackifier	25–40	25–40
Silane	0.25–5	1–3
Wax	5–54.75	5–49

A preferred procedure for preparation involves the use of a so-called melt mixing technique in which the wax (diluent) together with an antioxidant, if used, is maintained in a stirred molten state at a temperature between about 130° to about 230°C, preferably between about 140° to 170°C, during which time the tackifier is added followed by the addition of the ethylene-vinyl ester copolymer. Mixing is continued until a homogeneous mixture is obtained at which point the temperature is lowered to about 138° to about 144°C, preferably about 130° to about 137°C and the requisite amount of the silane is added and mixed at the stated temperature for about 10 to about 15 minutes. Generally, the total time for mixing an adhesive composition is from about 20 minutes to about 4 hours.

The adhesive composition is used to bond a low-energy plastic surface having a critical surface tension of about 24 to about 37 dynes per centimeter to metal surfaces. The process involves heating the low-energy plastic surface which is to be bonded to the metal surface to a temperature of at least 50°C, preferably about 60° to about 130°C, and the metal surface preferably to about 150° to about 170°C; followed by an application of the adhesive composition while the adhesive is at a temperature preferably about 160° to about 220°C. Then the coated surface is brought in contact with the other. After assembly the bond reaches maximum strength in a matter of minutes or few hours. No post-curing is required and, therefore, rapid assembly and handling of assembled parts is possible.

Adhesives for Laminated Film for Packaging

The packaging industry, particularly the food packaging area thereof, is currently utilizing large quantities of flexible films. Since all the properties desired in such films are not available in any one film, the industry generally employs laminates prepared from a combination of films. Most often, these laminates are formed from saran, or saran coated substrates, Mylar, nylon film or paper laminated onto a film or sheet of a corona treated polyolefin.

J. Sirota; U.S. Patent 4,137,276; January 30, 1979; assigned to National Starch and Chemical Corporation has formulated a curable two-part laminating adhesive which comprises:

(a) an alcoholic solution of a ketimine and/or amine terminated polyetherurethane-urea prepared by the addition of at least 50% of the stoichiometric equivalent of a diketimine having —NH functionality of 0, 1 or 3 to an isocyanate terminated polyalkylene ether diol urethane prepolymer; and

(b) a polyepoxide in a chain-extending and cross-linking amount of about 10 to 30% by weight based on the ketimine and/or amine terminated polyetherurethane-urea solids.

Such laminating adhesives may be stored indefinitely as a two-part system and combined with further dilution if desired, immediately prior to use.

In forming the laminates desired, conventional techniques are employed to apply the adhesive solution to one of the film webs. The adhesive coated web is then ordinarily passed through an oven to remove substantially all the solvent, and then heat nipped to form a bond with the other web.

The resultant laminate is characterized by the formation of a strong bond which cures to "tear" within 24 hours after formation, the bond being heat, water and chemical resistant, and, depending upon the particular substrate films used, capable of being employed in a variety of end-use applications. In the examples below, the following test procedures were used to evaluate the adhesives.

Peel Adhesion (Strength Test): A Suter Testing Unit was run at a rate of 12 inches per minute to pull apart a 1 inch wide strip of the laminate, and the force needed is shown in grams. Preferably, the adhesive should cure to tear, i.e., the base film should rupture instead of the film separating. Tests were done initially (for green strength), overnight and after one week storage.

Static Load Test (Using 1 Week Old Laminates): A one inch wide strip of the laminate is prepared in which no adhesive is applied to approximately the upper ½ inch. One substrate film is then attached vertically to a heavy cardboard card and a 35 g weight (including clip) is attached to the other film substrate. The assembly is mounted in an oven at 82°C for 15 minutes and then removed. The delamination (slippage or creep) caused by the weight is measured in inches. A fully cured, heat resistant bond will show no (0") delamination.

Boiling Water Test (Using 1 Week Old Laminates): A 2" x 3" swatch of lamination is placed in boiling water and removed after 15 minutes. A tunnel or blister would indicate failure of the adhesive due to water sensitivity or thermoplasticity.

Example 1: 60 g of poly(1,2-oxypropylene) glycol having a hydroxyl number of 110 and MW of 1,000 was allowed to react at 54°C for 2 hours with 14 g toluene diisocyanate (mixture of 80% 2,4-toluene diisocyanate and 20% 2,6-toluene diisocyanate) and 0.02 g dibutyltin dilaurate as catalyst. This prepolymer had 2.3% free NCO or 0.041 mol of NCO. Then 8 g of 50% diethylene triamine dimethyldiketimine in excess acetone (0.022 mol NH), further diluted with 23 g ethanol (0.50 mol) was added and mixed for 1 hour. In terms of equivalents, the diketimine was theoretically 54% of the available NCO (i.e., 0.022 mol NCO) with the remainder of the NCO (0.019 mol) capped via reaction with the alcohol. Thereafter, 3 g of water was added and the final product adjusted to a total solids content of 70% with alcohol.

To 110 g of this solution (containing 0.088 mol amine hydrogen still available from the diketimine) was added 17 g of an epoxide resin of the p,p'-isopropylidenediphenol-epichlorohydrin type (Epon 828) having an epoxide equivalent of 190. This represented 22 parts Epon resin on 100 parts polymer solids and 102% of the available amine hydrogens. This mix was then diluted to 25% solids with isopropanol and 1½ pounds per ream coatings applied and dried on ½ mil 50M nylon film. This was then immediately laminated to 2½ mil low density polyethylene (LDPE). Peel test values on 1" x 3" strips were as follows: immediate green strength = 620 psi, overnite values = 1,600 psi with tear at glue line, 1 week values = 1,800 psi with tear at glue line. Static Load Test gave 0 delamination and Boiling Water Test showed no failure.

Example 2: A prepolymer was prepared according to procedure of Example 1 using the same amounts and reagents employed therein. Then a mix of 5.53 g tetraethylene pentamine dimethyl diketimine and 1.82 g ethylenediamine dimethyldiketimine in 3.32 g excess acetone (0.0615 mol NH), further diluted with 24.72 g ethanol was added and mixed for 1 hour. In terms of equivalents, the diketimines represented theoretically 150% of the available NCO, i.e., a 50% excess and therefore no capping with alcohol occurred in this example. Then 3 g of water were added with additional ethanol to dilute product to 60% solids and a viscosity of 2,000 cp.

To 112.41 g of this solution (containing 0.134 mol amine hydrogen available from the diketimines) was added 20 g of a p,p'-isopropylidenediphenol-epichlorohydrin type (Epon 828) having an epoxide equivalent of 190. This represented 25.5 parts Epon resin on 100 parts polymer solids and 100% of the available amine hydrogens. This mix was then diluted to 25% solids with isopropanol and 1½ pounds per ream coating applied and dried on ½ mil 50M Mylar film. This was then immediately laminated to 2½ mil LDPE. Peel test values on 1" x 3" strips were as follows: green strength = 650 psi; overnite values = 2,100 psi with tear at glue line; 1 week values = 2,000 psi with tear at glue line. Static Load Test gave 0 delamination and Boiling Water Test showed no failures.

Improving Adhesion of Aluminum to Polysulfides

Aluminum, often with an anodized surface, is often used as a construction material for window frames, with which a polysulfide material is used as the glass-to-aluminum sealant. The adherence of the aluminum to the polysulfide usually tends to weaken upon exposure to water and heat and therefore there has been a particular need to improve the adhesiveness of such materials in a moist and warm environment.

T. Børresen and N.U. Harder; U.S. Patent 4,138,526; Feb. 6, 1979 have found that a strong and water-resistant bond between aluminum and a polysulfide material can be achieved in a very efficient, simple and inexpensive way. Namely, they have discovered that excellent adhesion can be achieved by applying to the aluminum metal a solution of a strongly basic-reacting inorganic alkali metal compound.

As the inorganic alkali metal compound or base there is preferably used one or more of the strongly basic reacting hydroxides of lithium, sodium or potassium. Also, sodium phosphate, potassium phosphate, sodium silicate and potassium silicate may be employed. Preferred inorganic bases also are carbonates of alkali metals, more particularly, lithium, sodium and potassium.

It is an advantage of the process that water can be used as a solvent. Also, lower alcohols can be used as a solvent for the inorganic base or, if desired, mixtures of such alcohols with water. The composition of the solvent is otherwise not critical, since the solvent is to be evaporated from the metal surface after priming.

The process can be used with very good results with electrolytically oxidized or chemically oxidized aluminum, but it is not necessary for the metal to be given such special oxidative treatment.

The concentration of base contained in the primer can be quite low and is desirably less than 5% by weight, preferably about 1% by weight. This concentration, however, is not critical for achieving good results. What is essential, however, is that the metal on priming and subsequent drying has a thin coating of the alkali metal compound contained in the primer.

Example: Aluminum extrusions which had been electrolytically oxidized were primed at room temperature by immersion in a solution as indicated below (% by weight). The last three tests were included for comparison with Tests 1 to 3 which were carried out according to the process described above.

Composition of Test Solutions

	. . . . Test No.					
	1	2	3	4	5	6
Water	98	99	–	98	–	–
Ethylene glycol	–	–	98	–	98	–
Na_2CO_3	2	–	–	–	–	–
NaOH	–	1	2	–	–	–
$Ca(OH)_2$	–	–	–	–	2	–
H_2SO_4	–	–	–	2	–	–
No priming	–	–	–	–	–	–

After drying for half an hour (in air at 28°C), the aluminum extrusions were joined to glass by means of a polysulfide jointing compound, PRC 408P, intended for the production of insulating windows. The test samples were then left to cure for one week at room temperature after which they were stored in water at 70°C for 4 weeks.

After this treatment the adherence between the jointing compound and the metal was tested.

Test Results

	. . . . .Tests					
	1	2	3	4	5	6
1 week in water at 70°C	K	K	K	A	K	K
2 weeks in water at 70°C	K	K	K	A	5A	K
3 weeks in water at 70°C	K	K	K	A	50A	10A
4 weeks in water at 70°C	10A	K	K	A	A	A

A: Rupture by adhesion failure to aluminum.
K: Rupture by cohesion failure of the jointing compound.
50A: 50% rupture by adhesion failure to aluminum.

For Use in Diazo Duplicating Material

Polyester sheets are used as a base for photographic films, drawing sheets or in duplicating materials. They are distinguished, after orienting and setting, by good mechanical properties, high chemical resistance and dimensional stability, which, in addition to other properties, explains their economic significance.

In the manufacture of coated polyester films a problem arises in obtaining good adhesion between the support sheet and the layers to be applied while still being able to roll and unroll the coated materials without blocking.

In diazo printing, alkaline development, for example, is necessary in the production of images after exposure under an original with a suitable UV-lamp in order to produce the diazo dyestuff at the areas that have not been exposed. In this development the sheets are subjected to elevated temperatures in an alkaline medium also in the presence of water or water vapor. It is particularly important here for there to be a flawless, strong bond between the polyester sheet and the applied layer or layers.

K. Thoese and K.-H. Jung; U.S. Patent 4,139,506; Feb. 13, 1979; assigned to Hoechst AG, Germany have provided a composition, suitable for use as an adhesive layer on a shaped structure of a polyester, which comprises a blend of a copolyester, a polyisocyanate and an organic polymer compatible with the blend of the copolyester and the polyisocyanate, the ratio of copolyester to the polymer being in the range of 20:1 to 0.5:1, preferably 10:1 to 1:1 by weight.

Example: A solution of the following composition is applied to a 100 μ thick, glass-clear sheet of biaxially oriented polyethylene terephthalate:

> 150 g of an 18% by weight solution of a commercially available copolycondensate of terephthalic acid with isophthalic acid and sebacic acid as acid components and ethylene glycol, diglycol and butanediol as alcohol components,

> 5 g of a film-forming polyisocyanate, dissolved in ethyl acetate, as a crosslinking component,

3 g of cellulose propionate having an acetyl content
of 3.6%, a propionyl content of 44.8% and a hydroxyl content of 1.8%,

332 g of acetone, and

110 g of toluene.

The solution dries to a clear film of 0.5 g/m^2 thickness and, even at high relative
humidity, can be stored on a reel and unwound from the reel when required
without blocking and sticking.

If the 0.5 g/m^2 thick layer does not contain the addition of cellulose propionate
then the sheet cannot be unwound from the reel without damage to the surface
even if the thin substrate has been dried at temperatures of 140°C.

To test the adhesion of the substrate with and without the addition of cellulose
propionate, a lacquer containing cellulose propionate as binder and with cou-
plers, stabilizing additives and diazo compounds typically used in the production of a blue dyestuff, was applied in a 6 μ thick layer and dried for three minutes at 110°C.

After exposure and treatment with water vapor and ammonia, to test the adhesion the layers were scored according to DIN 53,151 with a cross-hatch adhesion
test apparatus type GS 30. Then adhesive tape [Tesaband (R) No. 104] was
pressed onto the marked areas of the films and removed again with a jerk. The
adhesion of both films was good but in the case of extreme stress before the
adhesion test the adhesion to the substrate according to the described process
is better.

Laminates Having an Adhesive Barrier Layer

*D.D. Dixon and M.E. Ford; U.S. Patent 4,142,021; Feb. 27, 1979; assigned to
Air Products and Chemicals, Inc.* have developed laminated structures having
excellent resistance to moisture and oxygen vapor transmission which are
particularly adapted for the food packaging industry. Broadly, the laminate
structure comprises a base layer and an adhesive barrier layer bonded to the
base layer, the adhesive being a polyalkylene carbonate represented by the
formula

$$\left(\begin{array}{c} R_1 \ R_3 \quad\ O \\ \ | \quad\ | \qquad \| \\ -O-C-C-O-C- \\ \ | \quad\ | \\ R_2 \ R_4 \end{array} \right)_n$$

wherein:

R$_1$ is hydrogen, a halomethyl group or a hydrocarbyl
group having from 1 to 6 carbon atoms;

R_3 is hydrogen, a halomethyl group, or a hydrocarbyl group having from 1 to 6 carbon atoms;

R_2 and R_4 are hydrogen, or hydrocarbyl groups having from 1 to 6 carbon atoms; or

R_1, R_2, R_3 and R_4 are combined together constituting a carbocyclic ring having from 4 to 7 carbon atoms; and

n   is a number from about 250 to about 6,500.

In a preferred embodiment, the laminate has at least three layers with the base layer generally being a thermally formable, heat sealable polymer, the polyalkylene carbonate layer as the intermediate and adhesive layer, and a top layer which is a dimensionally stable polymer or paper. Advantages of this laminate structure include:

(1) excellent resistance to the transmission of oxygen and moisture vapor;

(2) excellent peel strength by virtue of the outstanding adhesion of the polyalkylene carbonate to the base layer and, in preferred situations, to the base layer and to the top layer or top coat;

(3) nontoxicity of the adhesive and barrier material;

(4) toughness and puncture resistance;

(5) excellent resistance to oil permeation and therefore suitability for packaging oil base products and

(6) it is readily manufactured to form a heat sealable package.

The polycarbonates which are employed as the adhesive barrier layer of these laminate structures are of the type which are normally solid at room temperature (70°F) and atmospheric pressure, and are formed by polymerizing an epoxide with carbon dioxide to form a polymer having alternating units of epoxide and carbon dioxide. The molecular weight of the polycarbonates generally ranges from about 50,000 to about 500,000, and preferably the molecular weight range is from about 75,000 to 150,000.

Example 1: A 5 g sample of poly(ethylene carbonate) formed by the procedure of Example 1 in U.S. Patent 3,900,424 having alternating units of ethylene oxide and carbon dioxide, a molecular weight of about 100,000, and a melting point of 190°-200°C was placed between two 6" squares of polyethylene film, the polyethylene film being 5 mils in thickness. The poly(ethylene carbonate) then was sandwiched between the polyethylene film by pressing at 5,000 psia in a Carver laboratory hydraulic press at a temperature of 140°-150°C. A laminate was formed and the oxygen transmission data was 0.1-0.15 cm^3/mil. The double layer control film of polyethylene had an oxygen transmission of 102-123 cm^3/mil.

The laminate has an advantage over many other laminates in that it is readily heat sealed utilizing conventional techniques.

Example 2:   Poly(propylene carbonate) having a molecular weight of about 100,000, melting point of $180°$-$190°C$ and alternating units of propylene oxide and carbon dioxide was prepared in a manner similar to that of Example 1, except propylene oxide was used in place of ethylene oxide. The poly(propylene carbonate) was dissolved in chloroform to form a 45% poly(propylene carbonate) solution. This solution then was cast over a 6" square of polyethylene. The solvent was evaporated to leave a film over the polyethylene base film. The resulting laminate had a vapor transmission rate of 2-22 cm^3/mil. The single layer of polyethylene film had a transmission rate of 336-405 cm^3/mil.

Utilizing Toluene Diisocyanate Residue

Diisocyanates, and especially toluene diisocyanates, are important industrial products useful in the manufacture of urethane polymers. They are produced commercially by phosgenation of diamines or mixtures thereof in the presence of a solvent followed by distillation to recover solvent and the diisocyanates or mixtures thereof. Unfortunately, distillation of crude toluene diisocyanates leaves a residue of material from which it is difficult to recover a large portion of the toluene diisocyanate and the material containing the TDI is discarded as a waste product. Toluene diisocyanate is manufactured in large quantities, thus disposal of the large quantities of residue wastes has become a serious problem.

As used herein TDI refers to toluene diisocyanate; and "TDI residue" as used herein is the residue remaining after vaporizing toluene diisocyanate from a reaction mixture, which results from substantially complete phosgenation of toluene diamine, until between about 12 and about 30% by weight of toluene diisocyanate remains in the residue. The residue includes from about 6 to about 25 weight percent of unreacted isocyanate groups, and is solid at $25°C$.

L.M. Zwolinski and J.W. Frink; U.S. Patent 4,143,008; March 6, 1979; assigned to Allied Chemical Corporation have found that the TDI residue can be utilized as a binder material when in finely divided form and also can be utilized to form adhesive, molding and binder compositions which comprise as an active ingredient up to about 95 weight percent of TDI residue and about 5 weight percent of a component which is free from toluene diisocyanate.

"Finely divided" as used herein means having an average particle size smaller than about 40 mesh. "Active ingredient" as used herein is intended to denote amounts greater than 0.5% by weight and, preferably, at least about 7% by weight of an ingredient which increases the cohesive or adhesive strength of the composition. The TDI free portion of the composition may be any inert substance adhered to or bound by the TDI residue, a solvent for the TDI residue or an active ingredient other than TDI residue.

The unreacted isocyanate groups in the TDI residue permit the formation of chemical bonds which create strong attachment of the residue to a substrate to be bonded.

Adhesives prepared by incorporating TDI residue can be used to bind a large variety of materials and may be either solvent adhesives, nonsolvent room temperature adhesives or hot melt adhesives. The adhesives may be used to bind a wide variety of materials including polyethylene, polypropylene, polyamides, paper, wood, cellulose, metals and glass. Particularly good adhesives for binding all of the above materials are made from a composition comprising rubber cement and TDI residue, wherein the TDI residue comprises between about 1 and about 20 weight percent of the adhesive composition.

Example 1: A rubber cement containing 40% by weight of butadiene rubber in toluene is prepared. Two 2" x 7" strips of cotton duck fabric are masked on one end with 2" x 2" strips of tape. The cement to be tested is spread thinly in equal amounts onto the exposed fabric on the taped side of each of the two strips of the cotton duck, so that the entire fabric surface is coated. The specimens are then air-dried for about 3 minutes until almost tack free and then joined together on the coated sides and rolled several times with a roller. The specimens are then cured overnight at 110°F. The adhesion of the strips is then tested. The rubber cement has a 5.0 lb/2" width strength.

To 90 parts by weight of the rubber cement prepared above is added 10 parts by weight of a 40% solution of TDI residue in o-dichlorobenzene to form an adhesive composition. Two 2" x 7" strips of cotton duck are prepared with the adhesive containing TDI residue using the above method. The strength of the bond is found to be 9.2 lb/2" width.

Example 2: A composition suitable for use as laminating adhesive, or a compression molding material for solid polymer structures is prepared by blending, 1 part by weight of finely ground, smaller than 40 mesh, TDI residue with 2 parts by weight of a high molecular weight hydroxy terminated polyester resin. The mixture is compressed at 2,000 psi at 300°F for three minutes. A solid, stiff, dry-feeling, amber colored urethane polymer film is produced. The film is flexible without cracking over a 90° radius and no odor of free isocyanate is detected. The composition prior to molding is found to be an effective hot melt adhesive.

Calcium or Magnesium Chloride Added to Polyolefin Adhesive

A. Mori, T. Okita, S. Kitamura, K. Kotani and M. Hama; U.S. Patent 4,144,113; March 13, 1979; assigned to Sumitomo Chemical Company, Ltd., Japan have found that the adhesion of plastic films to metals in salt water can be considerably improved by the addition of calcium chloride and/or magnesium chloride to the carbonyl-containing polyolefin adhesives used for bonding them.

The polyolefins containing carbonyl groups which can be used include (1) polyolefins modified with an acid anhydride and (2) bi- or multicomponent copolymers of ethylene and ethylenically unsaturated monomers containing a carbonyl group, and the saponified products thereof.

Anhydrous $CaCl_2$ is the most effective additive, and it may be used in a particle size of from 0.0001 to 0.2 mm and in an amount of from 0.1 to 10% by weight, based on the total weight of the composition.

Example: 5 g of anhydrous calcium chloride was mixed with 95 g of a vinyl acetate/ethylene copolymer (vinyl acetate content: 30 wt %; melt index (ASTM-1238-57T): 20 g/10 min), and the mixture was kneaded using a two-roll machine with the rolls being held at 50°C to form sheet-like thermofusible adhesive having a thickness of 1 mm. The resulting adhesive was interposed between a degreased iron sheet and a sheet of low density polyethylene [density (JIS-K6760-1966): 0.920 g/cm^3; melt index (ASTM-1238-57T): 7 g/10 min] , and the assembly was heated for 5 minutes at 200°C under a pressure of 50 kg/cm^2.

Results of evaluations showed that calcium chloride was very effective for improving the waterproof adhesiveness of the adhesive.

The salt water-resistance testing was carried out by immersing a test piece of a resin-coated metal, in which a slit leading to the surface of the metal had been made, in a 3% by weight salt water solution at 60°C for 10 days, and evaluating the resistance in terms of the area of the coated resin which was peeled off.

The salt crock-resistance testing was carried out by using the same test piece as a cathode and graphite as an anode, and passing a direct current of 3 V for 10 days through a 3% by weight salt water solution at room temperature (about 20°-30°C), and evaluating the resistance in terms of the area of the coated resin which was peeled off.

No peeling was evident in the testing of the formulation of the example, whereas when the $CaCl_2$ was not added to the adhesive, 4 cm^2 peeled in the salt water test and 36 cm^2 in the salt crock test.

For Thermosetting Layers Containing Mold Release Agents

W.D. Cottrell, Jr. and R.B. Jutte, Jr.; U.S. Patent 4,144,305; March 13, 1979; assigned to Owens-Corning Fiberglas Corp. have discovered that it is possible to adhere a thin sheet of particular kinds of thermoplastic materials having certain particular properties to molding compounds even though they contain mold release agents.

The thermoplastic material must be dissolvable at molding temperatures by the solvents and particularly the crosslinking monomers that are used in the molding compound. The layer of thermoplastic material need be only a few mils thick, and can adhere a second body of molding compound having mold release agent therein to the first body of molding compound through the thermoplastic sheet. The resulting interfaces, of course, cannot be precisely analyzed. However, the strong bonding of the thermoplastic sheets to the molding compounds seems to indicate that the molecules of monomers have in fact been utilized to adhere or bond the thermoplastic sheet to the crosslinked body of molding compound.

Example 1: A polyester resin was made from the following materials in parts by weight:

Materials	Parts by Weight
Propylene glycol	578.2
Maleic anhydride	674.4
Toluhydroquinone (25% solution in styrene)	4.8

The polyester was made by charging all the propylene glycol, one-third of the maleic anhydride and 1.6 grams of toluhydroquinone into a reactor using a continuous nitrogen sparge. The temperature was raised to 190°F, and after about 4 hours, the acid number was 35. Thereafter the remainder of the maleic anhydride was added at the rate of 100 to 150 parts per minute to control the temperature at 300°F.

Thereafter the temperature of the ingredients was held at 310° to 320°F for 60 minutes, following which the temperature was increased to 400°F. The material has an acid number of 29 to 32 and a sample cut 2 to 1 in styrene had a Saybolt viscosity of 21 to 25 seconds at 350°F. Thereafter the contents were cooled to 340°F. The polyester prepolymer when cut with styrene in a 90 to 10 ratio was stable at 120°C for 30 minutes before gelling.

In another tank 486.4 parts of styrene, 2.0 parts of MEHQ (methyl ether of hydroquinone) are mixed and held at a temperature between 130° to 145°F. Thereafter 1,138 parts of the polyester resin prepolymer produced as above described and at a temperature of 330° to 355°F was added with agitation to provide a thinned polyester syrup which was then cooled to a temperature of 180°F. The viscosity of the thinned polyester syrup was 1,500 to 2,000 Brookfield cp and had a water content of 0.08 and a monomer content within the range of 30 to 34% by weight.

A resin-filler paste is made of the following materials:

Materials	Preferred % by Wt
Unsaturated resin syrup described above	42.7
Dicumyl peroxide	0.85
2,5-dimethyl hexyl-2,5-di(peroxybenzoate)	0.09
Zinc stearate	1.71
Ca(OH)$_2$	1.28
Styrene monomer	4.27
Resin type (microethylene)	6.4
Nonresinous filler (CaCO$_3$)	43.7

The following mix procedure is used to make the resin-filler paste. The resin-filler paste is prepared by charging the resin to a Cowles type mixture. The 2,5-dimethyl hexyl-2,5-di(peroxybenzoate) dissolved in approximately half of the styrene is blended with the resin. Thereafter, the dicumyl peroxide and mold release agent are added. The resin type filler is then blended in, following which the nonresinous filler is likewise added and thoroughly dispersed. Immediately before the resin-filler paste is to be used, a slurry of the gelling agent in the other half of the styrene is added and mixed for approximately three minutes.

After aging for one week, a molded part approximately 0.100 inch thick was made using a male and female steel die set which was heated to 300°F. The female mold had a flat cavity 12 x 18 by 1 inch deep. Sufficient of the above molding compound was placed in the mold cavity to form a part 0.090 inch thick. A methyl methacrylate sheet 2 mils thick was placed on top of the molding compound, and the male die was brought down into engagement therewith with a force of 1,000 lb per square inch, and was held at this pressure and temperature for 90 seconds.

Thereafter the male die was raised to separate from the composite, and sufficient colored polyester molding compound was placed on top of the methyl methacrylate layer to form an additional thickness of 0.008 inch. The colored molding compound was obtained by mixing 1% by weight of carbon black with the molding compound described above. The male die was then brought down against the colored polyester layer with a force of 1,000 lb per square inch and this force was again held for 30 seconds. The male die was then raised and the composite removed from the mold and cooled. The top black layer could not be peeled from the base layer without destroying the composite.

Example 2:   The process of Example 1 was repeated excepting that the polymethacrylate film was replaced with a sheet of polystyrene. The composite so produced had substantially the same properties as did the composite of Example 1.

Example 3:   The process of Example 1 was repeated excepting that the polymethacrylate film was replaced with a sheet of poiyvinyl acetate. This composite so produced had substantially the same properties as the composite of Example 1.

Improvement of Adhesion of PVC to Metals

The resistance of polyvinyl chloride (PVC) or of its copolymers to degeneration in the presence of aggresive media has been known in the art for a considerable time and is widely used for the protection of metallic surfaces against corrosion, for adhering thin sheet metal constructions, and for sealing welded seams, particularly in the automotive industry.

For increasing the adhesion of these coatings, it has already been proposed to add adhesion-improving additives to the plasticized polyvinyl chloride, such additives being in the form of organic amine compounds or of esters of acrylic acid or methacrylic acid, or mixtures thereof. These two classes of materials nevertheless show serious disadvantages.

E. Bierwirth, N. Esper, C. Burba and R. Spickers; U.S. Patent 4,146,520; March 27, 1979; assigned to Schering AG, Germany have found that additives for PVC polymers can be made which do not have the disadvantages of the organic amines or acrylic esters. These adhesion-improving additives are condensation products prepared from

(A)   a polymerized fatty acid mixture having an increased content, X, in percent by weight, of tripolymeric and higher polymeric fatty acids and from

(B)   an excess of polyalkylene polyamines,

which condensation product has an imidazoline content, Y, in percent by weight. The additives are used in amounts from 0.5 to 5% by weight, based on the plasticized formulation, and the plastisols (polymers) are baked at temperatures from $90^{\circ}C$. If one of these values of X or Y falls below a limit of 40%, the value of the other component, to achieve a clearly recognizable effect, is at least 40 + Z (wherein Z is the difference between the smaller value and 40), but the value preferably should be 40 + 2Z.

The adhesion improvers are prepared from polycarboxylic acids and an excess of the amines. The condensation products suitably have an amine number between 200 and 600, preferably between 280 and 400.

The polycarboxylic acids can be prepared (a) by the polymerization of unsaturated fatty acids having from 12 to 22 carbon atoms, preferably 18 carbon atoms, and removal of unreacted monomeric fatty acids, or (b) according to a free radical mechanism in a directed reaction according to German Offenlegungsgeschrift 2,506,211. The following are typical compositions for the polycarboxylic acids:

Content	. . . . .Percent.	
	According to (a)	According to (b)
Monomeric fatty acid	0–5	1
Dimerized fatty acid	10–25	59
Trimerized and higher polymerized fatty acids	90–70	40

Amines suitable for reaction with the polycarboxylic acids are those amines capable of imidazoline formation, for example, polyalkylene polyamines such as diethylenetriamine, triethylenetetramine, tetraethylenepentamine, pentaethylenehexamine, and 3-(2-aminoethyl)aminopropylamine. The amidation reaction takes place at reaction temperatures between 210° and $260^{\circ}C$ optionally in vacuum. The reaction times are variable between wide limits depending on the imidazoline content desired. It is particularly advantageous that an imidazoline content greater than 40% be achieved.

A better understanding of the formulation and of its many advantages will be had by referring to the following specific examples. In the examples, the fatty acids have the following composition:

		Composition	**Percent**
(1)		Monomeric fatty acid	0
		Dimeric fatty acid	9
		Trimeric fatty acid and	
		higher polymeric fatty acid	91
(2)		Monomeric fatty acid	1
		Dimeric fatty acid	59
		Trimeric fatty acid and	
		higher polymeric, fatty acid	40
(3)		Monomeric fatty acid	9
		Dimeric fatty acid	75
		Trimeric fatty acid and	
		higher polymeric fatty acid	16
(4)		Monomeric fatty acid	1
		Dimeric fatty acid	25
		Trimeric fatty acid and	
		higher polymeric fatty acid	74
(5)		Monomeric fatty acid	1
		Dimeric fatty acid	95
		Trimeric fatty acid and	
		higher polymeric fatty acid	4

Example A: Polyaminoamide from Fatty Acid and Triethylenetetramine — 200.2 g of polymeric fatty acid (1) (saponification number, 196) are combined with 102.2 g of triethylenetetramine and reacted for 8 hours at a maximum temperature of 240°C. 18 ml of water distill over. The product has the characteristics: amine number, 350/351 and imidazoline content, 48%. A portion of the product is diluted to 80% with benzyl alcohol.

Example B: On reacting the remaining product with excess methyl isobutyl ketone, the ketimine of the polyaminoamide is obtained, having the following characteristic: amine number, 271/273.

Example C: Polyamide from Polymeric Fatty Acid and Triethylenetetramine — 292 g of polymeric fatty acid (2) (saponification number, 192.2) are combined with 146 g of triethylenetetramine, heated to 220°C, and held at this temperature for 3 hours. 22 ml of water are distilled off. The product has the characteristics: acid number, 2.1, amine number, 370/371, and imidazoline content, 33.2%.

Example D: In a further three hours at a maximum temperature of 245°C, an additional 10 ml of water distill off. The product has the characteristics: acid number, 1.1, amine number, 367/369, and imidazoline content, 67.2%.

Example E: On reacting product D with excess methyl isobutyl ketone, the ketimine of the polyaminoamide is obtained having the characteristic: amine number, 258/259.

Preparation of Plastisols: Two percent, by weight of the total mixture, of the aforementioned adhesion improvers A-E, or mixtures of these products with one another according to choice, is added to a plastisol comprising 45 parts by weight of polyvinyl chloride capable of being made into a paste and having a K-value of 70, 55 parts by weight of phthalic acid di(2-ethylhexyl) ester, 100 parts by weight of a filler mixture of 50% chalk and 50% barium sulfate, and 1.5 parts by weight of diisobutyltinisooctylthioglycolic acid ester.

The adhesion improvers can of course be added to other conventional plastisol formulations different from that given above in order to obtain the self-adhering plastisols according to this process.

The bonding strengths were determined according to DIN 53283 by measurement of the tensile shear strength. For this purpose, sheet metal bonding plates were used as bonding pieces. The size of the pieces to be joined: 2.5 cm x 10.5 cm x 0.15 cm (breadth x length x thickness).

The layer thickness of the plastisols in the adhesive joint was adjusted to 2 mm by means of spacers. The parts to be joined were heated at $160^{\circ}C$ for 30 minutes and bonded to each other with an overlap of 15 mm in this fashion. The following tensile shear strengths were measured.

Examples	Tensile Shear Strength (kg/cm^2)
A	14.5
B	26.8
C*	7.3
D	13.6
E	24.0
Comparison Example**	5.0

*This composition is presented for purposes of comparison.
**Polyaminoamide comprising dimeric fatty acid (3) and triethylenetetramine (amine number of 375, imidazoline content of 55%, trimer content, 15%).

Adhesive for Vinyl Film to Wood

The object of *O. Fogle, J. Cooley and M.E. Taylor; U.S. Patent 4,152,313; May 1, 1979; assigned to Champion International Corporation* was to provide an adhesive for preparing a vinyl film-wood product substrate laminate. The adhesive contains a major amount of a vinyl acetate-ethylene emulsion and a minor amount of a blend of toluene and N-lower alkyl substituted pyrrolidone.

In general, the adhesive contains about 85-95 weight percent of the emulsion and correspondingly about 15-5% of the blend, and preferably contains about 90% of the emulsion and 10% of the blend.

Vinyl acetate-ethylene emulsions are generally aqueous emulsions of the co-

polymer in which the vinyl acetate constitutes the major copolymerized monomer. The copolymer usually contains 75-90 weight percent vinyl acetate and
can also contain small amounts of other copolymerized material such as acrylic
acid or a sulfonic comonomer.

The solids content of the emulsion is adjusted such that the resulting viscosity
is appropriate for ease of coating and the solids content is usually between about
25 and 75% and preferably about 35 to 65 weight percent.

The other component of the adhesive formulation is a blend of toluene and an
N-lower alkyl substituted pyrrolidone. Lower alkyl encompasses 1 to 4 carbon
atoms and the preferred alkyl moiety is ethyl. The blend can be prepared in any
convenient manner using 1 to about 2 parts of one component per part of the
other component, i.e., in a weight ratio of about 2:1 to 1:2. Preferably, the two
components are combined in substantially equal weight amounts.

The adhesive composition is applied to the wood or wood product substrate by
conventional techniques such as spraying, doctoring, rolling, brushing, and the
like. In general, the adhesive is applied to provide a wet thickness of about 1 to
5 mils, and preferably about 2 to 3 mils. Thereafter the vinyl film is laminated
to the adhesive at elevated temperature and under pressure. The lamination temperature can be about 200° to 300°F, preferably about 240° to 260°F and pressures of about 8 to 40 kg/Lcm, preferably about 12 to 25 kg/Lcm are employed.

Example: An adhesive was prepared from 10 parts of a 50:50 blend of toluene
and N-methyl pyrrolidone and 100 parts of a vinyl acetate-ethylene copolymer
emulsion having a solids content of about 45%. The vinyl acetate constituted
about 20% of the copolymer. A lauan substrate was continuously conveyed
through several zones in a coating and laminating line. In the first zone, the
adhesive composition was direct roller coated on the lauan substrate at an application rate of 1 to 1.5 wet mils thickness. The second zone was a 30 foot long
oven maintained at a temperature of 100°-350°F and the coated substrate was
conveyed therethrough at a speed of 120 fpm to dry the first adhesive coat.

In the third zone, the coated substrate was direct roller coated with an additional 1 to 1.5 wet mil thick coating of the adhesive and the fourth zone was
a 40 foot long oven maintained at 100°-350°F through which the coated substrate was conveyed at a speed of 120 fpm.

In the next zone, the substrate with the dried adhesive coating was heated to a
surface temperature of about 250°F in order to reactivate the adhesive. A 2
mil thick polyvinyl chloride film was applied and the resulting composite was
then passed under a rubber roll maintained at 250°F in order to provide a
laminated composite about 150 mils thick.

The resulting board had a peel test rating of 7-8 lb/linear inch while the peel
test rating of the composite without the instant blend amounted to 2 lb/linear
inch.

For Polyvinyl Fluoride Sheets

Polyvinyl fluoride sheets (DuPont's Tedlar) are used, for example, as wall covering materials, since they have superior weather-, wear-, and solvent-resistance. An adhesive to be used for attaching the Tedlar film or sheet to the surface of a metal such as aluminum must have various requirements, for example, it should have high adhesiveness, high cohesiveness, a stability to hydrolysis and the like. If the adhesive does not meet these requirements, the film or sheet is apt to peel from the surface of the metal.

M. Nishimura, K. Naito, Y. Fujiwara and N. Kobayashi; U.S. Patent 4,158,725; June 19, 1979; assigned to Sony Corp., Japan have developed an adhesive which is made by copolymerizing a vinyl compound with a vinyl compound having an epoxy group and then reacting the copolymer thus produced with aqueous ammonia to aminate the same. The method for the manufacture of the adhesive comprises the following steps:

(1) Copolymerizing (a) vinyl compounds capable of copolymerizing with (b) vinyl compounds containing epoxy groups that produce a copolymer having an average molecular weight of from 50,000 to 500,000. Vinyl compounds (a) contain for every 100 parts by weight, at least 80 parts by weight of an acrylic or methacrylic acid ester of an alcohol containing 1 to 8 carbon atoms. The amount of epoxy compounds (b), is 1 to 10% by weight of the entire copolymer.

(2) Treating the resulting copolymer with aqueous ammonia in the presence of an amine catalyst in an amount of from 0.005 to 0.5 parts by weight per 100 parts by weight of copolymer to produce an aminated copolymer having an amine value of from 2 to 30. The amine catalyst must be present in sufficient amounts to substantially prevent gelatinization during the amination reaction.

The vinyl compounds with epoxy groups are selected from the group consisting of glycidyl methacrylate, glycidyl acrylate, glycidyl oxyethyl vinyl sulfide, glycidyl sorbate, glycidyl vinyl phthalate, glycidyl acryl phthalate, glycidyl acryl maleate, glycidyl acryl ether, butadiene monoepoxide, vinyl cyclohexane epoxide, glycidyl lycinolate and mixtures thereof, and the ratio of the vinyl epoxy compound to the entire copolymer is in the range of 2 to 5% by weight.

The aqueous ammonia is 5-30% by weight ammonia, and must be used in an amount sufficient to provide more than one molecule of ammonia for every epoxy group. The amination reaction is preferably carried out at a temperature of from 60°-80°C. A tertiary amine is the preferred catalyst, and the reaction time for the amination should be approximately 8 hours.

Metal Adhesives

EPOXY COMPOSITIONS

Primary Amine-Terminated Polyamides as Curing Agents

The use of epoxy adhesive technology as a substitute for spot welding, seam welding, and mechanical fasteners in assembly line processes to hold two substrates (particularly metal substrates) together may require an epoxy system with both an open time capability and the ability to be cured without resorting to ovens or oven-like equipment. Thus, the selection of a suitable epoxy system for a spot-welding approach to assembly-line bonding is by no means straightforward.

V.H. Batdorf; U.S. Patent 4,070,225; January 24, 1978; assigned to H.B. Fuller Company has found that when a suitable polyamide, dissolved in a suitable solvent is intimately blended with a suitable epoxide prepolymer, the solvent can be removed to form a uniform nonflowable curable adhesive mixture of epoxide and polyamide which has a very lengthy "open time." That is, the resulting, substantially solvent-free coating tends to be latent and may take days or even months to lose its ability to be cured with heat and pressure. Yet, the latent-curable coating or film can be cured sufficiently (e.g., with heat and pressure) to provide a minimum bond strength which increases with time.

By using conventional epoxides with epoxide equivalent weights greater than 60 and primary amine-terminated polyamide curing agents with a ball and ring softening point of 60° to 200°C a mixture can be formed which, upon the initiation of the epoxide cure reaction, gives sufficient immediate bonding strength to hold the substrates together and which will continue to cure after the initiation conditions have been removed; that is, under normal ambient conditions. The mixture can be coated as a film on at least one of a plurality of substrates which can later be mated under epoxide cure initiation conditions (e.g., a temperature above the ball and ring softening point of the polyamide) to form a bond between the substrates.

Stated another way, the process involves a method for forming an epoxy adhesive-bonded joint wherein: (a) a solution of the polyamide is applied to one or more surfaces, and this solution contains the epoxide prepolymer dissolved or dispersed therein; (b) most or substantially all of the solvent is removed from the resulting coating of adhesive (e.g., by evaporation); (c) the joint is formed by mating two surfaces and applying heat (and preferably also pressure) to provide epoxide cure initiation conditions; and (d) the epoxide cure initiation conditions are removed before the cure is complete. The cure then continues under normal ambient conditions, and can be completed under these conditions, if desired, e.g., in 1 to 40 days.

From the standpoint of effectiveness and commercial availability, an epoxide component typically preferred for use in this process comprises a polyglycidyl ether of a polyhydric alcohol, preferably a polyhydric phenol (polyhydroxy-aromatic compound). These polyglycidyl ethers of polyhydric phenols are readily available (D.E.N., D.E.R. and Epon).

A preferred polyamide for this process is the reaction product of a polyamine comprising a 1,4-bis-primary amino lower alkyl piperazine and a polyfunctional carboxylic acid or functional equivalent thereof. A preferred and commercially available 1,4-bis-primary amino lower alkyl piperazine is 1,4-bis(3-aminopropyl) piperazine which has a molecular weight of 200.34 (Jefferson Chemical Co.).

The ratios of the epoxide component to the polyamide component are similar to those commonly used in this art and can range from 0.5:1 to 10:1 by weight. Ordinarily, an excess of polyamide is preferred, e.g., at least 1.5:1. (These ratios assume 100% "solids" epoxide and polyamide components.)

The structural adhesives of this process have been found to work well on galvanized steel, phosphatized steel, aluminum, glass, wood, plastics, ceramics, and various vinyl films. Utility with metal substrates can be particularly important in assembly-line bonding procedures, due to the problems of spot-welding or the like.

Example: The polyamide component was prepared as follows: 500 parts by weight (57.2%) of polymeric tall oil fatty acid containing 96% of dimeric fatty acids (Empol 1014) was heated to 60°C under nitrogen with 125 grams (14.3%) of azelaic acid. Fifty parts by weight (5.7%) of polyoxypropylene amine (Jeffamine D-400, approximate molecular weight of 400), 155 parts by weight (17.7%) of 1,4-bis(3-aminopropyl) piperazine and 44 parts by weight (5.0%) of ethylenediamine were added and the temperature was raised to 235°C over 2 hours. The temperature was then held for three hours at 235°C. During the last two hours a vacuum of 25 mm/Hg was applied for improved removal of the water reaction. The polyamide obtained had a ball and ring softening point of 125°C and a total amine number of 70.

Forty parts of the polyamide were then completely dissolved in a solvent consisting of 27.3 parts ethanol, 35.0 parts toluene, and 4 parts 2-nitropropane and the mixture was blended until uniform.

Eight parts of diglycidyl ether-bisphenol A epoxy resin (Epon 828), 3 parts dimethylformamide, 6 parts Vagh vinyl, and 21 parts methylene chloride were uniformly mixed. This mixture was then uniformly dispersed in the polyamide solvent solution. The resulting adhesive mixture was 37.4% solids and had a pot life of approximately 8 hours at 25°C.

The adhesive mixture was applied to one or both of several pairs of substrates at thicknesses which would result in dry thicknesses of about 3 mils if only one substrate of a pair was to be coated and about 1½ to 2 mils if both substrates of a pair were to be coated. It was found that these films could be satisfactorily dried, i.e., the solvent sufficiently evaporated, after four hours at room temperature, or after 30 minutes at 65°C, or after 2 minutes at 150°C. Within a short time after the films were dry, pairs of substrates were mated and tested for T-peel and tensile shear strength as follows:

T-Peel Strength (all substrates acid-etched 2024 aluminum)—

    (a) Both substrates coated, mated for 1 minute at 93°C (200°F): 24 hr cure, 34 lb; 7 days cure, 36 lb; 14 days cure, 35 lb.

    (b) One substrate coated, mated one minute at 150°C (300°F): 24 hr cure, 25 lb; 7 days cure, 28 lb.

    (c) One substrate coated, mated one minute at 171°C (340°F): 24 hr cure, 20 lb; 7 days cure, 30 lb.

    (d) Both substrates coated, mated 15 minutes at 150°C (300°F) and cured for 7 days at 75°F: tested at 40°F, 11 lb; tested at 75°F, 21 lb; tested at 125°F, 18 lb; tested at 150°F, 11 lb; tested at 200°F, 6 lb.

Tensile Shear Strength—

    (1) 2024 aluminum, one substrate coated:

        (a) Both substrates wiped with methyl ethyl ketone only, mated 1 minute at 150°C: 24 hr cure, 1,500 psi; 7 days cure, 1,920 psi.

        (b) Both substrates acid etched, mated 1 minute at 150°C: 24 hr cure, 1,280 psi; 7 days cure, 2,850 psi.

    (2) Mild steel, methyl ethyl ketone wiped only, both substrates coated:

        (a) Mated 1 minute at 93°C (200°F): 24 hr cure, 940 psi; 7 days cure, 1,720 psi; 13 days cure, 1,970 psi.

(b) Mated 15 minutes at 150°C: initial strength, 2,500 to 3,000 psi.

(3) Acid etched 2024 aluminum: both substrates coated, mated 15 minutes at 150°C plus 7 days cure at 75°F.

Tested at –40°F, 4,000 psi; tested at 75°F, 3,000 psi; tested at 150°F, 1,000 psi.

With High Heat Distortion Temperatures and Longer Pot Lives

The substances described by *L.A. Hartmann; U.S. Patent 4,072,656; February 7, 1978; assigned to ICI Americas Inc.* are solid glycidyl polyethers which are prepared by reacting a 3,3',5,5'-tetraalkyl-4,4'-dihydroxybiphenyl (TADP) with a stoichiometric excess of halohydrin.

The starting compounds necessary to prepare these products are biphenols having the following formula,

where R_1, R_2, R_3, R_4, are alkyl whether straight or branched chain and may be the same or different. A specific starting compound which is particularly preferred is tetramethylbiphenol (TMBP).

Typical halohydrins which can be reacted with the biphenol to prepare the glycidyl ethers are polyhalohydrins, e.g., glycerol dichlorohydrin, bis(3-chloro-2-hydroxypropyl) ether, bis(3-chloro-2-methyl-2-hydroxypropyl) ether, 2-methyl-2-hydroxy-1,3-dichloropropane, 1,4-dichloro-2,3-dihydroxybutane, and epihalohydrins such as epichlorohydrin which is preferred.

An advantageous method for the preparation of these glycidyl ethers is to heat a dihydric phenol with epichlorohydrin in the presence of sufficient caustic alkali or other strong alkali, e.g., potassium hydroxide, to combine with the chlorine of epichlorohydrin. It is preferable to use an amount of alkali at least equivalent to the amount of dihydric phenol so as to insure the complete reaction of chlorine.

While the glycidyl ethers may be successfully cured with a variety of curing agents such as aliphatic secondary amines, aromatic amines or anhydrides and the like, bifunctional aromatic amines are preferred to obtain the highest heat distortion temperatures. Although such temperatures will vary depending on the ratio of glycidyl ether to curing agent, and the extent of the curing cycle, for best results, it is suggested to use the aromatic diamines generally in an equivalent ratio of 0.5 to 1.5. Specific curing agents to obtain high heat distortion temperatures of the resulting products are 4,4'-methylene dianiline (MDA) and m-phenylenediamine.

The glycidyl ethers and curable compositions made from them are useful in manufacturing structural parts requiring high heat distortion temperatures, insulation of electrical parts, and protective coatings. They can be cast or molded using simplified procedures to make numerous articles. They are particularly useful in making high-temperature structural parts for guided missiles and other high-speed aircraft, and tools and dies such as those used in the automobile industry for stamping fenders, hoods and other automobile body parts. The curable compositions are also particularly useful in making easily applied protective coatings and can be cured to hard, durable coatings which are tough, chip-resistant and resistant to attack by chemicals and which adhere tenaciously to surfaces of a wide variety of materials including glass and metals. The compositions are also useful as bonding agents in making adhesives, abrasive wheels, laminates and the like.

Example 1: Preparation of TMBP-Diglycidyl Ether — A glycidyl ether was made by reacting 545.3 grams TMBP with 2,081 grams epichlorohydrin. The TMBP-epichlorohydrin slurry was reacted with sodium hydroxide (189.6 grams) at 70° to 83°C for 2 hours and then 84° to 90°C for 1.5 hours. The product was cooled and filtered through a half-inch layer of SuperCel. A small amount of water (11.3 ml) was separated from the filtrate.

The product was vacuum-stripped at 50° to 70°C, finally for 20 minutes at 1 mm Hg. Methanol (1,500 ml) was added while crystallization took place. After 16 hours at 0°C, the product was filtered and washed with chilled methanol (600 ml). The product was dried 23 hours at 75°C under vacuum better than 1 mm. The yield was 73.5% (797.3 grams). The product melted at 105° to 107°C and had an epoxide equivalent of 187.5.

Example 2: Use of 85% of Stoichiometric Amount of Curing Agent — 22.5 parts 4,4'-methylene dianiline (MDA) (dried) and 100 parts of TMBP-DGE (mol ratio 0.113/0.267, 85% of stoichiometric amount of MDA) were ground in a mortar and stored at room temperature in a desiccator for 2 hours. 50 mg of the formulation was applied to aluminum strips. Bonding was accomplished by curing at 84° to 100°C (2 hours) and 150°C (2 hours). Reduced pressure (15 mm Hg) was applied for 10 minutes at 100°C (20 minutes after the start of cure). After cure the sample showed a lap-shear strength of 1,691 psi.

Containing Two Diglycidyl Ethers

During the past few years, a market for metal-plated polymer parts has grown rapidly as manufacturers have begun to appreciate the functional appearance of such parts when plated with bright, metallic finishes, and to take advantage of economies in cost and weight afforded by substituting molded polymeric parts for metal. Furthermore, such plated finishes are not as susceptible to pitting and corrosion because there is not a galvanic reaction between a polymeric substrate and a plated metal.

Because polymeric materials normally do not conduct electricity, it is common practice to provide a conductive layer or coating, such as copper, by electroless deposition so that an additional thickness of metals, particularly copper, nickel and chromium, can be electrolytically plated onto the electroless copper layer. Electroless deposition refers to an electrochemical deposition of a metal coating on a conductive, nonconductive, or semiconductive substrate in the absence of an external electrical source.

While there are several methods of applying this metallic coating by a combined use of electroless and electrolytic procedures, it was not until quite recently that processes were developed which can provide even minimal adhesion of the conductive coating to the polymer. This is because overall adhesion is governed by the bond strength between the polymer substrate and the electroless copper layer.

Polymers extensively employed and upon which electroless deposition is conducted, especially in the printed circuit industry are epoxy polymers resulting from curing uncured diglycidyl ethers of bisphenol A resins produced by a condensation reaction between bisphenol A and epichlorohydrin.

The surface of a cured or crosslinked epoxy article is hydrophobic and is therefore not wet by liquids having a high surface tension. Since electroless depositions usually employ aqueous sensitizing and activating solutions having metal ions therein, the surface will not be wet thereby. Since the sensitizing and activating solutions will not wet the surface, the catalytic species are not absorbed onto the surface and subsequent deposition of the metal ions cannot proceed.

In the present state of the art, various methods are available for rendering the surface of a polymer or a plastic material hydrophilic. These include mechanical roughening of the surface of the plastic material and chemical deglazing or etching. None of these techniques, however, has provided the necessary adhesion of metallic patterns to cured epoxies which are at least 5 lb/inch at a 90° peel and a peel rate of 2 inches/minute at 25°C for a copper thickness of 1.4 mils.

The object, therefore, of the process described by *C.A. McPherson; U.S. Patents 4,077,927; March 7, 1978; 4,091,127; May 23, 1978; and 4,121,015; October 17, 1978; all assigned to Western Electric Company, Inc.* is to produce a cured epoxy surface which gives good adherence without pretreatment.

It was found that such a cured epoxy polymer results from curing a mixture comprising (a) a diglycidyl ether based upon linoleic dimer acid, (b) an elastomerically modified epoxy resin blend, (c) a diglycidyl ether of bisphenol A (4,4'-isopropylidenediphenol) and/or bisphenol F (methylenediphenol) and (d) a suitable curing agent. The process will be illustrated by the following example.

Example: An epoxy resin-curing agent mixture (liquid) was prepared in the following manner. 150 grams (75 parts by weight per 100 parts by weight of the resin component of the mixture) of a commercially obtained epoxy resin, the adduct of 2 mols of the diglycidyl ether of bisphenol A (epoxide equivalent weight of 180 and 195) and 1 mol of linoleic dimer (epoxide equivalent weight of 650 to 750) was combined with 50 grams (25 parts by weight per 100 parts by weight of the resin component) of a commercially obtained, custom-synthesized elastomerically modified epoxy resin blend. The blend comprised the reaction product of a mixture of (1) 40 weight percent of a carboxyl terminated acrylonitrile/butadiene random copolymer having a structural formula

$$\text{HOOC–CH}_2\text{CH}_2\text{–}\underset{\underset{\text{CN}}{|}}{\overset{\overset{\text{CH}_3}{|}}{\text{C}}}\text{–[(CH}_2\text{CH=CHCH}_2\text{)}_5\text{CH}_2\text{CH–}]_{10}\underset{\underset{\text{CN}}{|}}{\overset{\overset{\text{CH}_3}{|}}{\text{C}}}\text{–CH}_2\text{CH}_2\text{COOH}$$
$$\underset{\text{CN}}{|}$$

containing 2.37% carboxyl groups, 18 to 19% bound acrylonitrile, having a number average molecular weight of 3,200 and a viscosity of 110,000 cp at 27°C and (2) 60 weight percent of a diglycidyl ether of bisphenol A having an epoxide equivalent weight of 180 to 195. The mixture had been at 300°F for 30 minutes to yield a reaction product or blend having an epoxide equivalent weight of 335 to 355, an acid value or number of less than 0.2 mg of KOH/g of reaction product, and a viscosity of Y-Z (Gardner-Holdt 80% NV in methyl Cellosolve).

To the resin component was added 40 grams of chlorendic anhydride (20 parts by weight per 100 parts by weight of the resin component) and 5 grams of phthalic anhydride (2½ parts by weight per 100 parts by weight of the resin component) to form the epoxy resin-curing agent mixture. The resultant mixture was then heated to liquefaction at 60°C for 30 minutes.

A commercially obtained epoxy-glass laminate was employed as a substrate. The liquefied epoxy resin-curing agent mixture was applied to one surface of it, using conventional means, to form an epoxy (partially cured) layer (0.5 to 2 mils thick). The coated substrate was then heated at 150°C for 30 minutes to obtain a fully cured epoxy layer as evidenced by infrared spectroscopy which revealed a disappearance of anhydride and epoxide functional groups.

The fully cured epoxy-coated substrate (virgin epoxy-coated) was then sensitized by immersion in a conventional aqueous hydrous oxide tin sensitizer (a 3.5 weight percent $SnCl_2 \cdot 2H_2O$ and 1 weight percent $SnCl_4 \cdot 5H_2O$ aqueous solution) for 1 minute at 25°C, activated in a 0.05 weight percent aqueous $PdCl_2$ solution by immersion therein for 30 seconds, water-rinsed for 2 minutes and then immersed in a commercially obtained electroless copper plating bath to deposit an electroless copper layer on the epoxy layer.

The electroless layer was then subjected to a conventional electroplating to obtain a 1.5-mil-thick copper layer and thereby form a laminar article. The electroplated deposited substrate was then heated or baked at 120°C for 10 minutes.

A peel strength measurement of the deposited metal was undertaken at a 90°
peel at a rate of two inches per minute employing a conventional peel testing
apparatus. The peel strength was 14.3 lb/linear inch at 25°C. At 90°C the peel
strength value was 1.1 lb/linear inch. For printed circuit manufacture a peel
strength of 5 lb/linear inch at 25°C is adequate.

Method for Sealing High-Pressure Piping

*D.A. Hubbard; U.S. Patent 4,104,354; August 1, 1978; assigned to Pont-A-
Mousson SA, France* has devised a method of assembling a joint between the
socket of a pipe and the male or smooth end of another pipe. The method
comprises, after application of layers of a first mixture which contains a first
resin which is slightly exothermic upon curing and has good adherence to the
parts being assembled, depositing a filler in a volume corresponding to the
annular space defined by the layers of the first mixture in the final assembled
state, and, after the filler has been deposited, injecting a second resin into the
filler which resin cures rapidly and is sufficiently exothermic to promote curing
of the first resin. The second resin has a poor adherence to the parts to be assembled but a good adherence to the first resin.

In its application to the assembly of a joint of electric cables, the method comprises, after application of layers of the first mixture to the surfaces of the electric cables and their connectors, disposing a mold around the assembly and
pouring the second mixture into the mold.

In its application to the assembly of a flange on a pipe, the method comprises,
after application of layers of the first mixture to the surfaces of the flange and
pipe adapted to be in confronting relation, applying on the layer on the pipe
and over the same area as the layer, a thick layer of the second mixture, then
placing on the thick layer, by making it extend beyond the latter, a winding
of a reinforcing material impregnated with the second mixture, placing the
flange in position on the pipe and compressing the winding in the annular space
defined by the flange.

Particularly advantageously, there is employed, as the first curable mixture, a
mixture of 50% of epoxy resin and 50% of hardener and, as second curable mixture, a mixture of 120 parts by weight of polyester resin, 150 parts of a granular
filler, 2.4 parts of an accelerator and 3.6 parts of a catalyst.

Containing Certain Polyether Diureide Additives

*H. Schulze and H. G. Waddill; U.S. Patent 4,110,309; August 29, 1978; assigned
to Texaco Development Corporation* have found that a specific diureide-ter-
minated polyoxyalkylene material having a molecular weight of from about
2,000 to 3,000, when employed as an epoxy additive, provides cured epoxy
resin compositions exhibiting outstanding strength of adhesion. Specifically,
epoxy resins incorporating these additives, upon curing with an amine, provide a material with high tensile shear strength and superior adhesion to substrates.

In accordance with a preferred embodiment, a diglycidyl ether of 4,4'-iso-propylidene bisphenol, a curing amount of a primary amine-containing curing agent consisting essentially of a polyoxypropylene polyamine having a molecular weight of from about 200 to 500, a piperazine-amino alkanol accelerator and an effective amount of a polyether diureide having terminal ureido groups and a molecular weight of about 2,000 are utilized to form a resin having superior adhesive strength.

Generally the vicinal polyepoxide-containing compositions which are amine cured are organic materials having an average of at least 1.8 reactive 1,2-epoxy groups per molecule. These polyepoxide materials can be monomeric or polymeric, saturated or unsaturated, aliphatic, cycloaliphatic, aromatic or heterocyclic, and may be substituted if desired with other substituents besides the epoxy groups, e.g., hydroxyl groups, ether radicals, aromatic halogen atoms and the like.

Preferred polyepoxides are those of glycidyl ethers prepared by epoxidizing the corresponding allyl ethers or reacting, by known procedures, a molar excess of epichlorohydrin and an aromatic polyhydroxy compound, i.e., isopropylidene bisphenol, novolak, resorcinol, etc. The epoxy derivatives of methylene or isopropylidene bisphenols are especially preferred.

The polyether diureide compounds are formed by the reaction of a ureido or monosubstituted ureido-forming compound with a polyoxyalkylene diamine having a molecular weight value such that the ureido-containing product has a molecular weight of from about 2,000 to 3,000 at temperatures in the range from about $25°$ to $150°C$ in a molar ratio of about 2 mols of ureido or monosubstituted ureido-forming compound for each mol of diamine. The diamines that are useful in forming the additives are polyoxyalkylene diamines of the formula:

$$[H_2N-(CH-CH-O)_n]_2-Z$$
$$X \quad H$$

wherein X is a hydrogen, a methyl radical or an ethyl radical; Z is a hydrocarbon alkylene radical having from 2 to 5 carbons; and n is an average number of from about 15 to 25. Preferred polyoxypropylene diamines are those wherein X is a methyl radical, n is an average number from 16 to 19, Z is a 1,2-propylene radical.

The ureido-forming compounds are generally those supplying the $O=C(NH_2)-$ radical. Urea is preferred. When urea is employed as a reactant, the reaction proceeds with the evolution of ammonia and the terminal primary amino groups of the polyoxyalkylenepolyamine are converted directly to ureido groups.

These amine-cured resins having superior adhesion are prepared in a conventional manner. The amine curing agent is admixed with the polyepoxide composition in amounts according to the amine equivalent weight of the curing agent employed. Generally the number of equivalents of amine groups is from about 0.8 to 1.2 times the number of epoxide equivalents present in the curable epoxy resin composition, with a stoichiometric amount being preferred. When using an accelerator, amounts from 1 to about 10 parts by weight based on 100 parts by weight of the resin are generally satisfactory. The exact amount of constituents in accordance with the above general requirements will depend primarily on the application for which the cured resin is intended.

The diureido additive is incorporated into the uncured resin by admixing. Preferably, the additive is first admixed with the curing agent and/or the accelerator prior to addition to the resin. The constituents forming the curable material are then intimately admixed by standard methods and degassed in the presence of a commercial defoamer and minute amounts of silicone oils to prevent voids and bubbles.

Epoxy resins based on aliphatic compounds are preferably not used exclusively. The presence of resins containing polyglycidyl ethers of polyhydric phenols in amounts greater than 50% by weight of the resin constituent, and most preferably 100% by weight has been shown to greatly enhance the desirable properties of the cured material, especially the adhesive strength.

Likewise, although many amine curing agents are generally useful to cure the resin, preferred amine curing agents are those polyamines having amine equivalent weights of from 20 to 70. Examples of such agents include polyoxypropylenediamines having a molecular weight in the range of 200 to 300 and polyoxypropylenepolyamines having a molecular weight of from about 400 to 600.

According to a greatly preferred embodiment, resins of the polyglycidyl ether of polyhydric phenol type are cured by incorporating therein about a stoichiometric amount of a polyoxyalkylenepolyamine having a molecular weight of about 230; from about 5 to 40 parts by weight of the polyether diureido terminated polyoxyalkylenepolyamine having a molecular weight of about 2,000; and from 1 to 5 weight percent based on 100 parts by weight of the resin of an accelerator comprising a 30:70 weight percent admixture of piperazine:triethanolamine. The composition is cured at room temperature (about 25°C) to produce products having superior adhesive strength in accordance with this process.

The described compositions can be used as impregnants, surface coatings, pottings, capsulating compositions, laminants, and most importantly, as adhesives for bonding metallic elements or structures permanently together.

Example: In this example a polyether diureido-terminated additive was prepared. Into a suitable reaction vessel, equipped with stirring apparatus, were added 1,980 grams (1 mol) of a polyoxypropylenepolyamine having a molecular weight of approximately 2,000 and an analysis of 1.01 milliequivalents primary amine per gram (Jeffamine D-2000) and 180 grams of urea (3.0 mols). The admixture was flushed with nitrogen and stirred under a nitrogen pad for 2 hours at 130° to 134°C. A second portion of Jeffamine D-2000 consisting of 990 grams (0.5 mol) was added over a three-hour period at a temperature of about 132°C. The reaction mixture was maintained at 134°C for another 70 minutes, during which time the admixture was vigorously stirred to continuously wash the sublimate on the upper surface of the reaction vessel. The crude reaction product was then stripped at 130°C/1.4 mm Hg to produce a viscous residue which upon analysis showed 2.54% N, 0.01 meq total amine per gram.

When various amounts of the product of the example were added to an epoxy formulation cured with a polyoxypropylenediamine having a molecular weight of 230, improved adhesion strength was easily demonstrated.

Containing Certain Polyether Diamide Additives

H. Schulze and H.G. Waddill; U.S. Patent 4,110,310; August 29, 1978; assigned to Texaco Development Corporation have found that polyether diamides having terminal amido groups and a molecular weight of from 2,000 to 3,000 can be used instead of the polyether diureides described in the previous patent to improve the adhesion of amine-cured epoxy resins. Polyepoxides and amine curing agents as described in the last patent may be used, and the diamines useful in forming the polyether diamide additives are also the same.

The amide-forming compounds are those which supply the acyl radical RCO–, wherein R is hydrogen; or a branched or straight chain alkyl radical of from 1 to about 10 carbons, and more preferably from 1 to about 6; or a monocyclic aryl, alkaryl or aralkyl having from 6 to 12 carbons, and more preferably 6 to 8; or, a branched or straight chain alkenyl or alkadienyl radical of from 2 to 10 carbons and more preferably 3 to 8.

Example 1: In this example a polyether diamide-terminated additive for use in accordance with this process, was prepared. Into a suitable reaction vessel, equipped with stirring apparatus, thermometer, reflux condenser, and Dean-Stark trap were added 971 grams (0.5 mol) of Jeffamine D-2000, 76.5 grams (1.5 mols) 90% by weight aqueous formic acid, and 200 ml toluene flushed with nitrogen and stirred under a nitrogen pad for two hours at reflux. An aqueous phase was separated in the Dean-Stark trap. The crude reaction residue was then stripped in a rotary evaporator at 97°C/0.4 mm Hg to produce a viscous residue which upon analysis showed 1.64% N, 0.07 meq total amine per gram.

To illustrate the advantage of the polyether amide additives, various epoxy formulations employing diglycidyl ether of 4,4'-isopropylidene bisphenol were cured with various known polyamine curing agents. Where indicated a commercial accelerator was utilized. Three drops of silicone fluid were added to each formulation to prevent formation of voids and bubbles. After degassing under vacuum, the formulations were cured under the conditions indicated. In appropriate examples, the cured products were subjected to standard ASTM tests for Izod impact strength (ASTM designation D-256), flexural strength and modulus of elasticity in flexure (ASTM designation D-790-66), tensile strength and elongation at break (ASTM designation D-638-64 T), deflection temperature (ASTM designation D-748-56) and hardness (ASTM designation 2240-64 T) and/or hardness Shore D, and peel strength (ASTM D-903). The tensile shear strength (ASTM D-1002-64) was measured on adhesive bonds. All substrates were aluminum panels (No. 2024-T-3 alloy, 16 gauge), degreased, then chromic acid etched prior to bonding. The abbreviations in the table, pbw, psi and g stand for parts by weight, pounds per square inch and grams, respectively.

Examples 2 through 5: In these examples epoxy resins were prepared wherein diglycidyl ether of 4,4'-isopropylidene bisphenol was cured with a polyoxy-propylenediamine curing agent of MW 230 having an equivalent weight of 58 to which were added the indicated amounts of the diamide prepared in Example 1. The resulting resins were used to bond aluminum to aluminum and the result-ant subjected to the ASTM tests herein described. The data, which are for com-parative purposes only, is presented in the following table.

| | . . . . Examples . . . . | | | |
	2	3	4	5
Epoxide (pbw) (eq 190)	100	100	100	100
Curing agent* (pbw)	30	30	30	30
Accelerator** (pbw)	10	10	0	0
Bisformamide***	0	30	0	30
Tensile shear strength (psi)	1,050	1,500	980	1,500

    *Jeffamine D-230.
    **A piperazine-triethanolamine admixture (30:70) known
      as Accelerator 398 (Jefferson Chemical Co.).
    ***The product of Example 1.

This example demonstrates the improved adhesion strength of the epoxy formu-lation when amounts of the bisamide are added to an epoxy formulation cured with a polyoxypropylenediamine having a molecular weight of 230.

Addition of Amino-Terminated Polyether Succinimides

In another patent from *H. Schulze and H.G. Waddill; U.S. Patent 4,115,361; September 19, 1978; assigned to Texaco Development Corporation,* the poly-ether diamides described in the previous patent are replaced by primary amine-terminated polyether succinimides having a molecular weight of about 4,000 to produce adhesives having outstanding adhesion strength. The following example shows how such additives are prepared.

Example 1: Into a suitable, clean, dry, reaction vessel, equpped with ther-mometer, stirring apparatus, reflux apparatus, and Dean-Stark trap were added 750 grams (0.39 mol) of a polyoxypropylenepolyamine having a molecu-lar weight of approximately 2,000, and an analysis of about 1.01 meq primary amine/gram (Jeffamine D-2000, Jefferson Chemical Co.), and 18.4 grams (0.183 mol) maleic anhydride and 50 grams of benzene. This admixture was heated at reflux (pot temperature $146°$ to $167°C$) for about 2¼ hours until azeotropic water removal (2.3 grams H_2O) had ceased. Benzene was then removed from the crude reaction mixture until a pot temperature of 205°C was attained. The remaining reaction product was then stripped at 180°C/3 mm Hg to produce a viscous residue which, upon analysis, showed 0.64 meq total amine/g, 0.43 meq primary amine/g, 0.009 meq acidity/g. IR analysis indicated the presence of the succinimide group.

To illustrate the advantage of these amino-terminated polyether succinimide additives, various epoxy formulations employing diglycidyl ether of 4,4'-isopropylidene bisphenol were cured with various known polyamine curing agents. Where indicated, a commercial accelerator was utilized. Three drops of silicone fluid were added to each formulation to prevent formation of voids and bubbles. After degassing under vacuum, the formulations were cured under the conditions indicated.

In appropriate examples, the cured products were subjected to standard ASTM tests for peel strength (ASTM D-903) and tensile shear strength (ASTM D-1002-64. The test results were measured on adhesive bond with substrates. All substrates were aluminum panels (No. 2024-T-3 alloy, 16 gauge), degreased, then chromic acid etched prior to bonding. The abbreviations in the table, pbw and psi stand for parts by weight, and pounds per square inch, respectively.

Examples 2 through 5: In these examples epoxy resins were prepared wherein diglycidyl ether of 4,4'-isopropylidene bisphenol was cured with a polyoxypropylenediamine curing agent of MW 230 having an equivalent weight of 58 to which were added the indicated amounts of the amino-terminated polyether succinimide prepared in Example 1. The resulting resins were used to bond aluminum to aluminum and the resultant subjected to the ASTM tests herein described. The data, which is for comparative purposes only, is presented in the following table. For the tests the cure condition was room temperature for 7 days.

	. . . . Examples			
	2	**3**	**4**	**5**
Epoxide (pbw) (eq 190)	100	100	100	100
Curing agent* (pbw)	30	30	30	30
Accelerator** (pbw)	10	10	10	10
Polyether succinimide***	0	75	14	29
Tensile shear strength (psi)	1,250	3,860	4,110	3,310
Peel strength (pli)	7.6	10.1	23.2	38.4

    *Jeffamine D-230.
     **A piperazine-triethanolamine admixture (30:70) known
       as Accelerator 398 (Jefferson Chemical Co.).
    ***The product of Example 1.

These examples demonstrate the improved adhesion strength of the epoxy formulation when amounts of the polyether succinimide are added to an epoxy formulation cured with a polyoxypropylenediamine of MW 230.

Diamides of Polyoxyalkylenepolyamine-Urea Condensates

In still another patent by *H.G. Waddill and H. Schulze; U.S. Patent 4,146,700; March 27, 1979; assigned to Texaco Development Corporation,* the addition of certain diamides of polyoxyalkylenepolyamine-urea condensates is recommended to improve the adhesion of epoxy compounds of the type described in the previous patents.

The diamides of polyoxyalkylenepolyamine-urea condensates can generally be depicted by the formula:

$$X(NHZ-NHX-R')_2$$

where R' is an alkyl radical, an aryl radical or hydrogen, and X is a C=O or a C=S radical, and Z is a polyoxyalkylene radical of molecular weight ranging from about 1,800 to 2,300.

The diamides of polyoxypolyamine-urea condensates are formed by the reaction of a carboxylic acid or a derivative thereof with a polyoxyalkylenepolyamine-urea condensate having a molecular weight value such that the diamide product has a molecular weight of from about 4,000 to 4,500. The polyoxyalkylenepolyamine-urea condensate is formed, for example, by the reaction of about 2 mols of a polyoxyalkylenediamine for each mol of urea.

The amount of additive effective in bringing about the increased adhesive property is somewhat empirical and will depend upon the resin, the amine curing agent, and the use of an accelerator. Generally, the diamide additive can be utilized in amounts from about 2 to 50 parts by weight based on 100 parts by weight of the resin constituent.

Preferably the adhesive properties of the epoxy resins are enhanced by addition of an effective amount of the diamide additive based upon the condensation of 2 mols polyoxyalkylenediamine with 1 mol of urea. The preferred resins comprise polyglycidyl ethers of a polyhydric phenol cured by incorporating therein a curing amount of a polyoxyalkylenepolyamine of molecular weight from about 200 to 500 and, optionally, an accelerator combination of piperazine and an alkanolamine, the combination having a combined weight ratio of between about 1:8 to 1:1.

The amine-cured resins are prepared in a conventional manner. The amine curing agent is admixed with the polyepoxide composition in amounts according to the amine equivalent weight of the curing agent employed. Generally the number of equivalents of amine groups is from about 0.8 to 1.2 times the number of epoxide equivalents present in the curable epoxy resin composition, with a stoichiometric amount being preferred. When using an accelerator, amounts from 1 to about 10 parts by weight based on 100 parts by weight of the resin are generally satisfactory.

The diamide additive is incorporated into the uncured resin by admixing. Preferably, the additive is first admixed with the curing agent and/or the accelerator prior to addition to the resin. The constituents forming the curable material are then intimately admixed by standard methods and degassed in the presence of a commercial defoamer and minute amounts of silicone oils to prevent voids and bubbles.

Example: This example demonstrates the preparation of a diamide(bisformamide) of a polyoxyalkylenediamine-urea condensate of about 4,000 MW. Initially a product, which is polyoxypropylenediamine of about 2,000 MW condensed with urea in a 2:1 molar ratio (Jeffamine DU-3000) is reacted to form a diamide as follows.

A mixture of Jeffamine DU-3000 (0.52 meq/g primary amine) 1,331 grams (0.31 mol), toluene (122 grams) and 90% formic acid (47 grams, 0.92 mol) was heated at reflux and with stirring under nitrogen. An aqueous phase (20.6 grams) was removed in a Dean-Stark trap. The product was stripped under vacuum (115° to 170°C/10 mm). The liquid product analyzed as follows: total amine 0.02 meq/g, total acetylatables 0.12 meq/g.

Epoxy-Amine Composition Curable at Low Temperatures

An epoxy adhesive compound is needed which can be brushed onto joint surfaces or formed into intricate shapes to match joint geometry, has a long shelf life, allows at least ½ minute working time after melt bonding, cures at low temperatures, and which will provide joint shear strengths of at least about 1,000 psi without requiring the extensive use of pressing equipment. The elimination of clamps in the curing process would be particularly advantageous in any continuous, commercial, production line operation requiring metal-to-metal bonding.

E.M. Petrie; U.S. Patent 4,113,684; September 12, 1978; assigned to Westinghouse Electric Corporation describes a solid reactive mass comprising a uniform dispersion of from about 1 to 6 parts reactive epoxy resin powder and 1 part of a solid curing agent consisting of a reaction product adduct consisting of solid epoxy resin and a stoichiometric excess of a liquid aliphatic amine or a solid aliphatic amine. The material is an unreacted mixture of epoxy resin powder and adduct, which can be used as an adhesive, and is capable of melting at temperatures of above about 60°C and forming a crosslinked adhesive material.

The adhesive has a long storage life and can be cold molded into pellets, rods, or intricate-shaped design configurations. The cold-molded adhesive can be melted on metal or glass substrates to be bonded, or applied with a hot melt gun apparatus with a zone heater in the nozzle, to form a single-phase system. A second substrate can then be placed in contact with the adhesive while it is in melted form.

This adhesive, in one embodiment, is melt applied and allowed to cool to room temperature, during which time it cures to a thermoset state, to provide initial joint shear strengths between the substrates of about 700 psi and joint shear strengths of about 1,100 psi after aging. In another embodiment, the adhesive can be mixed with water or other solvent to form a slurry, coated on metal surfaces to be bonded, and dried at room temperature to remove solvent. The coated metal may be stored, and when bonding is desired, the metal parts may be placed together, heated over about 60°C and allowed to cool to room temperature, during which time the adhesive cures to provide joint shear strengths between the substrates of about 2,000 psi.

The adhesive composition consists of: (A) a particulated solid reaction product adduct consisting of (1) one part by weight of a solid epoxy resin selected from the group consisting of diglycidyl ethers of bisphenol A epoxy resins, diglycidyl ethers of bisphenol F epoxy resins, and mixtures thereof, and (2) about 0.20 part by weight of about 1.5 parts by weight of an aliphatic polyamine, to provide a stoichiometric excess of amine; and (B) about one part by weight to about 6 parts by weight, for each one part of adduct, of a reactive solid epoxy resin powder having an average particle size range of from about 10 microns to about 420 microns.

The epoxy resin is selected from the group consisting of diglycidyl ethers of bisphenol A epoxy resins, diglycidyl ethers of bisphenol F epoxy resins and mixtures thereof, having a melting point of between about 55° to 225°C.

The composition is a solid two-phase reactive mass comprising a uniform solid dispersion of adduct (A) and epoxy powder (B), wherein the melting point of the adduct is within about 1° to 50°C of the melting point of the epoxy powder. The composition is capable of melting at temperatures of over about 60°C and forming a crosslinked adhesive material on cooling after melting.

Example: The following components were mixed together at 25°C in a reaction vessel: 20 parts by weight of a solid diglycidyl ether of bisphenol A having an epoxy equivalent weight of 475 to 575 (DER 661) ground to have an average particle size of about 75 to 210 microns; 16 parts by weight of water; and 20 parts by weight of a 50% solids polyfunctional aliphatic amine adduct solution.

The adduct solution was the reaction product of a solid diglycidyl ether of bisphenol A with an epoxy equivalent weight of about 450 to 575 and a stoichiometric excess of an aliphatic polyamide-amine. The adduct was in a 50 weight percent water solution and contained about 0.20 to 0.50 part polyamide-amine per one part by weight of epoxy resin known as Epi-Cure W-50-8535 (Celanese Resin Co.). This composition provided 2 parts epoxy powder per one part solid epoxy-aliphatic amine adduct.

This admixture was high shear mixed, for about five minutes at 25°C in a blender, until it was homogeneous and had a viscosity of about 10,000 cp at 25°C. The adhesive admixture slurry was brush coated onto abraded aluminum panels at 25°C and allowed to air dry for 16 hours at 25°C. The two coated aluminum panel surfaces were then mated, spring clamped, to keep proper alignment rather than to further bonding, placed in a forced air oven for five minutes at 125°C and allowed to cool to room temperature. The difference in melting point between the adduct and the epoxy powder was about 25°C. The melting point of the dried adhesive was between about 70° to 80°C. The completely cured aluminum-aluminum lap shear specimens were tested per ASTM D-1002 at 25°C, and the shear strength after complete cure was found to be 2,130 psi.

This adhesive is particularly useful due to its ease of application. After application, by brushing, spraying or other means, and drying, the coated parts can be stored for up to one year, melt bonded at temperatures over about 80°C under contact pressure in a suitable heating means and cooled to thermoset crosslink and cure the adhesive.

Containing Amino-Terminated Polyether Ureylene Additives

H. Schulze and H.G. Waddill; U.S. Patent 4,115,360; September 19, 1978; assigned to Texaco Development Corporation have found that the adhesive strength of amine-cured epoxy resins is enhanced by the addition of an effective amount of an amine-terminated polyether ureylene having a molecular weight of from 4,000 to 4,500.

In accordance with a preferred embodiment, a diglycidyl ether of 4,4'-isopropylidene bisphenol, a curing amount of a primary amine-containing curing agent consisting essentially of a polyoxypropylenepolyamine having a MW of from about 200 to 500, a piperazine-aminoalkanol accelerator and an effective amount of a primary amine-terminated polyether ureylene having a MW of about 4,000 are utilized to form a resin.

The preferred curable epoxy resin is prepared essentially in three steps. First, the polyether ureylene having terminal primary amino groups is prepared by charging a suitable reaction vessel with approximately 2.0 mols of a polyoxypropylenepolyamine having a MW of about 2,000 and consisting substantially of the polyoxypropylenediamine; and about 1.0 mol of urea. The mixture is then heated gradually to temperatures of about 180° to 200°C and maintained until the evolution of ammonia has ceased. The resultant reaction mixture is then stripped at temperatures of about 120° to 150°C at pressures of 1 mm Hg to form a viscous liquid.

In the second step, the viscous liquid obtained in the first step is admixed with a polyoxypropylenediamine having a MW of from about 200 to 250 in a ratio of from about 5:1 to 1:5 by weight to produce an admixture. To this admixture can be added a commercial accelerator if desired.

In a third step, the admixture obtained in the second step is added to a suitable amount of a diglycidyl ether of 4,4'-isopropylidene bisphenol such that the total number of equivalent amino groups is about equal to the number of equivalents of epoxide in the epoxy resin composition. The epoxy resin and the curing mixture are thoroughly admixed with the addition of about three drops of a silicone fluid to prevent the formation of voids and bubbles. The resulting formulation, after degassing under vacuum for about 2 to 5 minutes, is applied to substrates to be bonded and/or cast into molds. The resins are cured at room temperature and preferably post cured at temperatures from about 80° to 125°C. The cured products exhibit improved tensile shear strength; flexural strength, ultimate elongation; and, especially superior adhesion to substrates.

Preferred polyepoxides are those of glycidyl ethers prepared by epoxidizing the corresponding allyl ethers or reacting, by known procedures, a molar excess of epichlorohydrin and an aromatic polyhydroxy compound, i.e., isopropylidene bisphenol, novolak, resorcinol, etc. The epoxy derivatives of methylene or iso-propylidene bisphenols are especially preferred.

The described compositions can be used as impregnants, surface coatings, pottings, encapsulating compositions, laminants and, particularly and most importantly, as adhesives for bonding metallic elements or structures permanently together.

Aromatic Amine Curing Agent

The elimination of clamps during the curing of epoxy resins in metal-to-metal bonding would be very advantageous in continuous production line operations.

E.M. Petrie; U.S. Patent 4,120,913; October 17, 1978; assigned to Westinghouse Electric Corporation has formulated an adhesive composition which has a long storage life and can be cold molded into pellets, rods, or intricate-shaped design configurations. The cold molded adhesive can be melted on metal or glass substrates to be bonded, or applied with a hot melt gun apparatus with a zone heater in the nozzle, to form a single-phase system.

A second substrate is then placed in contact with the adhesive while it is in melted form. This adhesive, after melt application, can form a "thermoplastic" bond between the two substrates during cooling to room temperature. The adhesive is then fully cured to a thermoset state preferably by staged heating, to provide joint shear strengths between the substrates of about 1,900 psi. In most of the above curing processes, clamps or other fixturing equipment are not necessary.

The adhesive composition is a dispersed mixture of a reactive epoxy resin powder, and a liquid epoxy resin, and a stoichiometric excess of a liquid aromatic amine, based on the liquid epoxy resin.

The epoxy resin powder is selected from diglycidyl ethers of bisphenol A epoxies, diglycidyl ethers of bisphenol F epoxies, novolak epoxies, and their mixtures. The epoxy resin powder must have an average particle size range of from about 10 to 420 microns, with a preferred range of from about 37 to 230 microns. Over about 420 microns diameter there will be inadequate cross-linking between the epoxy powder and the adduct on cure.

The adduct must contain a liquid epoxy resin selected from diglycidyl ethers of bisphenol A epoxies, diglycidyl ethers of bisphenol F epoxies, and their mixtures, having viscosities ranging from about 500 to 22,500 cp at 25°C. Over about 22,500 cp the epoxy will not effectively react with the amine at low temperatures. The liquid aromatic diamine would preferably include a eutectic blend of about 60 to 75 weight percent meta-phenylenediamine and about 25 to 40 weight percent methylenedianiline.

The adduct curing agent will contain from about 0.20 to 1.5 parts liquid or liquefied solid amine for each one part by weight of epoxy resin in the adduct, to provide a suitable stoichiometric excess of amine.

Example: An epoxy composition was made from a mixture of epoxy powder and the adduct of a liquid epoxy and a liquid aromatic amine. The following reaction components were directly mixed together at 25°C in a reaction vessel: 80 pbw of a solid diglycidyl ether of bisphenol A resin having an epoxy equivalent weight of 450 to 550 and a Durran's melting point of 65° to 75°C (Epon 1001, Shell Chemical Co.), ground to have an average particle size of about 75 to 210 microns; 20 pbw of a liquid diglycidyl ether of bisphenol A resin having an epoxy equivalent weight of 175 to 195 and a viscosity at 25°C of 500 to 700 cp and containing 11 wt % butyl glycidyl ether diluent (Epon 815, Shell Chemical Co.); and 20 pbw of a liquid eutectic aromatic amine blend of 60 to 75 wt % of m-phenylenediamine and 25 to 40 wt % of methylenedianiline (Curing Agent Z, Shell Chemical Co.).

This provided a stoichiometric excess of amine, 1 part amine/1 part liquid epoxy, since for stoichiometric balance 20 pbw of Curing Agent Z (eq wt, 38) is required to react with 100 pbw of Epon 815 (epoxide eq, 175 to 195). This composition provided 2 parts epoxy powder per part combined liquid epoxy and liquid aromatic amine.

This mixture of liquid epoxy and adduct was milled for about 10 minutes at 25°C until it was homogeneous. The adhesive gelled after about 7 hours' standing at 25°C, and formed a glassy solid in approximately 16 hours. Before gelation the adhesive system had the consistency of a very thick, thixotropic paste which was formed in a variety of intricate shapes, as well as rods, pellets and balls. The exotherm generated by the Curing Agent Z and Epon 815 adduct reaction is so mild that a low-melting-point solid epoxy powder could be incorporated into the reaction without crosslinking and curing the epoxy powder.

A solid rod of adhesive was applied to a 150°C preheated 0.063 inch Type 6061 aluminum panel by wiping it across the area to be bonded. The panel was treated prior to bonding by abrasion with a 120X grit aluminum oxide sanding belt followed by a solvent wash with acetone. Another preheated abraded panel was immediately mated with the adhesive melt on the coated substrate and the joint was allowed to cool to room temperature without pressure on the bond or any clamping means. The tensile shear strength measured at 25°C before complete cure was 228 psi, and after complete cure it was 1,920 psi.

This adhesive was able to resist 80 psi loads at 60°C before final cure, exhibiting good creep resistance. This adhesive has sufficient shear strength and creep resistance to allow substantial handling throughout the cure cycle without the use of any fixturing.

Addition of Polymercaptans and Polyenes

Most epoxide resin adhesive compositions lack "green strength," that is to say, they are not tacky before they solidify, and surfaces to be bonded to each other by the adhesive must be held together by jigs, clips, presses, or other temporary fasteners while solidification takes place. Attempts have been made to overcome this disadvantage by dissolving certain high-molecular-weight polymers in one or more components of the composition to act as a tackifier. In general, such compositions suffer from the drawback that the blend of the added polymer and the epoxide resin is viscous and becomes tacky too rapidly, leading to difficulties in mixing it with other constituents of the compositions or in applying it in a sufficiently thin layer. To counter this, volatile solvents could be incorporated, but the solvent-containing compositions cannot always be spread in layers of the desired thickness, the solvent may attack the object to be bonded, and many types of solvents introduce flammability or toxicity hazards into the workshop.

E.W. Garnish and R.G. Wilson; U.S. Patent 4,126,505; November 21, 1978; assigned to Ciba-Geigy Corporation have found that the desired objective of producing an epoxide resin adhesive which can easily be mixed, dispensed, and spread, and which will become tacky in use, may be achieved, without the need to add a volatile solvent, by employing certain combinations of epoxide resins, polymercaptans, and polyenes with a curing agent for the epoxide resin. The curing agent may be one of certain aliphatic or cycloaliphatic polyamines which cure the epoxide resin by a crosslinking addition reaction, or it may be one of certain tertiary amines which cure the epoxide resin by catalytically induced polymerization.

The formulation accordingly provides a composition comprising: (a) an epoxide resin having, per average molecule, more than one 1,2-epoxide group of formula

$$-CH-\underset{\underset{R}{|}}{C}\underset{\underset{R^1}{|}}{\underset{\diagdown}{}}\underset{O}{\overset{}{\diagup}}\underset{\underset{R^2}{|}}{C}-H$$

directly attached to oxygen, sulfur, or nitrogen, where either R and R^2 independently of one another represent hydrogen, in which case R^1 denotes hydrogen or methyl, or R and R^2 conjointly represent $-CH_2CH_2-$, in which case R^1 denotes hydrogen; (b) a polymercaptan having at least two and preferably six mercaptan groups per average molecule; (c) a polyene having per average molecule at least two ethylenic double bonds, each β to an atom of oxygen, nitrogen, or sulfur, the sum of such ethylenic double bonds and of the mercaptan groups in (b) being more than 4, and preferably from 5 to 8; and (d) as curing agent for the epoxide resin, either a compound having at least three hydrogen atoms directly attached to aliphatic or cycloaliphatic amino nitrogen atoms or a tertiary amine having at least one nitrogen atom directly attached to carbon atoms of aliphatic or cycloaliphatic groups exclusively and at most two hydrogen atoms attached to amino nitrogen atoms.

The following examples illustrate the formulation. Unless otherwise indicated, parts are by weight. Temperatures are in $^{\circ}$C. Polymercaptan A is substantially of the average formula

$$CH_2 \left[\begin{matrix} CH_3 & OH \\ | & | \\ (OCH_2CH)_{b_1} OCH_2CHCH_2SH \end{matrix}\right]_3 \quad \begin{matrix}CH_2 - \\ | \\ CH - \\ | \\ CH_2 - \end{matrix}$$

where b_1 is an integer of average value 2.5 and was prepared as described in U.S. Patent 3,258,495.

Polymercaptan B is the commercially available polysulfide substantially of the average formula

$$HS-[CH_2CH_2OCH_2 OCH_2CH_2SS]_6-CH_2CH_2OCH_2OCH_2CH_2SH$$

Polymercaptan G is the tetrakis(3-mercaptopropionate) of pentaerythritol.

Polyene I is a polyester made in a conventional manner by the reaction of maleic anhydride (3.06 mols) with dipropylene glycol (3.21 mols).

Polyene II is a polyester prepared similarly from dipropylene glycol (2.1 mols), maleic anhydride (1 mol), and adipic acid (1 mol).

Epoxide resin I is a polyglycidyl ether of 2,2-bis(4-hydroxyphenyl)propane having a viscosity at 21°C in the range 200 to 400 poises; its epoxide content is 5.1 to 5.4 eq/kg.

Hardener I is a poly(aminoamide) obtained by reaction of diethylenetriamine with a mixture of dimerized and trimerized linoleic acid; its amine value is 210 to 220 mg KOH/g.

Accelerator I is 2,4,6-tris(dimethylaminomethyl)phenol. Shear strengths of joints were determined using aluminum alloy strips 1.63 mm thick obtained under the designation "2L 73 Alclad" which had been degreased, pickled by the process prescribed in the British Ministry of Technology Aircraft Process Specification DTD-915B, washed in running water, and dried at room temperature. Single lap joints 12 mm x 25 mm were prepared. T-peel strengths were determined by the procedure described in U.S. Military Specification MMM-A-132. Sheets of "2L 61 Alclad," 0.056 mm thick and 24 mm wide were used. Gelation times were determined by ascertaining when the mixture, stirred with a small wooden spatula, became a rubbery solid and/or formed "strings" when pulled.

Example 1: Mixtures were prepared, each containing 55 parts of Hardener I, 5 parts of Accelerator I, and 15 parts of Polymercaptan A. To a 5 gram portion of each mixture was added with stirring 5 gram portions of mixtures each comprising 100 parts of Epoxide resin I and 15 parts of a polyene (Polyenes I and II). In every case the admixtures gelled within 10 minutes at room temperature, and became tacky. If, however, either Polymercaptan A or the polyene was omitted, gelling did not take place within 30 minutes.

Example 2: Further compositions were prepared, each containing 100 parts of Epoxide resin I, 55 parts of Hardener I, 5 parts of Accelerator I, and certain polymercaptans and polyenes as indicated in the table below. The compositions were cured, either at room temperature for 24 hours or at 100°C for 30 minutes, and the lap shear strengths and T-peel strengths of bonds formed with the compositions were measured. All measurements were done at 22°C except where indicated.

Poly- mercaptan (parts)		Polyene (parts)		24 hr at 22°C Lap Shear Strength (MPa)	Cured for 30 min at 100°C		T-Peel Strength (N/mm)
					Lap Shear Strength (MPa)		
A	10	I	15	22.2	29.1	24.5*	1.3.
A	15	I	15	24.6	32.3	21.4	1.4
A	20	I	15	21.3	34.3	21.8	1.6
B	10	I	15	18.5	30.7	26.8	2.4
B	15	I	15	19.5	15.9	13.3	1.5
B	20	I	15	20.3	10.3	9.8	1.2
G	15	II	15	**	26.7	22.9	1.6

*These measurements were done at 60°C.
**Not determined.

In each case the compositions became tacky.

Having Long Pot Life

The formulations of *J.H. Kooi, R.M. Schure and J.M. Brown; U.S. Patent 4,129,607; December 12, 1978; assigned to Unitech Chemical Inc.* are for hot melt adhesives which have high tensile shear strength, exceptionally high peel strength, creep resistance at elevated temperatures, toughness and elasticity, and long pot life. The adhesives are blends of epoxy resins with certain hydroxy-terminated poly(ester/ether) block copolymers of the formula:

where R' and R'' are alkyl, alicyclic, acyclic, aryl, or arylakyl of from 2 to 12 carbons, p is a number of from 2.4 to 136.0, a is a number such that the "hard" segment within the first set of brackets makes up about 70 to 20% by weight of the copolymer, and b is a number such that the "soft" segment within the second set of brackets makes up about 30 to 80% by weight of the copolymer. The blends are formed by heating and mixing an epoxy resin and the copolymer shown until a compatible, thermoplastic mixture is formed. In use, the hot mixture may be applied to substrates and allowed to cool, thereby forming a thermoplastic bond of the substrates.

The hydroxy-terminated, substantially linear poly(ester/ether) block copolymer is a polymeric reaction product of: (1) one or more of an aromatic, aliphatic, or cycloaliphatic dicarboxylic acid or ester-forming derivative thereof; (2) one or more of a low molecular weight aliphatic, alicyclic, acyclic, or aromatic diol; and (3) one or more of a difunctional polyether, including the poly(alkylene ether) glycols.

The hydroxy-terminated poly(ester/ether) block copolymer includes two types of blocks, one being a "soft" segment that provides the polymer with a relatively low glass transition temperature and has an elastomeric character, the other being a "hard" segment that provides the polymer with a crystalline domain having a relatively high melting point to lessen chain slippage in the absence of elevated temperatures.

For example, the preferred hydroxy-terminated copolymer is prepared from dimethyl terephthalate, dimethyl isophthalate, 1,4-butanediol (tetramethylene glycol) and poly(tetramethylene ether) glycol. Both R' and R'' are $(CH_2)_4$. The "hard" segment has an average molecular weight of about 220 and the following structure:

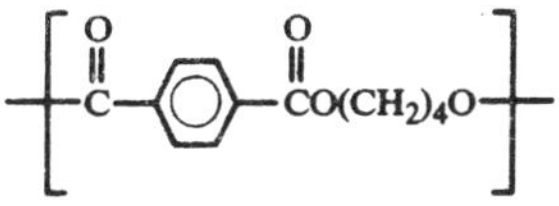

The "soft" segment has an average molecular weight of about 1,130 and a structure:

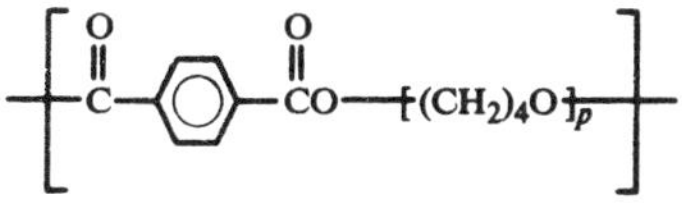

where p is an integer of from about 8 to 23. The "soft" segment makes up about 30 to 80% by weight of the total polymer, preferably about 40 to 70% by weight. The "hard" segment makes up about 70 to 20%, by weight of the total polymer, preferably about 60 to 30% by weight.

Epoxy resins that are suitable for forming the mixed blends of this process include those based on bisphenol A and epichlorohydrin that exhibit epoxide equivalents within the approximate range of from about 175 to 4,000 and average molecular weights of from about 350 to 3,800.

Example 1: A hydroxy-terminated poly(ester/ether) block copolymer having a MW of about 20,920 was made by reacting the following ingredients under nitrogen atmosphere in a two-gallon Ross mixer: 1,224 grams of polytetramethylene ether glycol (MW of about 1,000); 1,274 grams of dimethyl terephthalate; 546 grams of dimethyl isophthalate; and 1,102 grams of 1,4-butanediol.

In this formulation, the mol ratio of the dimethyl terephthalate to the dimethyl isophthalate is 70 to 30. This reaction was carried out in the presence of tetrabutyltitanate/magnesium acetate, an ester interchange catalyst, and octylated diphenylamine, an antioxidant.

Initially, the reaction temperature was held at 200°C until all methanol ceased distilling over, which was about one hour after the 200°C temperature had been reached. The pressure was then reduced to 6 mm Hg, and the temperature was increased to 250°C. These conditions were maintained for two hours. The reaction mixture was cooled to 200°C and the resulting hydroxy-terminated poly-(ester/ether) block copolymer was recovered. This polymer had a MW of about 19,670, a melting point of 140°C and is identified as "P-140" in the table below. It was then heated to its melting point along with various ratios of epoxy resins.

Two different resins were blended in amounts such that the final blend contained 10, 15 or 20 weight percent of the epoxy resin and 90, 85 or 80 weight percent, respectively, of the copolymer. One of the resins, identified as A, is the reaction product of bisphenol A with epichlorohydrin having an epoxide equivalent weight of about 900, while the resin, identified as B, is a similar product with an epoxide equivalent weight of about 5,000. The test results for the various thermoplastic products thus prepared are summarized in the table below.

Example 2: The procedure of Example 1 was repeated with the sole exception that the mol ratio of dimethyl terephthalate to dimethyl isophthalate was changed so that it was 80 to 20. This resulted in the copolymer having a MW of about 21,500 and a melting point of 160°C. It is identified as "P-160" in the table below.

Example 3: The procedure of Example 1 was repeated with the sole exception that the mol ratio of dimethyl terephthalate to dimethyl isophthalate was changed so that it was 90 to 10. This resulted in the copolymer having a MW of about 18,700 and a melting point of 180°C. It is identified as "P-180" in the table below.

Example 4: The procedure of Example 1 was repeated with the sole exception that the mol ratio of dimethyl terephthalate to dimethyl isophthalate was changed so that it was 100 to 0. This resulted in the copolymer having a MW of about 21,600 and a melting point of 195°C. It is identified as "P-195" in the table below.

LAB NUMBER OF BLEND	COPOLYMER	% EPOXY	EPOXY USED	TENSILE SHEAR STRENGTH	PEEL STRENGTH	CREEP at 300° F
P140	P140	0	—	990psi	16pli	Fail
P140-10-A	P140	10	A	940	110	Fail
P140-15-A	P140	15	A	1010	140	Fail
P140-20-A	P140	20	A	1040	138	Fail
P140-10-B	P140	10	B	1060	150	Fail
P140-15-B	P140	15	B	1170	145	Fail
P140-20-B	P140	20	B	1280	85	Fail
P160	P160	0	—	1070	16	Pass
P160-10-B	P160	10	B	1240	61	Pass
P160-15-B	P160	15	B	1340	80	Pass
P160-20-B	P160	20	B	1560	65	Fail
P180	P180	0	—	1070	10	Pass
P180-10-A	P180	10	A	1060	27	Pass
P180-15-A	P180	15	A	1240	34	Pass
P180-20-A	P180	20	A	1370	38	Pass
P180-10-B	P180	10	B	1320	51	Pass
P180-15-B	P180	15	B	1350	74	Pass
P180-20-B	P180	20	B	1590	62	Pass
P195	P195	0	—	980	2	Pass
P195-10-A	P195	10	A	1310	6	Pass
P195-20-A	P195	20	A	1420	8	Pass
P195-10-B	P195	10	B	1520	16	Pass
P195-20-B	P195	20	B	1710	11	Pass

Example 5: The P-140 copolymer alone and the present blends of P-140 with the epoxy resin A, all as prepared in Example 1, were tested for viscosity stability over time periods that would correspond to pot lives advantageous for commercial adhesives. Each sample was held at 400°F, and the viscosity of each was measured at various time intervals.

The unblended copolymer P-140 went through a marked viscosity decrease, resulting in loss of toughness and adhesive strength. The P-140-10-A and the P-140-20-A blends actually showed an increase in viscosity until the gel state was reached after about five hours. The P-140-15-A blend showed remarkable viscosity stability, with the viscosity remaining relatively constant over the eight-hour test period. No loss in toughness of the later-formed thermoplastic bonds was observed.

Containing a Rosin-Based Tackifier

K. Udipi and H.L. Hsieh; U.S. Patent 4,135,037; January 16, 1979; assigned to Phillips Petroleum Company describe adhesives useful for bonding together a wide variety of substrates, including metal ones, which comprise at least one epoxidized rubbery copolymer of a conjugated diene and a monovinylarene and at least one rosin-based tackifier.

The rubbery copolymers useful in this formulation generally include random, block, linear teleblock, and radial teleblock copolymers, including those containing random and tapered block segments, and mixtures thereof, having a conjugated diene/monovinylarene weight ratio of 45/55 to 95/5 and having a weight average molecular weight of about 15,000 to 350,000. A preferable range of conjugated diene/monovinylarene wieght ratios is from 50/50 to about 90/10. Especially preferred are linear teleblock copolymers which contain from 55/45 to about 85/15 conjugated diene/monovinylarene weight ratios and having a weight average molecular weight of about 50,000 to 150,000.

The conjugated dienes used for the preparation of the rubbery copolymers are those containing from 4 to 12 carbons per molecule, more preferably from 4 to 8 carbons per molecule. Especially preferred are 1,3-butadiene and isoprene.

The monovinylarene monomers can contain alkyl, cycloalkyl, and aryl substituents, and combinations thereof such as alkylaryl, in which the total number of carbons in the combined substituents is generally not greater than 12. Examples of aromatic monomers include styrene, α-methylstyrene, 3-methylstyrene, 4-n-propylstyrene, 4-cyclohexylstyrene, 4-dodecylstyrene, 2-ethyl-4-benzylstyrene, 4-p-tolylstyrene, 4-(4-phenyl-n-butyl)styrene, 1-vinylnaphthalene, 2-vinylnaphthalene, and the like. Styrene is preferred because of its availability and effectiveness.

Epoxidation can be effected by generally known methods such as by reaction with organic peracids which can be preformed or formed in situ during the reaction. Preformed peracids suitable for use include such as peracetic and perbenzoic acids, while in situ formation is exemplified by the use of hydrogen peroxide in the presence of a low molecular weight fatty acid such as formic acid, or hydrogen peroxide in the presence of acetic acid (or acetic anhydride) and a cationic exchange resin. In the latter case, the cationic exchange resin can be substituted by an alternate strong acid catalyst such as sulfuric acid or p-toluenesulfonic acid.

The epoxidation reaction can be conducted directly on the polymerization cement (polymer solution as formed), or alternately the recovered rubber polymer can be redissolved in an inert solvent such as toluene, benzene, hexane, cyclohexane, and the like, and epoxidation conducted on this new solution.

Tackifiers useful in this formulation are the modified and unmodified rosins and derivatives thereof. Examples include rosin itself, hydrogenated rosin, dehydrogenated rosin, dimerized and polymerized rosin, and the esters of rosin and the modified rosins with alcohols and polyols such as methanol, ethylene glycol, di- and triethylene glycols, glycerol, or pentaerythritol, and the like. Many such materials are commercially available.

The epoxidized rubbery copolymer, tackifier, and any other additives to be included in the formulation can be blended by any convenient method known in the art. The order of mixing is not critical. One well-known method is melt blending which comprises stirring of one or more of the materials into a melted batch of the other material using a mixer to assure adequate agitation for complete dispersion of the added components within the melted component. This composition can then be applied to the substrate by a method known in the art as hot melt application.

The tackifier level employed for hot melt applications is generally from 50 to 300 weight percent based on the epoxidized rubbery copolymer, but chosen preferably such that the initial viscosity of the adhesive formulation is between about 1,000 and 100,000 cp or higher, as measured at 400°F (204°C) using a Brookfield Viscometer equipped with a thermocell. In actual operation, these epoxidized rubbery copolymers and rosin-based tackifiers employed as adhesives are especially useful as hot melt adhesives which can be employed in any conventional manner.

The copolymers employed as adhesives also can be used as solvent cements, in which the polymers dissolved in a suitable solvent are applied in a conventional manner to substrates with the bond produced as the solvent evaporates or with solvent evaporation followed by a thermal bond-forming operation. These adhesives can be used to bond a variety of like or unlike substrates, e.g., metals such as aluminum, iron, steel, e.g., carbon steel, zinc, or tin; wood; paper; leather; plastics; and the like.

Example: A block copolymer of the structure styrene-butadiene/styrene (tapered)-styrene (20-54/6-20) was prepared and epoxidized by conventional polymerization and epoxidation procedures. Properties of this polymer are given in the table below.

Properties of Butadiene/Styrene Block Copolymer

	Unepoxidized	Epoxidized
Inherent viscosity	0.71	0.58
Molecular weight, $\overline{M}w$	52,000	73,000
Molecular weight, $\overline{M}n$	45,000	61,000
Gel, wt %	0	3*
Styrene, wt %	46.0	—
Block sytrene, wt %	37.8	—
Oxirane, wt %	~	3.72
Total oxygen, wt %	—	6.07

*Instead of gel, this may be organic sodium salt produced from the
  sodium hydroxide neutralization of the excess acid left from the
  epoxidation step.

Hot melt pressure-sensitive adhesive formulations based on the epoxidized butadiene/styrene block copolymer were tested for probe tack according to a modified procedure based on method ASTM D-2979-71. Composition 1, comprising a rosin acid tackifier, and comparative composition 2, comprising a hydrocarbon tackifier, were formulated as shown below.

These two compositions were melted on 0.002 inch (0.051 mm) Mylar film and the molten adhesive then spread to 0.001 to 0.003 inch (0.025 to 0.076 mm) film with a 0.0015 inch (0.038 mm) Bird applicator and allowed to cool to room temperature.

Composition 1

Components	Parts by Weight
Epoxidized butadiene/styrene copolymer	100
Hercolyn D (tackifier and plasticizer)	100
Pentalyn H (tackifier)	100
Irganox 1076 (antioxidant)	3

Composition 2

Components	Parts by Weight
Epoxidized butadiene/styrene copolymer	100
Wingtack 95 (tackifier)	150
Tufflo 6204 (plasticizer and extender oil)	100
Piccotex 120 (tackifier)	50
Irganox 1076 (antioxidant)	3

Results of probe tack testing are as follows: Composition 1, a probe tack of 1,508 g, and Composition 2, a probe tack of 0 g. This tackification was measured using a Polyken Probe tack tester (Testing Machines, Inc.). The test conditions were: probe speed, 1 cm/sec; dwell time, 1 sec; contact force, 100 g/cm^2. The test results were an average of five trials.

These data serve to illustrate the superiority of the composition based on rosin tackifiers (Hercolyn D and Pentalyn H) over the comparative composition based on hydrocarbon tackifiers (Wingtack 95 and Piccotex 120).

Addition of Compounds to Reduce Moisture Sensitivity

I. Hlavacek; U.S. Patent 4,148,778; April 10, 1979; assigned to Elitex, Koncern textilniho strojirenstvi, Czechoslovakia has found that the deleterious effects of water on epoxy resin adhesives are overcome by adding to the adhesives from 0.1 to 10 weight percent of water-soluble materials including the inorganic nitrites, alkalinized salts of hexavalent chromium, inorganic phosphates, silicates and carbonates. These may be used alone or in combination.

At the lower percentages of additive, since it is not dissolved in the adhesive, it has been found advantageous to employ a carrier for the additive to assure even distribution in the adhesive. Carriers selected for this purpose should evidence high absorption capacity. Typical carriers suitable for this purpose include carbon black, amorphous silica and the like. It is also important to avoid the use of an additive which might adversely react with the inorganic or organic filling agents such as titanium dioxide, insoluble silicates, metal powders and the like.

The epoxy resins suitable for use herein are selected from among the commercially available low molecular weight epoxy resins ranging in molecular weight from 350 to 4,000 which are terminated by at least one epoxy (i.e., ethylene oxide) group. These epoxy resins are typically prepared by alkaline condensation of 2,2-bis-p-hydroxyphenylpropane with epichlorohydrin or dichlorohydrin.

Example 1:

Components of Adhesive	Parts by Weight
Epoxy resin, MW 350 to 400	100
Aluminum powder	9
Carbon black	1.5
Microasbestos powder	3
Sodium nitrite	0.4
Diethylenetriamine	8

Example 2:

Components of Adhesive	Parts by Weight
Epoxy resin (modified), MW 500	100
Carbon black	0.5
Amorphous silica	11
Sodium metaphosphate	0.5
Aliphatic polyamine (amine number 800 to 1,100)	73

Studies of shear strength of commercial epoxy adhesives without additives indicated a range from 160 to 200 kp/cm^2. However, when exposing the synthetic adhesives of these formulations to water or a medium of high moisture content under ambient and elevated temperature, the basic shear strength remained unchanged whereas the adhesives of the prior art under these conditions evidenced a reduction of strength within the range of 20 to 40% of shear strength.

ANAEROBICALLY CURED COMPOSITIONS

Pressure Sensitive Adhesive

Anaerobic adhesive systems are typically supplied from a water-like liquid to a lightweight grease in consistency. One end-use application is to apply the adhesive to the threads of a bolt or mating nut which are then assembled. The adhesive fills the spaces between the threads which excludes oxygen and enables cure. In the normal situation, the metals present in the bolt or the nut accelerate cure.

A problem exists, however, in fixturing other surfaces together with initiation and completion of cure, and in providing a controlled quantity of anaerobic monomer to the surfaces to be bonded.

The composition described by *M. Douek, G.A. Schmidt, B.M. Malofsky and M. Hauser; U.S. Patents 4,092,376; May 30, 1978 and 4,118,442; October 3, 1978; both assigned to Avery Products Corporation and Loctite Corporation* may be instantly fixtured merely by finger pressure combined with sufficient cured strength to provide bonds of structural integrity. These anaerobic pressure sensitive adhesive compositions can be applied from or as sheets, tapes and the like to substrates to be bonded by cure upon the exclusion of oxygen.

The adhesives include a curable anaerobic resin system containing one or more anaerobic resins combined with a thermoplastic polymer system containing one or more high molecular weight thermoplastic polymers, the combination of which alone or upon inclusion of a tackifier, constitute a pressure sensitive adhesive system upon evaporation of essentially all of the solvent present. Further, there is provided in the anaerobic pressure sensitive adhesive composition an initiator system which is latent until made active by substantial exclusion of oxygen, preferably in combination with a suitable accelerator.

In one embodiment, if the anaerobic pressure sensitive adhesive system contains free transition metal ions, then at least the peroxy initiator may be encapsulated in microspheres which, upon rupture, and upon the exclusion of oxygen, will initiate cure. In another embodiment, a suitable metal accelerator may be encapsulated.

The thermoplastic polymers used in the preparation of the pressure sensitive anaerobic compositions are preferably of sufficient molecular weight so as to be elastomeric at room temperature. Further, they must be capable of being combined with the anaerobic resins and not greatly interfere with the creation of a crosslinked latticework of the anaerobic resins and prevent bending of the cured anaerobic polymer to the selected substrates to be joined.

In general, the amount of anaerobic resins combined with the thermoplastic polymer will range from about 35 to 99% by weight based on the total weight of the anaerobic resins provided and the thermoplastic polymer(s) with which it is combined, and, if present, a tackifier but exclusive of the amount of initiator system added. The preferred amount of anaerobic resin(s) combined in the thermoplastic polymer(s) is from about 55 to 95% by weight.

In addition, the thermoplastic polymer must be selected such that the composition alone or with tackifiers and upon the inclusion of an initiator system will form, after solvent evaporation, a curable pressure sensitive adhesive layer or film of sufficient cohesive strength to be applied to a substrate from differential release surfaces without disruption of the layer or film. The fully formulated, essentially solvent-free anaerobic pressure sensitive adhesive should be elastomeric at room temperature.

In addition, anaerobic pressure sensitive adhesive compositions should when applied to a surface, wet the surface and conform to the intricacies of the surface so that a uniform bond will be created upon cure and that cure will extend throughout the layer of applied anaerobic pressure sensitive composition to maximize cohesive bond strength.

To constitute a suitable pressure sensitive adhesive of this nature, the net composition when free of solvent should have, prior to cure, a static shear strength of at least about 2 minutes at a 250 gram load per 0.25 square inch and a $180°$ peel value of at least about 0.5 pound per inch, preferably at least about 1.0 pound per inch when using standard test methods.

Illustrative of the anaerobic resins which can be used in the preparation of these adhesives are polymerizable acrylate esters, including the α-substituted acrylate esters, such as the methacrylate, ethacrylate, and chloracrylate esters.

Of particular utility as adhesive monomers are polymerizable di- and other poly-acrylate esters since, because of their ability to form crosslinked polymers, they have more highly desirable adhesive properties. However, monoacrylate esters can be used, especially if the nonacrylate portion of the ester contains a hydroxyl or amino group, or other reactive substituent which serves as a site for potential crosslinking. Examples of monomers of this type are hydroxyethyl methacrylate, cyanoethyl acrylate, t-butylaminoethyl methacrylate and glycidyl methacrylate. Anaerobic properties are imparted to the acrylate ester monomers by combining with them a peroxy polymerization initiator as discussed more fully below.

One of the most preferable groups of polyacrylate esters which can be used in the adhesives disclosed herein are polyacrylate esters which are exemplified by the following materials: di-, tri- and tetraethylene glycol dimethacrylate, poly-ethylene glycol dimethacrylate, tetraethylene glycol diacrylate, diglycerol di-acrylate, ethylene dimethacrylate, trimethylol propane triacrylate, etc.

Yet another class of acrylate esters are those which are formed by the reaction of: (a) an acrylate ester containing an active hydrogen atom in the alcoholic moiety of the ester; with (b) an organic polyisocyanate.

Compositions including this general type of ester are disclosed in U.S. Patent 3,425,988. Preferably, the active hydrogen is the hydrogen of a hydroxyl or a primary or secondary amine substituent on the alcoholic moiety of the ester, and the polyisocyanate is a diisocyanate. Naturally, an excess of the acrylate ester should be used to ensure that each isocyanate functional group in the poly-isocyanate is substituted.

Typical polyisocyanates which can be reacted with the above acrylate esters to form polyacrylate monomers are toluene diisocyanate, 4,4'-diphenyl diiso-cyanate, dianisidine diisocyanate, cyclohexylene diisocyanate, 2-chloropropane diisocyanate, 2,2'-diethyl ether diisocyanate, and trans-vinylene diisocyanate.

Preferred anaerobic monomers are triethylene glycol dimethacrylate; the reaction product of hydroxypropyl methacrylate with methylene-bis-phenyl-4,4'-diiso-cyanate, a polymer formed by methacrylate capping of a 1:1 adduct of toluene diisocyanate and hydrogenated 2,2-bis-(4-hydroxyphenyl)propane as well as mixtures thereof.

The preferred thermoplastic polymers are polyvinyl chloride, polyurethanes, polyesters and acrylic based polymers. Typical of the peroxy compounds which may be employed as initiators are cumene hydroperoxide, methyl ethyl ketone hydroperoxide and the like. As accelerators there may be mentioned liquid and solid organonitrogen compounds.

Normally solid organonitrogen compounds are particularly preferred as they have the least effect on the viscosity of the resulting composition as well as the least tendency to migrate from the composition. Typical stabilizers are quinones, hydroquinones, and sterically hindered phenolic compounds.

Depending upon the amount of anaerobic resin system contained in the polymer system, the amount of initiator plus accelerator added will generally range from about 0.5 to 20% or more by weight based on the total weight of the polymer system plus anaerobic resin system, and if present, a tackifier.

As indicated, a tackifier may be employed to induce or enhance pressure sensitive properties. Typical tackifiers are rosins, rosin derivatives, terpenes, synthetic tacky resins, low molecular weight polyacrylates and the like as well as mixtures thereof. The tackifiers employed in general have a molecular weight less than about 5,000, preferably below about 1,000.

One way to prevent premature cure during shelf life or storage is to encapsulate the peroxy compound alone or with its accelerators. Then the active metals or metal ions can be left in the pressure sensitive adhesive composition without fear that premature cure or deactivation will occur.

When the anaerobic pressure sensitive adhesive is applied to differential release surfaces, it is required that it be removable from the release liner of maximum interfacial bond, typically a silicone-coated liner, for transfer to a substrate without cohesive disruption of the anaerobic pressure sensitive adhesive layer. It is desirable for any given application to have the coating as thin as conveniently possible when the surface(s) to which the anaerobic pressure sensitive adhesive is applied provides the active metal accelerator. Crosslinking will then rapidly occur throughout the anaerobic resin and the surfaces will be bonded together.

As an alternative, by employing microencapsulated accelerators within the pressure sensitive adhesive composition, greater cure rates and complete cure can be realized. Surface priming with accelerators may also be employed.

The types of products typically formed are the self-wound tapes, the surface of the supporting tape having differential release properties, sandwich constructions in which the anaerobic pressure sensitive adhesive composition is contained between two carrier liners having differential release surfaces, and similar products. All that is necessary is that the anaerobic pressure sensitive adhesive layer be transferable to a substrate and completely separated from its carriers to leave only an anaerobic pressure sensitive adhesive in contact with the substrate to be bonded to another substrate.

Of course, it is also understood that the liquid anaerobic pressure sensitive adhesive can also be applied directly to a surface to be subsequently joined to another surface, provided the solvent is removed before such joining.

In the following example, one or more of the following anaerobic resin systems were employed for the formulation of anaerobic pressure sensitive adhesive compositions:

Resin 1 consists of approximately 75% of a reaction product of two mols of hydroxypropyl methacrylate with one mol of methylene-bis-phenyl-4,4'-diisocyanate and 25% triethylene glycol dimethacrylate.

Resin 2 consists of approximately 66% of a polymer formed by hydroxypropyl methacrylate capping of a 1:1 adduct of toluene diisocyanate and hydrogenated 2,2-bis-(4-hydroxyphenyl)propane, 26% hydroxypropyl methacrylate, 7% acrylic acid and 1% methacryloxypropyltrimethoxysilane.

Example: To a heated flask equipped with a stirrer and a reflux condenser there was added 1,800 grams of toluene, 1,200 grams of Resin 1, and 300 grams of Resin 2. The mixture was heated with stirring at 70°C until a homogeneous solution was formed. To the stirred solution there was added 300 grams of a thermoplastic vinyl chloride copolymer (Vagh-2706, Union Carbide Corporation) and the mixture stirred until it again became homogeneous.

To the resultant mixture there was added with stirring 180 grams of an aqueous alcoholic solution containing a chelating agent for trace transition metal ions. The solution was held at a temperature between 40° and 50°C and stirred for three hours and the chelated transition metal ions removed.

To this solution there was added 70 grams of cumene hydroperoxide containing quinone, 37 grams of benzoylsulfimide and 37 grams of methylene-bis-dimethylaniline to form a catalyst system solution.

The anaerobic pressure sensitive adhesive solution was coated onto the release surface of a backing sheet fabricated from a plastic film and a paper having a silicone release coating. Coating weight after solvent removal was 28 grams per square meter. Another release sheet was applied to protect the anaerobic pressure sensitive adhesive.

A portion of the anaerobic pressure sensitive adhesive was tested for pressure sensitive properties. When applied to a paper support, the 250 gram static shear value was 5.5 minutes. The 180° peel test value on a Mylar support was 1.75 pounds per inch and failure was cohesive. Surface tack was about 2 inches.

Two aluminum alloy plates measuring 1" x 4" x $^{1}/_{16}$" were each etched on one end with a mixture of chromic and sulfuric acids to form a roughened surface. To one roughened surface there was applied a ½" x 1" layer of the anaerobic pressure sensitive adhesive. The roughened end of the other plate was placed on the anaerobic pressure sensitive adhesive in overlapping relation thereby excluding oxygen and initiating cure. The copper in the aluminum alloy accelerated cure and the anaerobic pressure sensitive adhesive bonded plates were allowed to cure for 24 hours at room temperature. The bonded plates were tested in an Instron tester and the bond was found to give a lap shear tensile value of 900 psi.

Containing a Monomer Mixture of Various Methacrylates

When anaerobically cured adhesives are used, certain disadvantages arise. These include the fact that the edges where the adhesive is exposed to air do not harden and have to be cleaned with solvent; if the anaerobic conditions are not perfect, the adhesive set-time is very long; the methacrylic acid added to various elastomers to improve the setting time causes corrosion of metals, particularly the copper wire used in electrical applications; etc.

To overcome these disadvantages, *I. Kishi, T. Nakano and H. Okai; U.S. Patent 4,096,201; June 20, 1978; assigned to Denki Kagaku Kogyo KK, Japan* have developed an adhesive composition, useful on many kinds of substrates, which does not require such strictly anaerobic conditions for curing, and which sets in a relatively short time.

The adhesive system consists of a redox system which comprises: (a) 5 to 30 weight parts of an elastomer of a copolymer of butadiene and acrylonitrile or a copolymer of butadiene and acrylonitrile and less than 5 weight percent of a functional monomer or a graft copolymer of butadiene and at least one of acrylonitrile, styrene and methacrylic acid ester; (b) 70 to 95 weight parts of a monomer mixture comprising 30 to 80 weight percent of 2-hydroxyethyl methacrylate or 2-hydroxypropyl methacrylate and 20 to 70 weight percent of a C_{1-4} alkyl methacrylate to total monomers having ethylenically unsaturated double bond; and (c) 0.1 to 10 weight percent of an organic hydroperoxide to total monomers having ethylenically unsaturated double bond.

The reducing agents used in the formulation include metallic soaps such as cobalt naphthenate, diethyl-p-toluidine, diisopropanol-p-toluidine or thioamides such as thiourea, acetyl thiourea, tetramethyl thiourea, ethylene thiourea, mercaptobenzoimidazole, etc. It is preferable to use thioamides from the viewpoint of high hardening speed. The formulation will be further illustrated by the example below, in which parts and percentages mean parts by weight and percentages by weight.

Example: In the example, various adhesive compositions were prepared by using various contents of 2-hydroxyethyl methacrylate (2-HEMA) and various contents of methyl methacrylate (MMA), as typical components and a reducing agent for a redox catalyst was used as the undercoat and the hardening time thereof was measured.

In the preparation of the adhesive composition, the specific amounts of MMA, 2-HEMA and butadiene-nitrile rubber were charged in a 1-liter glass flask and each mixture was stirred at 30°C for 30 hours to prepare each uniform translucent mixture and the catalyst and the additive were further added and the mixture was further stirred for 30 minutes to form each oxidizing system of the adhesive composition. The amounts of MMA and 2-HEMA in each sample are shown in the table below. These were reacted, in each case, with 15 parts by weight of butadiene-nitrile rubber (Hycar 1042); 0.3 part by weight of paraffin; and 8 parts by weight of cumene hydroperoxide.

Sample	. . Components (pbw) . .		MMA/2-HEMA
	MMA	2-HEMA	
a	68.0	17.0	80/20
b	59.5	25.5	70/30
c	51.0	34.0	60/40
d	42.5	42.5	50/50
e	34.0	51.0	40/60
f	25.5	59.5	30/70
g	17.0	68.0	20/80
h	8.5	76.5	10/90

The reducing agent for the redox catalyst was applied as a solution (a reducing system of the adhesive composition) prepared by dissolving 12 parts by weight of tetramethylene thiourea in 100 parts by weight of ethanol.

The hardening time of the adhesive composition was measured as follows. The solution was coated on a cold-rolled steel plate which was treated by the sandblast method, and was dried at the room temperature for 5 minutes (21°C, 55% RH) and then each adhesive composition (oxidizing system) was coated on it in a thickness of about 1 mm and the period from the coating to the initiation of hardening after losing finger tackiness was measured. The results were as follows:

Hardening Time of Adhesive Composition

	. . . . Sample					
	a	b	d	f	g	h
Hardening time (min)	30	20	15	5	20	*

*No hardening.

The results show that when the content of MMA is more than 80 wt % the hardening speed is too slow. When the content of 2-HEMA is 90 wt %, hardening in air could not be achieved.

Containing Reaction Products of Glycidyl Methacrylate and Half Esters of Dicarboxylic Acids

The anaerobically hardening adhesives, sealants, etc. developed by *B. Wegemund, W. Gruber and J. Galinke; U.S. Patent 4,096,323; June 20, 1978; assigned to Henkel Kommanditgesellschaft auf Aktien, Germany* are based on mixtures of methacrylic esters and organic peroxides, particularly hydroperoxides, which are preferably used in solvent-free form.

As essential components these systems contain reaction products of glycidyl methacrylate with linear half esters of dicarboxylic acids which are substantially free of epoxide groups. They can additionally contain methacrylic esters of mono- or polyhydroxy alcohols, and organic peroxides or hydroperoxides, as well as small amounts of additional polymerizable compounds, and optionally other auxiliary substances. Preferably the half esters have a molecular weight between about 200 and 1,800. To ensure sufficiently rapid hardening in the absence of oxygen, accelerators may be added to the system.

The half esters of dicarboxylic acids to be used as a starting material preferably have a molecular weight between about 200 and 1,800, more preferably between 400 and 1,200. They are prepared preferably by esterification of diols with dicarboxylic acids or dicarboxylic acid derivatives, particularly dicarboxylic acid anhydrides. To ensure that half esters are formed, a molar ratio of 1:2 of diol:acid anhydride must be used in the esterification.

The esterification is effected in known manner either in the melt or in inert solvents, with the possible addition of suitable catalysts, at temperatures between $50°$ and $150°C$, preferably between $80°$ and $120°C$, and optionally in an inert gas atmosphere, as for example, nitrogen.

The diol component of the half esters can be any short-chain, straight- or branched-chain, saturated or unsaturated compound. Particularly suitable are the aliphatic diols, among which the alkane-, alkene- and alkynediols are especially preferred. These diols can have 2 to 8, preferably 2 to 6 carbon atoms.

The above diols can be used alone or in combination with each other. The combination of an alkynediol with an alkanediol may be mentioned as one example.

The dicarboxylic acid-derived component of the half esters can be aliphatic, cycloaliphatic, aromatic, combinations of aromatic and aliphatic, and substituted groups thereof. The aliphatic and cycloaliphatic dicarboxylic acids and acid anhydrides can be saturated or unsaturated, preferably ethylenically unsaturated. The aromatic dicarboxylic acid or acid anhydride preferably contains at least one phenylene radical, most preferably one such radical. Cyclic anhydrides of dicarboxylic acids having five- or six-membered rings are preferred.

Examples of suitable dicarboxylic acids or acid anhydrides are maleic acid anhydride, succinic acid anhydride, succinic acid, phthalic acid anhydride, cyclohexane dicarboxylic acid anhydride, maleic acid, etc.

As in the case of the diol component of the half ester, the dicarboxylic acid-derived component can be esterified alone or in a combination of two or more of such components. Mixtures of maleic acid anhydride and phthalic acid anhydride, e.g., have been found to be extremely useful.

In the production of the reaction products of glycidyl methacrylate and half esters to be used for the adhesives and sealing compounds according to this process, the glycidyl methacrylate and the half ester are heated together, preferably in the presence of oxygen, as for example, by passing air through the reaction mixture, at temperatures between $60°$ and $120°C$, for about one-half to eight hours.

Since polymerization cannot be completely excluded in the production of the above reaction products, particularly at elevated temperatures, it is advisable to add known polymerization inhibitors, like hydroquinone. Furthermore, alkaline catalysts can be used to reduce the reaction time. Suitable here are those compounds which dissolve easily in the reaction mixture, for example, quaternary ammonium compounds, like trimethylbenzyl ammonium hydroxide.

In a preferred embodiment, the adhesive compositions consist of 70 to 80% of the dimethyacrylic ester and 20 to 30% monomethacrylates based on the total weight of the polymerizable portions of the composition. A favorable effect on the properties of a cemented joint has also been realized by addition to the mixture of small amounts of polymerizable carboxylic acids containing double bonds, such as alkenoic acids having 3 to 6 carbons, like methacrylic acid, acrylic acid, etc., in an amount of 0.1 to 5% by weight, based on the total weight of the polymerizable portions of the composition. Methacrylic acid has been found to be especially useful.

Another essential component of the anaerobically hardening compositions are the peroxide initiators. These are preferably hydroperoxides which derive from hydrocarbons with a chain length of 3 to 18 carbons. Especially suitable, e.g., is cumene hydroperoxide. The peroxides should be present in an amount of 1.0 to 10%, based on the total weight of the polymerizable portions of the compositions.

Commonly known accelerators may be used, e.g., organic tertiary amines, aromatic hydrazines, and organic sulfohydrazides. A sulfohydrazide accelerator combined with a tertiary amine and peracetic acid as stabilizer is a very appropriate combination, e.g., p-toluenesulfonic acid hydrazide with N,N-dimethyl-p-toluidine and peracetic acid.

These adhesive and sealing compositions are produced by mixing the components at room temperature. These compositions have excellent storage stability in air or oxygen. They are stable for months or years if they are kept in vessels that are permeable to air, like polyethylene bottles. They can further be stored in only partly filled bottles of glass, polyethylene, etc., without undergoing any change, a relatively low oxygen-partial pressure sufficing to inhibit polymerization. The bottles can also be colored to keep out shortwave light, which has an unfavorable effect on the stability.

The anaerobically hardening compositions are used in the industry for cementing metal sheets or metal parts of different materials, as e.g., for the cementing of screws and bolts in their threads, the sealing of screw-connections, nipples, etc., the cementing of plug connections, the sealing of flanges, the assembly of intricate metal shapes, sealing pipe joints, etc. Assemblies of metals such as iron, brass, copper and aluminum can be bonded to each other. Small quantities of the adhesive compositions are introduced between the surfaces to be bonded, after which the surfaces are contacted with each other sufficiently firmly or in another manner so as to exclude air or oxygen. Then the compositions of this process polymerize rapidly forming a firm bond. It is naturally also possible to accelerate the hardening with known means, as e.g., by heating the joint.

When the adhesives of this process are to be used for adhering or sealing glass or plastics or metals which are less catalytically active (for example, zinc, cadmium, high-alloyed steels and anodized aluminum), it is advantageous to pretreat these materials with metallic salt accelerators (for example, copper naphthenate and cobalt napthenate).

Example: 156.9 grams (1.6 mols) of maleic anhydride were reacted in a reaction vessel under nitrogen with 49.6 grams (0.8 mol) of ethylene glycol for 6 hours at 85°C. An acid number of 445 was attained. The yellow-colored viscous bis-maleic ester of ethylene glycol was then reacted at 80°C with 234 grams of glycidyl-methacrylate in the presence of 200 ppm hydroquinone and 6.4 grams of a 40% methanol solution of trimethylbenzyl ammonium hydroxide. Air was passed through the mixture during the reaction. After a reaction time of 6 hours the acid number was 5.2.

Using the dimethacrylic ester obtained as above, an anaerobically hardening cement of the following composition was prepared: 70 grams of dimethacrylic ester; 10 grams of methacrylic ester of 5,6-dihydrodicyclopentadienol; 10 grams of hydroxyethyl methacrylate; 0.5 gram of p-toluenesulfonic hydrazide; 0.5 gram of N,N-dimethyl-p-toluidine; 5 grams of a 70% solution of cumene hydroperoxide in cumene; 1 gram of a 40% solution of peracetic acid in acetic acid; and 3 grams of methacrylic acid.

This anaerobically hardening cement was tested in accordance with various testing methods described in full in the patent, and the following mean values were obtained: tensile shearing strength on steel, 270 kp/cm^2; tensile shearing strength on aluminum, 121 kp/cm^2; thermal stability, 300 kp-cm; and flexibility in bending test, 100°.

Stabilization of Acrylate Adhesives

It is desirable that adhesive compositions, typically containing vinyl unsaturation, in which the cure is initiated by peroxy compounds or ultraviolet radiation, be storage-stable for extended periods.

Many of these compositions are anaerobic, i.e., undergo cure when deprived of access to oxygen, but can be caused to cure by other means. It is desirable that such anaerobic compositions be preserved from curing when stored in an oxygen-poor environment, in order to undergo cure later, when required, by the use of other means.

Certain types of compounds are well known as inhibitors of premature polymerization for such compounds. Typically, these inhibitors are quinone- or hydroquinone-type compounds and their derivatives. Free radical scavengers of various types have also been used with some success.

None of these known inhibitors has proven fully satisfactory for all uses, however, especially in connection with acrylate and urethane-acrylate monomers such as those specifically described here.

D.J. O'Sullivan; U.S. Patent 4,100,141; July 11, 1978; assigned to Loctite (Ireland) Limited, Ireland provides inhibitor systems which provide advantageous improvements in stability. This stabilized, curable composition comprises: (a) a free radical-polymerizable monomer; (b) an effective amount for initiation of a peroxy or photosensitive initiator; and (c) an effective amount for stabilization of the composition of an inhibitor system selected from the group consisting of an allyl alkoxy hydroquinone and the combination of an allyl compound with an alkoxyl hydroquinone, the inhibitor system being dissolved in the monomer.

Preferred monomers are those prepared by reacting at a temperature between about 10° and 175°C (1) an organic polyisocyanate, and (2) an acrylate ester containing a hydroxy or amino group in the nonacrylate portion thereof, the ester being used in sufficient amount to react with substantially all of the isocyanate groups of the polyisocyanate.

Photo-initiated compositions are especially benefited since, by using increased levels of the inhibitor systems, improved stability is obtained without sacrifice of desirable cure properties.

With curable compositions which are inherently anaerobic in nature, the present inhibitor systems can effectively suppress the anaerobic cure characteristics, thereby allowing these desirable monomer compositions to be utilized in oxygen-starved aplications, such as the adhesive layer on tightly wound tapes or in oxygen-impermeable containers such as aerosol cans.

Generally, the initiator concentration will range from about 0.1 to 10% by wt of the total curable composition. As to the photosensitive (i.e., ultraviolet-sensitive) types of initiator systems, it is a particular advantage of these inhibitor systems that they may optionally be utilized in substantially increased concentration without significantly interfering with the ultraviolet-initiated cure properties.

Thus, the inhibitor system concentration may be increased by 4-5 fold, or more, over the lowest commercially useful level, thereby obtaining an outstandingly stable composition but yet without destroying its cure speed. In selecting photo-initiators, a degree of care should be exercised to avoid creating an interactive initiator-inhibitor combination which could impair the effectiveness of either. Thus, for example, the combination of benzoin ethers with the monoethyl ether of hydroquinone should be avoided since these have been found to interact.

Example 1: This example shows how the thermal stability of a polyethylene glycol dimethacrylate-based composition of the prior art (1) compares with that of a similar composition (2) according to the formulation shown below. These compositions are inherently anaerobic in nature, i.e., rapid cure is achieved only in the absence of oxygen. In order to test their anaerobic cure characteristics, a few drops of each composition were applied to the threads of mating black oxide nuts and bolts, the nuts and bolts were assembled (thereby excluding oxygen), and cure was allowed to proceed.

	..Compositions (pbw)..	
	1	2
Polyethylene glycol dimethacrylate	95.2	94.0
Saccharin	0.4	0.4
Cumene hydroperoxide	4.0	4.0
N,N-dimethyl-p-toluidine	0.4	0.4
Hydroquinone monomethyl ether (MEHQ)	Nil	0.2
Triallyl phosphate	Nil	1.0
Thermal stability at 82°C (min)	30	72

Both compositions developed satisfactory cured properties; however, it was found that the cure speed of composition (2), while commercially usable, was slower than that of composition (1) of the prior art.

Example 2: This example shows how the thermal stability of a trimethylol propane trimethacrylate-based composition of the prior art (3) compares with that of a similar composition (4) according to this formulation.

	..Compositions (pbw)..	
	3	4
Trimethylol propane trimethacrylate	95.0	93.8
Benzoyl peroxide	5.0	5.0
Triallyl phosphate	Nil	1.0
MEHQ	Nil	0.2
Thermal stability at		
55°C, hr	<8	>24
65°C, hr	<1	>4
82°C, min	50	>100

It was found that compositions (2) and (4) could be packed safely in high density polythene containers almost completely full, whereas the prior art compositions (1) and (3) had to be packed in partly filled low density polythene containers if premature polymerization due to oxygen exclusion were to be avoided.

Example 3: This example relates to a typical ultraviolet light initiatable curing composition of the prior art (5) and a similar composition according to this process (6). In both, a representative monomer A is used which may be obtained by reacting 1 mol (120 grams) of hydrogenated bisphenol A (i.e., 4,4'-dicyclohexanolyl dimethyl methane) with 2 mols (174 grams) of toluene 2,4-diisocyanate, and reacting the product with 2 mols (144 grams) of 3-hydroxypropyl methacrylate.

	..Compositions (pbw)..	
	5	6
Monomer A	93.0	91.0
Benzophenone	5.0	5.0
Acrylic acid	2.0	2.0
Triallyl phosphate	Nil	1.0
MEHQ	Nil	1.0

Composition (5) polymerizes either on deprivation of oxygen or on irradiation by ultraviolet light of suitable intensity, whereas composition (6) polymerizes only on irradiation, the anaerobic polymerization tendency having been substantially suppressed.

Thus the composition (6) can be applied to tapes which are afterwards coiled into rolls for use. This would not be possible for composition (5) unless expensive extra precautions were taken.

Containing Polycarbonates Terminated by Methacrylic Ester Groups

One of the objectives of the formulation described by *W. Gruber, J. Galinke and B. Wegemund; U.S. Patent 4,107,386; August 15, 1978; assigned to Henkel Kommanditgesellschaft auf Aktien, Germany* is the development of an anaerobically hardening adhesive and sealing compound based on methacrylic- or acrylic esters and organic peroxides or hydroperoxides, as well as small amounts of additional polymerizable compounds and optionally other auxiliary substances, containing as an essential component polycarbonates terminated by methacrylic ester groups which polymerizes at room temperature, i.e., which hardens to an adhesive layer or sealing compound, yielding after a short time a loadable bond which is strong and shows good thermal stability and flexibility, even for nonferrous materials, like aluminum and aluminum alloys.

The polycarbonates with terminal hydroxyl groups used contain diol-derived radicals from aliphatic, cycloaliphatic, and/or oxyaliphatic diols. Glycols containing both alkylene and cycloalkylene residues and those containing oxyalkylene residues are suitable.

Cycloalkylene glycols having five- or six- membered rings; short-chain diols, like butanediol; and short-chain ether diols, like di- and triethylene glycol, especially triethylene glycol, are preferred. The diols can be used alone or in combination with each other.

The polycarbonates with terminal hydroxyl groups preferably have an average molecular weight between about 300 and 10,000. In a preferred embodiment, α,ω-bishydroxy polycarbonates are used. An especially useful α,ω-bishydroxy polycarbonate can be prepared using several diols in widely varying mixing ratios to one another, as e.g., triethylene glycol and neopentyl glycol. Thus, very small amounts of a diol(s) can be used in combination with very large amounts of the other(s).

The α,ω-bishydroxy polycarbonates can be prepared according to known methods, e.g., by condensation of the diols with diphenyl carbonate or other readily accesssible carbonates.

The esterification or reesterification of these α,ω-bishydroxy polycarbonates with methyl methacrylate in the presence of acid catalysts, like cation exchangers, is effected by azeotropic distillation of the liberated methanol with methyl methacrylate.

The a,ω-dimethacrylates of polycarbonates thus obtained are liquid substances which are stable in the presence of oxygen at low temperatures.

The adhesives and sealing compounds made by this formulation can contain up to 90% by weight of the abovementioned polycarbonate dimethacrylic esters, based on the total weight of the polymerizable portions of the composition. According to a preferred embodiment, they additionally contain 10 to 40% by weight, based on the total weight of the polymerizable portions of cyclo-aliphatic, heterocyclic or aliphatic methacrylic esters. These cycloaliphatic, heterocyclic or aliphatic methacrylic esters can optionally contain free OH-groups.

In a preferred example, the adhesives consist of 70 to 80% of the polycarbon-ate dimethacrylic ester and 20 to 30% monomethacrylates, based on the total weight of the polymerizable portions of the composition. A favorable effect on the properties of a cemented joint has also been realized by addition to the mix-tures of small amounts of polymerizable carboxylic acids containing double bonds, such as alkenoic acids having 3 to 6 carbons, like methacrylic acid, acrylic acid, etc., in an amount of 0.1 to 5% by weight, based on the total weight of the polymerizable portions of the composition. Methacrylic acid has been found to be especially useful.

Another essential component of the anaerobically hardening compositions are the peroxide initiators. These are preferably hydroperoxides which derive from hydrocarbons with a chain length of 3 to 18 carbon atoms. Suitable, for example, are cumene hydroperoxide, tert-butyl hydroperoxide, methyl ethyl ketone hydroperoxide, and diisopropyl benzene hydroperoxide, especially cu-mene hydroperoxide.

The peroxides should be present in an amount of 0.1 to 20%, preferably 1.0 to 10%, based on the total weight of the polymerizable portions of the composi-tions. They are used mostly as phlegmatized (i.e., thickened) solutions or pastes, that is, with a relatively low content of inert substances, for example, dimethyl phthalate, cumene or the like.

The adhesives may also contain small amounts of auxiliary compounds such as are customarily present in adhesives of this type, such as stabilizers and, if de-sired, accelerators.

Example: 30 grams of an a,ω-bishydroxy polycarbonate, prepared from di-phenyl carbonate and triethylene glycol (MW, ~1,400), were reacted by ester exchange with 60 grams of methyl methacrylate in the presence of 15 grams of a commercial acid cation exchanger and 750 mg of hydroquinone by distilling off the methanol-methyl methacrylate azeotrope. During this reaction time the reaction mixture was maintained at the boiling point of the methyl methacrylate, and the ester was reintroduced to the reaction mixture after separation of the methanol. The dimethacrylic ester of the polycarbonate thus obtained was a colorless viscous liquid and is suitable for use as a component of anaerobically hardening adhesives.

Using the dimethyacrylic ester of the polycarbonate obtained as above, an anaerobically hardening cement of the following composition was prepared: 60 grams of polycarbonate dimethyacrylic ester; 30 grams of polyethylene glycol dimethacrylate (MW ~330); 1 gram of p-toluene sulfonic acid hydrazide; 0.5 gram of N,N-dimethyl-p-toluidine; 5 grams of a 70% solution of cumene hydroperoxide in cumene; 0.5 gram of a 40% solution of peracetic acid in acetic acid; and 3 grams of methacrylic acid.

This anaerobically hardening cement was tested in accordance with the testing method described below and the following mean values were obtained: tensile shearing strength on steel, 189 kp/cm^2; and tensile shearing strength on aluminum, 110 kp/cm^2.

Tensile Shearing Strength — The tensile shearing strength was measured (according to DIN 53283) using a tensile testing machine (feed = 20 mm/min) on steel plates (100 x 20 x 1.5 mm) which had been sandblasted at the place intended for the adhesive, and ahdered to one another by means of the composition of this process with a simple overlapping (DIN 1541/ST 1203) and on aluminum plates (DIN 1783, AlCuMg, 100 x 25 x 1.5 mm) which had been likewise adhered to one another with an overlapping length of 10 mm. The samples were tested as described above after hardening for 72 hours at room temperature.

ADDITIONAL FORMULATIONS

Butadiene Copolymers plus PVP plus Anionic Compounds

Mixed polymers of butadiene with styrene or acrylonitrile may be used in the form of their aqueous dispersions for gluing or bonding a wide variety of materials. The copolymers which have been used for this purpose are, in particular, those which additionally have groups such as carboxyl groups, or the like, which impart adhesion. Unfortunately, however, if such adhesives are used for gluing paper or cardboard to metals, and the laminates formed are exposed to hot water, the glued joint separates after a short period of time.

The objective of *K.-H. Kassner; U.S. Patent 4,076,661; February 28, 1978; assigned to Henkel Kommanditgesellschaft auf Aktien, Germany* is the production of an adhesive which overcomes this weakness to hot water.

The adhesive consists essentially of an aqueous dispersion containing from 20 to 50% by weight of adhesive solids consisting of a mixed polymerizate of butadiene with monomers selected from the group consisting of styrene and acrylonitrile, containing from 0.5 to 3% by weight of the mixed polymerizate of carboxyl groups derived from monomers polymerizable with butadiene, having a further content, based on the mixed polymerizate of: (a) from 0.5 to 8% by weight of a vinylpyrrolidone compound selected from the group consisting of polyvinylpyrrolidone and water-soluble copolymers of vinylpyrrolidone; and

(b) from 0.05 to 2% by weight of at least one carboxyl-group-containing, water-soluble, macromolecular compound and its water-soluble salts.

Suitable anionic water-soluble macromolecular substances containing carboxyl groups are, for example, carboxymethylcellulose (CMC) or its water-soluble salts, particularly the alkali metal salts. Suitable types of CMC have a viscosity, in accordance with Brookfield, of 6,000 to 25,000 cp at 25°C in a 2% aqueous solution. Furthermore, polyacrylic acids or their alkali salts, as well as poly-acrylic acids which are partially crosslinked, are suitable. It is also possible to use alginates and copolymers of styrene with maleic acid anhydride which are subsequently saponified with sodium hydroxide solutions and converted to a water-soluble form.

The adhesives described here are suitable for the bonding of metals to porous materials which can absorb the water in the adhesive, i.e., so-called absorbent substrates. Paper and cardboard are the most important materials in this re-spect. Among the metals to be glued, special reference may be made to alu-minum and copper which are frequently used in the form of foils. The adhesives can be applied in a conventional manner by means of rollers, nozzles or even by means of brushes and other suitable devices. The use of conventional laminating machines, which are equipped to rapidly discharge the evaporating water, has proved to be successful when effecting the gluing operation mechanically.

The adhesives in accordance with this process have satisfactory initial adhesion and, when used as laminating adhesives, enable the composite material to be rolled up immediately. The final strength is generally attained after a short stor-age period of 24 hours up to approximately 3 days. The bond between the paper and metal foils is distinguished by particularly favorable strength upon heating and when exposed to hot water. In addition, paper and polyethylene and poly-propylene can be bonded to one another in an adherent and durable manner if the surface of the plastics materials has been pretreated in a conventional man-ner.

Example: A latex of butadiene/styrene having a styrene content of approxi-mately 50% and a carboxyl group content of approximately 1.5% (by incorpo-rating acrylic acid by polymerization) was diluted to a 30% solids content. 1.5% by weight, based on the solid butadiene/styrene copolymerizate, of a polyvinyl-pyrrolidone having a K value of 90 was added under agitation, and 0.2% by weight, based on the solid butadiene/styrene copolymerizate, of carboxymethyl-cellulose in the form of its sodium salt was subsequently added. The sodium salt of the CMC had a viscosity, according to Brookfield, of 20,000 cp at 25°C.

A sodium kraft paper of 60 g/m^2 was laminated with a copper foil of 9 μ thick-ness on a commercially available laminating machine. The adhesive application was adjusted such that 1.5 grams was applied per m^2. A rigid bond was obtained after passing through the drying passage maintained at approximately 90°C. The samples were stored for 2 days at room temperature and then examined. It was determined that the sealing strength was 250°C/1 second.

Furthermore, the laminated material was suspended in hot water of 100°C. The test was discontinued, since no delamination phenomena were observed after 30 minutes.

Comparison Test — For the purpose of comparison, the same latex without an addition of polyvinylpyrrolidone and sodium salt of CMC was used to laminate the same materials. It was determined that after 2 days at room temperature, the sealing strength was only 180°C/1 second, and the laminated material was fully delaminated after it had been suspended for 5 minutes in hot water at 100°C.

One-Liquid Adhesive with Encapsulated Initiator

L.E. Wolinski; U.S. Patent 4,080,238; March 21, 1978; assigned to Pratt & Lambert, Inc. have formulated one-liquid adhesive compositions comprising a nonreactive thermoplastic polyurethane resin, an acrylic monomer, a polymerizable acid monomer, a free radical polymerization catalyst having dispersed therein pressure-rupturable microspheres encapsulating an activator for the catalyst.

Any polyester-based or polyether-based polyurethane resin may be used in the adhesive, so long as the resin is nonreactive and thermoplastic, as those terms are hereinafter defined. The term nonreactive means that the polymer contains no free, unreacted isocyanate groups, with the consequence that the polyurethane does not react to any notable degree with any component of the adhesive composition.

The term thermoplastic as employed herein applies to polyurethanes which are not gelled or crosslinked and which do not become crosslinked in the present adhesive formulations. Generally, such polyurethanes will be linear, or substantially so, although branching is not prohibited so long as the resultant polymer does not form an acrylic monomer insoluble gel.

Free radical addition polymerizable acrylic monomers are well-known and for purposes of this process conform to the general formula:

$$H_2C=\overset{X}{\underset{|}{C}}-\overset{O}{\underset{||}{C}}-Y$$

where X may be hydrogen, methyl, ethyl, or a halogen such as chlorine. Among these, it is preferred that X be hydrogen or methyl; Y in the above formula may be in turn represented by the general formulas:

$$-O-R, \text{ and } -N\overset{\displaystyle R}{\underset{\displaystyle R'}{}}$$

where R and R' each represent hydrogen, alkyl groups of up to 8, and occasionally more, carbon atoms and such alkyl groups substituted with hydroxyl, amino, halo, or aryl substituents.

Those skilled in the art will readily recognize that the foregoing formulas define acrylic and methacrylic esters, acrylic and methacrylic amides, and variously substituted variations thereof as preferred acrylic monomers. Acrylic acid and methacrylic acid are the preferred acid monomers for use in the formulations.

Since the adhesive is stable and noncuring in the absence of an activator or initiator for the catalyst, a separate formulation of an appropriate initiator is necessary for the use of the adhesive. Desirably, the activator component will be a tertiary amine, such as N,N-dimethyl aniline, N,N-diethyl aniline, N,N-dimethyl-p-toluidine, and the like. The activator may be supplemented by accelerators which function to increase the reaction rate of the adhesive cure. Such accelerators are most conveniently a source of a heavy metal, such as copper, iron, cobalt, manganese, lead, and the like, most desirably as an organometallic compound or salt wherein the heavy metal is oxidizable, i.e., not in its highest oxidation state. Generally a bead concentration of a broad range of 1.1 to 30.0 weight percent may be used with a preferred range of about 4.4 to 11.2 weight percent.

For any given adhesive composition, the proposed microsphere must be tested to determine its solubility therein. A microsphere having a wall composition consisting of a formaldehyde crosslinked gelatin and encapsulating an activator amine is sufficiently insoluble in adhesive compositions containing methacrylic acid as the sole polymerizable acid monomer to result in a one-liquid adhesive having a shelf life and pot life of more than 60 days.

The same adhesive solution having solely acrylic acid as the acid component in an amount of about 30%, with the same microsphere, has a shelf life of, at most, about 2½ hours. However, by reducing the amount of acrylic acid in the adhesive solution to less than 10%, a longer pot life of useful duration can be obtained. Bead sizes in the 200 to 600 micron diameter range are the most practical because of their ease of preparation, and because by suitable methods thin adhesive layers readily may be made from compositions containing beads of this size range. The following specific example illustrates the formulation. All parts are by weight.

Example: 33 parts of a polyester-based polyurethane resin (Estane 5712, B.F. Goodrich Co., Inc.), containing no free isocyanate groups was dissolved in a mixture of 33 parts of methacrylic acid and 34 parts of methyl methacrylate. After the resin dissolved, 5 parts of benzoyl peroxide and 0.1 part hydroquinone were stirred into the mix until dissolved. Thereafter, 5 parts of formaldehyde crosslinked gelatin microspheres encapsulating dimethyl-p-toluidine dissolved in tetrachlorethylene as the activator were added and the mixture stirred. The microsphere beads had essentially the same specific gravity as the adhesive solution so they remained in suspension therein.

The adhesive dispersion was coated on a steel plate to form a layer 7 mils thick and covered by a second steel plate. Pressure was then applied to crush the microspheres. The set time (i.e., the time when it was no longer possible to move the plates relative to one another by hand) was found to be 3.5 minutes, and the shear strength of the bond was 950 to 1,000 psi.

The one-liquid adhesive dispersion was allowed to stand for 21 days. At the end of this time, it had not gelled and was still pourable. The dispersion was poured on a steel plate, covered by a second steel plate and the microspheres in the dispersion were crushed by applying finger pressure to the plates. The set time was 3.5 minutes and the average shear strength was 950 psi.

The adhesive is capable of bonding a wide diversity of materials, attains bond strength rapidly, is effective with minimal or no surface preparation, and has a long shelf life.

Anaerobic Sealant

Sealant compositions which have anaerobic curing characteristics have been formulated for bonding closely facing metal surfaces such as threaded joints, for joining nuts to bolts without lock washers, gears to shafts for rotation therewith, and the like.

Some of the anaerobic compositions heretofore known cannot be used on passivated metal surfaces such as stainless steel, cadmium-coated steel, zinc-coated steel, and the like, while other such compositions, which are suitable for use with such surfaces, tend to promote corrosion when used on ordinary steel surfaces.

The formulation provided by *P.S. Patel and D.J. McDowell; U.S. Patent 4,090,997; May 23, 1978; assigned to Felt Products Manufacturing Company*, on the other hand, can be used with stainless steel and other passive metal surfaces as well as with ordinary steel to provide a high-strength bond and without promoting corrosion.

This high-strength anaerobic composition utilizes certain inorganic salts as polymerization initiators for a polymerizable polyacrylic ester monomer in combination with a N-nitrosoamine modifier and with a polymerization accelerator which can be a secondary or tertiary amine, a N,N-di(lower alkyl)amide of a monocarboxylic aliphatic acid, an organic carboximide of a polycarboxylic acid, an organic sulfimide of a carboxylic acid, or admixtures of the foregoing. Additionally, a quinone-type polymerization inhibitor is present in the composition in an amount sufficient to retard polymerization of the aforesaid monomer during storage of the composition in the presence of air.

The polyacrylic ester monomers suitable for use in compounding the anaerobic sealant compositions can be represented by the general formula shown on the following page.

In the formula R is a member of the group consisting of hydrogen, alkyl containing 1 to 4 carbons, inclusive, hydroxyalkyl containing 1 to 4 carbons, inclusive, and

$$-CH_2-O-\overset{\overset{O}{\|}}{C}-\underset{\underset{R'}{|}}{C}=CH_2$$

R' is a member of the group consisting of hydrogen, halogen, and alkyl containing 1 to 4 carbons, inclusive; R^2 is a member of the group consisting of hydrogen, alkyl containing 1 to 4 carbons, inclusive, hydroxy, and

$$-O-\overset{\overset{O}{\|}}{C}-\underset{\underset{R'}{|}}{C}=CH_2$$

m is an integer having a value of at least 1; n is an integer having a value of at least 1; p is an integer having a value of 0 or 1; q is an integer having a value of 0 or 1; and r is an integer having a value of at least 1.

Typical illustrative monomers within the purview of the foregoing general formula are the ethylene glycol dimethacrylates such as triethylene glycol dimethacrylate, ethylene glycol dimethacrylate, tetraethylene glycol dimethacrylate, polyethylene glycol dimethacrylate, and the like, 1,3-butylene glycol dimethacrylate, trimethylol propane trimethacrylate, neopentyl glycol dimethacrylate, ethoxylated bisphenol A dimethacrylate, propoxylated bisphenol C dimethacrylate, and the like. For improved heat resistance, particularly in retaining formulations, ethoxylated bisphenol A dimethacrylate is the preferred monomer and can be used alone or in admixtures with other monomers such as tetraethylene glycol dimethacrylate.

Preferably the acid number of the monomeric constituent in these sealant compositions is about 0.005 to 0.05. The inorganic salt initiator must be present in the anaerobic sealant composition in an amount sufficient to initiate polymerization of the monomer between two surfaces to be joined or bonded upon the exclusion of air, i.e., in the absence of a substantial amount of oxygen. The inorganic salt initiators are the persulfates or perchlorates of ammonium, an alkali metal, or an alkaline earth metal. Illustrative inorganic salt initiators are ammonium persulfate, ammonium perchlorate, sodium persulfate, sodium perchlorate, potassium persulfate, potassium perchlorate, lithium perchlorate, calcium perchlorate, and magnesium perchlorate.

Preferably the amount of initiator present in the anaerobic sealant formulations of this process can be about 0.005 to about 15 parts by weight per 100 parts by weight of the monomer.

The contemplated modifier is a N-nitrosoamine. These compounds are commercially available and can be a dialkyl-substituted amine, e.g., N-nitrosodipropylamine, N-nitroso-N-butyl-N-methylamine, N-nitrosodioctylamine, and the like, an aromatic amine, e.g., N-nitroso-N-phenyl-N-benzylamine, N-nitrosodibenzylamine, and the like, or a heterocyclic amine, e.g., N-nitrosopiperidine, N-nitrosoaziridine, N-nitrosopyrrolidine, N-nitrosopyrazole, N-nitrosocarbazole, etc.

The N-nitrosoamine modifier is present in the formulation in an amount of about 0.2 to about 0.9 part by weight per 100 parts of the monomer. Conventional polymerization accelerators may be used such as tertiary amines, organic carboximides of a polycarboxylic acid, etc.

However, these sealant compositions must contain a sufficient amount of a suitable stabilizer to retard polymerization of the monomer during storage of the sealant composition in the presence of air; thus sometimes the amount of stabilizer present in the commercial monomer compositions is supplemented. By the term "quinone-type stabilizer" as used herein is meant quinone and its derivatives sch as the benzoquinones, the naphthoquinones, the hydroquinones, and the like.

Example: Anaerobic Sealant Composition Containing N-Nitrosodiphenylamine Modifier — A sealant composition was compounded using tetraethylene glycol dimethacrylate monomer (100 parts by weight), hydroquinone (about 100 parts per million parts of monomer), potassium perchlorate (0.55 part by weight), benzoic sulfimide (0.5 part by weight), dimethyl-p-toluidine (0.3 part by weight), and N-nitrosodiphenylamine (0.2 part by weight). Initially an admixture of the monomer and potassium perchlorate was produced, benzoic sulfimide added, and the admixture then aged at room temperature for about 7 days.

The aged admixture was thereafter filtered, dimethyl-p-toluidine and N-nitrosodiphenylamine added thereto, and stirred. The produced composition was tested for set time, and for breakaway torque and runaway torque, using three-eighth-inch steel bolts and nuts. Gelling time at 180°F was also noted.

The set time was found to be about 15 minutes and the gelling time at 180°F about 5.5 hours. The breakaway torque was found to be 96 inch-pounds and the runaway torque was found to be 282 inch-pounds.

Block Copolymer Extended with an Anhydride

The object of the process devised by *R.M. Schure, J.H. Kooi and J.M. Brown; U.S. Patent 4,093,675; June 6, 1978; assigned to Unitech Chemical Inc.* is to provide both improved thermoplastic and improved thermoset adhesive compositions that have both high peel strength and excellent tensile strength.

The preparation of the high peel strength thermoset adhesive comprises the steps of: (1) forming a hydroxy-terminated poly(ester/ether) block copolymer by heating at a temperature of about 150° to 250°C and at a pressure of about 1 to 15 mm Hg, a reaction mixture of (a) a dicarboxylic acid or ester thereof, (b) a low molecular weight diol, and (c) a polyether that is hydroxy-terminated at both ends, until methanol distillation ceases, and further heating the reaction mixture at about 220° to 280°C under a pressure of about 1 to 15 mm Hg for an additional 1 to 6 hours.

(2) One mol of this block-copolymer was reacted at 100° to 220°C for 2 to 6 hours with 1 to 6 mols of the tetracarboxylic dianhydride plus an excess amount of a monoanhydride to form a carboxylated poly(ester/ether) block copolymer of the formula:

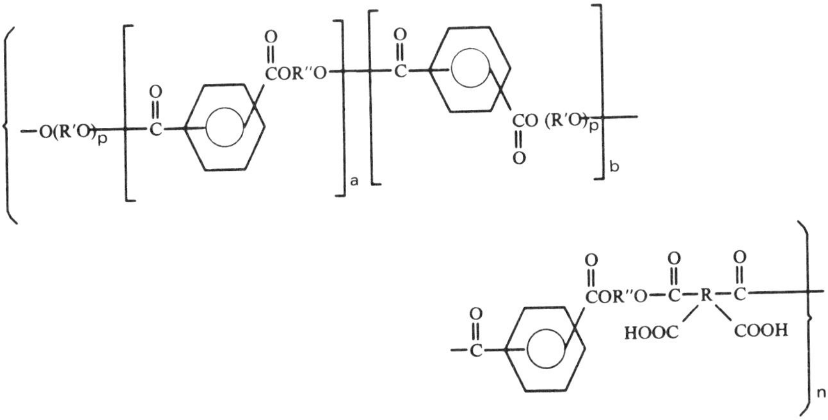

where R is alkyl, cycloalkyl or aryl, n is a number from about 1 to 2, R' and R'' are alkyl, alicyclic, acyclic, aryl, or arylalkyl, p is a number from 2.4 to 136.0, a is a number such that the repeating ester block segment within its set of brackets makes up about 70 to 20 weight percent of the polymer, and b is a number such that the repeating ether block segment within its brackets makes up about 30 to 80 weight percent of the polymer.

(3) The carboxylated poly(ester/ether) block copolymer was cured at about 300° to 450°F with a curing agent consisting of an epoxy resin, a phenol-aldehyde resin, or an amine-aldehyde resin to form a cured copolymer creep-resistant, high-tensile thermoset adhesive that simultaneously exhibits high peel strength.

Example: A hydroxy-terminated poly(ester/ether) block copolymer having a MW of about 17,000 was first made by reacting, under a nitrogen atmosphere, 135.8 grams (0.7 mol) of dimethyl terephthalate and 58.2 grams (0.3 mol) of dimethyl isophthalate with 117.5 grams (1.31 mols) 1,4-butanediol and 132.6 grams [0.13 mol of poly(tetramethylene ether) glycol, MW of 1,020] at 200°C in the presence of tetrabutyltitanate, an ester interchange catalyst, and octylated diphenylamine, an antioxidant. After all of the methanol had ceased distilling over, the pressure was reduced to about 12 mm Hg and the temperature increased to 250°C. These conditions were maintained for about 3 to 4 hours, with stirring to prepare a prior art hydroxy-terminated copolymer. This known thermoplastic adhesive exhibited a melting point of 140°C, a glass transition temperature of -53°C and a 15% degree of crystallinity. It was applied to a metal substrate, with heating, to form a thermoplastic adhesive bond having a tensile strength of about 1,000 psi on a 1 x 1 inch lap bond of unprimed steel at 77°F (25°C) and a peel strength of 40 pli on unprimed aluminum (approximately 20 mils thick) at 77°F. This polymer contained 44% "soft" segments and 56% "hard" segments.

One mol of the thus-prepared hydroxy-terminated poly(ester/ether) block copolymer was heated with one mol of 3,3',4,4'-benzophenone-tetracarboxylic acid dianhydride (BTDA) at 175°C for 4 hours. The resulting carboxylated poly(ester/ether) block copolymer had an acid value of 8.0. It was then applied to metal substrates, with heating, to form a thermoplastic adhesive bond having a tensile strength of 1,300 psi on unprimed steel and a peel strength of about 50 pli on unprimed aluminum at 77°F.

Polyesters Containing a Tertiary Amino Group

Conventional polyester adhesives such as those of polyethylene terephthalate-sebacate, polyethylene-tetramethylene terephthalate-adipate, polytetramethylene terephthalate-isophthalate, and polytetramethylene terephthalate-isophthalate-sebacate exhibit relatively low adhesive strength to metals. They have, consequently, been at a disadvantage since they cannot be used for building materials, mechanical parts, electrical devices and so on which demand high adhesive strength.

Y. Niinami and K. Etoh; U.S. Patent 4,096,123; June 20, 1978; assigned to Toyo Boseki KK, Japan have found that a polyester adhesive can be formulated which exhibits two or three times the adhesive strength of conventional polyester adhesives with respect to metals.

The adhesive has as its effective component a copolyester containing a tertiary amino group comprising: (1) a terephthalic acid residue; (2) at least one dicarboxylic acid residue selected from the class consisting of aliphatic dicarboxylic acid residues and aromatic dicarboxylic acid residues other than a terephthalic acid residue; (3) at least one alkylene glycol residue; and (4) at least one glycol residue or dicarboxylic acid residue selected from the class consisting of glycol residues and dicarboxylic acid residues each having the group represented by the formulas below.

(1) $-R_1-N\left\langle\begin{array}{c}-R_3\\ \\ -R_4\end{array}\right\rangle N-R_2-$ (2) $-R_5-\overset{\overset{\displaystyle R_6}{|}}{N}-R_7-$

(3) $-R_8-\overset{\overset{R_{11}}{|}}{N}-R_9-\overset{\overset{R_{12}}{|}}{N}-R_{10}-$

(4) $-R_{19}-\!\!\left[\!\!\bigcirc\!\!\right]\!\!-R_{20}-$ (pyridine ring with N)

(5) $-R_{13}-\overset{\overset{R_{15}}{|}}{\underset{\underset{R_{16}-N}{|}}{C}}--R_{14}-$ with $R_{16}-N\overset{R_{17}}{\underset{R_{18}}{\diagup}}$

In the formulas R_1, R_2, R_5, R_7, R_8, R_9, R_{10}, R_{13}, R_{14}, R_{16}, R_{19} and R_{20} are each an alkylene group of 1 to 15 carbons, R_6, R_{11} and R_{12} are each a hydrogen atom or an alkyl group of 1 to 10 carbons, R_{17} and R_{18} are either, independently, a hydrogen atom or an alkyl group of 1 to 4 carbons, or, in combination, make up a polymethylene group which may form a heterogeneous ring in conjunction with the adjoining nitrogen atom, R_3 and R_4 are each an alkyl group of 1 to 10 carbons, and R_{15} is an alkyl group of 1 to 3 carbons or:

$$-R_{16}-N\overset{\diagup R_{17}}{\diagdown_{R_{18}}}$$

The terms "acid residue" and "glycol residue" used mean the configurational forms of the indicated acid and glycol contained in the polyester, the "acid residue" referring to the group remaining after removal of a hydroxyl group from the relevant carboxylic acid and the "glycol residue" to the group remaining after removal of a hydrogen atom from each of the terminals of the relevant glycol.

Of the dicarboxylic acid and functional derivatives thereof which provide aliphatic dicarboxylic acid residues for the formulation, particularly desirable are those aliphatic dicarboxylic acids having 2 to 20 carbons and their functional derivatives, examples being oxalic acid, succinic acid, adipic acid, suberic acid, azelaic acid, sebacic acid, undecanedioic acid, dodecanedioic acid, brassydic acid, tetradecanedioic acid, and lower alkyl esters and chlorides of the dicarboxylic acids.

Examples of the carboxylic acids and functional derivatives thereof which provide aromatic dicarboxylic acid residues other than a terephthalic acid residue include isophthalic acid, ortho-phthalic acid, 2,6-naphthalene dicarboxylic acid, 2,7-naphthalene dicarboxylic acid and lower alkyl esters and chlorides of dicarboxylic acids.

Of the alkylene glycols which provide alkylene glycol residues, particularly desirable are alkylene glycols having 2 to 10 carbons, specific examples being ethylene glycol, propylene glycol, tetramethylene glycol, pentamethylene glycol, hexamethylene glycol, heptamethylene glycol, octamethylene glycol, nonamethylene glycol, decamethylene glycol and neopentyl glycol.

Examples of glycols which provide the glycol residues possessed of the groups of the generic formulas (1) through (5) are: (1) N,N'-bis(hydroxymethyl)piperazine; (2) diethanolamine; (3) N,N'-dimethyl-N,N'-bis(2-hydroxyethyl)ethylenediamine; (4) 3,5-dimethylolpyridine; and (5) 2-methyl-2-N,N-dimethylaminoethyl-1,3-propanediol. Examples of dicarboxylic acids and functional derivatives thereof which provide the dicarboxylic acid residues having the groups of the generic formulas (1) through (5) are: (1) N,N'-bis(2-carboxyethyl)piperazine; (2) N,N-bis(carboxymethyl)methylamine; (3) N,N'-dimethyl-N,N'-bis(2-carboxyethyl)-ethylenediamine; (4) 3,5-di(carboxymethyl)pyridine; and (5) 4-methyl-4-N,N-dimethylaminomethyl azelaic acid.

To produce the copolyester, terephthalic acid or a functional derivative and at least one dicarboxylic acid, or a functional derivative, chosen from the class of aliphatic or aromatic dicarboxylic acids other than terephthalic acid and functional derivatives are directly esterified or transesterified with an alkylene glycol; the product is reacted with a glycol or dicarboxylic acid having a tertiary amino group or a functional derivative thereof.

A wide variety of methods can be employed for actual use of the adhesive. Examples include a method which comprises the steps of shaping the adhesive in the form of a tape, ribbon, film, bar, rectangular sheet, strand or the like, inserting the shaped adhesive between the opposed surfaces of substrates desired to be joined and melting the inserted adhesive to cause desired adhesion of the substrates; a so-called dip-coat method which comprises melting the adhesive and applying the melt adhesive to the opposed surfaces of the substrates to cause the adhesion; and a method which comprises dissolving the adhesive in a solvent, applying the resultant solution to the opposed surfaces of the substrates, treating the applied solution so as to expel the solvent therefrom and thereafter melting the applied film to cause the adhesion. The polyester may also be cured by ultraviolet light.

Homopolyamide-Copolyamide Adhesive for Metal Cans

In the can manufacturing industry, so-called tin-free steels such as chromium-plated steels and steels electrolytically treated with chromic acid are used as can blanks. For side-seaming such new can blanks there has been developed a so-called side-lap-seam method where both side ends of the can blank are bonded to each other with an organic adhesive. Nylon-11, nylon-12 and copolyamides thereof have been used as nylon-type adhesives for this method because they have relatively low melting points and possess such merits as low water absorption and high impact strength. However, they are still insufficient in respect to high-speed bonding, bonding strength and processability of bonded can bodies. For instance, in the most advanced high-speed canning process conducted on a commercial scale, the operation must be conducted at such a high rate as 300 to

1,000 cans per minute, and for attainment of such high-speed operation, the step of bonding the side-seam portions, i.e., the cycle of melt adhesion and cooling of the adhesive, must be completed within such a short period as 20 to 200 milliseconds. Further, after packing of goods or contents the so-prepared can bodies must be subjected to the double-seaming operation at a temperature, for example, ranging from $-2°$ to $+5°C$ in the case of cans for beer and soft drinks. This double-seaming operation is accomplished at rates, for example, such as 1,600 cans or more per minute.

H. Ueno, S. Otsuka, T. Tsukamoto and A. Kishimoto; U.S. Patent 4,101,534; July 18, 1978; assigned to Toyo Seikan Kaisha Limited, Japan have therefore, directed their efforts to providing a metal adhesive which can give a metal-to-metal bonding of a sufficient bonding strength in a very short cycle of melt adhesion and cooling.

It was found that the following polymer composition is an excellent metal adhesive suitable for attaining this object: (a) 95 to 60% by weight, based on the total composition, of an aliphatic homopolyamide; and (b) 5 to 40% by weight, based on the total composition, of an aliphatic copolyamide, the homopolyamide (a) being a crystalline homopolyamide having up to 14 amido groups per 100 carbons, and the copolyamide (b) consisting of (1) 95 to 60% by weight of a polyamide having up to 14 amide groups per 100 carbons, and (2) 5 to 40% by weight of at least one polyamide whose recurring units are different from those of the polyamide (1).

In an especially preferable formulation, 95 to 60% by weight of the total amide recurring units of the copolyamide (b) are constructed of the same amide recurring units as those of the crystalline homopolyamide (a) and the remaining amide recurring units are constructed of other amide recurring units. The adhesive composition of this embodiment can give a desirable combination of an excellent bonding strength at a high-speed bonding operation and an excellent high-speed processability of the sealed portion.

For instance, if polylauryl lactam is used as the crystalline homopolyamide (a), 95 to 60% by weight of the total recurring units of the copolyamide (b) should be composed of the units of ω-amino dodecanoic acid.

The degree of polymerization of the copolyamide (b) is not particularly critical, but in general, it is preferred that the copolyamide (b) have a relative viscosity of from 1.8 to 3.5 measured in a solution of 1 gram polymer in 100 cc of 98% sulfuric acid at $20°C$.

In order to improve the high-speed processability of the bonded portion and the bonding ability of the adhesive as well as to make the high-speed bonding operation possible, it is important that the abovementioned crystalline homopolyamide (a) and copolyamide (b) are blended at a weight percentile ratio of (a):(b) = 95-60:5-40; preferably (a):(b) = 89-70:11-30.

Example: 75 parts by weight of granular solids of a polymer of ω-amino undecanoic acid (nylon 11; relative viscosity = 2.45), which is a crystalline homopolyamide, are mixed with 25 parts by weight of granular solids of a copolyamide
consisting of the amide recurring units of the polymer of ω-amino undecanoic
acid and the polycaprolactam recurring units at a weight ratio ranging from
100/0 to 30/70 having a relative viscosity of 2.45±0.05, and the blend is formed
into a film of a thickness of 50 μ by employing a film-molding apparatus. The
heat extrusion is conducted at a die temperature of 210°C and a screw rotation
rate of 40 rpm.

Steel strips having a length of 780 mm and width of 210 mm, are heated at about
250°C by high-frequency induction heating in both side end portions of a width
of about 7 to 8 mm along the longitudinal direction, and adhesive film tapes of
a 50 μ thickness having a width of 5 mm are pressed on both side end portions
of the strip for a period of 47 milliseconds by means of a press roll.

The strip is further cut into can blanks of 125 mm x 210 mm, and can bodies
are prepared from these blanks. At the time of adhesion, both side end portions
of blanks are heated at 250°C and the press is conducted for 43 milliseconds
under cooling.

Tests were made on the various weight ratios of the amide recurring units of the
polymer of ω-amino undecanoic acid and the polycaprolactam which showed
that when this ratio was kept between 95/5 and 60/40, the resulting bonding
had high strength and leakages in the double-seamed portion were greatly reduced. The bonding strength was greatest when the ω-amino undecanoic acid
was present in an amount of 90 to 70% by weight.

Adhesion Promoter for 2-Cyanoacrylates

Adhesive compositions based on 2-cyanoacrylate esters belong to a class of adhesives known as reactive liquid adhesives. 2-Cyanoacrylate adhesives are single-
part, low-viscosity adhesives which are characterized by features such as:
(1) their ability to polymerize at room temperature without the use of an added
catalyst when pressed between two substrates, (2) their rapid rate of cure, and
(3) the strength of the bonds produced with a wide variety of substrates.

Although adhesive compositions consisting of 2-cyanoacrylate esters and conventional stabilizers inherently yield high bond strength (as commonly measured by
the test of tensile shear strength), improvements therein would be desirable
particularly in cases where the substrate is of greater strength than the adhesive,
as in the case of many metal bonds.

Accordingly, the object of *J.E. Schoenberg and D.K. Ray-Chaudhuri; U.S.
Patent 4,125,494; November 14, 1978; assigned to National Starch and Chemical Corporation* is to provide an improved 2-cyanoacrylate adhesive having enhanced bond strength which does not significantly reduce its cure rate.

This objective is achieved in an improved adhesive composition comprising a mixture of: (a) a fully stabilized 2-cyanoacrylate adhesive consisting essentially of (1) a monomeric ester of 2-cyanoacrylic acid of the general formula

$$H_2C=\overset{\displaystyle O}{\underset{\displaystyle CN}{\overset{\displaystyle \|}{C}}}\text{--}C\text{--}OR$$

where R is an alkyl or alkenyl group having from 1 to 16 carbons, a cyclohexyl group or a phenyl group, and (2) an effective amount of an anionic polymerization inhibitor selected from the group consisting of acidic gases, protonic acids and anhydrides thereof, where the anionic polymerization inhibitor has a pK_a less than about 4; and (b) from 0.02 to 0.3% by weight, based on the total composition of acetic acid. The composition is characterized by a tensile shear strength of at least 80 kg/cm^2 as measured on brass.

The process for preparing such an improved adhesive composition comprises the step of dissolving the specified amount of acetic acid into the fully stabilized 2-cyanoacrylate adhesive described above.

If the adhesive composition is to be stored for an extended period of time, it may be desirable to add a free radical polymerization inhibitor to the composition for added storage stability. In addition, other optional ingredients which improve specific properties of the adhesive such as thickeners or plasticizers may be incorporated into the composition.

Example 1: This example illustrates the effect of carboxylic acids of varying strength on the cure rate of 2-cyanoacrylate adhesive compositions. Eight samples of 2-cyanoacrylate adhesive compositions were prepared by dissolving approximately equal amounts of a given carboxylic acid in a quantity of ethyl 2-cyanoacrylate obtained commercially containing 0.002% sulfur dioxide and 0.0075% hydroquinone as stabilizers. Each resulting composition was evaluated against a control containing no acid as to set time on phenolic chips. The results are given in the table below. The amounts of acid added are given in milliequivalents per kilogram. The numbers in this column indicate that comparable amounts of each carboxylic acid were used on an equivalent basis. (Equivalency is based on the number of acidic protons per molecule.)

Carboxylic Acid	pK$_a$ of Acid	Amount of Acid Added		Set Time (seconds)
		meq/kg	% by weight	
None (control)	–	0	0	5
Propionic	4.9	7.8	0.058	5
Isobutyric	4.8	8.3	0.073	10
Acetic	4.8	8.2	0.049	5
Benzoic	4.2	8.2	0.10	15
Formic	3.8	8.0	0.37	15
Salicylic	3.0	8.7	0.12	30
Malonic	2.8	8.5	0.044	60
Cyanoacetic	2.5	8.3	0.071	45

From the table it is evident that the weak acids such as acetic, propionic and iso-butyric acids have a minimal effect on the cure rate of the adhesive composition. In contrast, the stronger acids with pK_a less than about 4 act more effectively as anionic polymerization inhibitors and thus cause significant retardation in the rate of cure, the retardation being generally more pronounced as the acid strength increases.

Example 2: This example illustrates the effect of acetic acid concentration on the bond strength of the adhesive compositions of this formulation.

The 2-cyanoacrylate adhesive compositions were prepared by dissolving the given amount of acetic acid in ethyl 2-cyanoacrylate obtained commercially con-taining 0.0028% sulfur dioxide and 0.001% hydroquinone as stabilizers. The re-sulting compositions were evaluated against a control as to tensile shear strength. The results are given in the table below.

Acetic Acid	. . . .Tensile Shear Strength (kg/cm^2). . . .		
(% by wt)	Steel	Brass	Aluminum
None*	134	104	111
0.005	117	102	98
0.01	108	106	89
0.02	142	120	103
0.04	167	169	136
0.10	242	176	158
0.21	212	131	171
0.29	143	122	75
0.41	97	63	64
0.50	77	40	61

*Control

The results indicate that only a certain concentration range of acetic acid (about 0.02 to 0.3% by weight) is effective in enhancing the bond strength of the adhe-sive composition. Because of the reduction in set time at high concentrations of acetic acid, however, the preferred range is 0.03 to 0.1% to maintain optimum values for both bond strength and cure rate.

Example 3: This illustrates the use of acetic acid in adhesives based on three different 2-cyanoacrylate esters. Samples were prepared by dissolving 0.05% by weight of acetic acid in either methyl, n-butyl or allyl 2-cyanoacrylate, each con-taining a fixed amount of sulfur dioxide and hydroquinone as stabilizers. As con-trols, methyl, n-butyl and allyl 2-cyanoacrylate stabilized as above without any acetic acid incorporated, were used. Results of tensile shear strength tests of the compositions are given below:

Cyanoacrylate Ester	Adhesion Promoter	. . .Tensile Shear Strength (kg/cm^2) . .		
		Steel	Brass	Aluminum
Methyl	None (control)	179	180	166
	Acetic acid	177	214	171
n-Butyl	None (control)	63	64	55
	Acetic acid	118	108	86
Allyl	None (control)	87	90	69
	Acetic acid	131	113	90

In a further patent, *J.E. Schoenberg; U.S. Patent 4,139,693; February 13, 1979; assigned to National Starch and Chemical Corporation* has provided an adhesion promoter for 2-cyanoacrylate adhesives which does not significantly retard its cure rate which comprises a mixture of: (a) a monomeric ester of 2-cyanoacrylic acid of the general formula

$$H_2C=C-\overset{\displaystyle O}{\overset{\displaystyle \|}{C}}-OR$$
$$\underset{\displaystyle CN}{|}$$

in which R is an alkyl or alkenyl group having from 1 to 16 carbons, a cyclohexyl group or a phenyl group; (b) an anionic polymerization inhibitor; and (c) forms 1 to 30 mmols per kg of adhesive of an adhesion promoter of the general formula

where R' is hydrogen or an alkyl, aryl or cycloalkyl group having from 1 to 10 carbons.

The process for preparing the improved adhesive composition of this formulation comprises the step of dissolving into the stabilized 2-cyanoacrylate adhesive from 1 to 30 mmols per kg of adhesive of the adhesion promoter defined above.

The preferred 2-cyanoacrylate esters for use are those where the R group is an alkyl or alkenyl group having 1 to 4 carbons, with the ethyl ester being particularly preferred. The anionic polymerization inhibitor which is most preferred is sulfur dioxide. The preferred groups of adhesion promoters are those where R' is an alkyl, aryl or cycloalkyl group of 1 to 10 carbons, and most preferably an alkyl group having 1 to 6 carbons.

If the adhesive composition is to be stored for an extended period of time, it may be desirable to add a free radical polymerization inhibitor to the composition of this process to impart added storage stability. Other optional ingredients which improve specific properties of the adhesive such as thickeners or plasticizers also may be incorporated into the composition, if desired.

The adhesion promoter herein serves to enhance the bond strength of the adhesive composition when applied to many substrates which are stronger than the adhesive bond, such as most metal substrates. It is necessary that the adhesion promoter be present in the composition within a specified concentration range, and only the narrow range prescribed herein is effective in obtaining the high bond strengths characteristic of this process. To maximize bond strength while minimizing retardation in cure rate, the adhesion promoter is preferably employed in amounts ranging from 1 to 10 mmols/kg of total adhesive composition.

Example: This example illustrates the effect of the adhesion promoters of this formulation on the cure rate and bond strength of 2-cyanoacrylate adhesive compositions.

Four samples of 2-cyanoacrylate adhesive compositions were prepared by adding approximately equimolar amounts of the given adhesion promoter in a quantity of ethyl 2-cyanoacrylate obtained commercially containing 0.002% sulfur dioxide and 0.001% hydroquinone as inhibitors of anionic and free radical polymerization, respectively. The gallic acid and methyl gallate used as adhesion promoters were stirred to effect solution in the 2-cyanoacrylate ester, while, with the propyl and hexyl gallates, dissolution was effected only with shaking. Each resulting composition was evaluated as to set time and tensile shear strength against a control containing no adhesion promoter. The results are summarized in the table.

Adhesion Promoter	..Amount Added..		Set Time	Tensiie Shear Strength .....(kg/cm^2).....		
	ppm	mmol/kg	(sec)	Steel	Brass	Aluminum
None (control)	0	0	15	171	109	114
Gallic acid monohydrate	740	3.9	15	233	230	169
Methyl gallate	730	4.0	15	220	220	167
Propyl gallate	840	4.0	15	187	221	162
Hexyl gallate	990	3.9	15	198	187	159

It can be seen that gallic acid and the esters thereof increase the bond strength of the 2-cyanoacrylate adhesive on all three types of metal substrates. Furthermore, addition of these adhesion promoters does not adversely affect the cure rate of the adhesive.

Water-Based Sealants for Sealing Can Ends

D.L. Neumann; U.S. Patent 4,138,384; February 6, 1979; assigned to the Dexter Corporation has developed improved aqueous or water-based sealant compositions which are adapted for use in sealing a container end to a container body portion, and which are characterized by (1) having a relatively high level of total solids, (2) being substantially, virtually or entirely free of volatile organic solvents, (3) having desirable thixotropic properties, and (4) being capable of undergoing setting to a substantially solidified or relatively hardened state under air-drying conditions, and at ambient temperatures, and within requisite periods of time without sagging or slumping, and without developing undesirable or substantial void spaces, or structural blemishes or flaws during setting.

These aqueous or water-based sealant compositions are substantially, virtually or entirely free of objectionable or flammable organic solvents (e.g., more than

80% by volume of the total solvent or liquid vehicle content is water), and do not require oven-drying. The liquid sealant compositions, although being water-based or water-borne sealants, have a sufficiently high level of solids to be capable of setting effectively, on a production basis, under air-drying conditions, even within the setting time associated with air-drying, solvent-based sealants (e.g., up to but not more than 2 days).

Furthermore, the liquid sealant compositions have desirable thixotropic characteristics, which enable them to be effectively dispensed or ejected, on a production basis, from the nozzle of conventional dispensing equipment (e.g., automatic, high-speed, extrusion-spraying or sealant lining machines) under desirably controlled conditions onto the periphery of container ends and, thereafter, to stay in place on the container ends and not sag or slump during setting.

These sealants contain preferably from 67 to 70% by weight total solids. They consist of: (a) an acrylonitrile latex component having a synthetic, elastomeric, high molecular weight, acrylonitrile polymer component, such as acrylonitrile-butadiene copolymer; (b) a butyl latex component having a synthetic, elastomeric, high molecular weight, isobutylene polymer component, such as isobutylene-isoprene copolymer; (c) a compatible, thixotropic, mineral flow control agent; and (d) available water. Components (a) through (c) are distributed or dispersed substantially throughout the aqueous vehicle or component (d). In order to provide a water-based sealant that resists bacterial attack during shipping and storage and/or becoming unstable or nonhomogeneous as a result of bacterial growth, it is important to also include effective levels of a bactericide therein, such as paraformaldehyde.

The water-based sealants may be stored in drums or cans, and are substantially uniform, liquid blends or admixtures that may be readily poured therefrom, as desired or needed, and when taken from an appropriately sealed storage container or used, should have a Brookfield viscosity preferably in the vicinity of about 6,000 to 10,000 cp (LVF-5X, No. 3 spindle at 12 rpm and 70°F). The aqueous sealants of this process are stable in that they may be stored for at least 3 months.

Good results have been shown by using Hycar 1562X 155 Latex (B.F. Goodrich), as the acrylonitrile latex component, for use in cans to contain food products, or Hycar 1562X 117 Latex for cans which do not contain food products. The butyl latex component may be Exxon Butyl Latex 100.

Particularly effective results are achieved with fine aluminum silicate particles or siliceous pigment in the form of small hydrous clay particles (e.g., water-washed, nonhygroscopic, particulate kaolin clay) as the flow control agent.

Particularly outstanding results (e.g., thixotropy, thickening and stability) are provided by water-washed kaolin clay which has been pretreated to form a deflocculated suspension of high fluidity in water vehicles, such as ASP 102 of Engelhard Minerals & Chemical Corporation. ASP 102 essentially consists of small, particulate siliceous solids in the form of hydrous, complex aluminum silicate.

Example: A water-based sealant composition was prepared with the following materials used in the amounts indicated:

Materials	Parts by Weight
Exxon Butyl Latex 100	301.7
Hycar 1562X155 Latex	452.5
26% solution of ammonium hydroxide	10.0
91% solution of paraformaldehyde	0.5
ASP 102 clay	929.6
Added distilled water	174.4

In preparing the sealant, the Exxon Butyl Latex 100 and Hycar 1562X155 Latex products were added to a flask equipped with a stirrer and in which they were mixed. The ammonium hydroxide solution was added with mixing. The solution of paraformaldehyde and distilled water was added with mixing. The ASP 102 aluminum silicate clay was added very slowly with good mixing. After all of the clay had been added, the flask was sealed and a vacuum was applied (26 inches of vacuum), and mixing was continued until all of the detectable air had been removed from the sealant. The mixing was stopped and the vacuum was released. The resultant water-based sealant was poured into a container, which was then sealed, and it was ready for use, as desired.

The foregoing water-based sealant composition had 70% by weight of total solids and 30% by weight of water solvent, and a Brookfield viscosity of 3,000 cp (Brookfield LVF-5X, No. 3 spindle, 12 rpm at 76° to 79°F).

Adhesives
for Construction Materials

CONCRETE AND CEMENT ADHESIVES

Vinyl Chloride Resin-Coal Tar Pitch Joint Sealant

Sealants are in widespread use for sealing joints between adjacent slabs of concrete such as are found in aircraft runways, taxiways, aprons, highways, city streets, and parking areas. One type of related hot pour sealant composition heretofore in use on a limited basis has been a composition comprising coal tar and a minor proportion of a polyvinyl chloride (PVC) resin. Such a sealant composition is a heterogeneous mixture which is fluid prior to use and is subsequently heated, to disperse the resin, and flowed into a concrete joint where it forms an elastic adherent seal preventing the entry of foreign materials into the joint between the slabs.

While this composition is a very effective and highly satisfactory concrete joint sealant once in place between the concrete slabs, it has been found to have one serious defect which has limited its commercial use. Coal tar-vinyl chloride resin sealant compositions have been found to have an extremely short shelf life as a heterogeneous fluid and in fact, once mixed together, they will often gel within one day under normal storage conditions.

D.C. Payne and J.L. Chandler; U.S. Patent Reissue 29,548; Feb. 21, 1978; assigned to Superior Products Company, Inc. have discovered two factors which contribute to the production of a joint sealant which can more easily be stored and handled before pouring. First, they have found that the limited shelf life of the previous liquid coal tar-vinyl chloride composition is due to the presence in the coal tar of certain multiring aromatic compounds, notably naphthalene-based compounds, and that satisfactory shelf life could be obtained by limiting the amount of such compounds to not more than about 14% of the pitch or pitch fraction.

The second factor, contributing also to the overall quality of the sealant, is the selection of a particular fraction or fractions of the coal tar pitch. Normally bituminous coal tar pitch is a solid. However, applicants have discovered that there is a cut of fraction of the pitch lying between about 355° and 450°C which is liquid, although fractions lying on both sides of this liquid fraction are solid. This unique liquid pitch fraction, lying between about 355° and 450°C and herein referred to as D-pitch fraction, provides an excellent liquid composition with PVC and exhibits very good shelf life.

The following table illustrates the advantages of using the D-pitch fraction of coal tar. Compound ingredients are given in parts by weight.

	Approximate Equivalent Boiling Range, °C*	. . . . Examples						
		1	2	3	4	5	6	7
PVC homopolymer resin**		8	8	8	8	8	8	8
Ca(OH)$_2$		22	22	22	22	22	22	22
Aromatic petroleum plasticizer			70	17.5	17.5	17.5		40.5
Coal tar fractions:***								
A (liquid)	200 to 235			52.5				
B (liquid)	235 to 315				52.5			
C (solid)	315 to 355					52.5		
D-pitch fraction (liquid)	355 to 450						70	
E-pitch (solid)	450 up							29.5
Coal tar RT-5	200 up	70						
Viscosity,† poises								
Original		130	210	0.9	17	25	††	††
After storage at 122°F:								
1 day		†††	170	†††	†††	145	1,780	6,080
2 days			165				1,670	3,770
3 days			160				1,630	4,100
4 days						†††		
5 days								5,820
6 days			155				1,630	
8 days								4,600
9 days							2,080	
23 days			170					

*At atmospheric pressure.
**Average particle size, 15 microns.
***Bituminous coal source.
  †Measured at 77°F, except Examples 5 and 7, measured at 110°-115°F due to very high viscosity at 77°-78°F.
  ††Not measured.
  †††Gel.

The viscosity of each compound was measured periodically to determine stability during storage. Compound of Example 1 is typical of the prior art and is very unstable in storage, as shown by gelation in one day. Compound of Example 2 is not a sealant, but merely contains a high boiling, compatible aromatic petroleum plasticizer and is included to show that this plasticizer has negligible influence on storage stability and may be used to blend with coal tar fractions without adversely influencing the composition.

Compounds of Examples 3 and 4 result in rapid gelation on storage similar to that of Example 1. Compound of Example 5 is more stable than 3 or 4, but is considered marginally stable as gelation occurs in four days which limits its commercialization. Compound of Example 6 uses the aforementioned D-pitch, fluid at room temperature, and affords excellent stability.

Results with compound 7 are rather erratic due to the very high viscosity and difficulty in measuring viscosity at elevated temperatures. The significant result observed is that there was no gelation after many days' storage, which indicates the feasibility of using E-pitch with a suitable plasticizer. Accordingly, a wide range of combinations of pitch fractions D and E can be used. The higher the proportion of D-pitch, the less the need for plasticizer.

Styrene-Butadiene Interpolymer Latex

J. Peters, R.D. VanDell, R.D. Eash and L.F. Lamoria; U.S. Patent 4,086,201; April 25, 1978; assigned to The Dow Chemical Company describe a cement additive based on styrene-butadiene interpolymer latexes which gives significantly improved strength properties, workability and adhesion to cementitious substrates, e.g., to surfaces such as floors, pavements and bridge decks, cement blocks, bricks and the like.

The cement additive consists essentially of (1) a styrene-butadiene interpolymer latex containing up to about 60 parts by weight of interpolymer solids, and, based on 100 parts of the latex polymer solids, (2) from about 3 to about 10 parts by weight of a nonionic surfactant, (3) from about 0.3 to about 10 parts by weight of a polyelectrolyte having a number average molecular weight of less than about 3,000 and preferably from about 1,000 to about 2,000 and is comprised essentially of about 3 parts by weight of methyl methacrylate and about 1 part by weight of a sulfoester of α-methylene carboxylic acid or its salt having the formula $R-CO_2-Q-SO_3M$ wherein the radical R is selected from the group consisting of vinyl and α-substituted vinyl, and the radical $-Q-$ is a divalent hydrocarbon radical having its valence bond on different carbon atoms, and M is a cation, and (4) from about 0.1 to about 5 parts by weight of a polyorganosiloxane foam depressant.

A particularly preferred styrene-butadiene interpolymer consists essentially of about 63 weight percent styrene, about 32 weight percent butadiene and about 5 weight percent of acrylonitrile. Such an interpolymer can be prepared by mixing the monomeric ingredients in water containing an emulsifying agent or agents, and heating with agitation in the presence of a peroxide catalyst to initiate copolymerization.

Preferred nonionic surfactants are the polyoxyalkylene derivatives of propylene glycol having a molecular weight of at least about 1,000 to about 15,000; and the condensation products of ethylene oxide with alkyl phenols, particularly the dibutylphenoxynonaoxyethylene-ethanols. The above monomeric surfactants are advantageously used in concentrations of from about 3 to about 10 and preferably from about 4 to 5 parts by weight based on 100 parts of latex polymer solids.

The polyelectrolytes found to be particularly useful in this formulation are those polyelectrolytes having a number average molecular weight up to 3,000 and preferably in the range of 1,000 to 2,000, which are prepared by copolymerization of from 3 parts by weight of monomeric methyl methacrylate with 1 part by weight of a monomeric sulfoester of α-methylene carboxylic acid, or its salt, having the formula $R-CO_2-Q-SO_3M$ wherein the radical R is selected from the group consisting of vinyl and α-substituted vinyl, and the radical $-Q-$ is a divalent hydrocarbon radical having its valence bonds on different carbon atoms and M is a cation. A particularly preferred monomeric sulfoester is 2-sulfoethyl methacrylate or its sodium salt. The polyelectrolytes are used in concentrations of preferably from 3 to 7, based on 100 parts of latex polymer solids.

Polyorganosiloxanes are commercially available in several forms which are designated as "silicone fluids," "silicone emulsions" and "silicone compounds," the latter being siloxanes modified by the addition of a small percentage of finely divided silica or other inert divided solid. Any of these forms can be used.

Example 1: Preparation of Cement Additives — To a reaction vessel was added 1,655 g of deionized water and 55.5 g of the trisodium salt of N-hydroxyethyl-ethylenediaminetriacetic acid (as a 0.5% aqueous solution). The mixture was then adjusted to a pH of 3.5 by the addition of acetic acid. The reactor was then purged with nitrogen and heated to a temperature of 90°C. Thereafter, a monomer stream composed of 1,397 g of styrene (63 wt %), 713 g of butadiene (32 wt %) and 111 g of acrylonitrile (5 wt %) was added to the reactor over a 5-hour period along with a separate aqueous stream composed of (a) 670 g of a polyelectrolyte consisting essentially of 3 parts by weight of methyl methacrylate and 1 part by weight of 2-sulfoethyl methacrylate prepared by a homogeneous, continuous monomer addition solution polymerization technique substantially as set forth in U.S. Patent 3,965,032, (b) 18 g of sodium persulfate and (c) 9 g of sodium bicarbonate.

This aqueous stream was added to the reactor over a 7-hour period. The reactor was then held at 90°C over a period of about 2 hours and steam stripped to remove residual monomer. Thereafter, the following materials were post-added to the latex. Amounts are given in parts per 100 parts of latex polymer solids.

Material	Parts
Additional polyelectrolyte*	2
Nonionic surfactant**	4
Polyorganosiloxane foam depressant***	0.04

*10% aqueous solution
**di-tert-butylphenoxynonaoxyethylene-ethanol
***condensation product of dimethyl silane diol.

The resulting cement additive contained about 48% polymer solids, the solids having a particle size between about 1500-1800 A.

Example 2: A cement additive (1) of the composition of Example 1 was used to prepare a mortar according to the following recipe where the ingredients were mixed in a Hobart mixer.

Material	Amount (g)
Standard Ottawa crystal silica	975
Peerless brand type I portland cement	300
Piqua marble flour	150
Cement additive (solids)	48
Additional water	144

The above cement composition was tested for cross-brick adhesion (ASTM Test No. C-321, 7 days dry cure) wherein the cross-brick adhesion value is that value required to pull apart two bricks bonded by mortar at an angle of 90° to each other. For purposes of comparison, a cement additive (2) was also tested consisting of a latex of 75 parts by weight vinylidene chloride, 20 parts by weight vinyl chloride, 5 parts by weight ethylacrylate and 2 parts by weight of methyl methacrylate. Such cement additive is known as Sarabond latex mortar additive, which is conventionally used in the construction industry. The following table sets forth cement additive compositions and cross-brick adhesion values.

Cement Additive	Cross-Brick Adhesion (psi)	Vicat* (mm)	Water/Cement Ratio	Latex Solids/ Cement Ratio
(1) Process	158	65	0.48	0.16
(2) Comparison	150	63	0.55	0.20

*ASTM C-42

The above data illustrate the highly beneficial adhesion values obtainable with these cement additives.

Vinylidene Chloride Polymer Latex

In a similar patent by *L.F. Lamoria and R.D. Van Dell; U.S. Patent 4,086,200; April 25, 1978; assigned to The Dow Chemical Company,* a cement additive is formulated in much the same way as that of the previous patent, except that a vinylidene chloride polymer latex is used instead of the styrene-butadiene latex.

The cement additive consists of (1) a vinylidene chloride polymer latex, containing up to about 60% by weight of polymer solids, and based on 100 parts of the latex polymer solids, (2) from about 3 to about 10 parts by weight of a nonionic surfactant, (3) from about 2.5 to about 10 parts by weight of a polyelectrolyte having a number average molecular weight of less than about 3,000 and preferably from about 1,000 to about 2,000, the polyelectrolyte consisting essentially of about 3 parts by weight of methyl methacrylate and about 1 part by weight of a sulfoester of α-methylene carboxylic acid or its salt having the formula

R–CO$_2$–Q–SO$_3$M wherein the radical R is selected from the group consisting of vinyl and α-substituted vinyl, and the radical –Q– is a divalent hydrocarbon radical having its valence bond on different carbon atoms, and M is a cation, and (4) from about 0.1 to about 5 parts by weight of a polyorganosiloxane foam depressant.

In these portland cement compositions the vinylidene chloride polymer latex is present in an amount sufficient to provide from about 5 to about 20 parts by weight of latex polymer solids based on the weight of cement.

By the term "vinylidene chloride polymer latex" is meant any aqueous colloidal dispersion of an organic interpolymer composed of from about 35 to about 90 parts by weight of vinylidene chloride and from about 65 to 10 parts by weight of at least one other interpolymerized material of the general formula:

$$CH_2=\underset{\underset{R}{|}}{C}-X$$

wherein R is selected from the group consisting of hydrogen and the methyl group and X is selected from the group consisting of –CN, halogens of atomic numbers 9 to 35, and ester-forming groups, –COOY, wherein Y is selected from the group consisting of a primary alkyl group and a secondary alkyl group, each of the foregoing alkyl groups containing from 1 to 18 carbon atoms inclusively.

Representative types of water-insoluble vinylidene chloride polymers which have been discovered to be highly satisfactory as the latex components employed in these latex-modified portland cement compositions include those interpolymers designated in the following table.

Vinylidene Chloride Interpolymer Compositions

Organic Monomer Components	. . . . Parts by Weight.											
Vinylidene chloride	40	60	89	90	50	88	50	52	70	75	75	75
Vinyl chloride	–	–	–	–	–	–	40	35	20	20	20	20
Ethyl acrylate	60	40	–	–	40	7	10	–	10	5	–	5
Methyl methacrylate	–	–	–	10	10	–	–	–	–	–	–	2
2-Ethylhexyl acrylate	–	–	–	–	–	–	–	13	–	–	–	–
Acrylonitrile	–	–	11	–	–	5	–	–	–	–	5	–

Of particular benefit in the preparation of the exceptionally strong, latex-modified portland cement compositions is the interpolymer latex containing about 75 parts by weight of interpolymerized vinylidene chloride, about 20 parts by weight of interpolymerized vinyl chloride, about 5 parts by weight of interpolymerized ethyl acrylate, and about 2 parts by weight of interpolymerized methyl methacrylate.

The additives were made in the same way as those described in the previous patent and showed good properties of mechanical and shear stability, freeze/thaw stability and workability, and good adhesion to cementitious substrates.

Composition for Highway Marking

The following sets forth some of the requirements of a street or highway marking composition: adhesion to siliceous or asphaltic road surfaces; resistance to chemical attack by water and/or deicing salts; abrasion resistance with respect to rubber-tired vehicles, sanding, snowplowing, etc.; minimal solvent hazards during application of the coating, if possible; ability to adhere to or hold or retain a glass bead filler or glass bead overcoating, which filler or overcoating provides the required reflectorization; long-term weather resistance; ability to be applied under a wide variety of ambient temperature and road surface conditions, and the ability to become tack-free within a short time under any of these conditions; flowability or sprayability, e.g. adaptability for use with airless spray equipment; good wetting action with respect to the roadway surface; and flexibility (i.e., ability to move as road surfaces expand, contract, etc.); if possible, the ability to be applied without previous priming of the roadway surface.

R.S. Gurney; U.S. Patent 4,088,633; May 9, 1978; assigned to H.B. Fuller Co. have found that an outstanding method for marking a paved surface is within the scope of the capabilities (as well as the inherent limitations) of epoxy resin technology, provided that a combination of cycloaliphatic and aliphatic polyfunctional amines is used as the amine hardener (i.e., the cocurative or coreactant for the liquid vicinal epoxide composition). This combination of amines appears to provide a set of properties during application, and after cure, which is particularly well suited to the road surface marking art. The combination also provides properties that apparently cannot be achieved with either type of amine cocurative used alone.

Thus this method involves:

(a) supplying to the point of application the aforementioned blended cocurative agent;

(b) supplying to a point of application a suitable curable liquid vicinal epoxide composition, the proportioning of the epoxide composition and cocurative being controlled so as to provide a ratio of 3 epoxide equivalents to active hydrogen-bearing amine equivalents ranging from about 1:1 to about 1.5:1;

(c) applying the cocurative and the liquid epoxide to the paved surface (preferably after intimately mixing the cocurative and the liquid epoxide); and

(d) permitting the resulting mixture to cure in situ on the paved surface. It is ordinarily preferable to maintain the liquid epoxide and the cocurative in a moderately heated state prior to application (e.g., heated to a temperature ranging from about $40°$ to about $90°C$); however, once the epoxy and the curative have been blended and applied to the road surface (e.g., with airless spray equipment), rapid cures will ensue without any further application of heat and despite the lack of any special preparation of the asphaltic or siliceous road surface, even in moderately cold weather.

For example, a highway stripe or lane-marking applied in this way can be substantially tack-free and ready to accept automobile traffic in less than an hour after application under ambient temperature conditions ranging from $0°$ to $40°C$ or more.

Using the typical part A, part B terminology, a part A of this process typically comprises a 100% solids polyglycidyl ether of a polyhydric phenol. The typical part B comprises the aliphatic polyamine and the cycloaliphatic polyamine, blended in a weight ratio ranging from 10:90 to 90:10, a portion of this polyamine combination having been converted to an amine-epoxy adduct, the epoxy portion of this adduct being contributed by a diglycidyl ether of a dihydric phenol.

It is preferred that part B also contain a monohydric phenol and a tertiary alkanolamine. The preferred polyglycidyl ether for use in part A is derived from methylol-substituted bisphenol A and an epihalohydrin. Any diluents included in the composition are preferably reactive, so that the composition still remains substantially in the 100% solids category. For improved adhesion to road surfaces, silane adhesion promoters can be included in the epoxy coating composition, e.g., in part A.

Example: The following formulations illustrate a "fast cure" part A, a "slow cure" part A, and a part B. The part B can be used with either the slow cure or the fast cure part A. When pavement temperatures are below $45°F$, the fast cure part A is generally used alone. For warmer pavement conditions (pavement temperatures above $45°F$), a portion of the fast cure part A can be replaced with slow cure material. Under extremely hot pavement conditions, a 50:50 mixture of fast cure and slow cure can be used as the part A. Otherwise, the fast cure material can comprise the major amount of part A.

For optimum stoichiometry (i.e., a slight excess of epoxide over active hydrogen), the ratio of part A to part B is 2:1 by volume.

Fast Cure Part A

Percent by Weight	Ingredient
56.69	Apogen 104 (M&T Chemicals)
22.83	Epi-Rez 5044 (Celanese Corp.)
19.69	Unitane OR-600 (American Cyanamid Co.)
0.79	Silane A 187 (Union Carbide Corp.)
100.00	

Slow Cure Part A

Percent by Weight	Ingredient
58.77	DER-331 (Dow Chemical Co.)
12.24	P&G Epoxide No. 7 (Proctor & Gamble)
3.68	Dinonylphenol (Jefferson Chem. Co.)
0.82	Silane A 187 (Union Carbide Corp.)
20.40	Unitane OR-600 (American Cyanamid Co.)
3.13	Cab-O-Sil M-5 (Cabot Corp.)
0.96	Calidria RG-144 (Union Carbide Corp.)
100.00	

Part B

Percent by Weight	Ingredient
22.60	Trimethyl hexamethylenediamine (Thorsen Chem. Co.)
20.25	1,4-Cyclohexanebis(methylene) (Eastman Chemicals)
29.57	Nonylphenol (Jefferson Chem. Co.)
21.93	DER-332 (Dow Chemical Co.)
5.65	Triethanolamine (Union Carbide Corp.)
100.00	

Slurries Including a Polymer Latex

Certain "water reducers" have become available in the last few years. By admixing about 0.5-2.0% of such water reducers to portland cement, it is possible to produce practical mortars and concrete with significantly lower water content, without decreasing the slump or workability of the cement slurries. Examples of such water reducers are as follows:

Mighty (ICI United States). This consists of about 90% of a polymer of the sodium salt of naphthalene sulfonic acid partially condensed with formaldehyde, and about 10% sodium gluconate.

Melmont L-10 (American Admixtures) is similar to "Mighty" except that a melamine ring is used instead of a naphthalene ring in the polymer component.

Lomar D (Diamond-Shamrock) is identical to the 90% polymer component of Mighty.

FX-32 (Fox Industries) is similar to Melmont.

C.E. Cornwell and M. Plunguian; U.S. Patent 4,088,804; May 9, 1978 have found that when one of the above "super" water reducers are added to a cementitious formulation containing a relatively small percentage of a film-forming synthetic latex, a much tougher, adherent coating is obtained, even if the layer is only about two mils in thickness. Other important advantages of these new coatings are that they are nonburning and may be produced with mat, semi-glossy, or highly reflective glossy surfaces.

Fine mineral aggregates of about 100 mesh and finer are used in a ratio of about

15-100 parts by weight based on the weight of the hydraulic cement. A preferred aggregate is finely-ground calcite or calcium carbonate. Other aggregates are fine particle silicas, alumina trihydrate and aluminum oxide.

The synthetic latex, or emulsion, may be one of the numerous latexes which are stable when mixed with hydraulic cement and which give tacky films when dried at ambient temperatures. Examples of such latexes are the emulsions of vinyl acetate homopolymer, vinyl acetate-acrylic copolymer, internally plasticized and externally plasticized vinyl chloride copolymer, polyacrylic emulsions, styrene-butadiene copolymers, vinyl chloride-vinylidene chloride copolymers, vinyl chloride-vinylidene chloride-acrylic terpolymers, and others. The ratio of the latex (of about 50-55% total solids) to the weight of cement is about 5 to 30 parts by weight, or about 2.5-16 parts equivalent polymer.

It was found to be advantageous to use a small amount of a retarder in the cement formulation, of the order of about 0.05-0.5% on the weight of the cement. Some of the retarders used are glycerine, ethylene glycol, acetic acid, citric acid, sodium citrate, sugar, and zinc oxide.

The coatings may be made more waterproof by the addition of a small amount of calcium stearate, in the range of about 0.2-1.0% on the weight of the cement.

Pozzolonic fly ash may be added to react with the calcium hydroxide produced during curing of the cement and to provide better resistance of the coating to sulfate. Sodium nitrite is added as a rust preventive.

The preferred compositions contain the following components, in the following ratios, expressed in parts by weight:

Component	Parts by Weight
Powder	
Mineral cement	100
Mineral aggregate	15-100
Pozzolonic fly ash	0-15
Calcium stearate	0-1.0
Cement accelerator	0-3.0
Color pigments	0-5.0
Liquid	
Water	35-50
Latex (50-55% solids)	5-30
Water reducer (solids basis)	0.5-3.0
Cement accelerator (solids basis)	0-3.0
Retarder	0.05-0.5
Sodium nitrite	0.1-0.3

Epoxide Resin with Amine Hardeners

The formulation developed by *W. Huber-Nuesch; U.S. Patent 4,143,188; March 6, 1979; assigned to Ciba-Geigy Corp.* relates to the use of an aqueous solution of epoxide resins, which are completely water-soluble at room temperature and which have at least one N-heterocyclic ring in the molecule, and amines containing at least three hydrogen atoms bound to nitrogen, the solubility of which epoxide resins in water is at least 70% by weight, as binder and impregnating agent for the modifying, sealing or binding of substrates, especially as binder for loose aggregates and as sealing agent for porous materials. This aqueous solution is characterized in that as amines there are used those which satisfy the condition of the formula

$$\frac{a \cdot w}{c} < 14 \text{ (preferably 7 to 12)}$$

wherein a represents the H-active amine equivalent, w represents the sum of the amino groups, the hydroxyl groups and the ether groups, with two ether groups counting as one group, and c represents the number of carbon atoms in the molecule, or the water-soluble salts thereof. Basic primary, secondary and tertiary amino groups count as amino groups. The H-active amine equivalent of the amine compound is the molecular weight divided by the total number of hydrogen atoms bound to nitrogen of the amine compound.

After the dissolving of resin and hardener in water, there is formed from the clear solution within a few minutes up to two hours—depending on concentration and temperature—firstly a milky cloudiness (emulsion), which then precipitates as a liquid phase from the aqueous phase and, surprisingly, hardens or sets, in the presence of water, to form a compact, water-resistant, infusible substance. This reaction occurs even when the concentration of epoxide resin and polyamine is only 0.5% by weight. The concentration of these two constituents for application is generally 1-90% by weight, preferably 5-80% by weight.

Applicable liquid epoxide resins are, e.g., the following: N-glycidyl compounds of hydantoin and also glycidyl ethers of addition products of alkylene oxides with hydantoins. Also suitable are N-glycidyl compounds and glycidyl ethers of ethylene urea, of barbituric acid, of uracil, of cyanuric acid, of parabanic acid and of triacrylylperhydrotriazine. The following water-soluble resins are preferably used:

(a) N,N'-diglycidyl-5,5-dimethylhydantoin,

(b) N,N'-diglycidyl-5-methyl-5-ethylhydantoin,

(c) 1-glycidyl-3-(2-glycidyloxy-n-propyl)-5,5-dimethyl-hydantoin,

(d) a mixture of about 70% of polyepoxide (a) and 30% of polyepoxide (c),

(e) 1,3-bis-(1-glycidyl-5,5-dimethylhydantoinyl-3)-propanol-2-glycidyl ether,

(f) a mixture of three equal parts of polyepoxides (b), (c) and (e), and

(g) the glycidyl ether addition product of triacrylyl-perhydrotriazine of French Patent 1,267,432.

An applicable solid epoxide resin which is soluble at room temperature in water is:

> (h) triglycidyl ether of cyanuric acid.

Polyamines which fulfill the condition of the formula shown above are, for example, the following:

	Value of the Formula
1-methyl-4-(1,1-dimethylaminomethyl)-cyclohexylamine	8.5
2-aminomethylcyclopentylamine	9.5
2-oxo-1,3-hexahydropyrimidinedineo-pentylamine	9.6
3,5,5-trimethyl-3-(aminomethyl)cyclo-hexylamine	8.5
3-amino-1-cyclohexylaminopropane	11.6
3,3'-dimethyl-4,4'-diaminodicyclohexyl-methane	8.0
m-xylylidenediamine	8.5
trimethylhexamethylenediamine	8.8
1-cyclohexyl-4-amino-3-amino-methylpiperidine	13.3
1-phenylpropylenediamine	8.4
3,3-bis(γ-aminopropyl)-2-methylpiperidine	10.7
4a-(γ-aminopropyl)-8a-methyldecahydro-1,8-naphthyridine	13.3
1,3-diphenylpropylenediamine	7.5
2,2-bis(γ-aminopropyl)-propionaldehyde-neopentylglycol acetal	13.8
hexamethylenediamine	9.7
1,2-diaminocyclohexane	9.5
1,4-bis(aminomethyl)cyclohexane	8.9

The aqueous solutions are used as binders, preferably for loose aggregates. They are suitable for binding or sealing mineral or metallic aggregates, such as quartz flour, stone powder, kaolin, alumina, metal powders, fine sands, coarse sands, limestone, gravel, soil, aluminum shot, glass fibers, asbestos fibers, glass cloth, etc.; also for sealing or bonding organic materials, such as cork chips, sawdust, cotton fibers, paper, plastics chips, synthetic fibers, etc. Furthermore, the solutions can be used as adhesion promoters between old and new concrete.

SEALANTS

Silicon-Modified Polyurethane Polymers

L.R. Barron and H.M. Turk; U.S. Patent 4,067,844; January 10, 1978; assigned to Tremco Incorporated describe certain curable polyurethane polymers in which a proportion of the NCO terminations are reacted with certain amino silanes, or with the residual on reaction of a mercaptosilane with a monoepoxide, or with the residual on reaction of an epoxysilane with a secondary amine. Sealant compositions formed with these polymers are stable in storage, moisture curable and adhere well to various substrates such as glass, mortar, brick, concrete, steel, aluminum and other structural materials obviating the need for a primer.

In general, these polymeric materials are prepared by reacting a polyol material having a hydroxyl functionality in excess of 2, and preferably at least about 2.3, with a stoichiometric excess of a polyisocyanate; to provide an NCO terminated polyurethane prepolymer.

Thereafter, at least about 1%, and preferably from 2-25% of the NCO terminations are reacted with the silicon-containing material.

The polyol material, in addition to having a hydroxyl functionality in excess of 2, should have an average molecular weight within the range of about 1,000 to about 15,000. The material may be a triol, but is preferably a mixture of at least one diol and at least one triol.

The preferred class of polyols are polyalkylene ether polyols, including polyethylene ether glycol, polypropylene ether glycol, polybutylene ether glycol, polyethylene ether triol, polypropylene ether triol, polybutylene ether triol, etc.

A wide variety of organic polyisocyanates may be used, of which aromatic diisocyanates are preferred. An especially preferred material is 4,4'-diphenylmethane diisocyanate.

The polymers can be prepared by reacting the polyurethane prepolymer with a separately prepared intermediate in the form of certain aminosilanes, or the reaction product of a mercaptosilane with a monoepoxide, or the reaction product of an epoxysilane with a secondary amine. This is the the preferred order of addition of materials where the intermediate is in the form of an aminosilane. However, where the intermediate is in the form of the reaction product of a mercaptosilane with a monoepoxide, or an epoxysilane with a secondary amine, it is preferred to form the polyurethane prepolymer in the presence of the intermediates.

Useful intermediates in the form of aminosilanes include the reaction products of aminoalkylalkoxysilanes, such as gamma-aminopropyltrimethoxysilane and gamma-aminopropyltriethoxysilane, with ethyl acrylate, butyl acrylate, 2-ethylhexyl acrylate, Cellosolve acrylate, methyl methacrylate, butyl methacrylate, 2-cyanoethyl acrylate, glycidyl acrylate and acrylonitrile.

Other useful intermediates include the reaction products of mercaptosilanes, such as mercaptoalkylalkoxysilanes with monoepoxides. Specific examples include the reaction products of gamma-mercaptopropyltrimethoxysilane and gamma-mercaptopropyltriethoxysilane with butylene oxide, and styrene oxide.

Still other useful intermediates include the reaction product of an epoxysilane, such as an epoxyalkylalkoxysilane with a secondary amine. A specific example of this form of intermediate is the reaction product of gamma-glycidoxypropyltrimethoxysilane with di-n-butylamine.

Example 1: A prepolymer was made according to the following procedure:

Polyoxypropylene glycol (28.4 hydroxyl no.), g	4,938
Polyether triol (27.2 hydroxyl no.), g	5,156
Toluene (solvent), g	585
Diphenylmethane diisocyanate, g	1,040
Stannous octoate, ml	0.55

The first three ingredients were charged to a 12-liter reaction flask, and heated to 90°C at 150 mm Hg absolute pressure. The material refluxed, and the water azeotroped off was collected in a Barrett trap. After 2 hours the flask was cooled to 60°C, the vacuum replaced with a nitrogen blanket, and the remaining ingredients added. The batch was held at 90°C for 2 hours. The resultant prepolymer had an equivalent weight of about 3,800.

An intermediate material was prepared by placing the following ingredients into a test tube:

Gamma-aminopropyltrimethoxysilane, ml	4.42
Butyl methacrylate, ml	3.99
Toluene (solvent), ml	8.41

The test tube was stoppered and shaken, then allowed to stand for 1 day at 25°C. 13.46 ml of this intermediate was added to 760 g of the above prepolymer. The mixture was stirred and allowed to react for 3 days at 25°C. The resultant fluid material cured on exposure to the atmosphere to a rubbery elastomer with good adhesion to glass and to anodized aluminum.

Example 2: A prepolymer was made according to the following procedure:

Polyoxypropylene diol (38.3 hydroxyl no.), g	6,152
Polyether triol (26.7 hydroxyl no.), g	3,872
Toluene (solvent), g	590
4,4'-Diphenylmethane diisocyanate, g	1,275
Stannous octoate, ml	0.56

The first three ingredients were placed in a 12-liter reaction flask. A water-solvent azeotrope was refluxed for 2 hours at 90°C and 150 mm Hg absolute pressure, the water collecting in a Barrett trap.

The temperature was then reduced to 60°C, the flask was flooded with nitrogen at atmospheric pressure, and the remaining ingredients added. The batch was held at 90°C for 10 hours. The resultant prepolymer had an isocyanate equivalent weight of about 2,800.

An intermediate material was prepared by placing the following ingredients into a test tube:

Butylene oxide, ml	2.18
Gamma-mercaptopropyltrimethoxysilane, ml	4.08
Toluene (solvent), ml	6.26

The test tube was stoppered and shaken, then allowed to stand for 1 day at 25°C. 8.1 ml of this intermediate was added to 424 g of the prepolymer of this example. The mixture was stirred and allowed to react for 3 days at 25°C. The resultant fluid material cured on exposure to the atmosphere to a rubbery elastomer with good adhesion to glass and to anodized aluminum, as evidenced by cohesive failure.

Hot Melt Butyl Rubber Compositions

Hot melt butyl sealants having improved adhesion properties are provided by *J.E. Callan; U.S. Patent 4,092,282; May 30, 1978; assigned to Cities Service Co.* The compositions comprise (A) a butyl rubber component, (B) a resin selected from the group consisting of plasticizing and tackifying resins, and (C) a reacted mixture of (1) an alkaline earth metal hydroxide selected from magnesium, calcium, strontium, and barium hydroxides and (2) a solid polyamide resin having a molecular weight less than 10,000 and an amine value greater than about 50.

The butyl rubber component may be an unmodified butyl rubber but is preferably a butyl rubber that has been modified by crosslinking and/or compounding with conventional sealant additives, such as other elastomers, elastomeric resins, crystalline resins, etc. It ordinarily comprises at least about 20% by weight of the butyl rubber and up to about 80% by weight of additives.

The resin constituting the second component of the composition may be any one or more of the plasticizing and tackifying resins conventionally employed in sealant compositions. However, it is usually a hydrocarbon resin, such as a polystyrene, a vinyl toluene-alpha-methylstyrene copolymer, a polyterpene, a polybutene, a polyisobutylene, etc.; a phenolic resin, such as a modified alkylphenolformaldehyde resin, a thermoplastic terpene phenolic resin, etc.; a chlorinated bi- or polyphenyl; a coumarone-indene resin; natural rosin; a modified rosin, such as a glycerol ester of polymerized rosin, an ester of hydrogenated rosin, etc., and mixtures thereof. The concentration of this component is not critical but is ordinarily in the range of about 25-200 phr, i.e., about 25-200 parts per 100 parts of the butyl rubber component.

The polyamide-hydroxide reaction product which constitutes the crux of the formulation is a reaction product of (1) about 25-100 phr of a solid polyamide resin having a molecular weight less than 10,000 and an amine value greater than about 50 and (2) an amount of an alkaline earth metal hydroxide selected from the group consisting of magnesium, calcium, strontium, and barium hydroxides such as to provide a polyamide/hydroxide weight ratio of about 10-20/1.

In the following examples parts mentioned are parts by weight. Peel adhesion results given in the examples are measures of the adhesion to glass, obtained in accordance with Canadian Standard 19-GP-3a, paragraph 7.3.7, and show whether the type of failure obtained is "a" (the undesirable adhesive failure) or the more desirable "c" (cohesive failure) or "f" (film failure).

In the examples, compositions are prepared by (1) mixing the alkaline earth metal hydroxide with the high amine content polyamide in a sigma blade mixer at about 350°F for 20 minutes, (2) simultaneously mixing the butyl rubber component with 50% of the total amount of other ingredients in another sigma blade mixer at about 350°F for about 10-20 minutes, (3) adding the remainder of the ingredients to the second mixer and mixing for about 5 minutes, (4) adding half of the polyamide-hydroxide reaction product to the second mixer and mixing for 20 minutes, (5) adding the remainder of the polyamide-hydroxide reaction product, and (6) continuing mixing until the total mixing time is 60 minutes. Compositions in which a polyamide-hydroxide reaction product is not used are prepared similarly except that the first sigma blade mixer is not employed.

Example: Prepare three hot melt sealant compositions from the following recipe:

Ingredient	Parts by Weight
Slightly crosslinked butyl rubber having a Mooney viscosity (ML 1 + 3 at 260°F) of about 50-60	100
Terpene phenolic resin	100
Block copolyester resin*	100
Solid polyamide resin having a molecular weight less than 10,000 and an amine value greater than about 50	100-110

*A thermoplastic segmented copolyester elastomer having 15-30% by weight of ester units derived from butanediol and a phthalic acid, and about 85-70% by weight of units derived from polytetramethylene ether glycol and a phthalic acid, about 55-95% by weight of the phthalate units of the polyester being terephthalate units.

In the first composition, a control composition, the polyamide is 100 parts of unreacted resin; in the second and third compositions, it is the reaction product of 100 parts of the resin with, respectively, 5 and 10 parts of calcium hydroxide.

The compositions have respective normal peel adhesions to glass of 67 lb/in (f), 68 lb/in (f), and 80 lb/in (f) at 75°F and 35 lb/in (c), 36 lb/in (c), and 32 lb/in (c) at 150°F. After aging for three weeks at 150°F and 100% relative humidity, they have respective peel adhesions of 0 lb/in (a), 42 lb/in (f), and 23 lb/in (c) at 75°F.

In another patent *J.E. Callan and E.L. Scheinbart; U.S. Patent 4,127,545; November 28, 1978; assigned to Cities Service Company* describe other hot melt butyl sealant compositions which have good bond strength.

This is attained by the provision of a hot melt sealant composition (a) a slightly crosslinked butyl rubber compound having a Mooney viscosity (ML 1 + 3 at 260°F) of 55±5, (b) about 25-200 phr of a crystalline polymeric resin, and

(c) about 25-200 phr of a tackifying resin. This composition has a high adhesive bonding property, measured as a yield strength of about 60-350 psi at room temperature.

The slightly crosslinked butyl rubber compound used can be any such compound having a Mooney viscosity (ML 1 + 3 at 260°F) of 55±5. However, it is preferably one of the crosslinked butyl rubber compounds of U.S. Patent 3,674,735, e.g., a semivulcanized butyl compound obtained by mixing raw butyl rubber with a bromomethyl alkylated phenol-formaldehyde resin and polybutene in a Banbury mixer.

Example 1: A hot melt sealant composition is prepared from the following recipe:

Ingredient	Parts by Weight
Slightly crosslinked butyl rubber of U.S. Patent 3,674,735*	100
Ethylene-vinyl acetate copolymer**	100
Poly-beta-pinene tackifying resin	100

*Hereinafter designated as SCBR
**Hereinafter designated as EVA

Add the SCBR and EVA to a sigma blade mixer heated to 290°F. Mix for about 20 minutes to homogeneity. Then incrementally add the poly-beta-pinene tackifying resin over a period of 20 minutes, mix for an additional 10 minutes, and dump the batch. The resultant composition has the following properties:

Compression at 150°F	85 lb/in^3
Room temperature yield strength (glass-glass, applied at 350°F)	58 psi

Example 2: Repeat Example 1 except for replacing the 100 parts of poly-beta-pinene tackifying resin with 100 parts of thermoplastic terpene phenolic tackifying resin. The resultant composition has the following properties:

Compression at 150°F	192 lb/in^3
Room temperature yield strength (glass-glass, applied at 350°F)	80 psi

Example 3: Repeat Example 2 except for reducing the amount of EVA to 50 parts and increasing the amount of terpene phenolic resin to 150 parts. The resultant composition has the following properties:

Compression at 150°F	200 lb/in^3
Compression at 100°F	230 lb/in^3
Room temperature yield strength (glass-glass, applied at 350°F)	94 psi
Room temperature yield strength (glass-glass, applied at 300°F)	167 psi

As demonstrated above, excellent properties, including exceptional yield strength, are obtained with these formulations. Similar results are achieved when (1) the proportionation of components is varied within the permissable ranges of about

25-200 phr of crystalline resin and about 25-200 phr of tackifying resin, (2) the crystalline polymer of the examples is replaced by a saturated polyester elastomer, an amorphous polypropylene, an ethylene-vinyl acetate copolymer, an ethylene-ethyl acrylate copolymer, or any of the other crystalline polymers that are described as useful in the patent and/or (3) the tackifying resins of the preceding examples are replaced by polystyrene, a vinyltoluene alpha-methyl-styrene copolymer, a glycerin, methyl, or pentaerythritol ester of hydrogenated rosin.

One Pack Moisture-Curable Polysulfide

N.O. Price, H. Coates and C.S. Ely; U.S. Patent 4,096,131; June 20, 1978; assigned to Albright & Wilson Limited, England have provided a sealant composition which can be extruded and can be cured with moisture to give an elastomer. The composition comprises an SH containing reaction product of (a) a silanized polysulfide which is the product of reacting under anhydrous conditions a polysulfide of formula

$$HS[(CH_2(R')_bCH_2S_a)_c(R''S_a)_d(SH)_f]CH_2(R')_bCH_2SH$$

where a is 1 to 5, b is 0 or 1, c is 5 to 50, d is 0 to 0.5c and $0.05c \geqslant f \geqslant d$, $1 > f$ which is d times (no. of free valencies in $R'' - 2$), R' is O, S or a divalent saturated organic radical consisting of C, H and optionally O and/or S in COC, CS_aC or OH links, R'' is an at least trivalent saturated radical consisting of C, H and optionally O and/or S in COC, CS_aC or OH links, with a silane of formula Q_nSiX_{4-n} where n is 1 or 2, Q is a group capable of reacting under the reaction conditions with the SH groups of the polysulfide and at least 2 groups X are hydrolyzable groups and the remaining group X (if any) is alkyl or chloroalkyl, the silane having 1 or 2 groups capable of reaction with SH, and the silane reacting with 3 to 30% of the SH groups of the polysulfide, and (B) a disulfide of formula

$$\begin{array}{ccc} E & & G \\ \| & & \| \\ R_5-D-(C)_h-S-S-(C)_h-Y-R_6 \end{array}$$

where both h are 0 or both h are 1, and each of D and Y, which are the same or different, represents an $-NR_7$ group, each of E and G, which are the same or different, represents an oxygen or sulfur atom or an NR_8 group, and each of R_5, R_6, R_7 and R_8 which are the same or different, represents a univalent aliphatic, cycloaliphatic or heterocyclic group, an aryl group of 6 to 13 carbon atoms, or an aralkyl group of 7 to 19 carbon atoms, a cycloaliphatic aliphatic group or a heterocyclic aliphatic group, or at least one pair of R_5 and R_7, R_6 and R_8, R_5 and R_8 or R_6 and R_8 together represent a divalent aliphatic group of 2 to 8 carbon atoms, an arylene group, a cycloalkylene group, arylalkylene group, cycloalkylalkylene group or a divalent aliphatic group of 2 to 8 carbon atoms, which is interrupted by an oxygen or sulfur atom in an ether or thioether linkage respectively, or by an imino group of formula NR_9 where R_9 is hydrogen or an alkyl, aralkyl or aryl group, or, when both h are 1, each of D and Y, which are the same or different, may be an oxygen or sulfur atom.

The cured products obtained can be used as sealants, e.g., as building sealants. They usually have an extensibility preferably of 70-150% and especially 100-150%, though 50-120% and especially 70-120% may be suitable for some uses. The extensibility is measured, during elongation at 6 mm/min of a sample 50 x 12 x 12 mm, which has been cured by exposure to moisture for 7 days at 20°C and 50% RH. The conditions of curing and testing are basically those described in BS4254, or Fed. Spec. TTS-00230C. The cured products also have usually a recovery after 5 minutes extension by 50-200% of at least 75%, measured one hour after release of the extending force.

The formulation is illustrated in the following examples. The "gunnability" of a composition referred to therein is a measure of the fluidity of the mixture which is made by "gunning" material from a standard caulking gun through a standard nozzle for a fixed time under a fixed gas pressure. The conditions used are as follows: nozzle 2 x $\frac{1}{8}$ inch diameter at 50 psi for one minute, several measurements are made and an average value is recorded. This process may be repeated over several months as a measure of storage stability.

In the examples Calofort S is calcium carbonate, particle size 0.1 μ, coated with 3% stearic acid. Cereclor 56L is chlorinated liquid paraffins. The titanium dioxide used in the examples was Runa RP (Laporte Industries).

Example A: Polymer Preparation — A solution of 500 g of Thiokol LP12 in toluene (500 g) was azeotropically distilled to remove water (0.4 ml) for 4 hours. To this solution was added sulfur 5 g (1% by weight based on weight of polymer) and vinylmethyldimethoxysilane 15 g (3% by weight based on weight of polymer). The mixture was boiled under reflux for 15 hours. Analysis of the product showed that it contained 1.26% SH, i.e., 23.9% reaction.

Example B: Polymer Preparation — A solution of 500 g Thiokol LP-32, in toluene (500 g) was azeotropically distilled to remove water (0.5 ml) for 5 hours. To the solution was added 5 g sulfur (1% on polymer) and 15 g vinylmethyldimethoxysilane (3% on polymer). The mixture was boiled under reflux for 16 hours. Analysis showed that the product contained 1.24% SH, i.e., 25% reaction.

Filler Combinations

		Grams
(C)	Calofort S	17.5
	TiO_x	4.0
	Cereclor 56L	14.0
(D)	Calofort S	350
	TiO_x	80
	Cereclor 56L	350

Example 1: Toluene was stripped from 75 g of the product from Example A in vacuo at 100°C and 14 mm. A portion (25 g) was compounded with the filler combination C, together with 0.50 g of tetramethylthiuram disulfide TMTDS (2% by weight on the weight of polymer) and 0.50 g of diisopropyltitanium diacetyl acetonate DIPTDAA. The ingredients were intimately mixed on a 3

roll mill, and then gunned into test pieces 50 x 12 x 12 mm and allowed to cure for 7 days at 50% RH and ambient temperature ca 20°C. The samples were then extended at 6 mm/min. Cohesive failure occurred at 90% elongation. At 75% elongation a force of 14 lb was required, and the samples showed instantaneous recovery of 90%.

Example 2: Example 1 was repeated but with 0.25 g of tetramethylthiuram disulfide (1% on polymer). The blocks were cured and tested as before. At 100% elongation they were intact, requiring a force of 6½ lb. They were held at 100% elongation for 5 min, then released. Instantaneous recovery was 50%, and this increased to 75% in 15 min.

Example 3: Example 1 was repeated but with 1.25 g (5%) of tetrabutylthiuram disulfide (TBTDS) in place of the tetramethylthiuram disulfide. These samples needed a force of 4 lb at 100% elongation and recovered by 83% after 15 min. The samples were then placed in an oven at 50°C for 7 days and retested. A force of 30 lb was required at 100%, with an instantaneous recovery of 88%. Samples failed at 150% elongation with a force of 34 lb.

Example 4: Example 1 was repeated but with the polymer from Example B. At 100% elongation, 7¼ lb force was required. The instantaneous recovery was 54% which reached 75% after 15 min.

Example 5: 500 g of polymer of Example B were obtained in toluene solution, and this was added to the filler combination D, also in toluene. To the mixture was added 5 g (1%) of TMTDS, and the mixture was stripped free of toluene at 100°C under high vacuum (1 mm). The material was transferred to a Semco mixer/packager and 5 g of methyltrimethoxysilane (MTMS) were added, and a 5 min mixing cycle commenced. 10 g of DIPTDAA were added and a further mixing cycle of 10 min used. The material was packed into Schieferdecker cartridges and also test blocks were made up. A 10 lb force was needed at 100% elongation but cohesive failure occurred rapidly.

Thermoplastic Elastomer plus Mixed Modifying Resins

Rubbery materials are obviously ideally suited in their inherent characteristics for use in sealant compositions. However, because of unsaturation, rubber tends to become brittle with aging. It is readily apparent that it is desirable that the sealant adhere reasonably well to the sides of the joint or crack to which it is applied, yet after the sealant is cured it is undesirable for it to have a tacky exposed surface. Finally, it is desired that the sealant, in instances where a solvent base is used, have a relatively low bulk viscosity so as to facilitate the application; however, the use of increased percentages of solvent to achieve this low viscosity results in undesirable shrinkage when the solvent evaporates.

The objective of *R.C. Doss; U.S. Patent 4,101,484; July 18, 1978; assigned to Phillips Petroleum Company* is to provide a sealant formulation (1) having good flexibility and the ability to withstand weathering; (2) which has good adhesion and yet which, on curing, is not tacky; and (3) having low bulk viscosity in solvent-based formulations yet exhibiting low shrinkage.

In accordance with this formulation, there is provided a mixture of (1) thermoplastic elastomeric block copolymer of at least one monovinyl-substituted aromatic compound and at least one conjugated diene; (2) a poly(vinyl aromatic)-compatible component; (3) a polydiene-compatible component; and (4) a filler.

This formulation is based on a linear or radial teleblock copolymer of at least one conjugated diene and at least one monovinyl-substituted aromatic compound, this formulation containing a filler and a mixture of Resin A, at least one poly(vinyl aromatic)-compatible modifying resin, and Resin B, at least one polydiene-compatible modifying resin. Other possible ingredients in the sealant formulations include plasticizer, pigment, solvent and stabilizers.

The process sealant formulations exhibit lower bulk viscosities than comparable formulations containing a single poly(vinyl aromatic)-compatible modifying resin and decreased surface tack and increased canvas peel strength (adhesion) compared to formulations containing a single polydiene-compatible modifying resin. The lower bulk viscosities allow more convenient application of the sealant to the substrate. If the sealant is applied as a hot melt, a lower application temperature is required; if applied as a solvent-release system, lower solvent level is needed thus resulting in lower subsequent shrinkage of the applied sealant, less void formation and more economic application. The decreased surface tack of the formulations is important from the standpoint that tacky sealants tend to accumulate debris, dirt, insects, etc. on the surface and thus become aesthetically unattractive.

The sealant composition comprises

> (1) a thermoplastic elastomer having the structure ABA or $(AB)_nY$, wherein A represents a block of a polymerized monovinyl-substituted aromatic compound, B represents a block of a polymerized conjugated diene, Y represents a residue of a polyfunctional coupling agent or a polyfunctional initiator, and n is an integer having a value of 2 to 4, the thermoplastic elastomer having a weight ratio of the monovinyl-substituted aromatic compound to the conjugated diene within the range of 50:50 to 5:95, the elastomer having the unsaturation normally associated with such polymers;

> (2) 10-15 parts by weight per 100 parts by weight of the thermoplastic elastomer of a poly(vinyl aromatic)-compatible modifying resin component which is an aromatic hydrocarbon;

>   (3) 10-150 parts by weight per 100 parts by weight
>   of the thermoplastic elastomer of a poly(con-
>   jugated diene)-compatible modifying resin com-
>   ponent which is nonaromatic selected from nor-
>   mally solid modified and unmodified rosin and
>   rosin esters of polyhydric alcohols, normally solid
>   esters of polymerized rosin, polyterpene resins,
>   and resinous polyolefins derived from aliphatic and
>   cycloaliphatic olefins; and
>
>   (4) 25-250 parts by weight per 100 parts by weight of
>   the thermoplastic elastomer of a filler.

The monovinyl substituted aromatic compound may be styrene and the conjugated diene is 1,3-butadiene.

The poly(vinyl aromatic)-compatible modifying resin may be a vinyl toluene/α-methyl styrene copolymer present in the amount within the range of 25 to 125 parts by weight per 100 parts by weight of the thermoplastic elastomer. The poly(conjugated diene)-compatible modifying resin may be present in the range of 25 to 125 parts by weight per 100 parts of thermoplastic elastomer and may be selected from pentaerythritol ester of hydrogenated rosin, polymerized mixed olefins, polymerized olefins, polyterpene, mixed esters of polymerized rosin, and hydrocarbon-based polar resins.

The thermoplastic elastomer is a radial teleblock copolymer coupled with silicon tetrachloride.

Fast-Curing Phenolic-Acrylic System

B. Leffingwell; U.S. Patent 4,107,239; August 15, 1978; assigned to MPB Corp. has found a means for making storage-stable vinyl-polymerizable liquid sealants which provide both accelerated curing rates and superior adhesion.

Briefly stated, this formulation provides a composition comprising from about 5 to 90% by weight of the salt formed by reacting three mols of methacrylic acid with one mol of tris(dimethylaminomethyl)phenol in admixture with conventional olefinically unsaturated monomers and optionally with conventional compatible polymers.

Exemplarily, it has been found that a conventional vinyl-polymerizable sealant capable of curing in the presence of benzoyl peroxide within about 5-10 minutes is made to cure within about one minute by the admixture of as little as 5% or less, by weight, of the tri(hydromethacrylate) of tris(dimethylaminomethyl)-phenol. Likewise, when a conventional sealant already contains methacrylic acid, its curing is accelerated by the addition of up to one mol of tris(dimethylaminomethyl)phenol for every 3 mols of methacrylic acid. Changing the composition thus to include the tri(hydromethacrylate) of tris(dimethylaminomethyl)phenol also has the effect of substantially increasing shear strength.

The base, tris(dimethylaminomethyl)phenol whose methacrylic acid salt is used is obtainable by Mannich condensation of phenol, formaldehyde and dimethylamine. It is available as a commercial product under the proprietary name DMP-30 comprising substantially the 2,4,6-isomer, containing about 0.7% water, and boiling at $143°-149°C$ at 3 mm Hg.

The salt used is formed by reacting three mols of methacrylic acid with one mol of the base, tris(dimethylaminomethyl)phenol, corresponding to the use of 258 grams of pure methacrylic acid per 265 grams of pure base.

Monomers which can be used in admixture with the "tris" salt in these compositions include any liquid or soluble olefinically unsaturated compounds capable of polymerizing by free-radical catalysis, i.e., capable of vinyl polymerization. The lower alkyl acrylates and methacrylates are particularly suitable, including methyl acrylate, methyl methacrylate, ethyl acrylate, ethyl methacrylate, etc.

There can also be used styrene, acrylonitrile, vinyl acetate, vinyl propionate, vinyl chloride, chloroprene, acrylic acid, methacrylic acid or itaconic acid and the like. The preferred liquid monomer is methyl methacrylate.

Optionally, the composition can also include a polymer such as conventionally used in vinyl-polymerizable adhesive compositions for purposes of thickening or making more syrupy. Such a polymer can be a polyester or it can be an acrylic or vinyl-acrylic polymer or prepolymer.

In summary, the three main components of the compositions can be used in the following effective ranges of parts by weight per 100 parts.

	Permissible	Preferred	Most Preferred
Tris-salt	5-90	15-50	20-30
Monomer(s)	10-90	30-60	40-50
Polymer(s)	0-60	10-40	20-30

Compositions of this formulation have been kept in storage for as long as one year or longer without significant changes in properties. In order to become polymerized or cured they are placed into contact with a sufficient amount of conventional free-radical catalyst for vinyl polymerization, i.e., any substance capable of yielding by scission or under influence of a reducing agent a moiety having an unshared electron.

Such free-radical donors are exemplarily benzoyl peroxide, lauroyl peroxide, cumene peroxide or hydroperoxide, tertiary butyl peroxide or hydroperoxide, azobisisobutyronitrile or the like. The free-radical catalyst can be added to the composition either by itself or in a suitable solvent, just prior to application of the resulting adhesive to the surface or surfaces to be bonded. Characteristically such addition can be made in an amount corresponding to between about 1 and 5% by weight of catalyst, based on weight of composition used. Exemplarily a paste can be prepared from equal parts by weight of benzoyl peroxide and dibutyl phthalate, and 3% of this paste is used, based on the weight of the sealant.

Alternatively, one or both of the surfaces to be bonded is first primed with a solution or lacquer containing the catalyst in amount so as to effect the desired ratio of catalyst to sealant at the locus of adhesion. According to a well-known procedure in the prior art, benzoyl peroxide, for example, is dissolved in an appropriate solvent together with a compatible polymer in sufficient amount to thicken the solution to a lacquer which will stay in place on the primed surface while the solvent evaporates. Such a lacquer can be prepared, for example, from 10 parts benzoyl peroxide, 5 parts of polymethyl methacrylate, and 85 parts trichloroethylene or a 50/50 mixture of trichloroethylene and methyl isobutyl ketone. Alternate polymers can be selected from the acrylic or vinyl-acrylic polymers described above.

The sealants of this formulation can be used to bond a wide variety of substrates including metals, synthetic plastics and other polymers, glass, ceramics, wood and the like. After treatment and assembly of the surfaces to be bonded as above described, the assembly is permitted to stand. One of the principal features of the adhesives is that within a relatively brief period after application of the adhesive and joining of the parts to be bonded, the bonded assembly can be handled.

It is obvious that traces of grease, lacquers and the like as well as certain electroplated coatings in the case of some metal substrates, may retard polymerization or decrease the attainable bond strength. For this reason, it is desirable to remove such traces of grease or lacquer, conveniently by solvent treatment, before applying the adhesive and catalyst system.

Hot Melt High-Temperature-Resistant Composition

The term "sealant" is used to mean a high viscosity, high solids, elastomeric composition which is primarily used to fill gaps in structures to resist the passage of air, dust, and the like. In certain applications sealants may be required to have high tensile properties for applications requiring added structural strength between metal, glass, wood, and the like. In some of these higher performance applications these compositions have been referred to as structural adhesives.

In the field of adhesives, the trend to hot melt systems has been accelerated in recent years. Hot melt systems provide a 100% solids system containing no volatiles, thereby avoiding the air pollution problems encountered with solvent systems. Another advantage giving impetus to the development of hot melt sealant systems is their ability to develop final mechanical properties almost immediately after application. However, these final mechanical properties are limited because of the limitations required for the application of the adhesive in a melt state to a particular substrate. The high temperature final properties are even more limited as they approach the hot melt systems application temperature.

For example, adhesion and lap shear strengths at temperatures in service ranges of about 125°F (51.7°C) are generally with available hot melt sealant systems. To further complicate the problem these hot melt sealants should also exhibit good low temperature properties without detracting from ease of applicability or high temperature properties.

A hot melt sealant composition having both good low temperature and high temperature mechanical properties which can be applied to substrates with conventional hot melt application equipment has been developed by *A. R. Bullman; U.S. Patent 4,133,796; January 9, 1979; assigned to Union Carbide Corporation.* It comprises

(A) About 5 to about 50% (preferably from 15 to 20%) by weight of an ethylene copolymer selected from the group consisting of ethylene/vinyl acetate copolymers having a melt index of (a) about 1 to about 500 dg/min containing about 3 to about 40% by weight of vinyl acetate copolymerized therein and (b) ethylene/alkyl acrylate or methacrylate copolymers having a melt index of about 1 to about 500 dg/min containing about 3 to about 35% by weight of an alkyl acrylate or methacrylate copolymerized therein wherein the alkyl group contains about 3 to about 6 carbon atoms;

(B) About 5 to about 50% by weight of a rubber or elastomer;

(C) About 2 to about 40% by weight of at least one alkyl substituted phenolic resin;

(D) About 0.5 to about 10% by weight of an epoxy resin containing at least 2 epoxy units per molecule; and

(E) About 20 to about 70% by weight of filler all based on the weight of the total composition.

A preferred ethylene/alkyl acrylate copolymer is an ethylene/ethyl acrylate copolymer commercially available, having a melt index of about 20 dg/min and a softening point of about 128°C, containing about 23% by weight of ethyl acrylate copolymerized therein.

The following examples, in which parts and percentages are by weight, will serve to illustrate appropriate formulations.

Examples 1 through 5: Using a 1 gallon AMK kneader-extruder, a hot melt sealant composition was prepared by blending EVA-501 (an ethylene/vinyl acetate copolymer from Union Carbide), AFAX-800 (a low MW amorphous polypropylene from Hercules), Bakelite CK-1834 (a heat-reactive, oil-soluble alkyl-substituted phenolic resin from Union Carbide), Bucar 5214 (a crosslinked butyl rubber from Cities Service), Atomite (a $CaCO_3$ with an average particle size of 3 μ, from Thompson and Weiman) Bakelite CK-2400 (a non-heat-reactive, oil-soluble alkyl-substituted phenolic resin from Union Carbide), and Araldite 6010 (a bisphenol A epoxy resin from Ciba-Geigy) in the proportions shown in the table.

They were then evaluated by extruding them onto glass, aluminum, anodized aluminum, and steel substrates. The peel strengths of the resultant substrate-sealant composites were observed and rated subjectively as shown in the table where G is good, VG is very good, and EX is excellent.

In addition steel lap shear tests, using one square inch overlap, were also carried out bonding the steel specimens with each of the formulations at an extrusion temperature of 205°C and either a post heating of 30 minutes at 350°F (167°C) or no post heat treatment. The test was conducted by raising the temperature, at a rate of 1°/min, of a vertically hung pair of bonded specimens having a 500 g weight attached to the lower specimen. The temperature at which the weighted specimen yielded and dropped one inch was recorded as the failing point.

	. . . . Examples				
	1	2	3	4	5
Ingredients, wt %					
EVA-501	14	14	14	14	14
AFAX-800	15	15	14	14	13
Bakelite CK-1834	12	–	12	–	8.5
Bakelite CK-2400	–	12	–	12	4.5
Araldite 6010	–	–	2	2	2
Bucar-5214	19	19	18	18	18
Atomite	40	40	40	40	40
Test Results					
Adhesion to substrates*					
Anodized Al	EX	EX	EX	G+	EX
Aluminum	EX	VG	EX	G	VG
Steel	EX	EX	EX	G+	EX
Glass	EX	EX	EX	G+	EX
Steel lap shear**, °F					
No post heating	145	158	147	156	162
Post heating 30 min at 350°F	149	160	147	260	250

*After hot gun extrusion at 400°F (205°C).

**Temperature at which 500 g weight dropped, one square inch overlap, 400°F extrusion.

Sealant with Good Flexibility Able to Withstand Weathering

R.C. Doss; U.S. Patent 4,138,378; Feb. 6, 1979; assigned to Phillips Petroleum Company has provided a sealant formulation using a hydrogenated thermo-plastic elastomer and a low MW polyalkene plasticizer plus a modifying resin and a filler. The sealant has good flexibility, the ability to withstand weathering, good adhesion, but no tackiness on curing.

The thermoplastic elastomer used in the sealant formulation is a hydrogenated linear or radial thermoplastic elastomer of a conjugated diene and a monovinyl-substituted aromatic compound. The composition contains a low molecular weight polyalkene, which in combination with the hydrogenated thermoplastic elastomer results in an unexpected improvement in normally competing properties, for instance, improved hardness and improved elongation.

The hydrogenated thermoplastic elastomer copolymers of conjugated dienes and monovinyl-substituted aromatics generally useful in this formulation are of the structure ABA or $(AB)_nY$ wherein A represents a block of poly(monovinyl aromatic), B represents a block of poly(conjugated diene) before hydrogenation, Y represents a residue of a polyfunctional coupling agent or a polyfunctional initiator and n is a number having a value of at least 2 preferably from 2 to 4, these polymers being radial when n is greater than 2. A and B generally represent pure homopolymer blocks, but it is also within the scope of this formulation to include those block copolymers containing the well-known "tapered" blocks.

Of the monovinyl-substituted aromatic compound monomers useful in the preparation of the A blocks of the above-described copolymers, styrene is preferred. Useful conjugated dienes for the preparation of the B blocks of the above-described copolymers prior to hydrogenation are 1,3-butadiene and isoprene, 1,3-butadiene being most preferred.

ABA block copolymers are generally prepared by methods well known in the art such as by the sequential addition of monomers to a system utilizing an organo-monolithium initiator as described in U.S. Patent 3,639,521. $(AB)_n Y$ block copolymers are prepared using polyfunctional organolithium initiators or polyfunctional coupling agents by well-known methods.

Procedures for hydrogenating unsaturated polymers are well known in the art. An example of such a system which is convenient to employ is the catalytic hydrogenation of a polymer solution using a reduced nickel catalyst (such as triethylaluminum/nickel octanoate). Such procedures are described in U.S. Patent 3,696,088.

Low molecular weight (generally liquid at room temperature) polyalkenes useful as plasticizers in the formulations include homopolymers of ethylene, propylene, 1-butene, cis-2-butene, trans-2-butene, isobutylene, pentenes, etc. as well as copolymers and terpolymers thereof in all proportions and mixtures thereof in all proportions. These polymers generally possessing average molecular weights from 200 to about 3,000 and preferably from 300 to 2,000. Many are commercially available. Modifying resins which are useful in the sealant formulation are well known in the sealant art, such as modified and unmodified rosin and rosin esters, esters of polymerized rosin, polyterpene resins, terpene-phenolic resins, coumarone-indene resins, diolefin-olefin resins, phenol-aldehyde resins, aromatic resins, and the like. Examples of usable fillers include calcium carbonate, aluminum silicate, clay, talc, kaolin, silica, ground up polymers, etc., and mixtures thereof.

This sealant formulation is useful in sealing a wide variety of substrates. All common materials of construction, such as glass, aluminum, steel, concrete, brick, rock, ceramic, wood, etc., can be sealed by use of the sealant.

Example: The following materials were mixed to prepare a solvent-release sealant.

Ingredient	Parts by Weight
Hydrogenated thermoplastic elastomer*	54.25
VT/AMS** (modifying resin)	36.1
GEPR*** (modifying resin)	18.2
Plasticizer	42
Calcium carbonate	59.5
Talc	35.0
Titanium dioxide	10
Toluene	Variable

    *Hydrogenated 70/30 butadiene/styrene teleblock copolymer
      coupled with epoxidized soybean oil. Molecular weight
      (weight average) - 95,000.
    **Vinyltoluene/α-methylstyrene copolymer.
    ***Glycerol ester of polymerized rosin.

The sealant was compared to sealants made with similar formulations but containing either a chlorinated paraffin or a mixed dibutyl phthalate/methyl ester of rosin as plasticizer rather than the low MW polybutene. Much improved properties of tensile, modulus, elongation, hardness and peel strength were shown by the formulation given.

Gasket Material of Polyacrylated Polyurethane

Y.R. Bhatia; U.S. Patent 4,145,509; March 20, 1979; assigned to Dana Corp. has prepared polyacrylate-polyurethane copolymers for use as gasket sealants by reacting (a) a monoacrylated terminated active hydrogen-containing hydrocarboxy reactant, (b) an aromatic diisocyanate, and (c) a polyacrylated terminated hydrocarboxy reactant, in which the active hydrogen of reactant (a) reacts with at least one isocyanate radical of reactant (b) to form an isocyanate linkage, and the reaction product of reactants (a) and (b) reacts with reactant (c) in a molar ratio, respectively, of from 1:25 to 25:1.

Preferred reactants are a monoacrylated, lower alkylene glycol for reactant (a), phenylene diisocyanate or toluylene diisocyanate for reactant (b), and a diacrylated alkylene glycol for reactant (c).

The gaskets are prepared by applying a solvent-free reaction mix of the indicated reactants onto a base sheet of the gasket material and then bringing about the reaction as described to form preferably a crosslinked ultimate product.

The reaction may be initiated by heat alone, although preferably a peroxide catalyst is included in the reaction mixture to catalyze the reaction. Relatively heavy or thick coatings of a sealant on a base sheet can be realized in this process in one step. The resulting gasket has improved properties, particularly as to temperature resistance, compressibility, hardness, elongation, flexibility, and solvent resistance to improve the temperature resistance of the sealant and the gasket, metal powders such as aluminum powder may also be incorporated into the reaction mixture.

Examples of reactant (a) may have the formula

$$HO(CH_2)_n-O-\overset{\overset{\displaystyle O}{\|}}{C}-\overset{\overset{\displaystyle R}{|}}{C}=CH_2$$

in which R is hydrogen or methyl, and n is a whole number from 1 to 8. Some of the reactants that may be used for reactant (a) are: 2-hydroxyethyl methacrylate, hydroxypropyl methacrylate. The active hydrogen may be on an amine group as in 2-aminoethyl methacrylate.

Reactant (a) has basically an alkyl or alkylene linear stem which may have a molecular weight up to about 8,000 excluding the acrylate moiety. A preferred reactant (a) is monoacrylated ethylene glycol.

Useful aromatic diisocyanates may range from a simple single aromatic ring to a more complex diisocyanate arrangement containing 2, 3, 4 and even more aromatic diisocyanate rings ultimately joined together. A preferred aromatic diisocyanate is tolylene diisocyanate.

Reactant (c) is a polyacrylated terminated hydrocarboxy reactant such as alkylene glycol diacrylate, alkylene glycol dimethacrylate, a polyalkylene glycol diacrylate, and a polyalkylene glycol dimethacrylate, in which the alkylene group contains four to eight carbon atoms. The preferred diacrylated terminated reactant is diacrylated tetraethylene glycol.

Example: The following composition was prepared by mixing the ingredients with a mechanically driven stirrer:

	Grams
Polyacrylated urethane resin	28
2-Ethylhexyl acrylate	1
Hydroxypropyl methacrylate	9
Acrylic acid	2
Aluminum powder	8
Dicumyl peroxide	0.6
Calcium carbonate	0.9
Ethoxyethoxyethyl acrylate	2
Tetraethylene glycol dimethacrylate	5

The polyacrylated resin was a polyester based urethane acrylate Uvithane 782 of the Thiokol Corporation.

About 0.005 to 0.01 inch thick coatings were applied on both sides of phosphated 0.025 inch thick steel sheets with a brush, and baked for 10 minutes at 400°F to form a gasket. The coatings had good flexibility and adhesion and did not fail when bent around a 0.25 inch mandrel. Also, the coating did not chip off when heat aged for 2,000 hours at 425°F and then for an additional five hours at 500°F.

A nitrogen sealability test is carried out on a specimen having a 0.01 inch bead sealed nitrogen gas at a pressure of 60 psi. This is better than a conventional gasket specimen which has a sealant layer composed of 0.045 inch thick laminate of two perforated steel sheets and an asbestos sheet. Also, the specimen coated with the polyacrylated polyurethane coating prepared in this example sealed nitrogen gas at 60 psi completely with no leakage under a load on the gasket of only 3,850 pounds. But the conventional gasket specimen did not seal nitrogen completely even when a load of up to 20,000 pounds was applied.

Epoxy Composition for Improving Wear of Concrete Surfaces

J.A. Caramanian; U.S. Patent 4,153,743; May 8, 1979; has developed an epoxy resin sealing composition for treating concrete surfaces to render them tough and wear resistant yet flexible enough to expand and contract with the concrete.

The sealant is useful for the treatment of such concrete areas as runways, bridge decks, parking garages, sidewalks, curbing abutments, walkways, airport parking areas and highways.

The sealing composition preferably comprises two liquid components, (A) an epoxy resin base component and (B) an organic polysulfide modified curing and hardening agent. Both components comprise relatively low viscosity solutions, and when mixed during use provide a sealant solution which penetrates deeply into the concrete filling the voids and cavities therein.

Example: The parts and percentages given in the following formulation are by weight unless otherwise stated:

Component A

Ingredients	Parts
Glycidyl-ether modified 100% solids epoxy resin (Shell Epon 815)	34.35
Methyl ethyl ketone	17.20
Toluol	6.50
Butyl Cellosolve	8.92
Cellosolve acetate	21.01
Cyclohexanone	6.37
Hi Initial VM&P Naphtha	5.65
	100.00

Component B

Ingredients	Parts
Polysulfide elastomer (Thiokol polysulfide LP-3)	14.36
Diethylenetriamine	3.92

(continued)

Ingredients	Parts
Methyl ethyl ketone	21.41
Toluol	8.09
Butyl Cellosolve	11.11
Cellosolve acetate	26.15
Cyclohexanone	7.93
Hi Initial VM&P Naphtha	7.03
	100.00

Components A and B can be prepared and stored until ready for use. When ready for use, equal parts by weight or volumes of components A and B are mixed together to form a mixed treating composition having a viscosity of not less than 9 nor more than 25 seconds as measured in a Number 4 Ford Cup at 75°F. For high density air entrained concrete, such as the type used in runways and bridge construction, the viscosity of the mixed sealant composition preferably is between 10 and 12 seconds.

During use the mixed treating solution (A and B) is spread on the concrete surface with a brush or roller or may be sprayed on the concrete, preferably at a rate of approximately 1 gallon to 100-150 sq. ft. of concrete surface. The sealant, which comprises approximately 74% by weight solvent diluents, penetrates a substantial depth into the concrete, the depth of penetration varying somewhat with the consistency of the concrete penetration being about one-fourth to three-eighths inches in highway concrete.

CONSTRUCTION APPLICATIONS

Cementitious Facing Material for Smooth-Skinned Foam

In the preparation of composite panels for building construction comprising a cementitious facing material adhered to a generally smooth skinned foam material such as a styrene polymer foam, difficulty is experienced in obtaining an adequate and permanent bond between the contacting surfaces of the dissimilar materials. This is due at least in part to the fact that the prior known cementitious materials in setting, tend to shrink whereas the foam surface with which it comes in contact does not undergo a corresponding shrinkage. Also, the differences between the coefficients of thermal expansion between the two types of materials often provides a severe shearing stress which results in failure of the bond between the materials.

In order to overcome this problem, *W.J. McMillan; U.S. Patent 4,067,164; January 10, 1978; assigned to The Dow Chemical Company* has developed a cementitious facing material which consists essentially of an admixture of a portland cement, mineral aggregate, from about 5 to about 25% based on the weight of cement of a styrene-butadiene-1,3 copolymer having a styrene to butadiene weight ratio of about 30:70 to 70:30, water in amount of from about 25 to about 65% based on the weight of cement, and based on the weight of copolymer, (a) from about 2 to about 10% of nonionic surfactant, (b) from about 0.75 to 7.5% of anionic surfactant, at least about 15% of which is sodium alkyl sulfate in which the alkyl group contains 9 to 17 carbon atoms, and (c) from about 0.1 to 5% of a polyorganosiloxane foam depressant based on the weight of active polyorganosiloxane, the sum of (a) and (b) not exceeding about 11% by weight of copolymer and the weight ratio of (a) to (b) being within the range of about 0.7:1 to 10:1.

Particularly preferred results are obtained wherein the styrene-butadiene-1,3 copolymer modified portland cement is a shrinkage compensating portland cement which contains sufficient reinforcement to provide restraint against expansion.

Shrinkage compensating cements which may be utilized include the following:

Type K: This is a mixture of portland cement compounds, anhydrous calcium sulfoaluminate ($4CaO \cdot 3Al_2O_3 \cdot SO_3$), calcium sulfate ($CaSO_4$), and lime ($CaO$). The anhydrous calcium sulfoaluminate is a component of a separately burned clinker that is interground or blended with portland cement clinker. Alternatively, it may be formed simultaneously with the portland clinker compounds.

Type M: Either a mixture of portland cement, calcium aluminate cement and calcium sulfate or an interground product made with portland cement clinker, calcium aluminate clinker and calcium sulfate.

Type S: A portland cement containing a large tricalcium aluminate content and modified by an excess of calcium sulfate above usual amounts found in other portland cements.

If such shrinkage compensating cement compositions are not properly restrained, they literally expand themselves apart so that their potential strength is seriously impaired or totally lost. In general, any conventional reinforcing material such as, for example, deformed bar, rods, or wire mesh, in the proper amounts and properly installed will provide restraint sufficient to maintain compositional strength and integrity. Fiber reinforcing materials, such as steel fibers or alkali resistant glass fibers, also provide sufficient restraint. Fibrous types can be added to the composition during the mixing stage and hence will be evenly dispersed and become an integral constituent of the composition. These fibers are randomly oriented and will provide three dimensional restraint.

It has also been found that the combination of alkali resistant glass fiber reinforcement and latex modification creates an unexpectedly beneficial effect.

It has further been found that properly restrained modified shrinkage compensating cement compositions possess significantly increased freeze-thaw resistance, flexural strengths and water absorption characteristics.

The amount of water employed in preparing the shrinkage compensating cement compositions is also important with regard to providing compositions of optimum workability. In this regard at least 25% water, based on the weight of shrinkage compensating cement, is required with an amount from 35 to 65% being preferred.

It has further been found to be beneficial, for purposes of obtaining optimum adhesion of the cementitious facing material to the foam surface, to coat the foam surface with a substantially continuous coating of the styrene-butadiene-1,3 latex described above prior to the application of the cementitious facing material. In this regard, the latex coating is preferably not substantially dehydrated prior to application of the cementitious material.

Example: A number of individual blocks of closed cell, generally smooth skinned polystyrene foam measuring about 1¼ inches in thickness, 2 feet in width and about 4 feet long, were forwarded along a roll having individual projections thereon, which projections were about ½ inch apart in both directions and about ⅛ inch by ⅛ inch in cross sectional area and about ³/₁₆ inch in height. Such projections produced a plurality of indentations in the foam which were in the shape of a trapezoid with the short parallel side of the hole opening at the surface of the foam.

Thereafter, a ½ inch coating of a cementitious protective layer was cast on the surface of the foam. The cementitious protective material used was prepared by admixing a Type K shrinkage compensating cement with sufficient water to form water to cement ratios of 0.29 to 0.635, a sharp mason sand in amount to provide a sand to cement ratio of about 2.75:1 to 3:1, a styrene-butadiene latex composed essentially of an aqueous emulsion of about 48 weight percent of a solid copolymer of about 66% by weight styrene and 34% by weight butadiene-1,3; and based on the copolymer weight, about 4.65% of the nonionic surfactant di-tert-butylphenoxynonaethylene-ethanol; and about 0.78% of a mixture of anionic surfactants comprising predominant amounts of sodium lauryl sulfate and correspondingly lesser amounts of dodecyl-benzene sulfonate, in amount to provide about 15% latex solids based on the weight of cement, a polymethylsiloxane foam depressant in amount to provide about 0.4% by weight active silicon based on the weight of latex solids, and about 4 pounds of ½ inch long alkali-resistant glass fibers per 94 pounds of cement to furnish restraint.

The Type K compensating cement was a mixture of portland cement compounds, anhydrous calcium sulfoaluminate $(CaO)_4(Al_2O_3)_3(SO_3)$, calcium sulfate $(CaSO_4)$, and lime (CaO).

The cementitious protective layer was then vibrated to remove entrapped air and to seat a portion of such cementitious layer in the indentations present in the foam.

The so-formed panels were then cured under ambient temperatures. The cured panels were characterized by being exceptionally resistant to delamination. More particularly, delamination did not occur following 300 temperature cycles of from 15° to 85°F or following 500 temperature cycles of from 50° to 140°F. Further, the cured panels were characterized by a freeze-thaw value of greater than 300 cycles, as determined by ASTM Test No. C-666, i.e., such panels were not significantly deteriorated following such temperature cycling.

Weather-Resistant Polyethylene Terephthalate Adhesive Film

Y. Mitsuishi, S. Shiozaki and K. Hasegawa; U.S. Patent 4,115,617; Sept. 19, 1978; assigned to Teijin Limited, Japan describe an adhesive film having superior weatherability, ultraviolet transmission and transparency which comprises a polyethylene terephthalate film base that exhibits superior weatherability without using a large quantity of an ultraviolet absorbent.

They found that a polyethylene terephthalate film oriented in at least one direction and having (1) a refractive index, in the oriented direction, of 1.57 to 1.78, preferably 1.59 to 1.75, more preferably 1.61 to 1.72, (2) a refractive index, in the thickness direction, of 1.48 to 1.57, preferably 1.485 to 1.56, more preferably 1.49 to 1.55, (3) a density of 1.340 to 1.390 g/cm^3, preferably 1.350 to 1.382 g/cm^3, more preferably 1.355 to 1.379 g/cm^3, and (4) a thickness of 5 to 250 microns, preferably 10 to 150 microns, more preferably 15 to 50 microns is suitable for the base film.

The polyethylene terephthalate film having the aforesaid characteristics has an adhesive layer coated on at least one surface of it.

In one embodiment of the process, the polyethylene terephthalate film used as a base of the adhesive film (to be referred to hereinbelow as a base film or a film base) further has the property of shrinking (shrinkage) at least 4% in at least one direction thereof (therefore, in only one or in two directions, but preferably in two directions) when it is exposed in the relaxed state to hot air at 120°C for 1 minute. An adhesive film having a shrinkage of at least 4% as a base, when used for example as a wire insulating tape, gradually shrinks and becomes taut as it is exposed to sunlight.

Any known adhesives can be used in this process and for example, rubbery polymers, vinyl ether polymers, acrylic polymers, polyvinyl acetate, polyvinyl chloride, and an ethylene/vinyl acetate copolymer can be suitably used. Examples of the rubbery polymer are natural rubber, polyisobutylene, SBR (rubber composed of 1 mol of styrene and 4 to 5 mols of butadiene), ABR (rubber composed of 1 mol of acrylonitrile and 2 to 3 mols of butadiene), neoprene, polyisoprene (containing 70 to 80% of a trans-1,4 bond), a butyl rubber (containing 95 to 98% of isobutylene and 1.5 to 5% of isoprene), and a chlorinated rubber (a mixture of a trichloride and a tetrachloride). Examples of the vinyl ether polymer are polyvinyl butyral and polyvinyl isobutyl ether. Examples of the acrylic polymer include polyacrylates, and copolymers of acrylates and acrylonitrile.

When this adhesive film is to be used for controlling thermic rays, it is convenient to use a so-called water-activatable adhesive which becomes adhesive on addition of water. Examples of adhesives of this kind are polyvinyl alcohol, a partialy saponified product of polyvinyl acetate, polyethylene oxide, polyacrylamide, polyacrylic acid, polyvinyl pyrrolidone, hydroxyethylcellulose, methylcellulose, and carboxymethylcellulose.

Roofing and Surfacing Material

R.J. Graves; U.S. Patent 4,141,187; February 27, 1979 has developed a lightweight roofing and surfacing material comprising a water impermeable polymeric adhesive material as a base coat for adhering sheets of synthetic polymer substrate to a suitable surface. It may be used as a roofing material for providing either an original roof or for reroofing old structures or employed as a surfacing material for any concrete or other base, and provides a particularly efficacious method for repairing damaged concrete surfaces such as sidewalks, stairs, patios and driveways.

Even more particularly, this is a process relating to the building of a roof by covering the normal, e.g., wooden base used for a roof with, if desired, tar paper or other sheet backing material, and then completely covering and sealing this layer with, for example, a urethane polymer adhesive in a flowable condition. The urethane polymer adhesive is allowed to dry or cure to a tacky condition, wide rolls of polymeric sheet material, such as synthetic polymer sheet material, are applied as a top coat, and then, if necessary, rolled to provide the necessary adherence of polymeric sheet material to the adhesive undercoat. Optionally, an electrical heating grid may be disposed within or contiguous to the polymeric adhesive.

In the preferred embodiment, there is provided a synthetic polymeric substrate of 0.05 to 0.75 pounds per square foot, a water impermeable pressure-sensitive, preferably in situ applied, adhesive layer of a highly crosslinked polymer with a thickness of one-eighth to one-half inch, and optionally an electrical heating grid embedded in the adhesive membrane. Particularly preferred for their crosslinking ability are the polyepoxy or polyurethane adhesives.

The method for the installation of the surface covering comprises forming a smooth layer of water-impermeable polymeric adhesive on a suitable base in a flowable condition, curing the adhesive to a tacky condition, applying a sheet of synthetic polymeric material having a density from 0.05 to 0.75 pounds per square foot onto the adhesive layer, and firmly adhering the polymeric sheet onto the adhesive layer, the adhesive layer upon setting providing a water-impermeable membrane. This process may be used efficaciously for installing an original surface covering on any suitable base and as a particularly efficacious method for the repair of a damaged or leaky roof, or to repair damaged concrete.

Particularly preferred adhesives are the polyepoxides and epoxides since these adhesives have an exceptionally high bond strength and stability in the presence of water. Among the epoxides or polyepoxides which may be employed are the saturated, unsaturated, aliphatic, cycloaliphatic, heterocyclic, or epoxy compounds substituted by halogen, hydroxy groups, ether radicals, etc. Exemplary of the above compounds are epoxidized glycerol dioleate, 1,4-bis(2,3-epoxypropoxy)benzene, 4,4'-bis(2,3-epoxypropoxy) diphenyl ether, 1,3-bis(4,5-epoxypentoxy)-5-chlorobenzene, and the epoxy polyethers of polyhydric phenols obtained by reacting a polyhydric phenol with a halogen-containing epoxide or dihalohydrin in the presence of an alkaline medium.

Particularly preferred is an adhesive comprising an epoxide resin and a polyamine carbamate curing agent.

Cold Crosslinking Dispersion Adhesive

Adhesives in the form of aqueous dispersions are used in the building industry, for example, for the laying of floor coverings because, in comparison, with solvent adhesives, they have a series of advantages. Thus, they can be prepared with higher concentrations of the high molecular material which represents the actual adhesive, are not combustible, are not poisonous and are more economical.

G. Fuchs, H. Humbert and D. Pirck; U.S. Patent 4,141,867; Feb. 27, 1979; assigned to Deutsche Texaco AG, Germany provide aqueous dispersions of crosslinking adhesives which comprise:

> (a) a copolymer prepared from, in % by weight, 85-20 2-ethyl-hexylacrylate, 0-33 n-butylacrylate, 8-12 acrylonitrile, 2-6 acrylic acid, 0.5-2 acrylamide, and/or 0.5-2 hydroxyalkylmethacrylate and, based on the amount of the above stated monomers, 0.2-3% of 2-(α-chloroacetoxy) alkylacrylate or methacrylate;
>
> (b) 1-5% by weight, based on the copolymer, of a polyaminoamide comprising terminal amino groups and having a molecular weight of about 150; and optionally
>
> (c) rosin, filler and/or other adhesive additives, the dispersion having a solids content from about 40 to 60 by weight and a viscosity in the range of from 800 to 1,000 cp.

The copolymer which constitutes the adhesive proper contains an internal crosslinking agent, and in addition to this an external crosslinking agent is added to the copolymer dispersion. In this way it is possible to produce adhesive bonds having extremely good thermal stability and tensile shear strength, but at the same time to ensure an operating time between 5 and 60 minutes, given the condition that the formula be otherwise standard, i.e., it contains the usual quantity of filling materials; because as is known, the quantity of filling materials included in a dispersion adhesive has an important influence upon the open time.

A long working time is particularly desirable when laying polyvinyl chloride floor coverings over large areas.

Low molecular polyaminoamides are employed as the external crosslinking agent. These are the condensation products of unsaturated aliphatic acids with polyamines. The products are commercially available under the name Versamid.

Example: The examples were carried out with dispersion adhesives which differed from each other with respect to the proportion of 2-ethylhexyl acrylate (EH) and n-butyl acrylate (BU) in the copolymer, as well as in the type and the quantity of the crosslinking agent. There were also carried out, for the purpose of comparison, examples with 2-(2-chloroacetoxy)ethyl methacrylate as an internal crosslinking agent.

The dispersion adhesives were prepared in all the examples according to the same method. In each case there was prepared a monomer mixture A or B and, separately therefrom, an aqueous phase of the following compositions. All quantities are given in parts by weight.

	A	B
Monomer mixture		
Ethylhexyl acrylate (EH)	425	255
n-Butyl acrylate (BU)	–	170
Acrylonitrile	50	50
Acrylic acid	15	15
Acrylic acid amide	5	5
Hydroxyethyl methacrylate	5	5
EH/BU proportion	100/00	60/40

Aqueous phase	
Water (desalinized)	414.0
Anionic emulsifier	25.0
Nonionic emulsifier	5.0
Ammonium peroxydisulfate	5.0
Sodium disulfite	0.5
Defoamer	0.5
Soda lye* (15%)	50.0

*Added after the polymerization in two stages first stage pH ~5, second stage pH >10.

Preparation of Dispersion — One quarter of the monomer mixture (125 g), the internal crosslinking agent, where appropriate, (0.62 or 1.25 g) and of the aqueous phase (125 g), were placed in a 2 liter reaction vessel and heated to 75°-80°C while being strongly stirred (400 rpm). Within this temperature range, the polymerization reaction started up with a clearly marked heat of reaction.

The remaining three-quarters of the monomer mixture, the internal crosslinking agent, where appropriate, and the aqueous phase in a preemulsified condition, were slowly added over a period of 2 hours. In this operation, an internal temperature of about 80°C was established, which was also maintained for a further hour, after completing the addition, to complete the conversion.

From the polymerization reaction there resulted 1,000 g of a fine particle, co-agulate-free dispersion having a solid content of about 50%. The viscosity of the dispersion was below 1,000 cp, and the pH value amounted to about 5.

In the case where an external crosslinking agent was introduced, after preparing the dispersion, and in parallel experiments, the prescribed quantity was added by mixing in the appropriate formulation. In both of the parallel experiments, the results were always similar.

The adhesive dispersions obtained by means of the examples were used to secure polyvinyl chloride floor coverings onto Eternit sheets. The coated parts were stuck together after different exposure times, and in each case the stripping resistance was determined as a measure of the "open time" or available working time.

Tests were made to measure the thermal stability, the tensile shear strength, and the stripping resistance. It was found that the addition of polyaminoamide leads to a substantially more favorable relationship between the open time and the strength. The time interval between the application of the dispersion adhesive and the time when a bond becomes possible, is reduced from 7.5 to 5 minutes by the addition of 2 and 3% of polyaminoamide. The time interval between the application of the adhesive and the time when a bond can still be made initially increases with the increase of the amount of polyaminoamide (1 and 2%), but thereafter again slightly diminishes while the amount of this added substance becomes 3%. Nevertheless even when the added amount of polyaminoamide is 3% the working time is still satisfactory at 45 minutes; that is to say, such adhesives can be used to make direct bonds even over large areas.

For comparable available working times, the strength values of the adhesives made by the described formulation are always higher. It was found that the addition of polyaminoamide has the effect of producing a marked improvement in quality with different monomer mixtures.

In another formulation for the preparation of cold crosslinking adhesives with polyacrylic acid ester copolymers, *G. Fuchs, H. Humbert and D. Pirck; U.S. Patent 4,144,208; March 13, 1979; assigned to Deutsche Texaco AG, Germany* have used as the crosslinking agent compounds with an acetoacetic ester and a group having a polymerizable double bond. This crosslinking agent can then be incorporated into the copolymer via the double bond and functions as an internal crosslinking agent, such an example of crosslinking agent is the aceto-acetic ester of a β-hydroxyalkylacrylate or -methacrylate.

Preferably the crosslinking agent is a compound with two or more acetoacetic ester groups which is added to the dispersion at some point after completion of the polymerization and before applying the adhesive. Such compounds are, for example, 1,6-bisacetoacetoxyalkane in which the alkane group comprises from 2 to 6 carbon atoms, and 1,1,1-tris(acetoacetoxymethyl)alkane, the alkane groups of which comprise from 3 to 7 carbon atoms. These crosslinking agents may be produced in a simple manner. They may be prepared by the addition of diketones to the appropriate hydroxyl compounds. The polymerizable compounds are preferably prepared from β-hydroxyalkyl acrylates or methacrylates and diketones. Polyacetoacetyl esters are obtained by the conversion of polyvalent alcohols such as glycerine, trimethylol alkanes, hexanediol and other glycols with diketones.

The examples were carried into effect with dispersion adhesives, which are different from each other in respect of the proportions of 2-ethylhexylacrylate (EH) and n-butylacrylate (BU) in the copolymer, (monomer mixtures A and B of previous patent), as well as in the type and the quantity of the crosslinking agents. In addition to carrying out examples according to the process there were also carried out, for the purpose of comparison, examples with 2-(2-chloroacetoxy)ethyl methacrylate as an internal crosslinking agent, that is to say a crosslinking agent built into the copolymer.

The dispersion adhesives were prepared in all the examples according to the same method. In each case there was prepared a monomer mixture A or B and, separately therefrom, an aqueous phase of the composition described in the previous patent.

Examples: 1,000 g of dispersion (50%) was prepared as follows: $\frac{1}{4}$ of the monomer mixture (125 g) as prepared in the previous patent, the internal crosslinking agent, where appropriate, (0.62 or 1.25 g) and of the aqueous phase (125 g), were placed in a 2 liter reaction vessel and heated to 75°-80°C while being strongly stirred (400 rpm). Within this temperature range the polymerization reaction started up with a clearly marked heat of reaction.

The remaining $\frac{3}{4}$ proportion of the monomer mixture, the internal crosslinking agent, where appropriate, and the aqueous phase in a preemulsified condition, were slowly added over a period of 2 hours. In this operation an internal temperature of about 80°C was established, which was also maintained for a further hour, after completing the addition, for the purpose of finalizing the conversion.

From the polymerization reaction there resulted 1,000 g of a fine particle, coagulate-free dispersion having a solid content of about 50%. The viscosity of the dispersion was below 1,000 cp, and the pH value amounted to about 5.

In the case where an external crosslinking agent was introduced, after preparing the dispersion, and in parallel experiments, the prescribed quantity was added by mixing in the appropriate recipe. In both of the parallel experiments, the results were always similar.

The adhesive dispersions obtained by means of the examples were used to stick polyvinyl chloride floor coverings onto Eternit sheets. The coated parts were stuck together after different exposure times, and in each case the stripping resistance was determined as a measure of the "open time."

Other samples were stored as follows: Sample b was stored at 20°C for 7 days; Sample c was stored at 20°C for 1 day, at 50°C for 5 days, and then at 20°C for another day. Tests were then made of open time, thermal stability and tensile-shear strength. Characteristics which satisfied or exceeded the specifications were given by all the adhesive dispersions using the described esters as crosslinking compounds, as showed by the following table.

	Examples According . . . to the Process . . .			Comparison . Examples. .	
	1	2	3	4	5
Monomer mixture	A	A	B	B	A
Crosslinking agent*	I	II	III	IV	IV
Crosslinking agent (wt %)	2	2	2	0.5	0.5
Open time (min)	7.5-60	5-60	7.5-55	1-25	7.5-55
Tensile shear strength, kp/10 cm^2					
b	106	105	109	108	100
c	103	101	107	106	95
Thermal stability, min					
b	5.5	5.5	6	6	6.5
c	10	9	9	9.5	9.5

*I   = 1,1,1-Trisacetoacetoxymethylpentane
II = Bisacetoacetoxyhexane
III = 1,2,3-Triacetoacetoxypropane
IV = 2-(Chloracetoxy)ethyl methacrylate

In the above table, the specification minimum values were: tensile shear strength—b, 105; c, 100; thermal stability—b, 5; c, 8.5; and open time, 10–45.

For Installation of Wall and Floor Coverings

Dry-set portland cement mortar is normally used for installing wall and floor coverings such as ceramic tile, slate, marble, etc. The sag resistance of such mortars may be improved by the addition of asbestos fibers. It has now become desirable to find a substitute for such fibers and U.S. Patent 4,082,563 provides sag-resistant dry-set cementitious hydraulic compositions based on the use of certain salts which, under appropriate conditions, form gelatinous or hydrated hydroxides. These additives are nonfibrous, nontoxic and, in addition, provide sag resistance properties comparable to those imparted by asbestos fibers. The compositions of U.S. Patent 4,082,563 include a hydraulic cement and a water retentive agent or agents.

J.R. Ellis and H.B. Wagner; U.S. Patent 4,142,911; March 6, 1979; assigned to Tile Council of America, Inc., have found that gelatinous metal hydroxides and metal salts capable of forming a gelatinous hydroxide in the presence of a water-soluble alkaline material are also widely useful, even though they are nonfibrous, in providing sag resistance in a broad range of other room temperature hardening adhesive compositions, over and apart from the hydraulic dry-set mortar cementitious compositions. These other compositions include (a) hydraulic non-dry-set cementitious compositions comprising an hydraulic cement or plaster of paris, and (b) nonhydraulic adhesive compositions comprising a room temperature hardening polymer.

These compositions comprise in admixture;

> (A) an adhesive material selected from the group consisting of (a) materials capable of hardening at room temperature upon admixture with water, the materials being selected from among hydraulic cements and plaster of paris, and (b) room temperature hardening polymers; and

> (B) at least one nonfibrous compound selected from the group consisting of gelatinous metal hydroxides and metal salts capable of forming a gelatinous or hydrated hydroxide in the presence of a water-soluble alkaline material, the compound or compounds being present in an amount at least sufficient to provide the desired sag resistance.

In general, amounts of component (B) of at least about 0.2, and preferably from about 0.2 to about 5.0% by weight of the composition, are employed to provide the desired values of sag resistance.

The metal salts of component (B) can be water soluble, hydrated or anhydrous. The cations of these salts are selected from the groups consisting of aluminum cations, cations of the transition elements other than copper of Period 4 of the Periodic Table of the Elements: scandium, titanium, vanadium, chromium, manganese, iron, cobalt, nickel and zinc, and preferably chromium, manganese, iron, nickel and zinc; the cations of cerium and antimony; and the mixed cations of mono and trivalent metals, designated as M^+M^{+++} mixed salts, where M^+ is an alkali metal, e.g., sodium, potassium, or ammonium, and the like, and where M^{+++} is a trivalent metallic cation of Periods 3 or 4 of the Periodic Table of the Elements, preferably aluminum, chromium or iron.

The anions which are used in these salts are preferably selected from among the anions of strong mineral acids, e.g., sulfate, chloride and nitrate, as well as carboxylic acids, preferably alkanoic acids having up to 20 carbon atoms in the chain, including formates and oxalates. Especially preferred among the organic anions are oxalate and stearate.

Special mention is made of aluminum hydroxide and aluminum sulfate, especially powdered aluminum sulfate, as highly preferred compounds for use as the sag resistance imparting agent.

As mentioned above, these additive salts possess a common chemical characteristic in that all of them are capable of forming a gelatinous or hydrated hydroxide in the presence of an alkaline material, and more specifically, an aqueous solution of a base. The water-soluble alkaline materials with which the metal salts
are capable of coreacting to form gelatinous or hydrated hydroxides can be organic or inorganic. Examples of inorganic compounds include lime, i.e., calcium
oxide, as well as both monobasic water-soluble compounds such as ammonium
hydroxide and alkali metal and alkaline earth metal hydroxides or salts, e.g., sodium hydroxide, potassium hydroxide, soda ash and the like. Among these, lime
is preferred.

Examples of organic alkaline materials include aliphatic and cycloaliphatic primary and secondary amines such as diethylenetriamine, morpholine, diethanolamine, monethanolamine, 2-methyl-2-aminopropanol, and the like. In general,
enough of the alkaline material should be present in the aqueous composition,
i.e., after water has been added to the dry mix to provide a pH of at least about
7.1, preferably from about 8 to 13.

The sag resistance imparting agent of this process, component (B), is useful in a
wide variety of non-dry-set cementitious compositions capable of hardening
upon admixture with water, including portland cement-based patching plasters
and stucco compositions, as well as plaster of paris-based spackling formulations.

The nonhydraulic compositions in their broadest aspects, include any polymer-
based adhesive composition capable of hardening at room temperature to form a
connecting medium between surfaces and in which sag resistance is desired. Such
polymers include, e.g., urea-formaldehyde resins, melamine-formaldehyde resins,
polysiloxanes, phenolic resins, polyamides, polyesters, etc.

Example: A portland cement-based patching plaster is prepared using the following ingredients in dry admixture in percent by weight: white portland cement, 23.2; slaked lime, 5.8; sand, 69.8; and aluminum sulfate, 1.2. The sand
comprises about 95% by weight of particles 140 mesh or less, U.S. Standard
Sieve, and about 5% by weight of particles greater than 140 mesh.

Water in the amount of 21.9% by weight, based on the weight of the dry blend,
is added to obtain a workable viscosity, and the resulting aqueous composition
is trowelled onto a vertically disposed glass plate between guide strips ¼ inch
thick. A Type B tile, in accordance with American National Standard Specification A 118.1 (1967), is tapped lightly onto the surface of the trowelled patching
plaster immediately after application. The top edge of the tile is accurately
marked, and any downward displacement of the tile two hours after placement
is measured as the sag. No sag is observed after two hours.

By way of comparison, a control formulation containing the same ingredients
in the same amounts, but without aluminum sulfate, is tested and the tile is
observed to sag and fall off the glass plate within two hours.

Glass and All-Purpose Adhesives

ADHESIVES FOR GLASS

Aqueous Acrylate Dispersions

P.S. Columbus; U.S. Patent 4,077,932; March 7, 1978; assigned to Borden, Inc. describes adhesives useful for adhering a siliceous-containing material to itself or to a substrate which are aqueous dispersions of polyacrylates mixed with epoxy-silanes wherein the silicon atom of the silane is substituted with hydrolyzable alkoxy groups.

The aqueous adhesive composition has a pH of about 8 to 10, and contains from 25 to 75% by weight of water, the water having dispersed in it: (a) from about 25 to 75% by weight of the composition, of a water-insoluble linear polymer of interpolymerized monoethylenically unsaturated monomers in amounts by weight based on the linear polymer (1) from about 50 to 99% of an acrylic ester monomer of the formula

$$H_2C=\overset{\overset{\displaystyle R}{\displaystyle |}}{C}-\overset{\overset{\displaystyle O}{\displaystyle \|}}{C}-OR^2$$

or a mixture thereof, where R is H or methyl and R^2 is lower alkyl of 1 to 4 carbons, (2) from about 0.1 to 10% of acrylic acid or methacrylic acid, (3) from about 0.1 to 10% of a nitrogen containing monomer selected from the group consisting essentially of vinyl amines and salts thereof, vinyl ureido monomers, vinyl compounds having heterocyclic nitrogen containing groups, and mixtures thereof; and (b) 0.001 to 2% of a silane containing both alkoxy and epoxy functionalities.

These adhesives do not give off noxious odors, do not form release coatings, cure at ambient temperatures for example $10°$ to $25°C$, are easy to apply and

366

clean up, give final bonds with high tensile strength and water resistance. They do not contain organic solvents which evaporate off on drying, do not require baking to effect the cure, and are useful as adhesives for securing siliceous surfaces to other surfaces.

Example 1: Windshield Sealant — To a stainless steel tank with agitation was added 12 parts by weight of a 47% solids aqueous emulsion of Rhoplex AC-34, an aqueous acrylic emulsion (Rohm & Haas). The product has the following physical properties: white milky liquid in appearance; 46 to 47% solids content; 9.5 to 10.0 pH (as shipped); 8.9 lb/gal weight; 1.15 specific gravity of solids; and 9°C minimal film-forming temperature.

The solids are believed to be essentially a polymer of 58% ethyl acrylate, 38% methyl methacrylate, 1% of acrylic acid and 3% of a nitrogen containing monomer.

A thickening and suspending agent was prepared in a separate stainless steel tank by adding 96.70 parts of water then adding 3.0 parts of hydroxyethyl cellulose with agitation. The mixture was heated to 85°C and 0.15 part of methyl parahydroxy benzoate and 0.15 part of propyl parahydroxy benzoate are added. The mixing was continued at 85°C for approximately 20 minutes until smooth and homogenous. The mixture was cooled to 45°C.

4.75 parts of the suspending agent was then added to the acrylic polymer emulsion with agitation. Then an additional 5 parts of the Rhoplex AC-34 was added with agitation.

An additional 66 parts of Rhoplex AC-34 was then added. Then 0.75 part of propylene glycol and 0.11 part of gamma-glycidoxy-propyltrimethoxysilane were added with medium agitation. The agitation was continued for 20 minutes. The viscosity of the product should be 1,400 to 1,800 cp using a Brookfield LVF No. 4 spindle at 35°C at 60 rpm and 6,000 to 8,000 cp at 6 rpm, pH should be 9.0 to 10.0 and the percent solids should be 42±1.5%. The product is packaged in one fluid ounce wax-lined aluminum tubes.

A leaky automobile windshield was successfully sealed to its rubber gasket with the product of this example using the following procedure:

Surface areas must be clean, dry and free from dirt, oil, soap and wax as well as any other sealer used previously without success. Surfaces to be sealed should be at 60°F or higher.

Pressing tube lightly, apply sealer to problem area making certain that the sealer is applied as deeply as possible and that a continuous wet film is formed between surfaces to be sealed. The sealer may be considered dry when it turns from milky-white to crystal-clear and colorless (in approximately 15 minutes to several hours depending on thickness of application, temperature and humidity). Rather than applying one heavy coat, faster-drying may be accomplished by applying two light coats and allowing the first coat to clear before applying the second. Check dried clear film for continuity; if not continuous, apply another coat and let dry. Allow clear film to cure overnight before subjecting to rainy weather. Full cure is obtained in approximately 3 to 7 days depending on temperature and humidity. If subject to long continuous downpour, the sealer film will turn from crystal-clear to slightly translucent. Despite this color change, the sealing properties are not affected and the film will regain its crystal-clear appearance again on drying.

Example 2: To a stainless steel clean tank with agitation is added 12 parts of an aqueous dispersion containing 47% solids. The solids consist of an acrylic polymer containing 46% of ethylacrylate, 52% of methylmethacrylate, 1% of dimethylaminomethylmethacrylate and 1% of acrylic acid.

A thickening and suspending agent is prepared in a separate stainless steel tank by adding 96.70 parts of water then adding 3.0 parts of hydroxyethylcellulose, nonreactive with silane, with agitation. The mixture is heated to 85°C and 0.15 part of methyl parahydroxybenzoate and 0.15 part of propyl parahydroxybenzoate are added. The mixing is continued at 85°C for approximately 20 minutes until smooth and homogenous. The mixture is then cooled to 45°C.

To the acrylic dispersion in the tank is then added with agitation 4.75 parts of the suspending agent. Agitation is continued until the mixture is smooth and homogenous. An additional 82.40 parts of the acrylic polymer is then added slowly with agitation and the agitation is continued until the mixture is again homogenous.

Then 0.75 part of propylene glycol and 0.10 part of gamma-glycidoxypropyl-trimethoxysilane are slowly added with continued agitation. The agitation is continued for 20 minutes to insure that the mixture is homogenous. The mixture should now have a viscosity using a Brookfield No. 4 spindle at 25°C at 6 rpm of 2,500 to 3,500 cp. Viscosity of 60 rpm should be 500 to 700 cp. If the viscosity is too high, water can be added. The finished product has a pH of 9.0 to 10.0. The solids content is 44±1.5%. The product is to be marketed in one fluid ounce wax-lined aluminum tubes.

Broken china plates and crystal are mended using the abovedescribed adhesive, and after being mended and allowed to cure are washed in a dishwasher using a conventional dishwasher detergent such as Cascade or Electrasol and at a water temperature of 180°F. The adhesive withstands the washing cycle.

The adhesive is then tested for tensile strength by breaking the dishes. Once the adhesive film is fully cured, the dishes, in most cases, break in areas not mended by the adhesive.

Rapid-Curing Two-Part Acrylic-Based Adhesive

M.M. Skoultchi; U.S. Patent 4,081,308; March 28, 1978; assigned to National Starch and Chemical Corporation has found that acrylic-based adhesive compositions characterized by rapid curing and extended shelf life may be prepared by catalyzing an acrylic monomer with the inter-reaction mixture of copper salts, saccharin and sulfone or mixtures thereof. The two part systems comprise (a) the adhesive base and (b) an activator therefor. More specifically, part (a) comprises the acrylic monomer and a saccharin component present in the form of the copper salt of saccharin or as saccharin and a soluble copper salt or as a mixture thereof. Part (b) of the system comprises an alpha-hydroxy sulfone, an alpha-amino sulfone or mixtures thereof.

Ultraviolet Light-Curable Diacrylate Hydantoin Compositions

The formulation produced by *A. Fritz and B.L. Dunn; U.S. Patent 4,082,635; April 4, 1978; assigned to Ciba-Geigy Corporation* is for ultraviolet light-curable one-package diacrylate hydantoin adhesive composition comprising: (a) from 25 to 75 parts of a diacrylate of the formula

(b) from 25 to 75 parts of a diacrylate comprising (1) from 80 to 100 parts of a compound of the formula

(2) from 0 to 20 parts of a compound of the formula

(c) from 0.01 to 2 parts of a silane; and (d) from 0.01 to 3 parts of a light sensitizer wherein each of R_1 and R_2 is hydrogen, alkyl of 1 to 8 carbons, cycloalkyl of 5 to 6 carbons, or phenyl or together R_1 and R_2 is tetramethylene or pentamethylene; and R_3 and R_4 are hydrogen or methyl.

It is also necessary to have a photoinitiator present in the adhesive compositions. The photoinitiator absorbs the radiation to produce free radicals which initiate polymerization. Examples of the light sensitizers or photoinitiators used in this formulation are carbonyl compounds such as 2,3-hexadione, diacetylacetophenone, benzoin and benzoin ethers such as dimethyl, ethyl and butyl derivatives, for example, 2,2-diethoxyacetophenone and 2,2-dimethoxyacetophenone, benzophenone in combination with a catalyst such as triethylamine, N,N'-dibenzylamine, and dimethylaminoethanol, benzophenone plus Michler's ketone; nitrogen-containing compounds such as diazomethane, azo-bis-isobutyronitrile, hydrazine, phenylhydrazine, and trimethylbenzylammonium chloride; sulfur-containing compounds such as benzene sulfonate, diphenyl-disulfide and tetramethylthiuram-disulfide. Such light sensitizers are used alone or in combination with each other.

As noted, the compositions also contain a silane adhesion promoter. These are derivatives of trichlorosilane with the general formula $R'Si(OR)_n$, where R' is an organic group and preferably a reactive organic group, and n is 2 or 3. Organic functional groups representative of R' include vinyl, aminoalkyl, acrylatoalkyl, glycidoxyalkyl, and variations of these. The alkoxy substituents representative of OR are methoxy, ethoxy or β-methoxyethoxy.

These ultraviolet light-curable resinous compositions have long pot life so long as they are stored in a dark place. Also, it should be noted that these curable resinous compositions can be used as is, so they can easily be used in practice by the user. These adhesive compositions may easily be prepared by intimately mixing together components (a) through (d) in the specified amounts. The final viscosity of the adhesive composition is in the range of from 500 to 3,000 cp at $25°C$ and, preferably, from 1,000 to 2,000.

With respect to the amount of each component present in the adhesive composition, a preferred formulation contains an amount of component (a) which is in the range of from 40 to 60 parts; and amount of component (b) which is present in the range of from 40 to 60 parts; an amount of component (c) which is present in the range of 0.3 to 0.7 part; and an amount of component (d) which is present in the range of from 0.2 to 1 part.

As noted above, the one-package adhesive cures rapidly, and possesses a low viscosity which wets the bonding surfaces well. The adhesive composition cures to a transparent bond on glass surfaces and it is generally unaffected by long-term low intensity ultraviolet radiation, has good heat stability, water and weathering resistance. The adhesive composition, when exposed to ultraviolet radiation, attains maximum bond strength when exposed to high intensity ultraviolet radiation within 5 to 25 seconds. Further exposure is unnecessary but it does not appear to degrade the adhesive strength. At least one of the substrates to be joined must be glass or some other material transparent to ultraviolet light. Thus, this is a suitable adhesive for bonding glass to glass, glass to plastic, and glass to metal.

Example 1: A light-curable composition was prepared by mixing 48.8 parts by weight of the diacrylate of 1,3-bis-(β'-hydroxyethyl)-5,5-dimethylhydantoin, 48.8 parts by weight of the diacrylate of 1,3-diglycidyl-5,5-dimethylhydantoin, 1.9 parts by weight of dimethoxyphenylacetophenone and 0.5 part by weight of aminopropyltriethoxy silane. The adhesive was used to bond the following substrates: glass to glass (solvent wiped), glass to aluminum (etched), glass to stainless steel (solvent wiped), and glass to acrylic coated steel (solvent wiped).

The glass to glass specimens were prepared by bonding two glass pieces (1 x 4 x ¼ inch) with a ½ inch overlap. The glass to metal specimens were prepared by bonding a 1 x 2 x ¼ inch piece of glass onto two strips of 1 x 4 inch pieces of metal with a ½ inch overlap of glass to each metal strip. The thicknesses of the metal pieces were: Al, 64 mils; stainless steel, 50 mils; and the acrylic coated steel, 40 mils. The adhesive was cured by exposing the specimens to a mercury vapor medium pressure 400 watt lamp (360 mμ, 254 mμ) at a distance of 8 inches. The time for the adhesive to set was determined, as well as the tensile shear strength after 30 seconds and 1 minute exposure. The results are indicated in the following table.

	Aluminum	Stainless Steel	Acrylic Coated Steel	Glass
Time to set between substrate and ⅛" glass slide, sec	2-3	2-3	2-3	2-3
Tensile shear strength at 25°C*, psi				
after 30 sec UV exposure	290	340	650	660
after 1 min UV exposure	370	290	520	640

*Tested after 24 hr at 25°C.

Example 2: A similar composition was prepared, however, using 0.3 part by weight of dimethoxyphenylacetophenone rather than 1.9 parts by weight. Specimens were prepared wherein the substrates bonded were glass to glass and glass to aluminum. The specimens were of the dimensions described previously and were cured in the same manner. The time to set and lap shear strength were determined and are reported in the following table.

UV exposure set time, sec	14
Lap shear strength at 25°C*, psi	
Glass to glass	
10 sec	845
30 sec	840
60 sec	810
Glass to aluminum	
10 sec	650
30 sec	730
60 sec	720

*After 24 hr at 25°C.

Adhesive Films for Transparent Laminates for Aircraft Windshields

G.L. Ball, III, D.W. Werkmeister; and I.O. Salyer; U.S. Patent 4,137,364; Jan. 30, 1979; assigned to Monsanto Research Corporation have developed transparent or optically clear laminates and adhesive films for making these laminates having a temperature utility range far greater than those previously developed and which are especially useful for windshields of high speed aircraft which need such a broad temperature range. These transparent adhesive films have a temperature utility in the range of -65°F to at least 350°F in laminates and comprise an ethylene/vinyl acetate copolymer having a vinyl acetate content in the range of 40 to 60%, the copolymer being partially hydrolyzed to form a terpolymer having a hydroxyl content in the range of 1 to 6%, the terpolymer in film and/or laminate form being heat treated and crosslinked to give the adhesive film. The laminates are made from transparent layers of glass and/or plastic bound together with the transparent adhesive film.

Preferred ethylene/vinyl acetate copolymers have vinyl acetate content in the range of 45 to 55% and the preferred hydroxyl content in the terpolymer is in the range of 2 to 4%.

The particular ethylene terpolymer system, an ethylene/vinyl acetate/vinyl alcohol terpolymer, used experimentally to establish the utility of this process was made from an ethylene/vinyl acetate copolymer having about 50% vinyl acetate (available in pellet form commercially) which was hydrolyzed to contain 3.2% hydroxyl. The copolymer before hydrolysis analyzed 50±2% vinyl acetate (by saponification) with a melt index at 125°C of 1.6±1.2 g/10 mm based on sampling of a number of bags of the material. Unless differently stated, percentages are by weight.

The interlayer of this process had adequate strength to 350°F while retaining excellent high elongation down to -65°F.

The improved mechanical performance to 350°F was achieved through a controlled limited crosslinking of the basic ethylene terpolymer. It was shown that improved form stability could be provided by both chemical and radiation crosslinking. The crosslinking can be conducted during or after formation of the ethylene terpolymer interlayer sheet.

The two chemical crosslinking agents used were isophthaloyl biscaprolactam and vinyltriethoxy silane, and 0.3% of the caprolactam or 1% of the silane were determined to be adequate for the desired degree of crosslinking. These crosslinking agents are incorporated into the terpolymer in a Banbury mixer, on a mill roll, or other low (cold blending) temperature compounding equipment. The compounded terpolymer containing the crosslinking agent is then molded, calendered or extruded into a film at about 200°F and crosslinking is completed by heating for at least one hour at 350°F. The partially crosslinked sheet is then laminated between sheets of glass, glass-polycarbonate, polycarbonate, glass-acrylic, acrylic-polycarbonate, acrylic or other transparent plastics at temperatures of 250°F. By this process, there is obtained a laminate in which the interlayer material has significant strength and form stability up to a temperature of at least 350°F or higher.

In the electron bombardment crosslinking process, the ethylene terpolymer is first formed into film or interlayer of desired thickness. The sheets are then irradiated by electron bombardment at dosages in the range of 10 to 45 megarads at voltages of 180 to 300 kilovolts. Near optimum conditions for achieving the desired level of crosslinking involve a total dosage of about 45 megarads at 180 kilovolts. However, equivalent results can probably be obtained with a dosage of only 10 megarads at 300 kilovolts. The irradiated film, crosslinked as described above, is then placed between the desired glass and/or plastic layers, and laminated by conventional methods such as that described hereinabove at a temperature of about 250°F to obtain a laminate of similar properties to the laminate described above using a chemical crosslinking agent. These transparent adhesive films have a temperature utility in the range of -65° to 350°F in laminates.

GLASS-TO-PLASTIC ADHESIVES

Solar Control Film

In recent years reflective-transparent solar control film has achieved an important place in the economy of the United States and several other countries. Products of this type commonly involve a 1-mil biaxially oriented polyethylene terephthalate film, having a vapor-deposited aluminum layer about 25 to 125 microns thick on one face. The aluminum layer is contacted and protectively covered by a thin, transparent water-insoluble moisture-transmitting barrier coat.

Laminated windowpanes formed in this manner serve to substantially reduce heat transmission into the room on which the window opens without significantly reducing light transmission. Laminated windowpanes of this type can be formed by applying a water-soluble adhesive to either the glass or the protectively coated surface of the film prior to lamination. In either case, the slippery nature of the adhesive permits sliding the film to position it precisely against the glass. This feature is particularly important when it is considered that the film applied to a given windowpane may easily be 4 feet wide and 6 feet long.

There are certain disadvantages to these solar control films. When the interior surface of the window is exposed to high humidity, the film sometimes loosens in certain places greatly marring its esthetic qualities. It has also proved difficult to produce a film which can be readily handled, easily installed in large pieces on a window, easily positioned, quickly adhered, and readily removed.

D.R. Theissen; U.S. Patent Reissue 29,663; June 6, 1978; assigned to Minnesota Mining and Manufacturing Company have provided a solar control film which, although provided with a normally tacky and pressure-sensitive adhesive, can be handled, even in the form of extremely large sheets or wide rolls, without using a removable liner and without incurring the danger of premature adhesion to an undesired substrate or contaminating material. Applied to a windowpane, the resultant product is adherently bonded thereto, forming a laminate which is of high optical quality, i.e., free from haze, distortion, irregularities, bubbles, "fish eyes" and similar defects possessed by prior art products. In the preliminary stages of application, the product can be easily shifted on the windowpane to more correctly align its location, and if a small area inadvertently becomes adhered to the window, the film can be removed without transfer of the pressure-sensitive adhesive to the glass surface. At such time as it may be desired to remove the installed film, this can likewise be easily accomplished without leaving adhesive residue on the glass.

In its simplest terms, the solar control film is made by applying a normally tacky and pressure-sensitive adhesive layer over the aluminum vapor deposit-coated surface of a transparent polymeric film and then applying thereover a thin layer of a water-soluble film-forming material, thereby forming a water-activatable adhesive system. The thus-protected surface of the pressure-sensitive adhesive is not tacky, and thus is neither contaminated by dirt nor prone to stick to itself. Even wide rolls of this product can be readily unwound by one man without distorting the backing.

When the product is to be used, the water-soluble coating is moistened, thereby rendering the surface somewhat slippery without at the same time removing the water-soluble coating. When this slippery surface is placed against the windowpane, the solar control film can be slid to the desired location. A squeegee is then used to force the film against the glass and expel some of the excess water at the edges. Any excess water is then gradually transmitted through the film into the interior of the room on which the window opens, after which the film is found to be firmly anchored to the glass.

Example: A 1-mil biaxially oriented polyethylene terephthalate film 48 inches wide was vapor coated with aluminum by conventional techniques to a metal thickness of about 125 A, thereby reducing the light transmission at 5500 A wavelength to approximately 10% the value of the uncoated film, as measured by a Beckman DK-2 Spectrophotometer. Over the metal was then applied an 8% solution of a barrier coating consisting of 1:1 ethylene terephthalate:ethylene isophthalate copolymer containing 0.75% by weight of 2,2'-dihydroxy-4-methoxybenzophenone, known as Cyasorb (American Cyanamid Co.) in 1,1,2-trichloroethane, the total dry coating weight being 1 grain per 24 square inches. The solvent was then evaporated. The light transmission at 5500 A for the metallized barrier coated film was found to be 18% of the value measured for the clear film. The coated film was then slit to a width of 24 inches.

Over the barrier-coated film was applied a thin layer of primer in an amount equal to approximately 0.1 dry grain per 24 square inches. The composition of this primer was as shown below.

Components	Parts by Weight
50:47:3 isooctyl acrylate:vinyl acetate:acrylic acid terpolymer	4
Polymethyl acrylate, Acryloid 101 V, (Rohm & Haas)	1
Baking phenolic resin, BKR 2620, (Bakelite)	0.16
70:30 methyl isobutyl ketone:isopropanol solvent	8.8

The solvent was then evaporated from the primer and an ethyl acetate:heptane solution of a 96:4 isooctyl acrylate:acrylamide copolymer pressure-sensitive adhesive applied in an amount sufficient to leave 3 dry grains per 24 square inches after evaporation of the solvent.

A water-soluble film weighing approximately 0.10 to 0.15 grain per 24 square inches was separately prepared by casting a 4% aqueous solution of a vinyl ether:maleic acid copolymer, Gantrez AN-169, (General Aniline & Film Corp.), containing 1% wetting agent, Igepal CO 630, (General Aniline & Film Corp.), based on solids present, on a 1-mil temporary carrier sheet of biaxially oriented polyethylene terephthalate film, evaporating the water to leave an extremely thin, dry tack-free film, and winding the laminate on itself in roll form. The vinyl ether:maleic acid copolymer film was then transferred to the surface of the normally tacky and pressure-sensitive adhesive by placing it in contact therewith and removing the polyethylene terephthalate carrier sheet. The resultant product could be wound upon itself in roll form, unwound with very little effort, and handled conveniently with no detrimental effect on the product.

A 24 x 54 inch sheet of the material formed as described in this example was wet with water, applied to a clean plate glass window, positioned, and squeegeed to remove the excess water and water-soluble adhesive. The sheet material was in uniform contact with the glass but could be removed if desired. After drying for a period of approximately two hours, the edges of the product were firmly bonded. After three days, the entire product had developed a firm adhesion for the window. This product can be completely removed by simply peeling it from the window without separate steaming or soaking techniques, facilitating its replacement when desired.

Use of Silanes Having Capped Functional Groups

Good adhesion in glass- or glass fiber-plastic laminates is hard to attain if it is desirable to have the laminate be transparent.

J. Amort and H. Nestler; U.S. Patent 4,118,540; October 3, 1978; assigned to Dynamit Nobel AG, Germany, however, have found that the adhesion of glass and plastic can be achieved with a silane having a capped functional group.

The term, "silanes containing capped functional groups," as used herein, refers to compounds of the formulas shown below.

$$H_2C \underset{\underset{\underset{C}{\overset{\|}{O}}}{\diagdown}}{\overset{}{-----}} \underset{\underset{C}{\overset{\|}{O}}}{\overset{}{CH}} - CH_2 - X - (CH_2)_n - \underset{R'_a}{\overset{}{Si}} (OR')_{3-a}$$

$$\underset{R'-O}{\overset{R'-O}{\diagdown}} CH - CH_2 - NH - (CH_2)_n - \underset{R'_a}{\overset{}{Si}} (OR')_{3-a}$$

In the formulas n can be from 1 to 8, preferably 2 to 4, R represents an oxygen moiety or two hydrogen atoms, X represents oxygen or sulfur, R' represents identical or different alkyl moieties of 1 to 8 carbons which can be interrupted by an oxygen atom if desired, and a can be equal to 0 or 1.

The laminates made from organic binding agents and inorganic oxides or metals with the use of these silanes having capped functional groups display substantially improved mechanical strengths as a result of improved adhesion, as shown by wet strength measurements in bending experiments performed on standardized test specimens.

Of special technical interest is the pretreatment of glass fabrics and fibers and of glass surfaces and metal surfaces which are made into laminates with numerous organic polymers.

The commercial advantages of this process can be achieved either by a treatment of the inorganic metallic substrates with the claimed silanes by applying them either by spraying, dipping, atomizing or brushing, or by adding the claimed silanes to the polymers as adhesive adjuvants.

Example: Water-sized glass fibers are immersed for one minute in a 0.25% solution of $(C_2H_5O)_2CHCH_2NH(CH_2)_3Si(OCH_3)_3$ which has been acidified with acetic acid. Then, after allowing the excess solution to drain off, the fibers are dried for 15 minutes at $130°C$.

The glass filaments sized in this manner are then dipped in epoxy resin and made into round fiber-reinforced rods. The curing of these rods is performed at $130°C$ for 17 hours. The bending strength of the test specimens obtained in this manner is determined in accordance with DIN 53-452, one determination being performed directly after curing and the other after letting specimens stand in boiling water for 72 hours.

A test specimen prepared in the same manner, but reinforced with glass filaments sized with ω-aminopropyltriethoxysilane, was used for purposes of comparison. The blank specimen contains only water-sized glass filaments.

The results are given in the table below where the bending strength of the glass fiber-reinforced epoxy rods is given in kp/cm^2.

| Silane Used | . . . . . . Bending Strength . . . . . . | |
	Directly After Curing	After Standing in Water
None	9,800	3,500
$H_2N(CH_2)_3Si(OC_2H_5)_3$	9,500	7,900
$(C_2H_5O)_2CHCH_2NH(CH_2)_3Si(OCH_3)_3$	11,400	9,600

SEALANTS FOR GLASS

In Situ Curable Sealant for Multiple-Glazed Windows

G.H. Bowser; U.S. Patent 4,092,290; May 30, 1978; assigned to PPG Industries, Inc. describes a room temperature vulcanizable sealant useful in glazing applications which comprises an admixture of two rubbery components. The two components each comprise butyl rubber, polybutene, and polyisobutylene. The first component also includes a quinone dioxime, and the second component includes an oxidizing agent. The two components each have relatively low viscosities or consistencies, and can be extruded together at relatively low temperatures to form an initially uncured cold flowable mastic material in bead or ribbon form.

The individual components have unlimited shelf life. Curing only begins when the components are mixed together and the oxidizing agent comes in contact with the quinone dioxime, oxidizing it to the corresponding dinitro compound which cures or vulcanizes the rubbery composition. The extruded sealant composition is initially uncured and of relatively low viscosity, and since the curing is not accelerated as it would be if high temperature extrusion had been employed, as is necessary with high viscosity compositions, the low viscosity sealant composition readily cold flows and can easily be applied to double glazed units without the need for presssurizing equipment such as clamps, rigid spacers and the like.

After about 2 to 4 weeks at room temperature, while in situ in the window unit, the sealant composition fully cures in the multiple glazed window unit and forms a firm, resilient hermetic seal.

Example: A sealing composition comprising the following two components was prepared as follows:

| | Components (% by wt) | |
	A*	B**
Butyl rubber (Butyl 165CC, Exxon Chem. Co.)	19.2	19.2
Polyisobutylene (Vistanex LM, MS, Exxon)	16	16
Polybutene (Oronite 32, California Chem. Co.)	26.1	26.7

(continued)

	Components (% by wt)	
	A*	B**
Carbon black (Statex G, Columbia Carbon Co.)	14.1	–
Silica pigment (Hi-Sil 233, PPG Ind.)	8.8	8.8
Zinc oxide (XX4 grade, NJ Zinc Co.)	–	8.8
Silica pigment (TK 800, Degussa Inc.)	2.1	2.1
Zirconium orthosilicate (Superpax, National Lead Co.)	8.8	14.4
Lead dioxide (J.T. Baker Chem. Co.)	2.8	–
Paraquinone dioxime (GMF, U.S. Rubber, Naugatuck Div.)	–	1.4
Gamma-glycidoxypropyltrimethoxy silane (Z-6040, Dow Corning)	0.35	0.35
Butadiene polymer with terminal hydroxyl groups (Poly-BDR-45M, Sinclair Petrochemicals)	1.7	1.8

*Based on total weight of Component A.
**Based on total weight of Component B.

The ingredients were added in various increments over a period of 3 hours and thoroughly milled to a homogeneous mix on a rubber-compounding mill.

Mixing is conducted at a temperature of about $200°F$ and the time of mixing is given only as an indication that a good dispersion of each ingredient is needed before proceeding to the next material. Both components were formulated in this manner.

The two well-mixed and homogeneous Components A and B were cut into approximately equal strips and then fed into a 3½ inch rubber extruder having a length-to-diameter ratio of 4 to 1 where the individual components were blended together. The material was extruded into ribbon form directly onto a release paper from which point it was easily applied by hand to multiple glazed units. The sealant formed a hermetic seal and had excellent weathering properties as determined by accelerated testing. The physical properties of the sealant material immediately after it was extruded were as follows: tack, 100 g/in^2; initial Shore A-2 durometer hardness, 7; consistency, 150 m-g determined at $250°F$; and cold flow, final thickness 5 mils.

The material was then permitted to cure for two weeks at room temperature after which time it was completely cured. After curing, the sealant had the following physical properties: tensile strength, 82 lb/in^2; percent elongation at break, 900%; tensile product, 7.38; tack, 76 g/in^2; initial Shore A-2 durometer hardness, 12; consistency, 610 m-g determined at $250°F$ and at 15 minutes in the Brabender Plastograph; and cold flow, final thickness of 40 mils.

Based on Mercapto-Terminated Polymers and Quick Hardening

Mercapto-terminated polymers that can be oxidized with oxidation agents such as PbO_2, MnO_2, ZnO_2, cumene hydroperoxide, etc., into rubber-like elastic materials are used widely as caulking and sealing materials.

Those using such compounds are desirous of using one-component systems as often as possible; i.e., storage-stable mixtures which contain the polymer and the oxidation agent in the same system and which polymerize after use at ambient temperature are referred to as one-component systems. Such systems which harden upon contact with the moisture in air have the disadvantage that their hardening takes several days and at times even several weeks, so that they cannot be used for applications where an early strain on the seal is required.

Two-component sealing compounds in which the part containing the polymer is mixed with the hardener-containing paste prior to use also require one and often even several days for hardening.

In many cases, such as in the installation of window panes in automobiles, in the production of insulating glass units or in the sealing of expansion joints in a building, it is necessary that the seal be capable of withstanding stresses as early as possible; this cannot be solved satisfactorily with such hardening systems.

L. Hockenberger, M. Soyka and G. Wilhelm; U.S. Patent 4,100,148; July 11, 1978; assigned to Rütgerswerke AG, Germany have, therefore, provided a storage-stable, one-component adhesive sealant which hardens very quickly and permits stress to be placed on the seal soon after it has been made. This is achieved by employing the oxidizing agent necessary for the oxidation of the mercapto groups in a microencapsulated state; that is to say, the oxidation agent is present in the form of small pellets having a diameter of about 400 to 1,600 μ, which are surrounded by an inert wall of a capsule.

This wall prevents any reaction between the oxidation agent and the polymer during storage of the sealant. When it is used, the wall of the capsule can be destroyed by a suitable device. This results in the contents of the capsule being mixed with the polymer surrounding the capsule, and the resulting mass can be moved to the place that is to be sealed. These mixtures are very reactive and harden in less than one hour.

This process can be applied to all liquid (at room temperature, e.g., 20° to 23°C) polymers or mixtures of monomers, oligomers and polymer compounds having terminal mercapto groups, such as polysulfide polymers, thioether polymers, etc. Typical compounds are known as Thiokol Liquid Polymers (The Thiokol Corp.).

Inorganic dioxides, peroxides, or other compounds of heavy metals with a high valency, or organic hydroperoxides can be employed as oxidation agents, e.g., lead dioxide, manganese dioxide, hydrogen peroxide, cumene hydroperoxide, and others can be employed. Latent hardeners, such as calcium peroxide, barium peroxide, zinc peroxide and others, can also be employed whenever an activator is mixed into the mass.

The oxidation agents are made into a paste, effectively in a softener (plasticizer) compatible with the polymer, and are then provided with a mechanically destroyable wall forming a capsule according to processes well known in the art.

The wall of the capsule must be insoluble in the polymer employed as well as insoluble in the softener, and it must withstand slight mechanical stresses without breaking.

Example: The polymer employed in the example is a polysulfide having an average structure $HS\text{---}(C_2H_4OCH_2OC_2H_4SS)_{\overline{23}}\text{---}C_2H_4OCH_2OC_2H_4SH$ with about 0.5% of crosslinking. It has an average molecular weight of 4,000 and a viscosity of about 350 to 450 poises at 27°C. This polymer is commercially available from the Thiokol Corporation. In the example all quantitative data refer to parts by weight or percent by weight.

A mixture consisting of 1,000 parts polysulfide polymer; 450 parts plasticizer (benzylbutylphthalate); 350 parts chalk; 150 parts clay; 300 parts titanium dioxide; 2 parts sulfur, and 40 parts thixotropic agent (Aerosil, a finely divided SiO_2) was mixed for 2 minutes at a speed of 10 rpm in a planetary-type mixer with 200 grams of microcapsules.

The capsules consisted of a paste of lead dioxide:plasticizer in 1:1 ratio surrounded by a wall of gelatin in a ratio of paste:wall material of 12:1. The diameter of the capsules was 800 μ.

During the mixing process and during the subsequent filling of the mass into cartridges, no capsules were destroyed.

During the storage time of 12 months at standard climate 23/50, no increase of the viscosity of the sealant was observed, i.e., the sealant is storage-stable for at least 12 months.

A part of the sealant was sprayed out immediately after production with the help of a spray gun, which had a conveying worm and a chopper behind the spray nozzle. At the same time, the capsules were destroyed. The original white sealant was colored a uniform brown. After two hours, it had hardened into a nonadhesive, rubber-like elastic material. An analogous dispensing process was carried out after the sealant had been stored for 12 months. In this case too, the dispensed mixture hardened after two hours.

Nonaqueous Rubber Sealant for Use on Wet or Dry Substrate

The automotive repair industry has always used the same variety of sealant compositions which are used by the automotive manufacturing industry. Vehicles are often brought in directly from the rain for immediate service by the customer; or the vehicle may be temporarily parked in an outside lot and be exposed to rain before being brought in for a scheduled repair; or a mobile repair unit may have to make the repair in rainy weather at a remote location. In such situations, the area to which sealant has to be applied is wet and the sealant composition cannot be applied until the area is dried or is permitted to air dry, as satisfactory bond cannot otherwise be achieved.

Similarly, when the repair involves the detection and sealing of a water leak in the vehicle, it is often necessary to spray the suspected joint or seam with water to locate the leak. The area then has to be dried before sealant can be applied thereto. After the sealant has set, a water spray is again used to determine whether the leak has been stopped. If the leak persists, the area again has to be dried, sealant applied, a test water spray applied, and the process repeated until the leak is sealed.

It is a primary object of the formulation by *J.J. Moskal; U.S. Patent 4,101,483; July 18, 1978; assigned to Premier Industrial Corporation* to provide a nonaqueous rubber sealant composition which can be used on either wet or dry substrates with complete and satisfactory bond.

The sealant composition is a nonaqueous homogeneous mixture of the following constituents which are present in the composition as a percentage of the total weight of the listed composition in the range indicated in the table below.

Constituent	Percent by Weight
(A) Thermoplastic rubber base	5–20
(B) Liquid plasticizer	0.5–12
(C) Tackifying resin	1–20
(D) Antioxidant	0.1–3
(E) Cationic surfactant	0.25–0.70
(F) Inert filler	0–48
(G) Bodying agent	0–5
(H) Hydrocarbon solvent	20–70

The thermoplastic rubber base (A) is preferably one or more of the styrene block copolymers, such as styrene-butadiene-styrene or styrene-isoprene-styrene, but can be one of the other known rubbers such as butyl rubber or the like. It is utilized in particle form for better admixture and solution. Mixtures of different rubbers can be utilized for the base to achieve desired properties in the composition.

The plasticizer (B) is preferably an oxidation-resistant balsamic resin or could be one of the heavy aromatic hydrocarbons derived from petroleum and suitable for the purpose.

The tackifier (C) is preferably a thermoplastic ester derived from pentaerythritol or a polyterpene resin.

The antioxidant (D) for the rubber will ordinarily be one of the hindered phenols known to the art for this purpose.

The cationic surfactant (E) is an imidazoline derivative, preferably 1-hydroxy-ethyl-2-alkyl-imidazoline.

The filler (F) can be any suitable inert inorganic constituent in powdered form which may be utilized, if desired, to add mass to the composition. Its use in the composition is optional, but is often desirable in attaining desired consistencies or achieving cost reduction of product, as long as its proportion in the composition does not reduce the performance characteristics of the composition below a quality level which will be satisfactory for the particular purpose intended. A commonly used filler is calcium carbonate, but others such as magnesium carbonate could also be used.

Whether or not a bodying agent (G) should be introduced into the composition will depend upon the consistency which is desired for the composition; the extent to which such desired consistency is achieved by other constituents of the composition; the manner and environment in which the composition is to be applied to the substrate; and other factors. A bodying agent which is suitable and commonly used for this purpose is coatings grade asbestos. Other bodying agents, such as fumed silica with thixotropic tendencies, could also be used.

The hydrocarbon solvent (H) is any one or a suitable mixture of aliphatic or aromatic hydrocarbon liquids, such as toluol, naphtha or kerosene.

In addition to the foregoing constituents, it is frequently desirable for the sake of appearance, particularly in the automotive field, that pigment or coloring be added to the composition to achieve a particular color. For example, carbon black or any of the commonly available coloring agents may be added to the composition in such minor proportions as required to achieve the color intensity desired. Pigment or coloring is designated (I) in the table below.

All of the mandatory constituents of the composition, as well as those optional constituents selected for inclusion, are thoroughly mixed for a sufficient period of time to obtain complete solution of the soluble constituents and attain a homogeneous mixture of the desired consistency or viscosity.

The resulting sealant compositions can broadly be categorized as light-bodied or heavy-bodied. The light-bodied sealant compositions flow readily when brushed or flowed onto the joint or seam to be sealed and readily penetrate or seep into small openings, such as cracks and crevices, which could be a source of leakage through a substrate joint, for example, between a windshield and the automobile pinchwell or molding or even between a windshield and its original bed of sealant. These light-bodied sealant compositions have a Brookfield viscosity ranging from above 100 to about 5,000 cp.

The heavy-bodied compositions have a putty-like consistency and are considerably more viscous than the light-bodied sealants. They have a Brookfield viscosity ranging from above 75 thousand to about one million cp. They are generally applied by means of high-pressure dispensing devices to the substrate joint and tend to stay in situ rather than flow readily. After they set, these heavy-bodied compositions still retain an excellent degree of flexibility and resiliency; they have a Shore A durometer hardness of no more than 50, ranging down to about 30.

Six specific examples of representative sealant compositions are set forth in the table below.

Constituent, % by wt	. . . . Examples					
	1	2	3	4	5	6
(A)	8.5	10.0	10.5	12.6	9.3	164
(B)	6.6	6.4	3.5	1.7	3.6	6.3
(C)	7.0	2.0	12.0	12.7	17.5	14.0
(D)	0.1	0.1	0.1	0.1	0.1	1.5
(E)	0.6	0.5	0.6	0.6	0.3	0.25
(F)	37.8	40.5	44.1	46.7	–	–
(G)	3.8	5.0	3.4	3.6	–	2.7
(H)	35.6	30.8	23.7	22.0	68.7	58.85
(I)	–	4.7	2.1	–	0.5	–
Brookfield viscosity, cp	~100,000	~150,000	~800,00	~1,000,000	~135	~400
Shore A durometer	~45	~48	~50	~50	–	–

Examples 1 through 4 can be categorized as heavy-bodied. Examples 3 and 4 are of a consistency and character particularly adapated for use where an entire windshield or other fixed glass insert has to be replaced and reset in a vehicle. Examples 1 and 2 are of a consistency and character particularly adapted for use in rehabilitating an existing joint where the sealant has suffered cracking, embrittlement or other general deterioration.

Examples 5 and 6 can be categorized as extremely light-bodied. They can be flowed on easily for rapid penetration in localized areas of the joint, such as pin holes and the like, and can otherwise be utilized where relatively thin layers of flowable sealant are needed or desirable.

Using PTFE Fibers

Bearings made of polytetrafluoroethylene (PTFE) fibers have been made in the past. However, they are not suitable for sealing pumps and valves since bushes of this kind show in fact good sliding properties, but do not have any sealing effect.

It has, therefore, been the objective of *G. Beyer and G. Wölfel; U.S. Patent 4,157,328; June 5, 1979; assigned to Hoechst AG, Germany* to develop a packing which shows, besides good sliding properties, an especially good sealing capacity, a high resistance to pressure as well as an excellent thermal conductivity at the same time.

This objective was achieved by developing a sealing material consisting of PTFE short cut fibers which have been intimately mixed with graphite or with graphite plus oil or with graphite plus finely divided PTFE, or with graphite plus finely divided PTFE plus oil. Graphite having a purity of more than 95%, especially more than 99%, was used.

The sealing material has preferably been pressed into tablets. The PTFE short cut fibers have advantageously a staple length of from 1 to 20 mm, a length between 5 and 15 mm being particularly favorable. Good results are also obtained with lengths between 3 and 7 mm. The PTFE fiber portion is at least 50% by weight, calculated on the finished sealing material; the graphite portion is at least 10% by weight and at most 30% by weight; the portion of the finely divided PTFE is at least 5% by weight and at most 25% by weight. Lubricants may be added to the sealing material, for example, silicone oil, paraffin oil, mineral oils or polyhydric alcohols.

According to another preferred embodiment the sealing material consists of PTFE fibers which are prepared by machining, such as planing, milling, rasping, or turning. The PTFE block (residual block, block waste) used for machining is preferably sintered; it may show a round or angular cross section.

The PTFE fibers produced by machining are intimately mixed with graphite or with graphite plus oil or with graphite plus finely divided PTFE, or with graphite plus finely divided PTFE plus oil, or with graphite plus PTFE dispersion plus oil or with a graphite-PTFE dispersion plus oil.

The following sealant composition was made, using PTFE fibers having a length of 8 mm and obtained by machining, and pressed into tablets.

	Percent by Weight
Polytetrafluoroethylene fiber	70.0
Finely divided polytetrafluoroethylene	7.0
Graphite powder (purity 99%)	16.0
Propane-1,2,3-triol	3.5
Silicone oil	3.5

For determining the heat transmission, the temperature in the middle of the packing and in the outer case of the packing box was measured. The leakage was also measured. In order to carry out the test use was made of a centrifugal pump having the following technical data: shaft diameter, 32 mm; pressure, 0.2 to 4.0 bars; circumferential speed of the shaft, 2.2 to 4.4 m/sec; inner diameter of the packing box, 48 mm; medium pumped over, pure water.

The pump case and the shaft were made of chromium-plated special steel. The shaft was precisely centered. In order to apply full pressure onto the packing, the rear pressure reducing lamellas have been removed. The packing space of the pump was equipped with the required amount of packing tablets. The packing box gland was tightened by means of a dynamometric key to 15 cm x kp. The pressure of the pumping medium was 3.5 bars. The temperatures in the middle of the packing and at the case of the packing box were continuously measured by means of a thermoelement.

In the following table, the temperatures and leakages of the packing according to this process as compared with the conventional types of packing have been shown. These were found after reaching steady operation conditions. The conditions were identical for all tested packings in the pump.

| Type of Packing | . . . . . Temperature, °C . . . . . | | Leakage ml/hr |
	in the Packing	of the Case of Packing Box	
Commercial PTFE packing cord* with PTFE dispersion plus silicone oil	62	45	20
Commercial PTFE/asbestos (50/50%) kneading packing	70	50	0
Packing of the process	32	29	0

*(0–10 mm)

ADHESIVES FOR FIBER GLASS

Sandable Polyurethane Composition

Although polyester impregnated fiber glass vehicular bodies and boats have been made for some time by cementing the parts together, into the desired assembly, the resulting assembly frequently had unsightly joints or seam lines which were objectionable. For instance, the seam where two pieces of polyester sheet were bonded would shrink and the seam line would be readily noticeable even when painted. Therefore, it has been desirable to provide an adhesive that could be sanded to give a smooth finish and be painted without the seam line appearing through the paint coating.

A.J. Kieft, R.L. Cline and T.G. Rabito; U.S. Patent 4,097,442; June 27, 1978; assigned to The Goodyear Tire & Rubber Company have formulated an adhesive which will adhere to polyester-impregnated fiber glass parts and which can be sanded to give a joint which does not show the seam after coating.

The reaction mixture comprises: (a) about 1 mol of a reactive hydrogen containing polymeric material having a MW of about 500 to 3,000 selected from the class consisting of polyester polyols and polyether polyols; (b) about 2.5 to 7 mols and preferably 3.5 to 5 mols of a blend of organic polyisocyanates consisting of toluene diisocyanate (TDI), methane di(phenyl isocyanate) (MDI) and quasiprepolymers of MDI; (c) a filler composed of at least 50 to about 200 parts of silica composed of platelets, 50 to 200 parts of fine spherical particles of glass, a tertiary amine catalyst and a plasticizer; and (d) sufficient monomeric nitrogen containing tetraol of 200 to 1,000 MW to be equivalent to 50% of the excess polyisocyanate relative to the reactive hydrogen containing polymeric material but preferably 0.5 to 1.0 mol for each 1 to 2 mols of excess of the polyisocyanate.

Example: The adhesive composition was made by mixing composition A and composition B. Composition A was made by placing 73.5 parts of polypropylene ether glycol of 1,000 MW in an inert atmosphere jacketable reactor together with 16 parts of dioctylphthalate, 100 parts of silica essentially of the platelet type that passed 200 mesh Taylor sieve but stopped on 325 mesh sieve and 100 parts of fine spherical particles of glass that passed a 200 mesh Taylor sieve but stopped on a 325 mesh sieve and then the mixture was dehydrated at elevated temperature under a vacuum. To the dehydrated mixture 125.5 parts of a mixture of 48 parts of Mondur PF (Mobay Chemical Co.), a quasiprepolymer of MDI and dipropylene glycol, 35 parts of 80/20 isomeric toluene diisocyanate and 42.5 parts of methane di(phenyl isocyanate) referred to herein as MDI, was added to the reactor and reacted to form a prepolymer.

Composition B was made by mixing 85 parts of silica of platelet type indicated above in regard to composition A, 16 parts of fine spherical particles of glass of the type indicated above in regard to composition A, 50 parts of a polyol formed by condensing sufficient propylene oxide on an alkylene diamine to give a polyol having a MW of 250 to 1,000, 50 parts polyoxyalkylene tetraol of 500 MW and 0.2 part of a triethylene diamine as a catalyst.

Composition A (415.0 parts) and composition B (201.2 parts) were mixed immediately before use to form an isocyanate class adhesive. This isocyanate class adhesive was used to form a fiber glass polyester laminate that could be sanded and painted without a bond line showing through the paint.

The lamination operation can be performed in any of the well-known manners using the isocyanate adhesive of this process. Where butt joints are to be formed, it is preferred to grind or sand away the ends of specimens to be butt spliced to give a 45° angle to each end of the specimen. Then the specimens are placed in a bonding buck or clamp, preferably with 0.090 inch spacing between the ends to be bonded. The space between the ends of polyester fiber glass mat is filled with the adhesive and is allowed to cure at room or at elevated temperature. The excess adhesive is sanded away after the adhesive is fully cured to give a smooth finish to the two joined specimens. When the sanded smooth butt joined specimens are spray painted, the paint coat cures with no evidence of seam line being apparent through the paint coat as is experienced with the usual commercial adhesives.

ALL-PURPOSE ADHESIVES

Photocurable Compositions Containing Group Va Onium Salts

J.V. Crivello; U.S. Patent 4,069,055; January 17, 1978; assigned to General Electric Company has found that radiation-sensitive aromatic onium salts of Group Va elements, such as

$$\left[\left(\bigcirc\right)_3 \left(\bigcirc - \overset{O}{\underset{\|}{C}} CH_2\right) P\right]^+ \left[BF_4\right]^-$$

can be incorporated in epoxy resins to provide one-package radiation-curable compositions which do not require a stabilizer to minimize cure at ambient temperatures during the shelf period, and are free of all of the disadvantages of the aromatic diazonium salt compositions.

Included by the aromatic Group Va onium salts which can be employed are onium compounds of the formula,

$$\left[(R)_a(R^1)_b(R^2)_c\, X \right]_d^{+} \left[MQ_e \right]^{-(e-f)}$$

where R is a monovalent aromatic organic radical selected from carbocyclic radicals and heterocyclic radicals, R^1 is a monovalent organic aliphatic radical selected from alkyl, alkoxy, cycloalkyl and substituted derivatives thereof, R^2 is a polyvalent organic radical forming an aromatic heterocyclic or fused ring structure with X; X is a Group Va element selected from N, P, As, Sb and Bi; M is a metal or metalloid; Q is a halogen radical; a is a whole number equal to 0 to 4 inclusive; b is a whole number equal to 0 to 2 inclusive; c is a whole number equal to 0 to 2 inclusive; and the sum of a + b + c is a value equal to 4 or the valence of X, d = e–f, f = valence of M and is an integer equal to from 2 to 7 inclusive; e>f and is an integer having a value up to 8.

Complex anions included by MQ_e are, for example, BF_4^-, PF_6^-, AsF_6^-, SbF_6^-, $FeCl_4^=$, and $SnCl_6^=$, etc.

These curable compositions can be made by blending the epoxy resin with an effective amount of the Group Va onium salt. The resulting curable composition which can be in the form of a varnish having a viscosity of from 1 to 100,000 cp at 25°C or a free-flowing powder can be applied to a variety of substrates by conventional means and cured to the tack-free state within 1 second or less to 10 minutes or more.

Depending upon the compatability of the onium salt with the epoxy resin, the Group Va onium salt can be dissolved or dispersed therein along with an organic solvent such as nitromethane, acetonitrile, etc., prior to its incorporation. In instances where the epoxy resin is a solid, incorporation of the onium salt can be achieved by dry milling or by melt mixing the resin to effect incorporation of the onium salt. The onion salt also can be generated in situ in the presence of the epoxy resin if desired.

The curable compositions may contain inactive ingredients such as inorganic fillers, dyes, pigments, extenders, viscosity control agents, process aids, UV-screens, etc., in amounts of up to 100 parts filler per 100 of epoxy resin. The curable compositions can be applied to such substrates as metal, rubber, plastic, molded parts or films, paper, wood, glass cloth, concrete, ceramic, etc.

Some of the applications in which the curable compositions of this process can be used are, for example, protective, decorative and insulating coatings, potting compounds, printing inks, sealants, adhesives, photoresists, wire insulation, textile coatings, laminates, impregnated tapes, printing plates, etc.

Cure of the curable composition can be achieved by activating the onium salt to provide the release of the Lewis acid catalyst. Activation of the onium salt can be achieved by heating the composition at a temperature in the range of from 150° to 250°C. Preferably cure can be achieved by exposing the curable composition to radiant energy such as electron beam or ultraviolet light. Electron beam cure can be effected at an accelerator voltage of from about 100 to 1,000 kV. Cure of the compositions is preferably achieved by the use of UV irradiation having a wavelength of from 1849 to 4000 A and an intensity of at least 5,000 to 80,000 microwatts per cm^2. Several examples of such compositions used as adhesives are given for illustration.

Example 1: A curable composition was prepared of 97 parts of a mixture of a 60% epoxy novolak having an epoxy equivalent weight of 206 and 40% 4-vinylcyclohexene dioxide with 3 parts N-phenacylacridinium fluoroborate. The mixture was used to impregnate two 6 x 6 inch glass cloth squares, which were cut and stacked together. The resulting laminate was cured using a GE H3T7 lamp. Cure time was 1 minute exposure on each side. A completely dry rigid laminate was obtained which was integrally bonded together. The laminate could be used for the manufacture of circuit boards.

Example 2: There was added 18.35 grams (0.1 mol) 48% aqueous fluoroboric acid to 24.6 grams (0.2 mol) 2,6-lutidine-N-oxide dissolved in 100 ml absolute ethanol. A very pale yellow crystalline precipitate was formed on addition and after standing for 30 minutes was filtered and thoroughly washed with diethyl ether. Recrystallization from ethanol gave the pure salt.

To a 500 ml flask were added 26.5 grams (0.0793 mol) of the amine oxide acid salt in 64.3 ml (1.19 mols) of nitromethane. The solution was stirred at 30° to 40°C while 15.6 grams (0.159 mol) 1,2-epoxycyclohexane was added dropwise. After stirring for 1 hour, the reaction mixture was cooled to room temperature and poured into 500 ml diethyl ether. The white crystalline product was filtered and washed thoroughly with ether. After recrystallization from absolute ethanol, an 81% yield of product, MP 122° to 126°C, was obtained. Based on method of preparation and elemental analysis for $C_{13}H_{20}NO_2BF_4$, calculated: C, 50.5%; H, 6.47%; N, 4.53%; found: C, 50.7%; H, 6.51%; N, 4.50%; the product was a compound having the formula,

There was added 0.2 part of the above salt to 10 parts of an epoxidized buta-diene resin dissolved in 2 grams 4-vinylcyclohexene dioxide. After mixing the reagents thoroughly, the mixture was applied to a $\frac{1}{16}$ inch thick glass plate as a 1 mil coating. Another plate of glass was placed on top of the first and this assembly exposed to a GE H3T7 medium pressure mercury arc lamp having an intensity of 200 watts/in^2 at a distance of 3 inches. The total time of exposure was 1 minute. There was obtained a glass laminate.

Based on the characteristics of the resulting laminate, those skilled in the art would know that a similar procedure could be used to make a shatterproof auto-mobile windshield.

Chlorosulfonated Polyethylene

P.C. Briggs, Jr. and L.C. Muschiatti; U.S. Patent 4,112,013; September 5, 1978; assigned to E.I. Du Pont de Nemours and Company have provided an adhesive composition which can be formulated either as a two-part system or as a system utilizing a primer. In either case, the critical components of the adhesive compo-sition are a solution of a sulfur-bearing composition selected from chlorosulfo-nated polyethylene and a mixture of sulfonyl chloride with chlorinated poly-ethylene in a polymerizable vinyl monomer or a mixture of monomers and a polymerization catalyst.

The sulfur-bearing composition should contain about 25 to 70 weight percent chlorine and about 3 to 160 mmol sulfonyl chloride moiety per 100 grams of polymer and the polyethylene from which the chlorosulfonated or chlorinated polyethylene is prepared should have a melt index of about 4 to 500. The solu-tion can have a Brookfield viscosity of up to about 1 million.

The term "polymerization catalyst" as used here means at least one of the fol-lowing: (a) a free radical generator; (b) an initiator; (c) a promoter; and (d) an accelerator.

An initiator is a tertiary amine, for example, N,N-dimethylaniline, N,N-dimethyl-toluidine, N,N-diethylaniline, N,N-diisopropyl (p-toluidine) or a guanidine.

A promoter is an organic salt of a transition metal, for example, cobalt, nickel, manganese, or iron naphthenate, copper octoate, iron hexoate, or iron propio-nate.

An accelerator is an aldehyde-amine condensation product. In general, the con-densation products of aliphatic aldehydes with aliphatic or aromatic amines are useful.

A free radical generator can be an organic peroxide, an organic hydroperoxide, a perester, or a peracid. Suitable polymerizable vinyl monomers for the formula-tion include acrylic monomers and mixtures of monomers.

The chlorosulfonated polyethylene suitable in the formulation can be prepared by reaction of linear or branched polyethylene and sulfuryl chloride, or sulfur dioxide and chlorine. Chlorosulfonated polyethylene is also available commercially, for example, Hypalon.

Alternatively, sulfonyl chloride and chlorinated polyethylenes of suitable molecular weight can be used. The sulfonyl chlorides can be mono- or polyfunctional and can be C_{1-12} alkyl sulfonyl chlorides, such as methane or butane sulfonyl chloride, C_{6-24} aromatic sulfonyl chlorides such as benzene or toluene sulfonyl chloride. Some sulfonyl chlorides containing hetero atoms have also been found to work, such as diphenylether-4,4'-disulfonyl chloride.

The relative proportions of chlorosulfonated polyethylene and polymerizable vinyl monomer can vary within a rather broad range. In the case of acrylic polymers, the practical range is about from 25 to 2,000 parts by weight of the monomer per 100 parts of chlorosulfonated or chlorinated polyethylene. The preferred range is 50 to 500 parts by weight of the monomer per 100 parts of polymer, whether chlorosulfonated polyethylene or a mixture of sulfonyl chloride and chlorinated polyethylene is used.

These adhesive compositions require a polymerization catalyst to cause hardening of the composition within a practical time. The following catalysts or catalyst combinations are particularly suitable: (a) an initiator plus a promoter; (b) an accelerator; (c) a free radical generator and an initiator plus a promoter; and (d) a free radical generator and an accelerator. The preferred free radical generators are organic peroxides and hydroperoxides.

An amine initiator is usually used together with a transition metal compound as a promoter. The preferred proportion is about 2:1 by weight of the initiator to the promoter. The preferred initiator is N,N-dimethylaniline, while the preferred promoter is cobalt naphthenate.

The following concentrations of polymerization catalysts, as weight percent of the solution of polymer in monomer have been found to be practical: free radical generator up to 10%, 0.05 to 3% being preferred; accelerator up to 15%, 0.01 to 1.5% being preferred; initiator up to 5%, 0.01 to 1.5% being preferred; and promoter up to 5%, 0.01 to 0.75% being preferred.

The adhesive compositions can be formulated as a two-part system, wherein one part is a solution of polymer in a polymerizable vinyl monomer. The other part is the polymerization catalyst. Alternatively, it can be formulated as a primer system in which the polymerization catalyst is the primer and the adhesive solution is the polymer mixture. Usually, the catalyst will be either an accelerator or an initiator plus promoter. The curing rates can be increased by adding a free radical generator to either composition. In general, at any level of the polymerization catalyst, the rate of bond formation of a system employing an initiator plus promoter is lower than the rate of bond formation of a system employing only an accelerator.

The two-part system gives very strong bonds of 2,500 to 3,500 psi in shear. The two-part system using only an accelerator has a very short pot life. If an initiator and a promoter are used instead of an accelerator, the system's pot life can be extended. In practice, one or both surfaces to be joined are coated with the adhesive composition obtained by mixing both parts, and the surfaces are then placed in contact with each other.

In the primer system, a primer is first applied to one or both surfaces to be joined; then a solution of polymer in a vinyl monomer is applied to at least one of the surfaces. The solution can optionally contain a free radical generator. The primer is an accelerator, as defined above.

The primer system is operationally more convenient than the two-part system, and it gives a good bond strength of about 2,500 to 3,500 psi in shear when cured. Furthermore, it gives more rapid development of load-bearing strength.

The setting times for the two-part system and for the primer system will vary somewhat, depending on the nature of the catalyst, but usually will be about 5 to 10 minutes for the former and 0.05 to 5 minutes for the latter.

These adhesives offer several advantages. Thus, they are used at room temperature, no heat being required either for applying the compositions to the substrates or for curing. They can be used on porous surfaces, unlike those prior art adhesives which require the absence of air and thus cannot be used on surfaces containing air in their pores. The bonds containing elastomeric polymers such as chlorosulfonated polyethylene are flexible. The compositions do not require a careful surface preparation but can be used, for example, on bonderized or oily steel.

Other possible substrates which can be bonded by means of the compositions of this process include ordinary steel, etched aluminum, copper, brass, polar polymeric materials, (i.e., those having various functional groups, e.g., polyesters, polyamides, polyurethanes, polyvinyl chloride, etc.), wood, prepainted surfaces, glass and paper. In the following examples all parts, proportions, and percentages are by weight unless otherwise indicated.

Example 1: To a mixture of acrylic monomers comprising 85 grams of methyl methacrylate (containing 50 to 90 ppm of hydroquinone inhibitor), 15 grams of glacial methacrylic acid (containing 250 ppm of 4-methoxyphenol), and 2 grams of ethylene glycol dimethacrylate, there was added 100 grams of chlorosulfonated polyethylene, made from branched polyethylene having a melt index of 100 and containing 43% chlorine and 34 mmol sulfonyl chloride/100 grams of polymer. The mixture was rolled in a jar at room temperature until solution of the polymer was complete (24 to 48 hours).

An adhesive composition was prepared by stirring into 50 grams of the above solution, in turn, 1.5 grams of cumene hydroperoxide and 0.5 gram of N,N-dimethylaniline.

Lap shear bonds were prepared by pressing a small amount of the adhesive composition between 1 x 3 x 0.064 inch grit blasted and perchloroethylene vapor degreased hot rolled steel coupons in a mold so that a glue line of 0.010 x 1 x 1 inch was obtained. The specimens were stored for 18 hours in a nitrogen atmosphere (optional) and 14 days in air at room temperature. The specimens were tested in shear at a separation rate of 0.5 inch/minute and failed adhesively at 2,850 psi (ASTM D-1002-64).

Aluminum T peel specimens were prepared by pressing the adhesive composition between 1 x 10 x 0.018 inch etched aluminum strips so that a glue line thickness of 0.005 to 0.010 inch was obtained. The samples were stored as described above and tested in 180° peel at a separation speed of 10 inches per minute (ASTM D-1876-61T). The average peel strength obtained was 35 pounds per linear inch with mixed adhesive and cohesive failure.

Example 2: A second, shelf-stable adhesive composition was prepared by stirring 0.130 gram of cumene hydroperoxide into 65.3 grams of the chlorosulfonated polyethylene/acrylic monomers solution described in Example 1. Lap shear bonds were prepared by applying to steel coupons (treated as described in Example 1) an accelerator which was a mixture of butyraldehyde and aniline condensation products (known as Du Pont Accelerator 808). The accelerator was applied with a cotton swab and then wiped to a thin film with a piece of cloth. The adhesive was pressed between the treated steel coupons to give a layer 0.005 to 0.010 inch thick. The following rate of shear strength development was observed:

Time (min)	Lap Shear Strength (psi)
3	1,450
10	2,030
30	2,050
60	2,100

Modified with Certain Phenolic Resins

Nitrile rubber cement is a good adhesive for polar substances like polyvinyl chloride because it has strong polarity itself and its adhesiveness is improved by blending with phenolic resins.

A cement which is blended from 50 to 100 parts of phenolic resin and 100 parts of nitrile rubber is used for adhesion of flexible polyvinyl chloride leathers, flexible polyvinyl chloride films, flexible polyvinyl sheets, cotton cloth, leathers, wood and so forth as an excellent cement for general purposes, and a nitrile rubber cement which is blended with more than 100 parts of phenolic resin is also used as an excellent metal structural adhesive for aircraft, brake linings and shoes, and the like.

The biggest problem of the adhesive used for adhesion of flexible polyvinyl chloride leathers is the staining of polyvinyl chloride by the cement.

Y. Kako, T. Kikuga and A. Toko; U.S. Patent 4,112,160; September 5, 1978; assigned to Sumitomo Durez Company, Ltd., Japan have found an improved rubber cement which does not stain polyvinyl chloride, has excellent flexibility, bonding strength, and solvent resistance, and which can be heat cured.

It can be made by mixing the nitrile rubber with a phenolic resin produced by: (a) condensing a bifunctional phenol and an aldehyde in a molar ratio of about 0.5 to 3 mols of aldehyde per mol of bifunctional phenol under alkaline conditions, and (b) reacting the resulting condensation product with a polyfunctional phenol having at least 3 reactive positions on the phenyl nucleus in a molar ratio of about 0.05 to 2.5 mols of polyfunctional phenol per mol of bifunctional phenol in the condensation product in the presence of a sufficient amount of acid to provide acid catalysis of the condensation of the aldehyde and the phenol, to produce a phenolic resin having a number average molecular weight of about 700 to 1,900 and a melting point of about 80° to 160°C.

Example: 1 mol of para-octyl phenol, 1.8 mols of acetaldehyde, 150 weight percent of xylene to para-octyl phenol, and 7 weight percent of methanol to para-octyl phenol were charged into a flask, and the admixture was kept at 45°C, then 0.25 mol of 50% aqueous caustic soda solution was added very slowly drop by drop which took 1 hour until the pH of the mixture was 12.0.

Then also keeping the temperature at 45°C, there followed a methylol forming reaction for 8 hours, next the admixture was cooled and neutralized by adding 0.125 mol of 25% aqueous sulfuric acid until pH of 3.5 is indicated in the aqueous layer. To eliminate the production of neutralized salts therein, it was washed twice with water at 50° to 60°C. Then, 0.8 mol of bisphenol A and next 0.014 mol of 30% hydrochloric acid were added.

The pH of the admixture was 1.4, and it was heated 30 minutes at 100°C to continue the novolak-forming reaction, then dehydration was carried out under vacuum keeping the temperature of the admixture at 150° to 160°C. A novolak-type phenolic resin was obtained which showed a MP of 90°C and a pH of 3.4.

The resin was used in the following formulation and the comprised adhesive was tested against various adherents as shown in the table.

	Noncurable Cement (parts)
Hycar 1042	100
Phenolic resin (Example)	50
Methyl ethyl ketone	300
Toluene	150

Bonding strength tests of various kinds of substrates which were bonded by cements prepared by blending 50 parts of the abovementioned resin with 100 parts of Hycar 1042 were determined. The results in the table show such excellent bonding properties that it is evident that the cement is usable for general purposes. Measured values were obtained at 20°C, and 2 days after bonding.

| | Bonding Strength |
Substrates	(kg/in)
Canvas - canvas	10.2
Canvas - wood	11.5
Nitrile rubber - nitrile rubber	17.0
Nitrile rubber - sheet iron	2.5

Catalyst Activator as a Separate Ingredient

It is readily apparent that it would be highly desirable to obtain an adhesive which, in a single formulation, bonded a wide diversity of materials, with high strength joints, both with respect to shear strength and peel strength, which requires no solvent or other volatile ingredient, which attains bond strength very rapidly but has a long shelf life and pot life before the joint surfaces are mated, which is effective with minimal or no surface preparation, and which requires no elaborate or expensive equipment for use. To attain all these features and to attain a bond not overly sensitive to heat as well is more desirable still.

L.E. Wolinski and P.D. Berezuk; U.S. Patent 4,126,504; November 21, 1978; assigned to Pratt & Lambert, Inc. have attained these objects by dissolving a non-reactive elastomeric polymer in an acrylic monomer and a copolymerizable carboxylic acid-group-containing monomer and combining the solution thus formed with a dormant free radical polymerization catalyst to form an adhesive, and separately providing an activator for the dormant catalyst.

In use, the activator is applied to at least one of the surfaces to be joined, the adhesive is applied to at least one of the surfaces to be joined and the surfaces are mated and held in contact until the adhesive bond is formed. By such a technique, all the foregoing objects may be attained. Heat susceptibility of the adhesive bond is enhanced by the inclusion in the adhesive of a nonreactive epoxy resin. Any elastomeric polymer may be used, as long as it is nonreactive. Both natural and synthetic rubbers are included, as are grafted rubbers. The acrylic monomer includes acrylic and methacrylic esters and amides. Acrylic acid and methacrylic acid are the preferred acid monomers for use in the formulation. Catalysts of interest are those which have a half-life at 85°C of at least half an hour. A preferred catalyst is benzoyl peroxide.

The foregoing components are formulated into an adhesive by dissolving the elastomer and the catalyst into a mixture of the acrylic monomer component and the acid monomer component. In the solution thus formed, the elastomer component will comprise about 10 to 70, preferably about 10 to 30 weight percent; the catalyst will comprise about 0.1 to 5.0, preferably about 2 to 4 weight percent, and the balance will be the acrylic monomer/acid monomer blend. These two monomer components should be proportioned in such fashion that the acid monomer comprises at least about 5.0 weight percent of the total formulation, preferably about 5 to 35 weight percent. Since the acid monomer will generally be relatively expensive, it is ordinarily not preferred to use more than about 20.0 weight percent, but there is no reason greater amounts cannot be employed, up to as much as about 67% of the adhesive, or even more if desired. The acrylic monomer is preferably 30 to 50 weight percent.

The basic adhesive composition thus formulated is storage stable for considerable periods of time, but where prolonged shelf life is desired, it is preferred to add to the basic formulation a minor amount, usually on the order of about 0.1 to 1.0 of a polymerization inhibitor such as hydroquinone, a hindered phenol, acetylacetonate, or the like. In such fashion stable lives of up to as long as a year can be attained.

Since the adhesive is stable and noncuring in the absence of an activator or initiator for the free radical catalyst system, a separate formulation of an appropriate activator is necessary for use of the adhesive. Desirably the activator component will be a tertiary amine, such as N,N-dimethyl aniline, N,N-diethyl aniline, N,N-dimethyl-p-toluidine, and the like. The activator may be supplemented by accelerators which function to increase the reaction rate of the adhesive cure. Such accelerators are most conveniently a source of a heavy metal, such as copper, iron, cobalt, manganese, lead, and the like, most desirably as an organo-metallic compound or salt wherein the heavy metal is oxidizable, i.e., not in its highest oxidation state.

The activator and (optional) accelerator selected for use will be formulated into a composition suitable for application to a substrate to be bonded with the adhesive. Activation of the free radical catalyst occurs upon contact of the adhesive with the activator, and by the use of such a system, no mixing or compounding of ingredients is required at the time and place of use.

Generally, it will be preferred to formulate the activator (and accelerator) as a solution or dispersion in a volatile liquid carrier medium to facilitate application to the substrate. The solution or dispersion of the bonding activator or initiator in the solvent then can be applied, as by brushing, spraying, or the like, upon at least one surface to be bonded, and the solvent allowed to evaporate leaving a deposit of bonding activator (and, optionally, accelerator) on the surface. Because of the extremely rapid cure speed attainable from such a technique, it is usually preferable to apply the activator to only one of each pair of mating surfaces to be bonded, affording a slightly longer opportunity to manipulate and adjust the parts.

It has been found preferable to use an activator/accelerator concentration in the solvent of between about 0.01 and 10% by weight, and preferably between about 0.2 and 5.0% by weight.

The most highly preferred method of applying the bonding activator/accelerator dissolved in a solvent to the surface is from an aerosol container.

Bond strengths attained by these cured adhesives are exceptional, both in shear strength and peel strength, on a wide diversity of substrates. In many cases it has been found possible to attain bonds having shear strengths greater than the coherent shear strength of the substrate. This is particularly true in the case of wood, glass, natural and synthetic rubber, and polyvinyl chloride. In some circumstances such results have been attained with other substrates, such as mild steel and the like. Bond shear strengths as high as about 4,000 psi and peel strengths of as high as about 90 pli have been attained, and in some cases have been even greater.

Among the substrates specifically evaluated to date, there may be mentioned steel, both clean and oily, as received from a mill, neither cleaned nor sanded, aluminum, wood, glass, polyvinyl chloride, nylon, polystyrene, glass-reinforced polyester, polyester films, such as Mylar, surface activated polyolefins and polytetrafluoroethylene, such as Teflon, ABS, natural rubber, SBR rubber, neoprene rubber, hot galvanized steel, and electrogalvanized steel. Bond strengths in shear qualify as very high with all these materials except hot galvanized steel although results attained with this substrate are better than is normally attained in the art.

Peel strength of the adhesive bonds is also exceptionally high, and ranges from 25 to 90 pounds per linear inch, at a rate of 0.1 inch per minute.

Example: 33 grams of a high acrylonitrile/butadiene rubber, Hycar 1431 (B.F. Goodrich Co.), was dissolved in a mixture of 33 grams of acrylic acid and 34 grams of methyl methacrylate. Thereafter 5 grams of benzoyl peroxide and 0.1 gram of hydroquinone were dissolved in the solution.

The foregoing solution was applied at a 10 mil thickness to an oily steel specimen (as received from the mill, not cleaned or sanded). Dimethyl aniline was applied to a similar steel specimen in an amount to form a layer 0.05 mil thick. The two pieces of steel were then joined with a light pressure by placing one on top of the other, the weight of the steel being 4 g/in^2. The time was noted when it was no longer possible to move the two pieces relative to one another by bond, and was designated as the "set time" for the bond.

The set time was 60 to 75 seconds, the shear strength was 2,900 psi (0.2 in/min) and the peel strength was 25 to 35 pounds per linear inch, at a rate of 6 inches per minute.

Instead of the two-liquid adhesive composition described above, wherein the adhesive is applied to one of the surfaces to be joined and the activator is applied to a mating surface, a one-liquid adhesive composition may be formulated having the same active components. In the one-liquid formulation, the activator is encapsulated in an insoluble rupturable microsphere and dispersed in the liquid adhesive. Only one bonding surface is coated and on mating with the other surface and applying pressure, the microspheres rupture thereby releasing the activator. Strong bonds rapidly develop with high shear and peel strength values.

In the above patent, the compounds recommended as separate components to activate the catalyst for the necessary polymerization of the monomers in the adhesive were certain tertiary amines, but other amines could also be used.

L.E. Wolinksi and P.D. Berezuk; U.S. Patent 4,155,950; May 22, 1979; assigned to Pratt & Lambert, Inc. found that, due to toxic reactions which the tertiary amines might give, other, nontoxic amines should preferably be used. Reaction products of N-methyl aniline with epoxy compounds were therefore substituted for the tertiary amines.

Example: A 1-liter reaction flask equipped with a mechanical stirrer and reflux condenser was loaded with 297.5 grams (2.1 mols) of glycidyl methacrylate, 214 grams (2 mols) of N-methyl aniline and 100 ppm (0.05 grams) of hydroquinone. The reaction mixture was heated for a total of 24 hours at 98°C with agitation. Air was maintained over the reaction at all times to prevent premature polymerization. The product, N-phenyl-N-methyl-amino-2-hydroxy-propyl methacrylate, $CH_2=C(CH_3)COOCH_2CHOHCH_2N(CH_3)C_6H_5$, is a light green liquid with a refractive index of 1.5418, a vapor pressure of 2 mm at 162° to 165°C and of less than 0.05 mm at room temperature.

The material was applied to a clean steel panel and the adhesive base of the example of the previous patent was applied to a second panel. The two were then brought together and allowed to set. Setting time was 10 to 12 minutes at room temperature. The bond strength in shear was 2,400 pounds per square inch and peel strength was 25 pounds per linear inch. Thus, an acceptable curing rate was obtained.

Carboxylic Acid or Ester plus Metal Salt plus Olefin Polymer

The rheological properties of adhesive and coating materials based on carboxylic acids and their derivatives, such as unsaturated polyester resins, are ordinarily altered with mineral fillers and/or thixotropic materials such as colloidal silica or bentonite in order to attain stable formulations.

Various disadvantages accrue from the use of such thixotropic or filler materials in the adhesives including limited shelf life, poor gel-like character, high shrinkage, crack formation, high material cost, and the fact that such products must be supplied as two-component systems.

Microencapsulation of certain of the reactive materials has also been tried, but without significant success.

R. Hinterwaldner; U.S. Patent 4,154,774; May 15, 1979 has, therefore, formulated a single component thixotropic material suitable for adhesive, sealing, and coating purposes, consisting of the reaction product of: (a) a hardenable component selected from the group consisting of unsaturated monomeric, oligomeric, or polymeric carboxylic acids and esters thereof having an acid number in excess of 0.1; with at most a stoichiometric amount of (b) a salt forming component of a metal of the second group of the periodic system; and (c) a pulverulent olefin polymer selected from the group consisting of polyolefin copolymers with acrylates or vinyl acetate and a grafted polyolefin copolymer grafted with acrylic or methacrylic compounds, the olefin polymer being present in an amount from 0.1 to 50% by weight based on the weight of the reaction product. The composition further contains a hardening agent. The reaction product, the hardening agent or both may be contained in rupture protective sheaths (i.e., may be micro-encapsulated) to prevent their reaction with one another until rupture thereof.

The hardenable component may be an unsaturated polyester or a monomer or polymer of acrylic or methacrylic acid or their esters. The metal salt is preferably an oxide or hydroxide of magnesium, zinc or calcium.

The hardening agents may include a wide variety of materials including reaction initiators such as inorganic or organic peroxides; reaction accelerators such as organometallic compounds, amines such as tert-butylamine, azo catalysts such as azobisisobutyronitrile, or mercaptans like lauryl mercaptan; and coreactants like styrene or vinyl benzoate.

Example 1: An adhesive material was prepared with the following formulation: 100 pbw unsaturated polyester resin (viscosity of 800 cp, DIN 53 015; 30% styrene; and an acid number of 16, DIN 53 042); 8 pbw microencapsulated benzoyl peroxide in a plasticizer at a concentration of 40%; 2 pbw magnesium oxide, light, DAB; 4 pbw polyethylene powder, particle size $<$50 μm; 50 pbw coated chalk; and 38 pbw quartz sand, particle size $<$2 mm.

After 2 days, the adhesive material had reached its final viscosity of 3×10^6 cp and was thixotropic.

The adhesive material was reactivated by extended mastication by hand (2 to 3 minutes). Aluminum strips were glued with the reactivated material. The tensional shear strength was 50 kg/cm^2 after 2 hours, and 70 kg/cm^2 after 24 hours.

Two small concrete blocks were glued together with the reactivated material. A bending strength test (DIN 1164) was carried out after 24 hours, in which the break occurred outside of the glue line.

Example 2: A metal adhesive below had the following composition: 100 pbw unsaturated polyester resin (viscosity of 2,200 cp; 20% styrene; and an acid number of 28); 20 pbw 2-hydroxyethyl methacrylate (2% free methacrylic acid); 10 pbw microencapsulated benzoyl peroxide, in a plasticizer at a concentration of 40%, particle size $<$200 μm; 2 pbw magnesium oxide; 4 pbw zinc oxide; and 14 pbw grafted olefin copolymer powder, with 6.5% methacrylic acid, particle size $<$50 μm.

Steel/steel and aluminum/aluminum bondings were carried out with this adhesive. The steel surfaces were sandblasted and the aluminum surfaces were pretreated by pickling (sulfuric acid - sodium chromate). The test pieces were prepared according to DIN 53 281. To reactivate the adhesive material, the glued surfaces were pressed in a gluing device. The pressure amounted to about 5 to

8 kg/cm^2. The following values were determined after a pressing time of 1 hour followed by storage at 20°C for 4 hours: steel/steel bonding—124 kg/cm^2; and aluminum/aluminum bonding—117 kg/cm^2.

Anaerobically-Cured Compositions

The phenomenon of anaerobic polymerization is one that is well known in the art and one which is easily effected in commercial practice by interposing an adhesive composition between two opposing substrate surfaces that are to be bonded together with such adhesive composition. Upon such substrate surfaces being pressed together, air is thereby excluded, giving rise to rapid polymerization of the adhesive composition which, in turn, causes the two substrates to be bonded via the resultant polymerized adhesive composition.

Such bonding processes have suffered because they often fail to produce adhesives with uniform bond strength, or because the bond strength decreases over a moderate time span, or because they require the use of solvents to promote bonding.

J.L. Azorlosa; U.S. Patent 4,158,647; June 19, 1979; assigned to GAF Corporation has formulated a solventless, fast-setting anaerobic adhesive composition which comprises two monomer parts which are adapted to be copolymerized anaerobically at ambient temperature between opposing surfaces to form a high-strength bond. One part comprises a multifunctional acrylate monomer and the other an N-vinylamide monomer. The formulation also includes a redox system consisting of oxidizing and reducing agent components for initiating polymerization between the monomers.

The oxidizing agent of the redox system may be present in either one of the parts, preferably with the multifunctional acrylate. The reducing agent is present in the other part, usually the N-vinylamide. Upon anaerobic admixing of the respective parts of the composition at ambient temperature, the oxidizing and reducing agents react to generate free radicals which initiate copolymerization of the monomers to form the desired high-strength adhesive bond to the surfaces.

Part A of the adhesive comprises a multifunctional acrylate monomer which has at least two acrylate groups per molecule, such as a diacrylate or a triacrylate. Preferred multifunctional acrylates are those which are formed by esterification and condensation of polyols with acrylic acids. Thus, suitable multifunctional acrylates include, 1,6-hexanediol diacrylate, trimethylolpropane trimethacrylate, diethylene glycol diacrylate, dipropylene glycol dimethacrylate and diglycerol dimethacrylate.

Part B of the formulation contains a monomer which will copolymerize anaerobically with the multifunctional acrylate component at ambient temperature to form a crosslinked copolymer of high-strength. In accordance with the process, this monomer is an N-vinylamide which polymerizes through its ethylenic group. Suitably N-vinylamides include N-vinyl carboxylic amides and N-vinyl sulfonamides. Preferred N-vinyl carboxylic amides are the N-vinyl lactams. Exemplary of such N-vinyl lactams are N-vinyl-2-pyrrolidone and its alkylated analogs, such as N-vinyl-5-methyl-2-pyrrolidone, N-vinyl-3, 3-dimethyl-2-pyrrolidone, N-vinyl-2-piperidone, N-vinyl-6-caprolactam, N-vinyl-hexahydro-phthalamidine and N-vinyl-morpholidone.

The oxidizing and reducing agents are present in an amount sufficient to generate the necessary free radicals to initiate polymerization. Usually about 0.1 to 5% by weight of the part is sufficient, preferably about 2 to 3% by weight.

Preferred Redox Systems

Oxidizing Agent	Reducing Agent
Organic hydroperoxide	Polyamine, $H_2N(-CH_2CH_2NH-)_nH$*
tert-Butyl hydroperoxide	Triethylenetetramine
Organic peroxide	N-alkylated aromatic amine
Benzoyl peroxide	N,N-diethylaniline
Organic hydroperoxide	Metalloorganic
Cumene hydroperoxide	Cuprous acetonyl acetones

*$n = 1$-8.

Preferably, parts A and B are admixed so as to effect copolymerization between the monomers to form a crosslinked copolymer having about a 1:1 molar ratio of acrylate group to N-vinylamide. The molar equivalent of an acrylate group is calculated by dividing the molecular weight of the multifunctional acrylate monomer by the number of acrylate groups present therein, e.g., two, three, etc. This bond developed with this composition is produced within the fastest setting times.

To improve the bond strength and to speed up the polymerization, a quantity of aluminum powder may be added to either or both parts of the formulation. Generally about 10% by weight of aluminum will result in the improvement desired with respect to these parameters.

Generally, the setting time of the adhesive composition of this process is less than 60 seconds, and often as low as 5 seconds. Occasionally, the setting times may be somewhat longer particularly if the surfaces to be bonded are rather porous.

Example: The following solutions were made up: part A contains 7 grams cellulose acetate butyrate, EAB-500-5 (Eastman); 93 grams 1,6-hexanediol diacrylate; and 3 grams tert-butyl hydroperoxide (70% by wt). Part B contains 7 grams EAB-500-5; 93 grams N-vinylpyrrolidone; and 3 grams triethylenetetraamine.

One-half ml of part A was placed on a glass plate; one-half ml of part B was placed on another glass plate. The two glass plates were pressed together so that the two components were pushed together to give a thin glue line. Within 60 seconds a film bond resulted.

Adhesives for Medical and Dental Uses

DERMATOLOGICAL ADHESIVE TAPES

Containing Tretinoin

It has been demonstrated that prolonged topical application of tretinoin (trans-retinoic acid, or Vitamin A acid) is effective in the treatment of such dermatological disorders as warts and localized acne vulgaris.

However, cream formulations containing tretinoin possess some undesirable attributes, such as difficulty in uniformly applying sufficient amounts of the active ingredient to the lesion being treated to be effective and at the same time avoid local excesses, surface spread or pooling into facial creases, the naso-labial folds and corners of the mouth where the cream may cause erythema, stinging and itching.

Another undesirable attribute of cream formulations of tretinoin is their relative instability, often necessitating the use of refrigeration or antimicrobial preservatives to prevent microbiological contamination, as well as special additives to maintain physical stability.

J.R. Marvel and J.A. Mezick; U.S. Patent 4,073,291; February 14, 1978; assigned to Johnson & Johnson have found that tretinoin can be incorporated in a wide range of concentrations in commonly used, dermatologically acceptable pressure-sensitive adhesive masses, such as the acrylic based adhesive masses, that are compatible with tretinoin. When these masses are then spread on a flexible backing to form an adhesive tape and the tape is applied to the skin, the tretinoin retains its biological activity and is readily released from the tape for absorption through the skin for sustained periods of time.

The pressure-sensitive adhesives most generally used for skin application are the rubber based pressure-sensitive adhesives, the polyvinyl alkyl ether pressure-sensitive adhesives and the acrylate pressure-sensitive adhesives. These adhesives are well known to those skilled in the pressure-sensitive adhesive art and differ from each other primarily in the type of base polymer used in preparing the same.

While the concentration of the tretinoin in the adhesive mass may vary widely, depending on the condition to be treated, its severity, etc., it is generally desirable to maintain the concentration equivalent to that obtained by spreading on the skin a cream vehicle containing a concentration of from about 0.05 to 0.5 wt % tretinoin. This is equivalent, in terms of amount per unit area, to about 23.3 and 233 mg/m^2, respectively.

Example: (A) 0.5% Tretinoin Tape – 200 mg of tretinoin were added to 70.875 grams of wet acrylate pressure-sensitive adhesive mass (C) and mixed thoroughly.

(B) 0.05% Tretinoin Tape – 20 mg of tretinoin were added to another 70.785 gram portion of the same wet acrylate pressure-sensitive adhesive mass as used for Sample (A), again with thorough mixing.

(C) The Acrylate Pressure-Sensitive Adhesive Mass – The wet adhesive mass described in (A) and (B) was prepared by reacting 300 parts by weight of 2-ethylhexyl acrylate, 125 parts by weight of vinyl acetate, and 75 parts by weight of diacetone acrylamide in a solvent comprising 373.5 parts by weight of ethyl acetate and 373.5 parts by weight of cyclohexanone, using 1.65 parts by weight of benzoyl peroxide as the polymerization initiator, at a temperature of about 80° to 85°C.

Separate samples of wet mass (A), containing 200 mg of tretinoin, wet mass (B), containing 20 mg of tretinoin, and wet mass (C), the control to which no tretinoin was added, were cast on silicone treated release paper in such a manner that, when the solvent was evaporated, one ounce of dry mass covered one square yard of release paper. It was calculated that the dry mass thus cast from Sample (A) contained 0.71% by weight tretinoin and that cast from Sample (B) contained 0.071% by weight tretinoin.

Then the cast masses were transfer coated onto polyethylene film backings to form tretinoin tapes (A), (B) and (C), respectively.

Four one inch square samples of tape (A) were separately extracted with 50 ml of acetone and the tretinoin content was determined by means of a Beckman DB-G spectrophotometer. The results were as follows.

Sample	Milligrams Tretinoin per Square Inch Tape
1	0.094
2	0.103

(continued)

Sample	Milligrams Tretinoin per Square Inch Tape
3	0.105
4	0.107
Average	0.102

Acrylic Ester Polymers with Improved Cohesion

The formulation of a pressure-sensitive adhesive given by *L.W. Takanen, K. Palmius and G. Myrthil; U.S. Patent 4,079,030; March 14, 1978; assigned to Salve SA, Switzerland* is one for use on a flexible backing, preferably an adhesive plaster or strip in contact with the skin, which at the same time is suitable for use in securing and retaining objects to be used in autoclave sterilization, such as tubes, catheters, lids, etc., and which can also be removed from the skin or a surface at elevated temperature without the pressure-sensitive adhesive agent partially remaining.

The pressure-sensitive adhesive composition comprises (1) an acrylic ester adhesive including the molecular grouping

$$-\overset{|}{C}=CH-COOC\overset{<}{<}$$

(2) a crosslinking agent which is at least one tetrahydro-4H-1,3,5-oxadiazine-4-one of formula

$$\begin{array}{c} O \\ \parallel \\ C \\ ROCH_2-N \diagup \diagdown N-CH_2OR \\ | \qquad\qquad | \\ H_2C \diagdown\quad\diagup CH_2 \\ O \end{array}$$

where R is an alkyl group with straight hydrocarbon chain containing 1 to 4 carbons; or 5-alkyltetrahydro-s-triazine-2(1H)-one or -thione of the formula

$$\begin{array}{c} X \\ \parallel \\ C \\ ROCH_2-N \diagup \diagdown N-CH_2OR \\ | \qquad\qquad | \\ H_2C \diagdown\quad\diagup CH_2 \\ N \\ | \\ C_nH_{2n+1} \end{array}$$

where n is 1 to 5 and R is H or an alkyl group with a straight chain containing 1 to 4 carbons and where X is oxygen or sulfur, or a mixture thereof; and (3) a catalyst.

In a preferred form of the process, the crosslinking agent is present in an amount of 0.1 to 2.0% by weight based on the weight of the composition.

The catalyst may be an inorganic acid such as hydrochloric acid or sulfuric acid, an organic acid such as toluene sulfonic acid or methane sulfonic acid, or a Lewis acid such as boron trifluoride, e.g., in a quantity of 0.01 to 0.5% calculated on the weight of the acrylic ester. Catalyst 1010 (American Cyanamid) may also be used as catalyst.

The composition shows cohesion of the adhesive compound through crosslinking. When a substrate coated with the adhesive composition is removed from the skin or a surface, even at elevated temperature, no remnants of adhesive remain on the skin or surface.

After storage testing, it has been ascertained that the abovementioned crosslinking agent does not split to form formaldehyde which is a great advantage, particularly in the case of products to be used in contact with the skin, as formaldehyde can cause contact eczema and allergies.

Example:

	Grams
Vinyl acetate	19.4
n-Butylacrylate	31.5
2-Ethylhexylacrylate	46.8
2-Hydroxyethylmethacrylate	2.3

The above ingredients were dissolved in 31 g ethylacetate and 47 g cyclohexane, to which 0.25 g benzoylperoxide was added. One quarter of this solution was placed in a reaction vessel with reflux coolers and heated to 78°C. Thirty minutes after this temperature had been reached, the rest of the solution was added over a period of 4 hours. The temperature was maintained at 82° to 85°C during the rest of the polymerization process which was permitted to continue for a total of 12 hours.

0.13 g 3,5-bis(methoxymethyl)tetrahydro-4H-1,3,5-oxadiazine-4-one dissolved in 5 g methanol, and 0.05 g p-toluene sulfonic acid dissolved in 5 g methanol, were added to the polymer solution and mixed well.

A strip consisting of a plasticized PVC foil having a weight of 95 g/m^2 was coated with an adhesive compound made as described above. The solvent evaporated and the strip was kept at a temperature of 65°C for 6 minutes. The weight of the adhesive coating was 32 g/m^2.

By adhesion ability is meant the force required to remove an adhesive strip having a width of 25 mm at an angle of 180° from a polished steel surface. The test was performed in accordance with method PSTC-1, 180° peel adhesion. The test was performed at room temperature. The abovementioned adhesive strip had an adhesion ability of 840 g/25 mm.

The adhesive strip had satisfactory adhesion ability on the skin and could be removed after 24 hours without remnants of the adhesive compound being left on the skin.

No formaldehyde content could be found in the adhesive compound, which was tested as described above. The adhesive strip could be removed from a steel plate heated to 70°C without leaving remnants.

Highly Tacky Pressure-Sensitive Adhesive

R. Korpman; U.S. Patent 4,080,348; March 21, 1978; assigned to Johnson & Johnson has found that superior "finger tack" and skin adhesion, as well as the ability to adhere to oily surfaces, can be attained with pressure-sensitive adhesive formulations employing particular types of A-B and A-B-A block copolymers in particular proportions.

More specifically, the adhesive composition comprises a thermoplastic elastomeric component and a resin component and the thermoplastic elastomeric component consists essentially of about 60 to 75 parts of a simple A-B block copolymer, and about 25 to 40 parts of a linear or radial A-B-A block copolymer. The A blocks are derived from styrene or styrene homologues in both the A-B and A-B-A block copolymers.

In order to attain the desired properties of the adhesive, it is important that the total styrene-derived A block content of the copolymers be not above about 20% by weight of the total block copolymers, and that the A blocks of the A-B block copolymers constitute about 10 to 18%, preferably 12 to 16% by weight of the A-B copolymers. It also is important that the B blocks of the A-B copolymers be derived from isoprene either alone or in conjunction with small proportions of other monomers.

In these A-B copolymers, the number average molecular weight of the individual A blocks should be about 7,000 to 20,000 and the total molecular weight of the block copolymer generally should not exceed about 150,000.

The elastomeric component of the adhesive may include small amounts of other more conventional elastomers but these should not exceed about 25% by weight of the elastomeric component. These include natural rubbers, synthetic rubbers based on butadiene, isoprene, butadiene-styrene, butadiene-acrylonitrile and the like, butyl rubbers, and other elastomers.

The adhesive composition includes about 20 to 300 parts, preferably 50 to 150 parts, of the resin component per 100 parts by weight of the thermoplastic elastomeric component. The resin component consists essentially of tackifier resins for the elastomeric component. In general, any compatible conventional tackifier resin or mixture of such resins may be employed. These include hydrocarbon resins, rosin and rosin derivatives, polyterpenes, and other tackifiers.

Examples 1 through 3: The following table gives the adhesive composition formulations for Examples 1 through 3. In the examples all proportions are expressed in parts per 100 parts by weight of the total elastomeric component unless otherwise indicated.

Ingredients and Characteristics	. . . . .Examples.		
	1	2	3
S-I simple block copolymer, A	70	—	—
S-I simple block copolymer, B	—	80	—
S-I simple block copolymer, C	—	—	80
S-I-S linear block copolymer (15% S)	30	—	20
S-B-S linear block copolymer (30% S)	—	20	—
Wingtack 95 tackifier resin	70	—	—
Wingtack 76 tackifier resin	—	60	—
Piccolyte S-115 tackifier resin	—	—	55
Zinc dibutyldithiocarbamate	2	2	2
2,5-Di-tert-amylhydroquinone	1	1	1

The S-I simple block copolymers A, B and C are as follows.

Copolymer	$\overline{M}n$	Percent Styrene
A	110,000	15
B	150,000	12
C	110,000	17

The S-I-S and S-B-S linear block copolymers each have a number average molecular weight of 125,000. As indicated in the table, the former contains 15% styrene and the latter contains 30%.

The above adhesive formulations are coated onto unified creped paper backing sheets and then dried. The resulting coated sheets are slit into one inch widths to form pressure-sensitive adhesive tapes. Each of the adhesives possesses superior finger tack, i.e., the ability to stick to the finger when the finger is pressed against the adhesive side of the tape.

In fact, the tapes exhibit superior skin adhesion when adhered to other parts of the body and in general are able to adhere to oily surfaces. Thus, they are suited for a variety of applications where adhesion is necessary despite the presence of some oil on the application surface.

With Improved Long Term Skin Adhesion

Several types of pressure sensitive adhesives are known to be useful as the adhesive component in adhesive bandages, self-adherent surgical drapes and the like. Acrylate pressure-sensitive adhesives, that is, pressure-sensitive adhesives comprising random interpolymers derived from at least 50% by weight of an alkyl acrylate wherein the alkyl group has from 4 to about 12 carbons, are well known for their hypoallergenic nature and good cohesive strength. However, pressure-sensitive adhesives comprising acrylate polymers sometimes suffer from the disadvantage that adhesion to human skin is poor under hot, humid conditions such as would occur if the wearer were sweating after vigorous exercise.

C.H. Beede and T. Blumig; U.S. Patent 4,082,705; April 4, 1978; assigned to

Johnson & Johnson have provided surgical pressure-sensitive adhesive compositions having excellent long term skin adhesion characteristics which comprise from about 82 to 97½ parts by weight of a pressure-sensitive adhesive and, correspondingly, from about 18 to 2½ parts by weight of a nontacky, crosslinked, hydrophilic random interpolymer which can absorb significant amounts (for example, from about 10 to 125 times its own weight) of water.

Nontacky, crosslinked, random hydrophilic interpolymers which have been found to improve the long term skin adhesion of polyacrylate, rubber and polyolefin pressure-sensitive adhesives are derived from the polymerization of a mixture of monomers comprising an ester of an α,β-olefinically unsaturated carboxylic acid and a monohydric or polyhydric alcohol having a terminal quaternary ammonium group; an α,β-olefinically unsaturated comonomer; and a crosslinking agent comprising a difunctional monomer derived from an α,β-olefinically unsaturated carboxylic acid.

The monomeric ethylenically unsaturated esters useful in the preparation of the crosslinked, hydrophilic random interpolymers include esters of an α,β-olefinically unsaturated carboxylic acid and a monohydric or polyhydric alcohol having a terminal quaternary ammonium group.

Such esters are exemplified by 2-methacryloyloxyethyltrimethylammonium methyl sulfate and 2-hydroxy-3-methacryloyloxypropyltrimethylammonium chloride, the latter being preferred.

The preferred difunctional monomer for preparing the crosslinked interpolymers is N,N'-methylenebisacrylamide. The preferred olefinically unsaturated comonomers are acrylic acid, methacrylic acid, acrylamide and methacrylamide.

Example 1: This example shows the preparation of a crosslinked, hydrophilic random interpolymer comprising 45 parts by weight 2-hydroxy-3-methacryloyloxypropyltrimethylammonium chloride, 10 parts by weight acrylic acid, 45 parts by weight acrylamide, and 0.05 part by weight N,N'-methylenebisacrylamide.

A 5 liter multineck flask, equipped with a nitrogen inlet, mechanical stirrer, thermometer, reflux condenser and an addition funnel was charged with the following reagents

	Grams
2-Hydroxy-3-methacryloyloxypropyl-trimethylammonium chloride (Sipomer Q-1)	108
Acrylic acid	24
Acrylamide	108
N,N'-methylenebisacrylamide	0.12
Water	2,700

After purging the system with nitrogen for 100 minutes, the contents were heated until the temperature reached 55°C. Then a solution of 2.4 g of ammonium persulfate in 10 ml of water was added. When the temperature reached 68°C, and the viscosity of the reaction mixture became such that the stirrer began to labor, 240.0 g of methanol were added over a 10 minute period and the reaction mixture cooled. The colloidal dispersion obtained had a solids content of 7.76%, equivalent to 100% conversion.

The polymer was obtained as a brittle film by drying down the dispersion at 100°C. The dried polymer was charged into a ball mill and pulverized. The material was sieved to obtain material less than 244 microns in diameter.

Example 2: Solutions of acrylate pressure-sensitive adhesive polymers having the following composition and characteristics were prepared.

	A	B
Ingredients, parts by weight		
2-Ethylhexyl acrylate	60	60
Vinyl acetate	25	25
tert-Butylacrylamide	—	15
Diacetoneacrylamide	15	—
Silane	0.05	0.035
Cyclohexane	150	—
Ethyl acetate	—	150
Characteristics		
Williams plasticity, mm	2.8–3.2	2.25
Percent solids	40	40

Reaction mixtures were prepared by mixing the indicated monomers and solvent in a polymerization flask and adding thereto 0.02 g/100 g of monomer of benzoyl peroxide. The reaction mixture was heated to reflux and maintained there for 6 hours. The resulting polymer solution was cooled to room temperature after which the solids content of the solutions and Williams plasticity of the polymers were determined in accordance with standard techniques.

Example 3: Four pressure-sensitive adhesive solutions were prepared in accordance with the following general procedure. The designated amount of the crosslinked, hydrophilic random interpolymer of Example 1 was wetted with 10 ml of cyclohexane and added to the designated amount of the various polyacrylate pressure-sensitive adhesive solutions of Example 2. To this mixture was then added 0.5 g of dibutyltin dioctoate (an organic catalyst) predissolved in 2.0 ml of cyclohexane. The entire mixture was then stirred until homogeneous. Adhesive sheets were then prepared by coating the pressure-sensitive adhesives on various backing materials. Control samples were also prepared and tested.

The results of the long term skin adhesion tests clearly demonstrated that the addition of the random hydrophilic interpolymer improves the long term skin adhesion characteristics of polyacrylate pressure-sensitive adhesives.

With Total Adhesive Mass Being Microporous

One medical tape in use is composed of an extremely thin tissue-like nonwoven fabric backing composed of short discontinuous, randomly laid, highly flexible thin fibers sized with an acrylate water-repellent binder and having a thin acrylate adhesive layer coated on one surface thereof. The selection and treatment of the fibers and the amount of sizing in the formation of the backing is such that the resulting product is highly porous to air and has a light cloth-like hand and drape, resulting in high patient comfort, as contrasted, for example, with a heavy cloth or crinkly paper.

The water-insoluble acrylate adhesive is of sufficient thinness that the passage of moisture vapor therethrough is sufficiently unimpeded that when adhered to the skin of a patient the underlying skin does not macerate even though the tape may remain in place on the skin for days or even weeks.

F. Salditt and W.L. Hansen; U.S. Patent 4,112,177; September 5, 1978; assigned to Minnesota Mining and Manufacturing Company have found that the valuable properties of porous pressure-sensitive adhesive tapes can be maintained and the adhesive properties thereof enhanced by providing the adhesive as a plurality of layers. Further, this can be done even though the thickness of the adhesive is increased beyond the approximately 20 to 30 g/m^2 generally found on the single layer adhesive coated porous pressure-sensitive adhesive tapes.

In Figure 9.1 the tape is seen to comprise a backing **12**, a backing contacting adhesive layer **14** and a skin contacting adhesive layer **16**. Layers **14** and **16** are provided with microporously sized pores **18** and **20** therethrough.

One way to produce this tape is to place a thin layer of pressure-sensitive adhesive on a smooth surfaced release liner, dry it, next apply a second thin layer of an adhesive over the first layer, then laminate the multilayered adhesive to the porous backing with subsequent removal of the release liner.

The layer that contacts the surface of the tape backing during lamination (the second layer on the release liner) is of a kind and of a thinness which ordinarily develops pores after such lamination, while the other layer (the first on the liner), which in the finished tape provides the exposed adhesive surface (the overlying layer), will likewise develop pores. The first layer is preferably somewhat thinner and softer than the second, although there may be tape constructions where this relationship need not be followed.

For consistency and simplicity, the layer first coated on the liner, and contacting the skin when the tape is used, will be referred to herein as the skin layer, while the other layer which contacts the backing when the tape is in use will be referred to as the backing layer.

Acrylate adhesives are capable of being formulated to provide rubbery, polymeric, hydrophobic adhesives throughout a very broad spectrum of properties.

Figure 9.1: Porous Medical Adhesive Tape

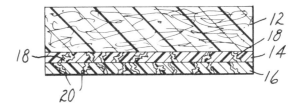

Source: U.S. Patent 4,112,177

They may be crosslinked, i.e., by covalent bonding or by hydrogen bonding, or they may be free from crosslinking. In general, the greater the crosslinking, the more firm the adhesive. These adhesives may be formulated to provide very firm or very soft rubbery layers with varying degrees of adhesive and cohesive properties, but like all pressure-sensitive adhesives they are viscoelastic materials. Unlike rubber based pressure-sensitive adhesives, the acrylate adhesives may be used significantly free of additives which may cause skin irritation.

Example: A specific porous pressure-sensitive medical adhesive tape of this process is one wherein the backing adhesive layer comprises a relatively firm acrylate adhesive coated to a dried layer weight of about 23 g/m^2 with a much softer skin adhesive layer having a dried layer weight of about 4 g/m^2. The backing is that described in Example 1 of U.S. Patent 3,121,021 and the adhesive layers are provided from siliconized release paper following the general procedures of that example.

The backing layer is an acrylate copolymer comprising 94 parts by weight isooctyl acrylate and 6 parts by weight acrylic acid, dispersed in a mixture of 70 parts by weight n-heptane and 30 parts by weight isopropanol, to about 20% solids by weight.

The skin layer comprises the same basic adhesive as above, with the addition to the adhesive of 40% by weight of a tackifier resin, e.g., glycerol ester of hydrogenated rosin (Foral 85), resulting in an adhesive dispersion of about 28% solids. The addition of the tackifier provides a demonstratable improvement in adhesion to skin.

The tape of this example has a porosity, measured in the Gurley densometer, of 1,455 seconds and adhesion values, as measured by procedures set forth in *Test Methods for Pressure Sensitive Tapes,* Pressure Sensitive Tape Council, Glenview, IL, of 525 g/2.54 cm width to a steel plate, 680 g to glass, and 45 g to human skin.

In the following table the adhesive compositions are represented by letters having the following significance: (A) the backing adhesive of the example, (B) the skin adhesive of the example and (C) a water-based thermoplastic polymeric adhesive containing 65% n-butyl acrylate and 35% n-butyl methacrylate (Rohm & Haas N 560).

	. Experiment Number .	
	1	2
Backing layer, g/m^2	*	20C
Skin layer, g/m^2	27A	4B
Total coating weight, g/m^2	27	24
Porosity, seconds	180	14
Adhesion/steel, g/2.54 cm	539	624
Adhesion/glass, g/2.54 cm	652	454
Adhesion/skin, g/2.54 cm	73	58

*Single layer construction.

The benefits of this process can be seen from the above results. Gurley densometer readings of greater than about 4,000 seconds are considered substantially nonporous for purposes of this process.

Experiment 1 is a standard microporous tape, such as described in U.S. Patent 3,121,021. Experiment 2 is a tape made by this process where the backing adhesive is, as noted hereinbefore, a water-based acrylate. The use of water-based adhesives is highly desirable, eliminating as it does the waste of hydrocarbon solvents, and lessening pollution problems.

Further, such adhesives are generally more firm than solvent-based adhesives and thus are believed to maintain porosity more readily over extended periods of time. The very thin skin layer, however, coated from solvent, while nonporous upon microscopic examination, becomes porous after contacting the backing layer and drying of the backing layer in contact with the nonwoven adhesive tape backing.

Vapor-Permeable Substrate

British Patent 1,110,051 describes the production of net from a polymer film where one surface is provided with a set of parallel grooves or channels and the other surface is also provided with a set of parallel grooves or channels, the direction of which intersects the direction of the first set. In such a film the channels are of such depth that the combined depth of the channels (on the two surfaces of the film) is a small amount less than the thickness of the film so that a thin continuous membrane is left at the points of intersection.

R. Lloyd, A.G. Patchell, W.O. Murphy and P.J. Herbert; U.S. Patent 4,135,023; January 16, 1979; assigned to Smith & Nephew Plastics Ltd., England have found that if a certain type of such film is stretched exclusively or predominantly in one direction it has unexpected properties, and in particular provides a material which can be readily torn in either of two directions at right angles. Such a material is particularly useful as a substrate for a strapping tape as used in a medical context.

In one aspect therefore, the product consists of a melt-embossed polymer film having on one surface a plurality of primary parallel ribs and grooves extending in a first direction and on the other surface a plurality of secondary parallel ribs and grooves extending in a second direction intersecting with the first direction at an angle of not less than 30°, the combined depths of the grooves being equal to the film thickness, or only a small amount less than the film thickness, so that a thin membrane is left at the areas of intersection, the secondary ribs and grooves being more closely spaced than the primary ribs and grooves, and the film having been subjected to exclusively or predominantly uniaxial stress in a direction at right angles to the direction of the primary ribs and grooves.

Preferably, the polymer film is produced by melt-embossing and the first direction may in such a case be the longitudinal or machine direction. It is, however, possible in such an instance for the primary grooves to be transverse and the secondary grooves to be longitudinal. Suitable polymer materials include polymers of olefins, particularly of ethylene, and copolymers of olefins. High density polyethylene is particularly suitable.

The material described above is coated on at least one surface thereof with an adhesive, preferably a physiologically acceptable gas-permeable and water-vapor-permeable adhesive.

Usually this is made available in the form of a tape. The tape is preferably one having adhesive on one side only. A tape having adhesive on both sides may be used for attaching medical appliances to the skin or for joining two or more porous sheets to form a laminate that is also porous. Such a laminate may be used, for example, in clothing.

Adhesives for the tapes of this process are preferably pressure-sensitive adhesives. However, adhesives of other kinds may be used, for example, solvent-responsive or heat-sealing adhesives.

Example: A high density polyethylene of melt flow index 6 was extruded as a molten film and embossed between two cooperating grooved rollers cooled to about 60° to 70°C. One roller was provided with 100 circumferential grooves, triangular in cross section, per transverse inch; the other with 250 grooves, triangular in cross section, per circumferential inch, the solidified and embossed film being taken off over this latter roller.

The solidified grooved sheet tore easily in the machine direction but only with difficulty in the transverse direction. However, after stretching in the transverse direction by 116% it was possible to tear the resultant net in the transverse direction; 200% transverse stretch gave even better transverse tear characteristics.

The net described was converted into an adhesive tape by direct coating with an emulsion based adhesive. The adhesive used was N 580 (Rohm & Haas), thickened with Collacryl VL. This adhesive was chosen to give adequate adhesion to the net together with an acceptable unspooling tension.

The net was "Corona" discharge treated before spreading to give improved keying of the adhesive to the nets. Subjectively the adhesive shows improved keying to the styrene based net. This observation was confirmed by the results of the wash trial where less adhesive is left on the skin than from the tape based on the all-polyethylene net.

Reduction of Skin Damage on Tape Removal

E. Schonfeld; U.S. Patent 4,140,115; February 20, 1979; assigned to Johnson & Johnson has found that skin damage, i.e., the stripping of tissue cells from the stratum corneum, by removal of the backing material which has been adhered to skin by a pressure-sensitive adhesive can be greatly reduced by including an unreacted polyol uniformly dispersed in the water-insoluble pressure-sensitive adhesive composition used to coat the article for adherent application to skin.

Such water-insoluble pressure-sensitive coatings are well known in the art and are collectively referred to by the term "adhesive mass." It is necessary that such pressure-sensitive adhesive compositions be water-insoluble in order to prevent their peeling when they come in contact with water.

Although the adhesive masses based on rubber elastomers which include a polyol are quite satisfactory insofar as reduction of skin damage is concerned, the preferred adhesive masses are based on acrylic systems. Such acrylate masses are based upon an adhesive polymer of monomers which consist essentially of a major amount of a medium chain length alkyl acrylate monomer and preferably minor amounts of cohesion-inducing short chain monomers, plus a very small amount of an alkoxy silyl crosslinking monomer polymerizable in the acrylate system.

The medium chain length alkyl acrylate monomers of this process generally are those averaging about 4 to 12 carbon atoms in the alcohol moiety and include: butyl, hexyl, 2-ethylhexyl, octyl, decyl and dodecyl acrylates and the like, alone or in combination with one another or with higher and lower alkyl acrylates. The medium chain length acrylate monomer is present in this adhesive copolymer in a major amount by weight of the total monomers, preferably about 55 to 85 parts by weight of the monomers. The term "parts" used above and hereinafter means parts per 100 parts by weight of the total monomer solids, unless otherwise indicated.

The cohesion-inducing short chain monomers generally are selected from vinyl acetate, methyl acrylate, acrylic acid, diacetoneacrylamide, N-tert-butylacrylamide, and the like, and preferably are present in the total monomers in the amount of about 10 to 30 parts, preferably above about 15 parts.

The alkoxy silyl crosslinking monomer of this process comprises an alkoxy silyl alkyl group and an unsaturated functional terminal group copolymerizable with the other monomers. This functional terminal group preferably is an acrylate or substituted acrylate group.

A preferred silyl crosslinking monomer is 3-methacryloxypropyltrimethoxy silane.

The polyol is physically incorporated in the adhesive mass in a quantity range of preferably 9 to 13% by weight. Polyols which are suitable for inclusion in adhesive masses include the polyoxy(C_{2-4})alkylene glycols, the polyoxy(C_{2-4})-alkylene sugars or sugar alcohols such as sorbitol and the C_{12-18} fatty acid esters of polyoxy(C_{2-4})alkylene sugars or sugar alcohols, such as polyoxyethylene sorbitol monolaurate. The preferred polyol is polyoxyethylene glycol having an average molecular weight of about 400 (Carbowax 400).

Two batches of adhesive, (A) and (B), were prepared from a mixture of 120 g of 2-ethylhexyl acrylate, 50 g of vinyl acetate, 30 g of N-tert-butylacrylamide, 0.07 g of 3-(trimethoxysilyl)propylmethacrylate and 200 g of cyclohexane. Each batch is heated to reflux under nitrogen, 0.6 g of benzoyl peroxide is added, and the reaction mixture refluxed 4 hours. The viscous solution is then diluted by stirring in an additional 100 g of cyclohexane and the reaction mixture cooled. To each of the batches is added 0.1 g dioctyltin maleate (Thermolite 813).

30 g of polyoxyethylene glycol, Carbowax 400, was added to batch (B). Batch (A) had no additive. Each adhesive was then coated onto a polyvinyl chloride film and cured for two minutes at 340°F. One by three inch samples of the two tapes were applied to the backs of human volunteers. Each day for seven days the tapes were removed and reapplied. Visual determination of skin damage showed very little for the adhesive (B), containing the polyol, and severe damage for adhesive (A), which contained none.

Showing Improved Adhesion in Water

A particularly important and desirable characteristic of a pressure-sensitive adhesive designed for application to the skin is its adhesion time when worn in or exposed to water. The failure of adhesive bandages, adhesive tapes and the like to adequately hold when immersed in or exposed to water, such as washing dishes, doing laundry, swimming or the like, has long been an undesirable problem.

I.J. Balinth; U.S. Patent 4,147,831; April 3, 1979; assigned to Johnson & Johnson has formulated a pressure-sensitive adhesive which comprises from about 30 to 50% of an elastomeric mixture consisting of natural rubber and polyisobutylene, from about 5 to 20% of a liquid plasticizer component and from about 30 to 50% of a solid tackifier component.

The term "natural rubber" as used in describing this process includes both the naturally occurring form of rubber, i.e., cis-1,4-polyisoprene, and synthetically prepared cis-1,4-polyisoprene.

The elastomeric mixture of the adhesive composition should be preferably from about 34 to 40% by weight to achieve the desired results.

The elastomeric mixture gives the adhesive composition its pressure-sensitive adhesive characteristics and high temperature stability. This high temperature stability permits the sterilization of the products, such as adhesive bandages, to which the adhesive composition is applied. The elastomeric mixture consists of natural rubber and polyisobutylene in a ratio by weight of from about 1:1 to 3:1.

The liquid plasticizer component of the pressure-sensitive adhesive compositions should be from about 5 to 20% by weight of the total composition, preferably about 8 to 14% by weight to achieve the desired results. The liquid plasticizer component controls the tackiness of the adhesive compositions and can be selected from the group consisting of isomeric liquid polybutenes; mineral oils; low molecular weight polyterpenes such as Wingtack 10, Piccolyte S-55, Piccolyte A-40, and Zonarez 25; and low viscosity rosins such as Stabelite Ester #3; and mixtures thereof.

When mineral oils are utilized as the liquid plasticizer component, they should comprise no more than 50% of the liquid plasticizer component to avoid any loss of adhesive characteristics and they should therefore be combined with one of the other abovementioned liquid plasticizers.

The solid tackifier component of the adhesive compositions should be preferably from about 38 to 48% by weight of the total composition to achieve the desired results.

The solid tackifier component can be selected from the group consisting of normally solid polyterpenes, solid rosins and mixtures thereof, such as Piccolyte S115, Wingtack 115, Escorez 115 and Nirez 1115 and Nirez 1125. Specific solid rosin tackifiers include Pentalyn A and Polypale Ester No. 10.

It has further been found that to obtain pressure-sensitive adhesive compositions with the desired characteristics including the improved water adhesion time, it is necessary for such compositions to have a Williams plasticity measurement of preferably about 1.8 to 2.2 mm.

In the preparation of pressure-sensitive surgical sheet materials for application to the skin, such as pressure-sensitive adhesive tapes, adhesive bandages, surgical drapes and the like, these pressure-sensitive adhesive compositions are coated onto a flexible backing material in accordance with known techniques. Suitable flexible backing materials include polymeric films, paper, woven and nonwoven fabrics or other similar flexible sheet materials.

Example 1: A pressure-sensitive adhesive composition is prepared by placing 160 lb of natural rubber and 58 lb of polyisobutylene of a molecular weight range of from 64,000 to 81,000 in a Banbury mixer for a period of 5 minutes. Thereafter, 50 lb of isomeric liquid polybutenes are added over a period of 10 minutes while maintaining the temperature between $220°$ and $290°F$. The resultant mixture is placed on a two-roll sheeter mill (Farrell-Birmingham Co.).

The temperature of the rolls is adjusted to 200°F and mixing is commenced for 5 minutes. Thereafter, 250 lb of a solid tackifier, such as Wingtack 115, are added and the temperature is raised to 250°F on the front roll and 300°F on the back roll for a period of 10 minutes to produce a pressure-sensitive adhesive of the following composition.

	Percent by Weight
Natural rubber	30.89
Polyisobutylene	11.20
Isomeric liquid polybutenes	9.65
Solid tackifier	48.26

Examples 2 through 5: Four pressure-sensitive adhesive compositions were prepared substantially in accordance with the process of Example 1 and had the following general composition.

	Percent by Weight
Natural rubber	28.00
Polyisobutylene	10.10
HiSil 233 filler	4.60
Isomeric liquid polybutenes	8.74
Primol 355 mineral oil	2.80
Agerite stalite	1.03
Ionol	1.03
Wingtack 115	43.70

The four compositions differed only in the molecular weight ranges of the polyisobutylene as follows.

	Polyisobutylene (MW)
Example 2	64,000–81,000
Example 3	81,000–99,000
Example 4	99,000–117,000
Example 5	117,000–135,000

These pressure-sensitive adhesive compositions were coated by means of calender coating techniques on a vinyl plastic backing with an absorbent pad to form an adhesive bandage. These bandages were then applied to the middle three fingers of thirty individuals who then placed their hands in a dishwashing solution consisting of a commercial dishwashing product mixed with hot water.

The fingers were flexed in the solution for 15 minutes and then withdrawn and the adhesive performance ranked according to the Friedman Statistical Ranking Test. The resultant statistical data has been converted to a rating system of excellent, good, fair and poor. The ratings for the molecular weight ranges are summarized below.

Example	Polyisobutylene (MW)	Rating
2	64,000–81,000	Excellent
3	81,000–99,000	Excellent
4	99,000–117,000	Fair
5	117,000–135,000	Fair

The results show that as the molecular weight range of polyisobutylene goes outside the range of 64,000 to 99,000 of the adhesive compositions of this process, the performance characteristics in water of the resulting adhesive compositions are unsatisfactory.

ADHESIVES FOR DENTAL USES

Dental Restorative Composite

H.L. Lee, Jr. and J.A. Orlowski; U.S. Patent 4,107,845; August 22, 1978; assigned to Lee Pharmaceuticals provide an adhesive dental composite based upon particular sized filler fibers, or "whiskers," which are suspended in liquid polyacrylate resin blends having viscosities of less than 5,000 cp. The particular fibers utilized comprise from 30 to 70% by weight of the total composite. The composites are unusually adhesive to etched enamel and dentin.

Utilizing the composites of this process, developmental enamel defects, hypocalcified lesions and eroded or abraded enamel or dentin may be repaired, without the need for conventional cavity preparations and the resultant sacrifice of sound tooth structure. Additionally, when properly formulated, these composites may be used without operative procedures for coating stained enamel and for masking other restorative materials, such as unsightly amalgam fillings.

The fibers must be from 1 to 100 microns in length and have diameters which average from one-twentieth to one-fifth the length of the fibers. Inorganic fibers having a hardness on the Mohs scale of 3.5 to 6.0 are preferred. Particularly preferred are calcium silicate fibers; one, commercially available, has an average fiber length of 5.5 microns, with 97% by weight of the fibers being less than 20 microns in length and 94% by weight less than 10 microns in length, with diameters on the average of one-fifteenth to one-thirteenth the lengths.

A number of compounds for possible use in the polyacrylate resin composition are given in the patent. However, a preferred formulation comprises from 25 to 90% by weight of a polyethylene glycol diacrylate and from 10 to 75% by weight of an aromatic or alicyclic polyacrylate compound.

An especially preferred composite is one wherein the resin comprises 40 parts triethylene glycol dimethacrylate to 60 parts bisphenol-A-bis(3-methacrylato-2-hydroxypropyl) ether having suspended therein calcium silicate fibers having an average length of 5.5 microns, and in which 97% by weight of the fibers have a length less than 20 microns, and 94% a length less than 10 microns. The calcium silicate fibers comprise 55% by weight of the composite.

Example 1: 30 parts by weight of bisphenol-A-bis(3-methacrylato-2-hydroxy-propyl) ether are blended with 70 parts by weight triethylene glycol dimethacrylate. To 100 parts of the resin blend are added 100 parts by weight of calcium silicate fibers having an average length of 5.5 microns with 97% by weight less than 20 microns, 94% by weight less than 10 microns, and diameters from one-thirteenth to one-fifteenth of the length. The fibers remain suspended in the resin blend over extended periods of storage.

To 100 parts of the resin-fiber mixture are added 1.5 parts by weight N-bis(2-hydroxyethyl)-p-toluidine, 0.6 part benzoyl peroxide, and 0.05 part 3-tert-butyl-4-methylphenol. The composite sets in approximately 100 seconds at room temperature and hardens in about 4 minutes from the initial mixing.

A human tooth is etched for 2 minutes with a 50% solution of phosphoric acid, washed and dried. The composite is applied to the tooth and allowed to harden. The tooth is then submerged in water at $37°C$ for 24 hours. The bond strength of the composite to the enamel of the tooth is then measured as 700 psi. Examination of the sectioned tooth also shows tag-like extensions of the resin composite penetrating the enamel.

Example 2: The teeth of a patient which exhibit areas of eroded dentin and enamel are treated with an 85% solution of phosphoric acid for 2 minutes, washed, dried and then treated with a 5% solution of hydrolyzed γ-methacryloxypropyltrimethoxysilane.

100 parts of the suspension of calcium silicate fibers in the resin blend described in Example 1 is divided into two equal portions of 50 parts each. 0.025 part of 3-tert-butyl-4-methylphenol is added to each portion. The 0.6 part benzoyl peroxide is added to one portion and the 1.5 parts N-bis(2-hydroxyethyl)-p-toluidine added to the other portion. Equal parts of each portion are then mixed together and applied to the prepared teeth. The applied composite is allowed to harden for about 20 minutes and is then finished and polished. Examination after six months shows 100% retention and the arrest of erosion and elimination of hypersensitivity.

Allyl 2-Cyanoacrylate Adhesive for Orthodontic Brackets

D.M. Stoakley and J.R. Dombroski; U.S. Patent 4,134,929; January 16, 1979; assigned to Eastman Kodak Company have developed an allyl-cyanoacrylate-based adhesive that can be rapidly cured in a thick layer which provides and maintains good adhesion of metal or polycarbonate brackets to the enamel surface of incisors and molars. The adhesives are further characterized as convenient, quick-setting allyl-cyanoacrylate-based compositions which exhibit their adhesive properties over a $4°$ to $60°C$ temperature range for extended periods of time as well as exhibit excellent hydrolytic stability.

Accordingly, there is provided an adhesive composition comprising Components (A), (B) and (C) as follows:

(A)   A polymerizable monomeric allyl 2-cyanoacrylate-containing portion comprising:

   (1)   From 75 to 92 parts by weight of allyl-2-cyano-acrylate;

   (2)   From 1 to about 12 parts by weight of a difunctional monomer diester of an acid from the group consisting of acrylic and methacrylic acid and an aromatic diol; and

   (3)   From 3 to 20 parts by weight of a thickening agent selected from poly(methylacrylate-acrylonitrile) co-polymers (sometimes referred to hereinafter as MA/AN);

(B)   From 0.1 to 3.0 weight percent based upon the weight of Component (A) of an hydrophobic cyclic imino initiator selected from 2,4,6-tri(allyloxy)-s-triazine, tri(alkyloxy)-s-triazines having 1 to about 20 exocyclic carbon atoms and a benzoxazole having the formula

   wherein each R is the same or different and represents hydrogen, phenyl, alkyl of 1 to about 20 carbon atoms, chloro or bromo; and

(C)   An amount of an organic peroxide free radical-providing-compound sufficient to cause crosslinking of the difunctional monomer diester with the allyl 2-cyanoacrylate.

This adhesive composition is prepared by sequential blending of preferred proportions of the three Components (A), (B) and (C). After carefully mixing for 30 to 60 seconds, the adhesive may be either applied to the bracket base or to the tooth or both. The bracket is positioned on the tooth with polymerization occurring within 2 to 7 minutes from the time of mixing. The bracket is then firmly adhered to the tooth after about 10 minutes and the dental wires may be inserted into the brackets.

Example 1: Into a three inch diameter aluminum cup is added 0.3 g of solution previously prepared from 8.5 g allyl 2-cyanoacrylate and 1.5 g 60/40 poly-(methylacrylate-acrylonitrile) copolymer. 12 mg of bis(hydroxyethylterephthalate)diacrylate and 1.5 mg of benzoyl peroxide are added and carefully stirred until dissolved. One drop (~0.02 g) of initiator solution, previously prepared by dissolving 5 g of 2,4,6-tri(allyloxy)-s-triazine (also called triallyl cyanurate) in 5 g of dioctyl phthalate, is added and stirred for 30 to 60 seconds with the other components. The base of a metal orthodontic bracket is dipped into the adhesive, and the metal bracket is placed on the acid-etched enamel surface of an extracted incisor.

The bracket is held in position on the tooth for about 1 minute. Dental wires may be placed through the brackets within 10 to 15 minutes after application of the bracket to the tooth. After 6 hours in 60°C water and 1 hour of cycling 1 minute each between 4° and 60°C water, tensile strengths for this composition with metal brackets were determined to be ~700 psi. Tests conducted in the same manner on a commercial bisphenol A/glycidyl methacrylate system gave tensile strengths of ~600 psi.

Example 2: 0.4 g of solution previously prepared from 8.5 g allyl 2-cyanoacrylate and 1.5 g of 60/40 poly(methylacrylate-acrylonitrile) copolymer is added to a three inch diameter aluminum cup. 2 mg p-xylylene glycol bisacrylate and 0.5 mg benzoyl peroxide are added and carefully stirred until dissolved. One drop (~0.02 g) of initiator solution, previously prepared by dissolving 1.5 g triallyl cyanurate in 5 g dioctyl phthalate, is added and stirred about 30 seconds with the other components.

The base of a polycarbonate bracket is dipped into the adhesive and the bracket is placed on the acid-etched enamel surface of an extracted incisor. The bracket is held in position for about 1 minute. Dental wires may be placed through the brackets within 10 to 15 minutes after application of the bracket to the tooth. Tensile strengths were determined to be about 600 psi. Tests conducted in the same manner on the commercial bisphenol A/glycidyl methacrylate system gave tensile strengths of ~600 psi.

In another process by *J.R. Dombroski and D.M. Stoakley; U.S. Patent 4,136,138; January 23, 1979; assigned to Eastman Kodak Company,* the use of a similar 2-cyanoacrylate adhesive is described for use as a dental sealant to fill pits and fissures, thus preventing dental decay. Additionally, the compositions find use as adhesives for sealing cavity margins, for attaching crowns, caps, inlays and pins, for restoration of incisal fractures and for bonding to dentin. The adhesive composition comprises Components (A), (B) and (C) as follows:

(A) A polymerizable monomeric 2-cyanoacrylate containing portion comprising:

   (1) From 60 to 90 parts by weight of allyl 2-cyanoacrylate;

   (2) From 3 to 25 parts by weight of an alkyl 2-cyanoacrylate having the formula

$$\begin{array}{c} CN \\ | \\ CH_2\!=\!C\!-\!COOR \end{array}$$

wherein R is alkyl having 6 to 10 carbon atoms;

(3)   From 0.25 to 5 parts by weight of a difunctional monomer diester of an acid from the group consisting of acrylic and methacrylic acid and an aromatic diol; and

(4)   From 3 to 15 parts by weight of a 2-cyanoacrylate-compatible polymeric thickener;

(B)   From 0.05 to about 1.5 weight percent based upon the weight of Component (A) of a cyclic imino initiator selected from 2,4,6-tri(allyloxy)-s-triazine, 2,4,6-tri(alkyloxy)-s-triazines having 1 to about 20 exocyclic carbon atoms, benzoxazole and substituted benzoxazoles; and

(C)   An amount of an organic peroxide free radical-providing-compound sufficient to cause crosslinking of the difunctional monomer diester with the 2-cyanoacrylates.

The above compositions are prepared for use by the sequential blending of the preferred portions of the three Components (A), (B) and (C). After mixing for 30 to 60 seconds, the adhesive is applied to the molar surface with a small brush or spatula applicator. Within 3 to 7 minutes from the time of mixing, the adhesive hardens to a resistant surface that will not chip or flake or be dislodged by applying pressure with a sharp dental explorer.

Example: Into a three inch diameter aluminum cup is added 0.2 g of a solution previously prepared from 7.2 g allyl 2-cyanoacrylate, 1.8 g octyl 2-cyanoacrylate and 1.0 g poly(methylmethacrylate). 1 mg of benzoyl peroxide and 4 mg of p-xylylene glycol diacrylate are added and stirred until dissolved. One drop (about 0.03 g) of initiator solution, previously prepared by dissolving 1.5 g of 2,4,6-tri(allyloxy)-s-triazine in 5 g of dioctyl phthalate, is added and stirred for 30 to 60 seconds with the other components.

An extracted molar tooth was previously prepared by first washing the tooth with a stream of tap water for 1 minute and then lightly swabbing the surface with a cotton roll saturated with ethanol. A light stream of air was then directed on the tooth surface for 15 seconds to facilitate removal of excess water. The fully compounded adhesive described above was applied to the molar surface with the aid of a small brush to obtain the desired thickness.

After completing the application, the adhesive hardened within 2 minutes to a clear, hard surface. The coated tooth was placed in boiling water for 20 hours. Subsequent examination of the tooth under a microscope showed no cracks in the coating and the adhesion remained very good.

Polycarboxylate Cement

S. Crisp and A.D. Wilson; U.S. Patent 4,143,018; March 6, 1979; assigned to National Research Development Corporation describe a process for the preparation of a polycarboxylate cement which comprises mixing a water-soluble polycarboxylic acid or a precursor thereof with an aluminosilicate glass in the presence of water, the ratio of acidic to basic oxides in the glass having been selected such that the glass will react with the polycarboxylic acid to form a polycarboxylate cement.

The glass compositions are described here in the conventional manner as containing alumina, silica, calcium oxide, sodium oxide and other oxides, though it is to be understood that these oxides are chemically combined in the matrix of the aluminosilicate glass, and are not present as free oxides. The proportions of oxides quoted for the glass compositions refer to the amounts of these oxides (added in some cases as the corresponding carbonates) added to the glass frit.

The weight ratio of the acidic oxides to basic oxides in the aluminosilicate glass is usually chosen such that the polycarboxylate cement stiffens within a relatively short period, termed the working time, which is usually less than 10 minutes. It has been found that the rate of reaction increased with increasing basicity of the glass and thus the ratio of the oxides can be chosen in order to allow adequate working time to form the cement into a desired shape before it has set.

For many applications it is preferred to attain a working time of about 5 minutes, or less, and then to have the shortest possible setting time in which the set cement hardens and attains an appreciable compressive strength. Preferably, the ratio by weight of acidic to basic oxides in the glass is from 0.1 to 3.0 and most preferably from 0.2 to 2.5.

The principal acidic oxide in the aluminosilicate glass is silica, although the glass may also contain minor amounts of phosphorus pentoxide, and boric oxide. The principal basic oxide in the glass is alumina, which, although it has amphoteric properties, can be considered for these purposes solely as a basic oxide. Particularly preferred aluminosilicate glasses fall within the composition range of 10 to 65% w/w silica, and 15 to 50% w/w alumina.

The aluminosilicate glass desirably contains at least one other basic oxide, preferably calcium oxide, which may be present in the glass composition in an amount of from 0 to 50% w/w. The calcium oxide may be partly or wholly replaced by sodium oxide or another basic oxide or mixture of basic oxides, although in some applications the presence of sodium oxide may be undesirable as this oxide tends to increase the solubility of the resultant cement.

Preferred glasses for use in this formulation are the gehlenite and anorthite glasses, and in general glasses falling within the composition range of 10 to 65% w/w silica, 15 to 50% w/w alumina and 0 to 50% w/w calcium oxide.

These aluminosilicate glasses may be prepared by fusing mixtures of the components in the appropriate proportions at temperatures above 900°C, and preferably between 1050° and 1550°C. The mixture is preferably fused for 1 to 4 hours. Silica and alumina may be included in the mixture as oxides, but it is convenient to add calcium oxide and sodium oxide as calcium carbonate and sodium carbonate respectively.

The glasses may be readily obtained in fine powder form. The degree of fineness of the powder should preferably be such that it produces a smooth cement paste which sets within an acceptable period when mixed with the polycarboxylic acid in the presence of water. Preferably the degree of fineness of the powder is such that it will pass through a 150 mesh BS sieve and, most preferably, a 350 BS sieve. Mixtures of differing glasses may be used if desired.

The preferred polycarboxylic acids are those prepared by the homopolymerization and copolymerization of unsaturated aliphatic carboxylic acids, for example, acrylic acid, itaconic acid, mesaconic acid, citraconic acid and aconitic acid, and copolymerization of these acids with other unsaturated aliphatic monomers, for example, acrylamide and acrylonitrile. Particularly preferred are the homopolymers of acrylic acid, and copolymers thereof, in particular copolymers of acrylic acid and itaconic acid.

It is also possible to use a precursor of a polycarboxylic acid which will be transformed into the polycarboxylic acid on contact with water, for example, a polycarboxylic acid anhydride or other suitable polymer. The polycarboxylic acid anhydride may be a homopolymer of an unsaturated carboxylic acid anhydride, or a copolymer with a vinyl monomer, and particularly a vinyl hydrocarbon monomer. Good results may be obtained using homopolymers of maleic anhydride and copolymers thereof with ethylene, propene, butene and styrene.

The polycarboxylic acid or precursor thereof is preferably linear although branched polymers may also be used, and most preferably has an average molecular weight of 5,000 to 100,000.

The cement packs of this process preferably comprise the polycarboxylic acid in the form of an aqueous solution containing from 20 to 65% by weight of the polycarboxylic acid. The cement pack may be a two-part pack in which the weight ratio of aluminosilicate glass to liquid in the two parts is from 0.5:1 to 5:1 and preferably 1.5:1 to 4.5:1, so that when the entire contents of the pack are mixed together a rapidly hardening cement is obtained.

The aluminosilicate glass is preferably from 15 to 85% by weight, the polycarboxylic acid is preferably from 3 to 50% by weight, and the water is preferably from 5 to 70% by weight, based on the total weight of the components.

The polycarboxylate cements may be made up in the conventional manner. Thus the materials in the one or two-part pack are brought together and mixed forming a plastic mass which can be cast, molded, or otherwise formed in the required shape during the brief period in which the mixture retains its plastic properties. The components can be mixed quite rapidly to give a uniform mass which commences to harden in a few minutes and is usually set within 10 minutes of mixing. The rate of hardening and strength of the final product are partly determined by the glass/liquid ratio which is preferably as high as possible and compatible with adequate working time.

These polycarboxylate cements may find application in dentistry, and also in orthopedic surgery where they may be used to assist in the resetting of fractured bone material and in the production of water hardenable surgical dressings. They are particularly useful for cementing in moist environments and may find application as grouting cements. In addition they may be useful as binders, for example, in foundry sand casting techniques.

ADHESIVES FOR MEDICAL USES

Bio-Event Electrode Material

The problem of providing a suitable electrode for applying to skin for measuring bio-electric events is a difficult one. The cardiac patient whose heart action must be monitored and who must wear electrodes for protracted periods, and astronauts who must wear electrodes for days, are all concered with the irritation which those electrodes produce. More than discomfort is involved. The skin may be damaged and discolored.

The object of a process by *J.A.R. Kater; U.S. Patent 4,094,822; June 13, 1978* is realized by the provision of a combination adhesive and electrolyte mixture performing the entire ionic transfer and adhesive function. The combined adhesive and electrolyte substance is formed by mixing a salt in an adhesive which includes, or has added to it, a solvent for the salt.

In the preferred form, the adhesive in the mixture is water-soluble. The primary solvent should have a low vapor pressure so that it does not evaporate away completely. If the adhesive is also water-soluble, perspiration will not interfere with attachment of the electrode to the skin, and cleaning after removal of the electrode is simplified.

The salt should include a metal salt and the adhesive-electrolyte, in preferred form, includes current collectors made of the same metal that is included in the salt. Not all of the salt need be metal salt, and, to some extent, salt content can be reduced by the addition of current collecting metal. All that is required is to provide a reversible ion transfer path and that is done by including metal, salt of what should ordinarily be the same metal, and a solvent for the salt.

In addition to serving as an adhesive, the adhesive component in the adhesive-electrolyte material serves substantially the same function as does the gel in prior art electrodes. Thus, it serves as a filler in which the concentration of salt can be varied and by which the salt is dispersed more or less uniformly. It serves as the vehicle for containing or retaining the solvent so that it too will be dispersed relatively uniformly throughout the body of electrolyte.

The adhesive is a nonconductor of electricity, but the dissolved metal salts are ions and they and the metal powder are conductive. The resistivity exhibited by the electrode measured from its lower face to the connector varies with the proportion of salt and powdered metal to adhesive. The proportions are not critical, so long as ionic conductivity is maintained. Total resistance is lowered by increasing the area of contact between skin and the adhesive-electrolyte material. The increased area permits a reduction in salt concentration and a consequent reduction in irritation.

The metal powders that are dispersed in the adhesive-electrolyte may be any of the materials that are customarily used in reference electrodes and in bio-electrodes. Silver is a good choice, and in that case, the salt would be an alloy of the silver. The standard salt is AgCl. Use of the adhesive-electrolyte eliminates adhesive tapes, permitting a larger area to be engaged by the adhesive-electrolyte.

As a consequence, the number of parallel flow paths is increased and resistance is lowered. Some of the materials that are ordinarily not selected for use in electrolyte gels because of high resistivity are excellent choices for use in this process. That is particularly true of a combination zinc and zinc-carbonate, and zinc and zinc-citrate. The salts, zinc-carbonate and zinc-citrate, are less an irritant when applied to human and animal skin than are the conventional salts.

A good remoistenable adhesive for surface electrode has the following formula: 15 to 25% polyvinyl alcohol, degree of polymerization 1,700, 88% hydrolyzed; 5 to 7.5% boric acid; 1.5 to 2.5% carboxymethylcellulose; 5 to 10% glycerol; and the balance is water.

Borax, or boric acid is used to increase the viscosity whereby wet tack is increased. Other thickening agents such as carboxymethylcellulose may be used either as a substitute for borax or boric acid, or for addition to those substances. Viscosity can also be increased by using fillers such as clay and fumed silica. To maintain wet tack during storage, a plasticizer such as polyethylene glycol or glycerol may be used.

If a more moisture resistant adhesive is desired, the formula may be adjusted so that it has the following proportions of ingredients: 8 to 15% polyvinyl alcohol, degree of polymerization 1,750, 98% hydrolyzed; 5 to 7% boric acid; 1.5 to 2.5% carboxymethylcellulose; 5 to 10% glycerol; and the balance is water. In either of these, bacterial growth may be prevented by adding 0.5% sodium benzoate.

Adhesives for Electrical and Photographic Uses

FOR ELECTRICAL USES

Hydantoin Diglycidyl Compounds

R. Seltzer and D.A. Gordon; U.S. Patent 4,071,477; January 31, 1978; assigned to Ciba-Geigy Corporation have succeeded in formulating hydantoin diglycidyl resins which are liquid at room temperature, possess excellent water resistance after curing, and are easily processable for laminating and casting.

The resins have the following formula:

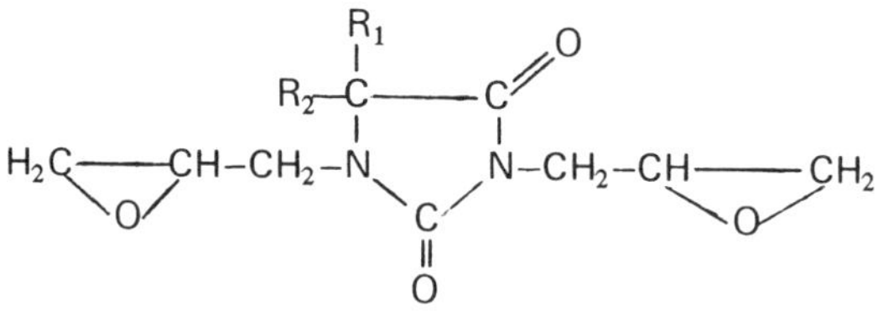

wherein R_1 is hydrogen, alkyl containing 1 to 8 carbon atoms or cycloalkyl containing 5 to 6 carbon atoms; and R_2 is alkyl containing 5 to 8 carbon atoms or cycloalkyl containing 5 to 6 carbon atoms.

The alkyl group employed herein includes both straight- and branched-chain alkyl groups which are methyl, ethyl, propyl, isopropyl, butyl, pentyl, neopentyl, amyl, sec-amyl, isoamyl, hexyl, octyl, and the like. The cycloalkyl groups include cyclopentyl and cyclohexyl.

Preferably, R_1 is hydrogen, or alkyl containing 1 to 8 carbon atoms; and R_2 is alkyl containing 5 to 8 carbon atoms.

426

Most preferably, R_1 is H, or alkyl containing 1 to 6 carbon atoms and R_2 is alkyl containing 5 to 6 carbon atoms.

The curable epoxides may be employed in the fields of surface protection, the electrical industry, laminating processes and the building industry. More specifically, when combined with the appropriate curing agents, they may be used as insulating compositions for electrical parts, to prepare printed circuit boards and can coatings, and further, for the preparation of structural laminates and flooring.

The diglycidyl hydantoin resins react with the customary acid and basic type curing agents for epoxide compounds for both room temperature and heat-curing systems. However, the basic room temperature type curing agents are especially preferred.

As suitable room temperature type basic curing agents, various kinds of amines may be used, aliphatic, cycloaliphatic, or heterocyclic, primary and secondary. Examples in the patent use triethylenetetramine, 2,4,4-trimethylhexamethylenediamine, benzyldimethylamine, etc.

The Preparation of 5-sec-Amyl-5-Ethylhydantoin: To a slurry of ammonium carbonate (865 parts), sodium cyanide (180 parts) in water (1,200 parts) was added 5-methyl-3-heptanone (385 parts) in ethanol (1,200 parts) at ambient temperature with stirring. The reaction mixture was heated to 55°C over a period of 30 min and maintained at 55°C for 6 hr. After cooling to ambient temperature, chloroform (1,000 parts) was added and the mixture stirred for 10 min.

The reaction mixture was filtered and the filter cake washed with additional chloroform (500 parts). The organic phase was collected and the aqueous phase washed with additional chloroform (1,000 parts) in two portions. The combined organic phase was evaporated to dryness yielding crude product. The resultant white solid was slurried in water (2,000 parts), filtered and dried to constant weight to afford 5-sec-amyl-5-ethylhydantoin (560 parts, 94% yield, MP 151° to 156°C).

The Preparation of 1,3-Diglycidyl-5-sec-Amyl-5-Ethylhydantoin: A mixture of 5-sec-amyl-5-ethylhydantoin (397 parts), epichlorohydrin (1,575 parts) and tetramethylammonium chloride (10 parts) was heated slowly to 80°C over a period of 1 hr and maintained at 80°C for 2.5 hr. The reaction mixture was cooled to 60°C and a reflux was established by reducing pressure. 50% aqueous sodium hydroxide (416 parts) was added dropwise over 2.5 hr while water was removed azeotropically by a circulatory distillation. The reaction mixture was cooled to 40°C, filtered and the filter cake washed with additional epichlorohydrin (500 parts).

The filtrate was treated with activated charcoal (2 parts) and filtered through a pad of filter aid. The filtrate was concentrated to near dryness to yield the crude resin. The resin was diluted with chloroform (2,000 parts) and washed with water (1,000 parts). The organic phase was dried with magnesium sulfate (250 parts), filtered and concentrated to dryness to afford 584 parts (94% yield) of 1,3-diglycidyl-5-sec-amyl-5-ethylhydantoin as a pale yellow resin; epoxy value of 6.18 eq/kg (96% of theory); Cl, 0.75%.

The hydantoin glycidyl resins made by such a method are then cured to produce casting or laminating compositions.

Addition of Adduct of Epoxide with a Novolak Resin to Polyurethane

Polyurethane adhesives or polyisocyanate adhesives are often used where heat resistance is required. However, the polyurethane adhesives hitherto used for such adhesive bondings, for example, of electrically insulated materials, will withstand a long-term thermal load of only $100°$ to $130°C$. This corresponds to what is called heat class "E".

K.-H. Kassner; U.S. Patent 4,080,401; March 21, 1978; assigned to Henkel KGaA, Germany has devised a formulation for polyurethane and polyisocyanate adhesives to increase their heat resistance which comprises the addition of 5 to 30 parts by weight of an adduct of an epoxy compound having more than one epoxide group in the molecule with a novolak resin, to 100 parts by weight of a solvent-free polyurethane or polyisocyanate adhesive, and thermally hardening the adhesives.

The multifunctional epoxy compounds are those epoxy compounds with more than one epoxide group in the molecule which are free of other reactive groups. Such compounds are the polyglycidol ethers of polyols such as the diglycidol ethers of diphenylolmethane and diphenylolpropane, the polyepoxidized aliphatic or cycloaliphatic polydienes, such as epoxidized cyclohexadiene, or also glycidyl esters of isocyanuric acid.

A novolak resin is the reaction product of a slight excess of a phenol and formaldehyde in the presence of an acidic catalyst. The resulting product has essentially no methylol groups and a diphenylmethane type of structure. The adducts of the multifunctional epoxy compounds with a novolak resin are ordinarily prepared by heating the two components to temperatures above $100°C$. The components of the adducts are preferably chosen so that the adduct has an epoxide oxygen content of from 3 to 6%.

The polyurethane adhesives employed are, in particular, those which have been produced from commercial polyesters still containing free OH groups, and which are crosslinked with multifunctional isocyanates. The polyesters still containing free OH groups often contain as the polycarboxylic acid component, alkanedioic acids having from 4 to 40 carbon atoms, such as adipic acid; or benzenepolycarboxylic acids, such as phthalic acid; and as the polyol component, ethylene glycol, diethylene glycol, or triethylene glycol, and in minor quantities, glycerin or trimethylolpropane.

Also, it is possible to use polyethers of higher molecular weight containing OH groups; for example, polyethylene glycol or polypropylene glycol, as OH group-containing component together with the isocyanates. The named hydroxyl compounds should have a hydroxyl number of between about 40 and 120. The multifunctional isocyanates employed are the benzene polyisocyanates, alkylbenzene polyisocyanates, alkanepolyisocyanates, as well as trimerization products thereof and the adduct of toluylenediisocyanate to trimethylolpropane in the molar ratio of 3 to 1.

Preferred adducts are those of crystalline triglycidylisocyanurate to novolak in the weight ratio of 30:70 to 60:40. It has proved particularly advantageous to use adducts which have an epoxide content of 3 to 6%.

The adhesives are produced by mixing the named components advantageously with the use of organic solvents.

The adhesives can be applied from their solutions on the materials to be bonded, or films can be produced from the solutions by applying solutions of the adhesive to silicone paper or materials otherwise finished antiadhesively and detaching the film after evaporation of the solvent. In this way, coatings can be produced on materials to be bonded, as well as adhesive films which are detached from the support advantageously only just before their use. During the evaporation of the solvent, which advantageously occurs under the action of heat at temperatures of from $40°$ to $110°$, a prereaction takes place, while the actual hardening occurs only in the course of the adhesive bonding.

The bonding itself requires only low pressures of from 1 to 10, more particularly 2 to 5 kp/cm^2 at temperatures of between about $120°$ and $180°$C for about 2 to 30 min.

Good bonds can be produced between a variety of materials, such as copper, aluminum, bronze, stainless steel (V2A or 18/8). Also thermostable plastics can be bonded, such as polyimides, linear polyesters, polyhydantoins, polyamides, glass fibers, and veneers impregnated with phenol/formaldehyde resin or melamine/formaldehyde resin.

Example: 45 g of a liquid commercial polyester having a hydroxyl number of 58, and a molecular weight of 2,000, which had been prepared from adipic acid and isophthalic acid (molar ratio 1:1) and diethylene glycol, were reacted with 10 g of an adduct in equal parts by weight of crystalline triglycidylisocyanurate to novolak (epoxide oxygen content 4.8%) dissolved in 45 g of methyl ethyl ketone.

The mixture was warmed to about $40°$C. To this reaction mixture, 20 g of the addition product of toluylenediisocyanate to trimethylol propane, in a molar ratio of 3:1, were added. From this mixture films of about 10 g/m^2 were produced on silicone paper by evaporation of the solvent at a temperature of about $70°$C. At room temperatures the films produced had a storage stability of about 6 months.

With the films, steel sheets of a width of 2.5 cm were bonded with an overlap of 1 cm at a pressure of 3 kp/m^2 in 5 min at 160°C. The average tensile strength of the bond was 80 kg/cm^2 and the peeling strength was 11 kg/cm^2.

Atactic Tape for Repairing Seals in Secondary Batteries

The tape developed by *P. Bernstein and J.P. Coffey; U.S. Patent 4,088,628; May 9, 1978; assigned to ESB Incorporated* is useful as a sealant and for insulation, but it is particularly designed for repairing defective seals in secondary batteries.

Casings for secondary batteries consist of containers and covers. Usually the containers and covers are molded of a thermoplastic material, such as polystyrene, polymethacrylate, polyamide, polyolefin, polyvinyl chloride, polyformaldehyde and the like. After the battery is assembled, the space between the container and cover is sealed with a sealant compound.

Most sealant compounds are not entirely satisfactory because they are not able to provide a crack-free seal during the total useful life of the battery. This is particularly true for secondary batteries exposed to vibration, very low or very high temperatures and used for applications requiring a rugged long life device. Over the course of years the sealant compound may, for a variety of reasons, become brittle and often breaks, forming cracks and holes which allow solvent to evaporate and electrolyte to escape. For such batteries to remain useful the holes and cracks, i.e., the defective seal, must be repaired.

The known compounds used for repairing defective seals are utilized in processes that require the battery to be disconnected, removed from operation, and possibly moved to a different site for the repair to be done safely. A tape of this formulation can be used in a process done with complete safety while the battery is in operation and on site.

The tape is prepared by mixing an atactic polypropylene with a nonconductive filler and then forming the mixture into a tape.

The composition of the tape is 60 to 95% by wt based on the weight of the total composition atactic polypropylene and 5 to 40% by wt based on the weight of the total composition nonconductive filler. Preferably, the percentage by weight atactic polypropylene will be 65 to 90%, more preferably 75 to 85%, and the balance nonconductive filler.

Useful atactic polypropylene is, at room temperature, a noncrystalline, waxy, slightly tacky solid. It becomes softer and more tacky with increasing temperature and gradually becomes molten. Atactic polypropylene does not have a sharp melting point, but over a temperature of 120° to 175°C, it becomes a viscous liquid that can be pumped by conventional means. The typical molecular weight of atactic polypropylene is 3,000 to 10,000.

A preferred atactic polypropylene will have a softening point determined by ASTM-E28-51 of about $100°$ to $110°C$, more preferably $105°$ to $107°C$; a viscosity determined by Brookfield Thermosel system of 5,000 to 9,000 cp at $149°C$, more preferably 5,500 to 8,500 cp, and a density determined by ASTM-D792 at $23°C$ of 0.82 to 0.90 g/cc, more preferably 0.85 to 0.87 g/cc.

Examples of such atactic polypropylene are Eastobond N5K, N5W, N500S, and N510S.

Nonconductive filler useful in the formulation can be asbestos, talc, titanium dioxide, and carbon black. The nonconductive filler must be a material that is nonleachable under operative temperatures and hence not interfere with the operation of the object sealed or insulated. Addition of the filler improves the flow characteristics and extrudability of the tape.

Preferably, the filler will be a carbon black having a medium particle size, i.e., able to pass through a 70-mesh sieve. A preferred carbon black is thermal black. More preferably, the thermal black will have a specific gravity of 1.80, and a moisture content of 0.5% maximum. Such a medium thermal black is Floform Thermax ASTM-N-990. The filler having large particles is ground to reduce the particle size to the required size.

The tape, once formed, will be low-temperature melting, tacky, inert, and unaffected by secondary battery electrolyte. It will adhere to thermoplastic materials, such as those which are utilized to construct containers and covers of battery casings, especially thermoplastic materials such as polystyrene, polyethylene, and crystalline polypropylene.

The tape may be used for sealing and/or insulating any object so long as it will, in operation, not be exposed to temperatures over $75°C$. The tape can be stamped or shaped into ropes, washers, plugs, sheets, channels, half rounds, preform or whatever shape is required for a specific purpose.

High Temperature Refractory Adhesive

Recent years have witnessed an increasing growth in the use of ceramic fibers for numerous types of high temperature applications. However, most commercially available refractory adhesives generally have physical and chemical properties which make them unsuitable for use with ceramic fiber products. Often such adhesives contain binders, such as inorganic silicates, which can attack the ceramic fibers and/or become electrically conductive at high temperatures. Such undesirable characteristics severely limit the areas of application of such adhesives leaving alternatives which are economically unattractive.

In spite of such drawbacks, the tendency for favoring inorganic adhesives may be appreciated when one has used and encountered the problems inherent with employing many of the organic adhesives at high temperatures. Carbonization and volatilization are representative of the drawbacks of such compositions.

What is needed is an inorganic adhesive which is not only useful at high temperatures, but displays numerous desirable properties so as to make it suitable for a broad spectrum of applications. Such a composition should not only retain its insulating value at high temperatures, but should also be able to retain sufficient rigidity to remain in place when subject to stresses and other forces at high temperatures.

D.L. Keel and R.A. Waugh; U.S. Patent 4,090,881; May 23, 1978; assigned to The Babcock & Wilcox Company have developed a low-cost high temperature inorganic adhesive including mullite, alumina, ceramic fibers, colloidal silica and a binding agent blended within certain specified ratios to give desired characteristic properties to the mixture including nonconductibility, workability, ability to bond metallic elements to nonmetallic elements at high temperatures, improved dry strength, and ability to be used in a broad spectrum of applications.

Because of such properties, a long life composite of the type encountered, for example, in electrically heated support pads can now be produced, minimizing the deterioration encountered because of the undesirable drawbacks of the prior art adhesives.

The formulation on a dry weight basis consists essentially of 33 to 80% mullite, 10 to 43% ceramic fiber, 5 to 20% colloidal silica, 5 to 20% alumina and 3 to 13% bentonite clay. Such a composition displays desirable electrical properties along with, among its other properties, suitable strength, shelf life and viscosity.

The ceramic fiber is a determinative factor in composing this refractory composition. While many commercially available ceramic fibers may be used in making the adhesive, varying results may be obtained. Ceramic fibers, such as Kaowool, and a composition of polycrystalline alumina fibers can be used in formulating the refractory-adhesive composition. Of course, mixtures of the various ceramic fibers are also useful. The high temperature electrical insulating properties of this composition are believed to be attributed to the fact that constituents, such as zirconia, are absent.

The colloidal silica can be supplied to the mixture from a number of different sources, but a preferred form is as found in the composition Ludox HS, an aqueous colloidal sol containing about 30% by wt of the dispersion as silica.

The composition of this process is extremely well suited as an adhesive. An adhesive may generally be thought of as a substance used to bond by surface attachment two or more solids so that they act or can be used as a single piece. This composition displays excellent refractory properties.

This composition is characterized by its ability to retain its insulating qualities even at those temperatures where other refractory adhesives become conducting. It also maintains its structural integrity and resists cracking and crumbling that could be occasioned by thermal stresses resulting from changes in temperature and is compatible with a wide variety of materials giving it a broad spectrum of useful applications.

In addition to being inexpensive, this adhesive is relatively simple to make. Preferably the raw materials, alumina, mullite and the ceramic fibers are milled to a particle size of less than 325 mesh; that is less than 44 μ. It has been found that the most stable and strongest cements result from such a milling step. Wet ball milling is appropriate and gives a smooth homogeneous product. Alternatively, the adhesive can be prepared by ball-milling the dry ingredients and later adding the liquid colloidal silica binder and water in a blending operation, such as a Hobart mixer.

In the preferred range, on a dry weight basis, mullite will be present in an amount of 33 to 43%, ceramic fiber in an amount of 33 to 43%, alumina in an amount of 8 to 12%, colloidal silica in an amount of 8 to 12% and the clay in an amount of 3 to 5%.

Besides electrical insulation, other uses for the adhesive are in attaching heating wires to glass ceramic surfaces, bonding layers of fiber insulation, bonding different vacuum formed fiber shapes, etc.

Fluid-Permeable Adhesive Tape for Binding Electrical Conductors

In the electrical industry, coils of transformers, electrical appliances or electrical equipment are provided, after winding, with impregnating fluids by immersion or dropwise. These impregnating varnishes or resins serve to maintain the functional capacity of electric materials over a lengthy period, since they represent additional electrical insulation and offer protection against mechanical injuries of wires and thermal and chemical effects. Prior to treatment with impregnating fluids, the coiled wires must be bound or fixed so that the relative positions of the individual wires are maintained.

If an adhesive tape is used to wrap the coiled conductors, such a tape must not only retain its adhesive power at high thermal stress and show good stability against organic materials, but must be able to be thoroughly impregnated with the varnishes and resins used for insulating the coils.

E. Bräunling, K.D. Kuhlmann and P. Lutz; U.S. Patent 4,096,007; June 20, 1978; assigned to Beiersdorf AG, Germany have provided an impregnating-fluid-permeable, pliant adhesive tape, which can be covered with an adhesive-repellant separating paper before it is used, for binding coiled electrical conductors. It contains an impregnating fluid-permeable carrier material; the adhesion side of which is at least 50% free of adhesive compound. The wrapped coils together with the tape can then be impregnation varnished; the time required for this process being substantially reduced.

The support for the tape can be an open-meshed woven fabric or textile, a coarse pored nonwoven fabric, or a stamped nonwoven fabric with a netlike structure, and the adhesive compound is coated on the support material in fine strips or particles in such a manner that at least 50% of the side to which the adhesive gluing compound is applied remains free of the adhesive so that impregnation with the necessary varnish or resin can be carried out more readily and thoroughly.

The adhesive compound can be based on a conventional natural rubber compound, a synthetic rubber compound, polyacrylate or silicon rubber, or heat crosslinked compounds thereof.

Example 1: The adhesive tape consists of a coarse meshed polyester fabric with a mesh space of 0.3 mm, the meshes of which are fixed at intersection points by a polyester resin, and self-adhering acrylate based adhesive. The striped coating of this compound covers 50% of the fabric area. The width of the strips is about 1 mm, the coating thickness about 50 g/m^2.

Example 2: A coarse, nonwoven polyester fabric with netlike structure, and which has been punched with ellipse shaped holes, has the bore diameter (of the ellipses) where a is 4 mm and b is 2 mm. Its weight per unit area is 60 g/cm^2, the thickness 0.25 mm. It is provided with about 1 mm wide strip lines of a heat crosslinked, natural rubber compound, the strips having 1 mm spacing from each other. The adhesive coating quantity is about 50 g/m^2.

Electrically Conductive Sulfur-Bearing Adhesive

J.H. Aumiller and J.F.W. Becker; U.S. Patent 4,127,699; November 28, 1978; assigned to E.I. DuPont de Nemours and Company describe a curable adhesive composition comprising:

(1) A sulfur-bearing component selected from chlorosulfonated polyethylene and a mixture of a sulfonyl chloride with chlorinated polyethylene, the sulfur-bearing component containing about 25 to 70 wt % of chlorine and about 3 to 160 mmol of sulfonyl chloride moiety per 100 g of chlorosulfonated or chlorinated polyethylene, the chlorosulfonated or chlorinated polyethylene being made from polyethylene having a melt index of about 4 to 500;

(2) At least one polymerizable vinyl monomer;

(3) At least one member of the group of conductive particles selected from silver, carbon and titanium carbide in an amount which will produce an electrically conductive adhesive composition when cured; and

(4) A polymerization catalyst.

The adhesive compositions are cured by polymerization catalysts, i.e., ingredient (4). The catalysts can form one package of a two package system wherein the other package consists of components (1) (2) and (3).

The chlorosulfonated polyethylene useful in the formulation can be prepared in a manner well known to those skilled in the art by reaction of linear or branched polyethylene and a sulfuryl chloride, SO_2Cl_2, or sulfur dioxide and chlorine. Chlorosulfonated polyethylene is also available commercially, for example, under the tradename Hypalon. In practice, the chlorosulfonated polyethylene can be a chlorosulfonated copolymer of ethylene with small proportions of propylene or other olefins.

Alternatively, a mixture of a sulfonyl chloride and chlorinated polyethylene can be used. The sulfonyl chlorides can be mono- or polyfunctional sulfonyl chlorides having 1 to 24 carbon atoms and 1 to 2 chlorosulfonyl groups; for instance, C_{1-12} alkyl sulfonyl chlorides, such as methane and butane sulfonyl chloride; and C_{6-24} aromatic sulfonyl chlorides, such as benzene or toluene-sulfonyl chloride; and sulfonyl chlorides containing hetero atoms such as di-phenyl ether 4,4'-disulfonyl chloride.

Polymerizable vinyl monomers which can be used include acrylic monomers, and polyalkylstyrenes. Acrylic monomers, especially lower alkyl acrylates and meth-acrylates and ethylene glycol diacrylate and dimethacrylate are preferred.

The particles for use include metallic-conductive silver, TiC particles, and con-ductive carbons which generally have low levels of volatile materials present. These three types of particles are referred to herein as "conductive" for the sake of simplicity.

Carrier particles which are coated with silver, carbon or TiC are also useful. Silver is generally more conductive than TiC and carbon, and, for this reason, the preferred conductive particles are silver. The most preferred form of silver is a flake form. However, for bonding aluminum or iron, compositions con-taining silver particles tend to give bonds whose resistivity increases with time. Therefore, TiC particles and carbon particles are preferred for these substrates.

Operable catalysts to facilitate polymerization of the adhesive can easily be selected by those skilled in the art and include amines, transition metal com-positions, various aldehyde and amine condensation products and such free-radical generating catalysts as organic peroxides.

In the following examples all parts are by weight unless otherwise noted.

Example 1: To a mixture of acrylic monomers comprising 678 g of 2-ethyl-hexyl methacrylate (containing 50 to 90 ppm of hydroquinone inhibitor), 22 g of glacial methacrylic acid (containing 250 ppm of 4-methoxyphenol), and 11 g of ethylene glycol dimethacrylate, there was added 400 g of chlorosulfonated polyethylene (Hypalon 20 Synthetic Rubber) made from branched polyethylene having a melt index of 10 and containing 29% chlorine and 1.4% sulfur.

The mixture was rolled in a container at room temperature until solution of the polymer was complete. To a portion was added in turn, with stirring, 0.5% by wt of cumene hydroperoxide and 6 g of titanium carbide powder (20 μ average particle size) per gram of solution.

Several 1 in^2 pieces of aluminum sheet were primed on one surface with a com-mercial butyraldehyde-aniline condensation product such as 808 Accelerator.

The curable adhesive composition was then applied between two facing primed aluminum surfaces. The adhesive set within 15 min at room temperature and the resistance was determined to be about 300 ohms. Another amount of curable adhesive composition was then applied in the same way between facing primed aluminum surfaces. The resistance of the cured adhesive was determined to be 200,000 ohms.

Example 2: A mixture was prepared as in Example 1 except that replacing the 2-ethylhexyl methacrylate as monomer was methyl methacrylate similarly inhibited with hydroquinone, the resultant percentages by weight of components in the rolled solution being 36.8% Hypalon 20 Synthetic Rubber, 51.65% methyl methacrylate, 10.5% glacial methacrylic acid, and 1.05% ethylene glycol dimethacrylate.

The preparation of conductive adhesive compositions was made without addition of cumene hydroperoxide. Portions of the rolled solution were loaded respectively with 5, 4, 3 and 2 parts of finely divided silver powder per part of solution. This powder contained at minimum 99.1% Ag, at maximum 0.02% chlorides, and is available as Conductive Silver 4569.

Each solution into which silver powder has been stirred was applied as in Example 1 between 808 Accelerator primed aluminum pieces. Each set within 15 min and curing continued for 40 min altogether, at which time electrical resistances and lap shear strengths between the bonded pieces were measured. Resistances were less than 1 ohm for the 5, 4, and 3 part Ag bonds and 1,000 ohms for the 2 part Ag bond. Lap shear strengths of the 2 and 3 part Ag bonds were greater than 1,000 psi, and were 2 to 3 times greater than the 4 and 5 part Ag bonds.

For Connecting Flexible Heating Resistors to Electrodes

M. Leiser, K.-H. Wegehaupt and W. Marsch; U.S. Patent 4,130,707; Dec. 19, 1978; assigned to Wacker-Chemie GmbH, Germany have formulated an adhesive composition which consists of a vinyl-containing organopolysiloxane having the formula:

$$R_2(CH_2{=}CH)SiO(R'_2SiO)_nSi(CH{=}CH_2)R_2$$

wherein R is selected from the group consisting of methyl and phenyl radicals, R' is selected from the group consisting of methyl, vinyl and phenyl radicals, n is a number sufficient to provide an organopolysiloxane having a viscosity of at least 100,000 cp at 25°C in which 0.1 to 1 mol % of the diorganopolysiloxane units are vinyl methyl siloxanes and at least 90% of the remaining organic radicals consist of methyl radicals; an organopolysiloxane containing at least three silicon bonded hydrogen atoms in which the silicon bonded hydrogen atoms are present in an amount of from 0.01 to 1.7 wt %; and a catalyst which promotes the addition of the silicon bonded hydrogen atoms to the vinyl groups, intimately contacting the coated surfaces and thereafter heating to a temperature of from 100° to 200°C.

The catalyst preferably consists of platinum or platinum compounds, which are preferably employed in amounts of from 5 to 100 ppm, calculated as Pt and based on the total weight of the composition.

Reinforcing fillers may also be used in the composition, and must have a surface area of at least 50 m^2/g. Examples are pyrogenic silicon dioxide, silicic acid hydrogels which have been dehydrated while their structure has been maintained and precipitated silicon dioxide.

The adhesives that are obtained by curing these compositions exhibit excellent adhesion to electrodes and area heating resistors which are generally composed of metals, such as copper and aluminum, even though these metals have not been pretreated with a primer. These adhesives exhibit excellent adhesion on organopolysiloxane elastomers, organic plastics, such as olefinic addition polymers, and condensation polymers containing carbonyl groups, such as polyethylene terephthalate and polyimides.

One of the preferred embodiments is to use these compositions, especially when they are electrically conductive, for connecting electrodes to area heating resistors which contain organopolysiloxane elastomers, or organic plastics, such as olefinic addition polymers and condensation polymers which contain carbonyl groups as binding agents.

For this reason, the composition is particularly suited for connecting surface heating resistors to insulating films which contain organic polymers obtained from the olefinic addition and/or carbonyl groups in order to insulate the resistor material.

Example: About 100 parts of a diorganopolysiloxane, which contains vinyl-dimethylsiloxy groups as terminal units (with 0.2 mol % vinyl methyl siloxane units and a viscosity of approximately 3×10^6 cs at 25°C with the remaining organic radicals being methyl groups) are mixed on a rolling mill with 60 parts of acetylene black, 15 parts of pyrogenically produced silicon dioxide, 2 parts dimethylsiloxanols, which on the average contain 4% by wt of silicon-bonded hydroxyl groups, and 3 parts of a mixed hydrolysate of methyldichlorosilane and trimethylchlorosilane at a weight ratio of 1,000 to 51.5 with a viscosity of about 30 cs at 25°C and about 1.6% by wt of silicon-bonded hydrogen.

The acetylene black used has an average particle size of 56 mμ, a BET-surface area of approximately 70 m^2/g and a specific electrical resistance measured at room temperature and a pressure of 200 kp/cm^2 or 200 kg/cm^2 of 0.14 ohm cm. The silicon dioxide used has a BET-surface area of approximately 130 m^2/g.

After 8 days of storage the mixture obtained is dispersed in 700 parts of trichloroethylene. The obtained 15% dispersion is mixed with 0.05% based on the weight of the dispersion of chloroplatinic acid as a 1% solution in isopropanol.

Two opposing edges of a fiberglass resistor material for area heating resistors, which has been coated with electrically conductive organopolysiloxane elastomer and which measures 10 x 10 cm and has a square resistance of 432 ohms, measured with two 12 cm long retention electrodes, which were held by constant spring pressure and which on the two inner sides were equipped with smooth massive silver sheetmetal, are coated with the catalyzed dispersion whose preparation was described above.

Two copper wire webs which measure 12 cm x 5 mm, which were previously immersed in the abovementioned catalyzed dispersion are pressed upon the coated edges as electrodes. They are then heated for 5 min at 120°C and for 3 min at 180°C. Good adhesion is achieved between the electrodes and the resistors. The connection is extremely flexible. The square resistance of the obtained area heating resistor is now 430 ohms and does not change even when the electrically conductive connections between the resistors and the electrodes are bent.

For Bonded Covering of Fire Extinguishing Materials in Appliances

Fire-extinguishing materials contained under adhesive bonded covers in electric appliances, such as ranges, may be released under emergency conditions to extinguish fires caused by failure to turn off the ranges, or fires caused by cooking oils spilled on the ranges. Heat generated by these fires is utilized to release or separate the adhesive bonded covers and thereby free the fire-extinguishing materials.

Moreover, adhesive bonded automobile windshields and adhesive bonded emergency exit areas for vehicles, such as airplanes, may also be released by utilizing the heat generated from accidentally caused fires to separate the adhesive bonded windshields or exit areas. In this manner, emergency exits may be provided for occupants trapped inside these vehicles. Also, adhesive bonded jackets on space reentry vehicles may also be released by utilizing the heat induced through atmospheric reentry conditions.

However, present adhesives available for surface bonding are limited in ability to provide a bonded joint having a controlled release function, particularly a release function which separates a surface bonded joint and to withstand severe environmental conditions.

P.W. Juneau, Jr. and M.M. West; U.S. Patent 4,145,328; March 20, 1979; assigned to General Electric Company have devised a bimetallic adhesive for surface bonding comprising a mixture of titanium and boron powders in the ratio of one atom of titanium and two atoms of boron and which further includes a resinous adhesive binder. The resinous adhesive binder may consist of epoxy, polyamide, polyvinyl and resorcinol resins, acrylic latexes, polyesters or a combination of these adhesives with or without polytetrafluoroethylene. The mixture of titanium and boron powders is employed in place of a filler material and provides a controlled release function to separate a bonded joint.

The mixture is triggered by heating the mixture at a reaction temperature above 600°C. The heat causes the mixture to exothermically react to effect complete deflagration of the mixture. Moreover, the adhesive is nonexplosive and possesses excellent shock, temperature, moisture and spark resistance characteristics.

The mixture of bimetallic powders utilized consists of 4.8 parts by weight of titanium powder and 2.2 parts by weight of boron powder.

Example: The formulation is illustrated with reference to the following compositions:

Composition 1

Ingredients	Percent by Weight (±0.1%)
Epon 815	17.5
Pentamid 840	7.5
Acetone	25.0
Titanium	34.3
Boron	15.7

Epon 815 (Shell Chemical Co.) is a phenylglycidyl ether modified diglycidyl ether of bisphenol A, having an epoxy equivalent of 160 to 220 and having a viscosity of 2,000 cp. The phenylglycidyl ether can be present to the extent of 15 to 30% by wt. Pentamid 840 (Ciba-Geigy Corp.) is a polyamide which is the reaction product of a dimerized fatty acid and an excess of ethylenediamine, so that amino end groups are available for reaction. A viscosity of 3,000 to 5,000 cp at 25°C is typical.

Composition 2

Ingredients	Percent by Weight (±0.1%)
Formvar 15/95E	4.46
Dioxane	53.5
Toluene	35.71
Titanium	4.28
Boron	1.96

Formvar 15/95E (Monsanto Co.) is polyvinyl formal consisting of 95% cyclic structures and 5% hydroxyl groups.

Composition 3

Ingredients	Percent by Weight (±0.1%)
Polyvinyl butyral	4.59
Methanol	87.24
Titanium	4.41

(continued)

Ingredients	Percent by Weight (±0.1%)
Boron	2.02
PTFE*	1.74

*Polytetrafluoroethylene.

All three formulations are mixed mechanically in a conventional manner to produce a smooth even blend, with all solid material having particle sizes from 2 to 30 μ. The metals in powder form are mixed by tumbling together in the dry state to ensure thorough mixing in all formulations. Other additives, such as powdered PTFE (up to 30%), may be included at this state of mixing.

Then a dilute dispersion (5 to 10% solids) of the resin is added, and the mixture is stirred with a low shear mixer or a spatula until throughly mixed. It should be noted that varying the amount of solvent has no effect on the properties of the cured material, since the solvent evaporates and is not present during the operation of the composition as an adhesive bond. The final composition may consist of 30 to 50% resinous binder, with the rest of the composition being the mixture of boron, titanium and PTFE powders.

Electrically and Thermally Conductive Adhesive

The formulation described by *J.M. Shaheen and L.J. Quintana; U.S. Patent 4,147,669; April 3, 1979; assigned to Rockwell International Corporation* relates to electrically and thermally conductive adhesives containing gallium and other metals dispersed in a resin. A composite is formed by producing a substantially liquid eutectic of gallium and another metal, such as tin, indium or the like.

The eutectic is mixed with a powdered metal, such as gold, copper, silver, or any other electrically conductive material to form an alloy which is then dispersed in a resin, such as an epoxy with a curing agent or the like. The composite is typically produced in the form of a paste which cures to a relatively homogeneous solid mass to form a conductive bond between two items.

In a preferred embodiment, gallium (89%) and tin (11%) are first mixed together to form a eutectic mixture. Typically, the gallium is heated to 30°C and is in the liquid state while the tin is in powder form. The gallium is found to be desirable because it has the ability to be mixed and to form a eutectic mixture which remains liquid at room temperature. Generally, the eutectic does not require heat to react with other metals. Moreover, the eutectic remains in the liquid form at room temperature.

A powdered metal, such as gold, silver, copper or the like, is then mixed with the eutectic mixture to form an alloy. The alloy is generally formed of a 50/50 weight proportion of the eutectic and the powdered metal. However, the proportions may vary from 20/80 to 80/20 of the same materials. The alloy is initially a pastelike mixture. The wetness of the paste is a function of the materials used and the proportions thereof.

Next, a resin or resin with curing agent is mixed with the alloy. The resin and the alloy harden independently of one another but form a relatively homogeneous mass capable of forming a relatively homogeneous junction between a die and a carrier to which it is bonded. The resinous portion of the composition used can comprise a mixture of one or more epoxy resins.

The desired characteristics of the epoxy resin include the requisite adhesiveness between the composition and the substrate surface. The epoxy resins are available commercially. The amount of curing agent that is used in conjunction with the epoxy resins in the gallium alloy composition is based on the weight of the resin. The amounts of epoxy resin in the composition can vary up to 25% by wt. Amounts from 1 to 10% by wt are preferred.

Example: A specific adhesive formulation cured at room temperature and at 125°C may be made as follows.

	Percent by Weight
Gallium eutectic	49
Gold powder	48
Delta epoxy resin (ER11-201A)	2.99
Delta epoxy curing agent (ER11-201B)	0.01

This adhesive possesses excellent adhesion to silicon, alumina, gold plated alumina, gold plated Kovar and rolled annealed copper. The electrical resistivity measurements remained relatively constant (about 55×10^{-6} ohm cm) through a wide range of temperatures and for a prolonged period at high temperatures (e.g., 60 hr at 350°F). The thermal conductivity of this formulation, which was found to be brittle, is comparable to metal filled epoxies (i.e., 0.015 ± 0.002 W/in °C). The coefficient of expansion was well within the expansion requirements of conductive adhesives used in hybrid packaging (6.5×10^{-5} below Tg; 3×10^{-4} above Tg when Tg is below 15°C). A thermal gravimetric analysis performed on this formulation revealed a weight loss well under the requirements of hybrid packaging (0.3% wt loss at 250°C).

ADHESIVES FOR PHOTOGRAPHIC USE

Adhesive Layer for X-Ray Intensifying Screen

X-ray intensifying screens for medical and industrial radiography generally consist of the following layers:

A rigid or flexible support layer consisting of cardboard or a foil of polyethylene terephthalate, cellulose acetate, polyvinyl chloride or a copolymer of vinyl chloride/vinyl acetate and other polymers;

An adhesive layer containing a binder, such as a vinyl chloride copolymer, polyvinyl butyral, chlorosulfonated polyethylene or a similar polymer;

Optionally an additional reflective or radiation absorbent layer containing pigments, such as titanium dioxide, magnesium oxide, barium sulfate or carbon black, together with the same binder as in the adhesive layer or a similar binder;

A fluorescent layer containing fluorescent substances, such as calcium tungstate, zinc sulfide, zinc cadmium sulfide, lead barium sulfate, a rare earth metal compound or a mixture of such compounds in a binder and a transparent layer as protection against dirt and mechanical damage.

The object of *W. Brandstätter, G. Haschka and H. Heindze; U.S. Patent 4,078,108; March 7, 1978; assigned to AGFA-Gevaert, A.G., Germany* is to provide an x-ray intensifying screen in which the adhesive layer bonds the fluorescent layer more firmly to the support.

The problem is reduced or substantially solved in an x-ray intensifying screen consisting of a flexible support layer of polyethylene terephthalate with an adhesive layer, optionally a reflective or radiation absorbent layer, a fluorescent layer, and optionally a protective layer. The binders in the adhesive layers are polyesters or copolyesters of isophthalic acid with aliphatic diols, optionally in combination with saturated dicarboxylic acids, with a molecular weight of at least 4,500 and 2 to 35% by wt of sulfonated castor oil based on the dry weight of the binder.

Excellent results are obtained with a copolyester of isophthalic acid, terephthalic acid, ethylene glycol and 1,4-butanediol, in which the acid component consists of 70% by wt of isophthalate and 30% by weight of terephthalate groups. Esters with an isophthalate content of 35 to 75% by wt are generally preferred.

The polyesters or copolyesters used in the material are either commercial products, for example, the isophthalic acid ester mentioned above which is obtainable under the name Vitel PE 200, or they can easily be prepared by well-known polycondensation processes.

The surprisingly advantageous effect of the described adhesive layers is achieved by the addition of Turkey red oil. So-called Turkey red oil available commercially is a castor oil which has been treated with sulfuric acid and contains sulfonic acid esters in which the OH group of ricinoleic acid is esterified with sulfuric acid. The Turkey red oil is added to the composition of the adhesive layer in a quantity of from 2 to 35% by wt, preferably 8 to 15% by wt, based on the dry weight of the polyester.

Example: A polyethylene terephthalate foil 250 μ thick was coated with an adhesive layer of the following composition by the immersion process: 4,000 g of a 25% solution of the above described copolyester of isophthalic acid, terephthalic acid, ethylene glycol and 1,4-butanediol in chlorobenzene and ethyl acetate (1:1); 50 g of oleic acid glyceric ester; 50 g of Turkey red oil; and 6,500 g of chlorobenzene.

The dispersion was milled in a ball mill for 96 hr, diluted with a mixture of chlorobenzene and ethyl acetate in proportions by weight of 1:1, filtered through a filter with a pore size of 0.001 to 0.005 mm and applied.

The thickness of the dry layer was 2 μ. A fluorescent layer of the following composition was applied to the adhesive layer: 10,000 g of calcium tungstate; 2,000 g of ethyl acetate; 800 g of cellulose acetobutyrate; 280 g of polyethyl acrylate; 200 g of toluene; 100 g of methyl glycol acetate; and 1,600 g of methyl ethyl ketone.

The thickness of the calcium tungstate layer when dry was 200 μ. The bond between the layers obtained was very firm but was insufficient if no Turkey red oil was added to the adhesive layer of otherwise the same composition.

Amorphous Polyester Adhesives

In certain formats for photographic film units the image-receiving layer containing the photographic image for viewing can be separated from the photographic layers after processing. In some embodiments, it can remain permanently attached and integral with the image-generating and ancillary layers present in the structure when a transparent support is employed on the viewing side of the assemblage.

Typically, adhesives are used in photographic film units to bond supports, subbed or unsubbed, to cover sheets or other layers around the edges of the units. Adhesives can also be used to bond subbing layers to supports or cover sheets, photosensitive layers to image-receiving layers, and the like. It has been also desirable to provide high strength adhesives for bonding spacer rail substrates to other layers, such as barrier-timing layers, supports, photosensitive layers and the like. These spacer rail substrates generally separate the photosensitive element of the film unit from a cover sheet which contains the barrier timing layer.

It is evident that there is a need in the photographic arts for amorphous adhesives which are heat-activatable and which have high bonding strengths over a wide range of temperatures.

M.A. Sandhu and K.L. Tingler; U.S. Patent 4,113,493; September 12, 1978; assigned to Eastman Kodak Company have found that certain amorphous, organic solvent soluble condensation polymers are useful in adhesive compositions for radiation sensitive materials and particularly for photographic film units. These polymers have high bonding strengths at moderate to high temperatures and can be used to bond hydrophilic surfaces to hydrophobic surfaces.

One adhesive composition comprises an amorphous polyester with an acid component which has at least one phthalic acid derivative, and a glycol component comprising from about 15 to 85 mol % of at least one linear aliphatic glycol having the structure HO–R–H where R is selected from the group:

$$-H_2CH_2CO-\bigcirc-OCH_2CH_2O-$$

and $+CH_2CH_2O+_n$ wherein n is an integer or from 2 to 4, and from about 85 to 15 mol % of at least one branched aliphatic glycol having the structure:

$$HOCH_2-\underset{\underset{R^2}{|}}{\overset{\overset{R^1}{|}}{C}}-CH_2OH$$

wherein R^1 and R^2 are alkyl; wherein the polyester is less than 1% crystalline, is capable of remaining amorphous at temperatures above about 20°C, is soluble in halogenated solvents, and has a glass transition temperature from about −30° to +50°C.

The phthalic acid derivative should have at least two condensation sites. Examples are terephthalic acid, isophthalic acid and hexahydroterephthalic acid.

Exemplary linear aliphatic glycols include diethylene glycol, triethylene glycol, tetraethylene glycol, 1,3-bis(2-hydroxyethoxy)cyclohexane, 1,4-bis(2-hydroxyethoxy)cyclohexane and the like.

Exemplary branched glycols include neopentyl glycol, 2,2-diethyltrimethylene glycol, 2,2-dipropyltrimethylene glycol, 2,2-dibutyltrimethylene glycol, 2-methyl-2-ethyltrimethylene glycol, 2-ethyl-2-propyltrimethylene glycol and the like.

Example: Preparation of Poly[Oxydiethylene-co-2,2-Dimethyl-1,3-Propylene (50:50) Terephthalate] − A mixture of dimethyl terephthalate (97.0 g, 0.5 mol), diethylene glycol (45.6 g, 0.43 mol), 2,2-dimethyl-1,3-propanediol (31.2 g, 0.3 mol), zinc acetate dihydrate (60 mg) and antimony trioxide (20 mg) was heated with stirring under a slow stream of nitrogen at 200°C for 2 hr. The temperature was then raised to 240°C and stirring was continued for 2 hr. The mixture was then stirred at 260°C for 1 hr.

The polymerization was performed at 280°C/0.2 mm for 2.5 hr to give a light amber, amorphous polymer of inherent viscosity of 0.57 as measured in a mixture of phenol:chlorobenzene (50:50) at a concentration of 0.25 g/100 ml at 25°C. The glass transition temperature was 50°C.

5 g of the adhesive were dissolved in 25 ml of methylene chloride. After solution was complete, the homogeneous solution was coated at a wet coating

thickness of 4 mils on 2.5- or 4.0-mil poly(ethylene terephthalate). The coating was then dried for 2 hr at 85°C, heat sealed under pressure to various substrates, and tested for peel strength. (Testing procedures are fully described in the patent.) Other adhesives were tested in the same manner. The adhesives made by the described formulation showed overall improvement in bonding strengths to all substrates, though some make better bonds to photosensitive elements than to cover sheets.

Film Adhesive Layer Containing No Gelatin

In order to ensure a perfect adhesion of hydrophilic layers, such as photographic light-sensitive emulsion layers to dimensionally stable polyester film supports, i.e., polyester films that have been biaxially stretched and heat set, it is known to apply several intermediate layers between the support and the light-sensitive emulsion layer. In most cases, two intermediate layers are needed. The first one is an adhesive layer which is usually hydrophobic in order to exhibit good adhesion to the polyester film and, at the same time, possesses good attaching properties in respect of the second layer, called the subbing layer, which usually is formed to a great extent of a hydrophilic colloid such as gelatin.

Although such an adhesive layer attaches itself very tenaciously to a dimensionally stable polyester film support, a supplemental subbing layer is still needed. Indeed, the adhesion of a photographic gelatin-containing layer directly to the hydrophobic adhesive layer leaves much to be desired. This confirms the general rule that a good adhesion of a gelatin layer to a hydrophobic film surface can only be obtained when the surface of the hydrophobic film has been covered with a subbing layer that contains a sufficient amount of gelatin.

It is an object of the process described by *A.J. Van Paesschen and L.J. Van Gossum; U.S. Patent 4,123,278; October 31, 1978; assigned to Agfa-Gevaert N.V., Belgium* to provide a single adhesive layer improving the adhesion of hydrophilic layers to dimensionally stable polyester films; the coating composition for the single adhesive layer being free of gelatin.

A coated film is therefore provided comprising a dimensionally stable polyester film support as defined hereinafter, the film support having a hydrophilic coating adherent thereto by way of an intermediate adhesive layer strongly adhering to it, the adhesive layer consisting essentially from 25 to 60% by wt of a chlorine-containing copolymer, from 15 to 40% by wt of a butadiene copolymer and from 2 to 10% by wt of a water soluble melamine-formaldehyde or hydantoin-formaldehyde resin and being free of gelatin.

By an adhesive layer strongly adhering to the film support is meant that when a pressure sensitive adhesive tape is pressed onto the adhesive layer and then torn off at an acute angle, the adhesive layer should be left undamaged, even after the adhesive layer had been scratched several times with a knife to facilitate the tearing off.

In a method of forming a dimensionally stable polyester film, e.g., of polyethylene terephthalate, the polyester is extruded in the melt and quenched immediately on a cooling cylinder to form an amorphous film. This amorphous film is then stretched longitudinally and transversely, i.e., first longitudinally and then transversely, or first transversely and then longitudinally, or in both directions in a single operation, i.e., biaxially, the stretching being performed at 80° to 90°C to form a crystalline film having its dimensions increased by 3 to 5 times. Subsequently, the film is heat set at 180° to 225°C while being kept under tension in both directions.

The chlorine-containing copolymers of the adhesive layer are preferably copolymers comprising 70 to 95% by wt of vinyl chloride and/or vinylidene chloride, 0.5 to 10% by wt of a hydrophilic monomer, and 0.5 to 25% by wt of at least one other copolymerizable monomer.

Suitable copolymerizable monomers are the esters of acrylic or methacrylic acid; further vinyl esters such as vinyl acetate; and the vinyl ester of versatic acid, which is sold under the trade name Veova 10.

Other suitable copolymerizable monomers are acrylonitrile, N-tert-butylacrylamide, acrolein, and vinyl sulfofluoride.

The hydrophilic monomer may be selected from the following acids, acrylic, methacrylic, crotonic, maleic, fumaric, itaconic; the unsubstituted amides of acrylic and methacrylic acid; the monoalkyl esters of maleic acid; and vinyl pyrrolidone.

The chlorine-containing copolymers are added in latex form to the coating composition for the adhesive layer. This latex is obtained by the emulsion polymerization of the different comonomers according to known emulsion polymerization techniques. The so-called primary dispersion directly formed upon emulsion polymerization is used as such or after adaptation of its concentration. Greatly varying concentrations can be used depending on the final concentration and viscosity needed in the coating composition.

The butadiene copolymer preferably comprises 30 to 70% by wt of monomeric butadiene units. The balance is formed by units deriving from other ethylenically unsaturated hydrophobic monomers, such as acrylonitrile, styrene, acrylic acid esters, methacrylic acid esters and acrolein. Moreover, at most, 5% of the comonomers may be formed by hydrophilic monomeric compounds.

Example 1: A substantially amorphous polyethylene terephthalate film having a thickness of approximately 2.2 mm was formed by extrusion of molten polyethylene terephthalate at a temperature of about 280°C on a quenching drum and was chilled to a temperature of about 75°C and then stretched in the longitudinal direction over a differential speed roll stretching device to 3.5 times the initial dimension at a temperature of 84°C.

The thus stretched film was covered with a layer at a rate of 70 m^2/l of the following coating composition.

Ingredients	Amount (ml)
20% latex of co(vinyl chloride/vinylidene chloride/n-butyl acrylate/itaconic acid) (63:30:5:2% by wt)	300
20% latex of co(butadiene/styrene) (50:50% by wt)	300
10% aqueous solution of Ultravon W*	30
Glycerol (50% in water)	15
40% aqueous solution of dimethyl trimethylol melamine-formaldehyde resin	25
10% aqueous solution of the sodium salt of sulfosalicylic acid	300
Water	30

*Ultravon W is a dispersing agent.

After drying of the coating the film was stretched in a tenter frame to 3.5 times in the transverse direction at about 87°C. The final thickness of the film was about 0.180 mm.

The surface resistance of the dried layer at a relative humidity of 60% proved to be 3.60 x 10^{11} ohms per square.

The film was then conducted into an extension of the tenter frame, where it was heat set while kept under tension at a temperature of 200°C for about 1 min. After heat setting the coated film was cooled and wound up in the normal way.

The thus heat set film was provided with a gelatin silver halide emulsion layer as used in photographic x-ray material. The layers of the photographic material thus obtained showed a very good adhesion to the polyester film support in wet as well as in dry state.

The adhesion in dry state was checked before and after the processing. The gelatin silver halide emulsion layer was scratched crosswise by means of a sharp knife, whereafter an adhesive tape that was pressed over the scratches was torn off at once. The quality was approved only if very small pieces of the photographic layer were torn off.

The adhesion in wet state was checked by scratching the material superficially and trying to rub off the gelatin layer with a finger after each step of the photographic processing (development, rinsing, fixing, rinsing). The gelatin layer should not be damaged during this rubbing.

Example 2: The procedure of Example 1 was repeated except that longitudinally stretched polyester film was covered with the following coating at a rate of 70 m^2/l.

Ingredients	Amount (ml)
20% latex of co(vinyl chloride/vinylidene chloride/n-butyl acrylate/itaconic acid) (63:30:5:2% by wt)	300
40% latex of co(butadiene/acrylonitrile) (55:45% by wt)	150
20% aqueous solution of polyvinyl alcohol	100
10% aqueous solution of the sodium salt of sulfosalicylic acid	300
40% aqueous solution of dimethyl trimethylol melamine-formaldehyde resin	37.5
Glycerol (50% in water)	5
10% aqueous solution of Ultravon W	7.5
Water	100

The surface resistance of the dried layer at a relative humidity of 60% was 2.40×10^{11} ohms per square.

After transverse stretching and heat setting as described in Example 1 a gelatin silver halide emulsion layer as used in photographic x-ray materials was applied to the adhesive layer.

The layers showed a good adhesion to the film support in dry as well as in wet state.

Low Temperature Adhesives for Film Units

J.R. Dann; U.S. Patent 4,126,464; November 21, 1978; assigned to Eastman Kodak Company has found that mixtures of certain block copolymers and co-polymers formed from ethylenically unsaturated monomers are useful as adhesive compositions in radiation sensitive materials. These adhesives have high bond strengths at low temperatures, such as below 75°C and as low as 5°C or lower, and can be used to bond various layers of photographic elements, and particularly barrier timing layers to other layers in photographic film units.

The non-pressure-sensitive, heat sealable adhesive composition having high bond strength at temperatures below about 25°C consists essentially of a mixture of from about 1 to 99% by wt of a block copolymer formed from about 10 to 50% by wt of at least one vinylbenzene and from about 90 to 50% by wt of at least one conjugated diolefin; and from about 99 to 1% by wt of a copolymer formed from about 40 to 85% by wt of at least one alkene and from about 60 to 15% by wt of vinyl acetate.

Example 1: Adhesive Composition Containing Linear Block Copolymer – An adhesive composition was prepared by mixing two commercial materials, namely, 8 g of Borden's HA6164 ethylene-vinyl acetate copolymer adhesive with 2 g of Kraton 1102 (linear block copolymer of polystyrene and polybutadiene) in 50 ml of toluene. This mixture was then heated to 60°C and stirred until all components were dissolved (about 2 hr).

Example 2: Comparison of Superior Bonding Strength of Adhesive over Individual Components — Spacer rail substrates made of polyethylene terephthalate subbed with poly(acrylonitrile-co-vinylidene chloride-co-acrylic acid) (weight ratio 14.1:79.9:6) were coated with the adhesive composition of Example 1 and with similar adhesives having different proportions of copolymers with a 0.004" (0.01 cm) coating knife. After drying at room temperature for about 16 hr, then under vacuum at about 50°C for 1 hr, strips of these coated substrates were bonded to cover sheets having barrier timing layers made of dried latex terpolymers such as those of U.S. Patent 4,056,394.

Two other spacer rail substrates were coated with Borden's HA6164 adhesive and Kraton 1101, respectively, and bonded to barrier timing layers to serve as controls. Bonds were formed by use of heat and a pressure roller on a drum.

Adhesive strength was determined by peel tests using an Instron testing machine. Samples bonded on the drum were peeled at a 90° angle at various rates and temperatures.

Peel strength tests were made for a series of adhesives of varying composition at a peel temperature of 5°C and a peel rate of 12 in/min when bonded to the so-called "single timing layer" of the cover sheet.

These tests clearly showed the strength exhibited by the described adhesives at low temperatures over the individual components. In fact, the improved strength is greater than the sum of the individual component strengths.

Example 3: Adhesive Composition Containing Radial Block Copolymer — 1 g of Solprene 406, a radial block copolymer, was mixed with 9.0 g of the Borden HA6164 ethylene-vinyl acetate adhesive in 50 ml toluene at 60°C until all materials were dissolved. The resulting adhesive was tested as in Example 2 to determine its bonding strength. Peel test strength at 5°C and a 12 in/min (30.5 cm/min) rate was about 1,200 g/in (472 g/cm), also considerably greater than the strength of the control adhesives of Example 2.

A comparison of the bonding strengths of the described adhesives with control adhesives at temperatures from 5°C to 70°C showed that the adhesives of this process exhibit far superior bonding strengths over the formulations of the prior art at temperatures less than 70°C and particularly below 25°C. It is also seen that compositions comprising 10 to 30 wt % of the block copolymers are exceptionally strong adhesives.

Heat-Activatable Adhesives with High Bonding Strengths Above 50°C

J.M. Noonan; R.C. McConkey and M.J. Hanrahan; U.S. Patent 4,150,217; April 17, 1979; assigned to Eastman Kodak Company describe certain water-soluble condensation polymers which are useful in adhesive compositions for photographic materials and laminates. These polymers have high bonding strengths at elevated temperatures and can be used to bond hydrophilic surfaces to hydrophobic surfaces.

One adhesive composition is a water-soluble polyester which comprises a glycol component having at least 50 mol % of an aliphatic diol selected from the group consisting of HO−R−H wherein R is $-(CH_2CH_2O)_n$, and

$$HO-(H_2CH_2CO)_n \underset{\bigcirc}{} (OCH_2CH_2)_n-OH$$

wherein n is an integer from 1 to 4; and an acid component comprising greater than 15 and up to about 35 mol % of at least one dicarboxylic acid having an iminosulfonyl moiety containing a monovalent cation as an imino nitrogen atom substituent, and from about 65 to 85 mol % of one or more other diacids.

In a preferred embodiment, the dicarboxylic acids have an iminosulfonyl moiety represented by the formula:

$$\begin{array}{c} O \qquad\qquad O \\ \parallel \qquad\qquad \parallel \\ -C-Y-Q_p-C- \\ \mid \\ Q'_m \end{array}$$

wherein m and p are integers whose sum equals 1; Q is defined by the formula:

$$\begin{array}{c} O \quad O \\ \parallel \ M^+ \parallel \\ -S-N^--S-Y- \\ \parallel \quad \parallel \\ O \quad O \end{array}$$

Q' is selected from the group consisting of the following:

$$\begin{array}{c} O \quad O \\ \parallel \ M^+ \parallel \\ -S-N^--S-Y' \\ \parallel \quad \parallel \\ O \quad O \end{array} \qquad \text{and} \qquad \begin{array}{c} O \\ \parallel \ M^+ \\ -S-N^--Y' \\ \parallel \\ O \end{array}$$

wherein Y is arylene or arylidene, preferably having from 6 to 12 carbon atoms, such as phenylene, naphthylene, phenylidine, naphthylidyne, and the like; all of which may be substituted with alkyl having from 1 to 4 carbon atoms, such as methyl, ethyl, and the like; halide, such as fluoride, chloride, bromide and the like; and other substituents known to those in the art which will not interfere with the desired properties of the resulting copolyesters. Y' is substituted or unsubstituted aryl, preferably having from 6 to 12 carbon atoms, such as phenyl, naphthyl, tolyl, and the like; or substituted or unsubstituted alkyl, preferably having from 1 to 7 carbon atoms, such as methyl, ethyl, propyl, methylphenyl, and the like; and M is a solubilizing cation and preferably a monovalent cation such as an alkali metal, ammonium cation or hydrogen.

A preferred iminosulfonyl compound is 3,3'-[(sodioimino)disulfonyl] dibenzoic acid or an equivalent benzoate such as the dimethylbenzoate.

From about 65 to 85 mol % of the acid component of the polyesters useful in the process is one or more other diacids or functional derivatives thereof. Exemplary of such diacids are aromatic dicarboxylic acids, such as phthalic, isophthalic, terephthalic and the like; aliphatic dicarboxylic acids, such as malonic, succinic, glutaric, adipic, sebacic and other higher homolog dicarboxylic acids which may be aryl- or alkyl-substituted; carbocylic dicarboxylic acids, such as 1,3-cyclohexanedicarboxylic acid, and the like; heterocyclic dicarboxylic acids, such as 1,4-piperazinylenedicarboxylic acid and the like; and light sensitive ethylenically unsaturated dicarboxylic acids, such as cinnamylidenemalonic acid, etc.

These adhesives can be used in any layer of a photographic element wherein an adhesive is useful. One use would be as subbing layers to render hydrophilic proteinaceous compositions of emulsion layers adherent to hydrophobic element supports, as described in U.S. Patent 3,658,541. Other uses can be determined from general knowledge of the photographic art possessed by a worker of ordinary skill in the art. Typical coating concentrations of the adhesives in various applications would be from about 8 to 24 g/m^2 of support.

A preferred embodiment of the process is an image transfer unit comprising:

A photographic element comprising a support having thereon at least one photographic silver halide layer;

At least one layer comprising an adhesive comprising a water-soluble polyester described hereinabove;

An image-receiving layer; and

Means containing an alkaline processing composition adapted to discharge its contents with the unit.

The polyester adhesives disclosed herein can be used anywhere in image transfer units where there is a need to bond two layers of materials. Preferably, the adhesives are used to close the entire film unit, i.e., to bond the subbed or unsubbed support to a cover sheet during manufacture of the film unit. The adhesives are also useful when bonding hydrophilic subbing layers to hydrophobic layers such as film support surfaces, including polyethylene terephthalate, cellulose acetate butyrate and the like. However, the adhesives disclosed herein also bond strongly to metals, such as aluminum, steel, lead, tin, copper and the like; glass, ceramics, wood and other plastics known to those skilled in the photographic arts. Still another use for the polyester adhesives disclosed herein is to reseal edges of image transfer units after dispensation of the processing composition within the unit and separation of the unit from the means adapted to discharge the processing solution. Still another use of these adhesives is to bond a suitable barrier or timing layer to a support material which may be either the photographic element support or a cover sheet for the entire film unit.

Specialty Applications

FOR SHOE MANUFACTURE

Carbon Blacks Added to Adhesive Used for Shoe Cap Stiffening

In making shoes, certain thermoplastic resins are often used as binders for stiffening the toe cap. When these resins are joined to the leather shoe uppers, using UHF or microwaves, caution must be used to prevent damage to the leather uppers.

P. Kremer, B. Gora and C. van Amsterdam; U.S. Patent 4,069,602; January 24, 1978; assigned to Deutsche Gold- und Silber-Scheideanstalt Vormals Roessler, Germany have found that such damage can be prevented by the inclusion of certain substances in the adhesive coating used.

Stiffening material charged in this manner permits the heating with microwaves in a few seconds to the softening or activation temperature of the binding or adhesive agent without the other materials connected or bonded to them being heated to any appreciable extent. By virtue of these additions it is now possible to use thermoplastic synthetic resins having relatively high softening temperatures for the production of the new stiffening materials.

These shoe-capping materials consist of at least one backing or layer of fibrous material and at least one thermoplastic synthetic resin as a binding or stiffening agent, optionally containing one or more fillers, plasticizers, dyes, pigments and/or stabilizers against light, heat and/or mechanical influences, either individually or in combination. The stiffening material is optionally provided with an adhesive coating based on a thermoplastic synthetic resin on one or both sides (surfaces). The stiffening material is characterized in that the thermoplastic synthetic resin contains at least one active substance in an amount from 3 to 40 parts by weight per 100 parts by weight of synthetic resin.

452

As active substances, the most important are carbon blacks. The active substances also include carbon black preparations and carbon black compounds, as for example the gray pigments which are produced according to British Patent 1,139,620. For example, there can be used the gray pigment (made as described in the above patent) by heating 50 parts by weight of lampblack (Flammruss 101) under reflux for 15 minutes at 60°C with 10 parts by weight of silicon tetrachloride. The excess silicon tetrachloride is then evaporated off in vacuo. The increase in weight is 2.3 parts by weight.

A suspension of 50 parts by weight of aluminum oxide in 1,200 parts by weight of water is heated to 83°C in an open agitation vessel. The pretreated carbon black is introduced into the suspension over a period of 1 hour, followed by stirring for 30 minutes at 83°C and then by stirring with the heat off for another 3 hours. The product is then suction filtered, washed thoroughly with water and dried. 92 parts by weight of gray pigment are obtained.

Furnace blacks with a BET surface area in the range from 50 to 150 m^2/g per 100 pbw of thermoplastic binding agent (solids content) also permit the stiffening materials to be heated to the working temperature in a conventional UHF oven of 2.5 kW capacity in 3 seconds, while the material connected thereto or in contact therewith, for example, leather, remains practically cold. A further group of active materials are the graphites.

Polyurethane Adhesive for Bonding Shoe Soles to Uppers

B.E. Bailey and A.J. Bunker; U.S. Patent 4,098,747; July 4, 1978; assigned to Interox Chemicals Limited, England have formulated an adhesive composition which comprises a polyurethane having (a) units derived from a diisocyanate; (b) units derived from a poly(ε-caprolactone) diol having a hydroxyl value of from 25 to less than 30; and (c) units derived from an alkane diol chain-lengthening agent containing at least 2 carbon atoms, the mol ratio of alkane diol:-poly(ε-caprolactone) diol being from greater than 0.7:1.0 to less than 1.2:1.0. A hydroxyl value of from 25 to less than 30 for the poly(ε-caprolactone) diol represents a molecular weight range of from 4,480 to 3,750.

To improve the storage stability of the polyurethane adhesives an isocyanate blocking agent may be added thereto. The isocyanate blocking agent is preferably a 5- to 7-membered lactam, most preferably ε-caprolactam. The blocking agent may be used in an amount of from 0.1 to 10% by weight, preferably 0.05 to 5.0% by weight, based on the weight of the polyurethane.

In general, there is no discernible difference between the ultimate properties of the adhesives which do contain the isocyanate blocking agent and those which have otherwise identical formulations but do not contain the isocyanate blocking agent. The only difference is that the polyurethane adhesives which do contain the isocyanate blocking agent remain usable for a longer period, that is, have a longer storage life.

The poly(ε-caprolactone) diols suitable for use in this formulation are the polyester reaction products of ε-caprolactone and glycols containing from 2 to 10 carbon atoms, for example, ethylene glycol, trimethylene glycol, tetramethylene glycol and hexamethylene glycol or mixtures thereof. Most preferably tetramethylene glycol (butane-1,4-diol) is used. Most preferably the reaction between the glycol and ε-caprolactone is carried out at a temperature not exceeding 160°C, in the presence of a conventional catalyst.

The diisocyanates which may be used include, for example, diphenylmethane-4,4'-diisocyanate (MDI), tolylenediisocyanates (particularly dimerized tolylenediisocyanates), diisocyanate-dicyclohexylmethane, hexamethylene-diisocyanate and naphthalene-1,5-diisocyanate. MDI is particularly preferred.

Any alkane diols or mixtures thereof which contain at least 2 carbon atoms are suitable for use as chain-lengthening agents and include, for example, ethane-1,2-diol, propane-1,3-diol, butane-1,4-diol, pentane-1,5-diol and hexane-1,6-diol. Butane-1,4-diol is particularly preferred.

The polyurethane adhesives according to this formulation may be used in the conventional manner. For example, they may be applied to surfaces to be adhered together from solutions in suitable solvents. Solution in methyl ethyl ketone is preferred.

The adhesive coated surfaces are preferably heated to about 80° to 90°C to evaporate the solvent and to activate the adhesive. The surfaces may then be pressed together immediately.

The polyurethane adhesives generally have excellent adhesive properties immediately on contact and readily crystallize to form the desired hard but flexible structural adhesive, reaching initial hardness within 24 minutes of being heat softened.

A number of materials, for example, paper, cardboard, leather, wood, glass, metal, thermoset and thermoplastic materials, and rubber (natural, synthetic and urethane rubbers including poromeric) can be bonded very firmly with the polyurethane adhesives according to this formulation. The adhesives are particularly useful for bonding rubber materials, PVC (particularly plasticized PVC) and leather to one another, especially for bonding shoe soles made of these materials to shoe uppers made of leather or of synthetic materials which simulate leather.

In a different formulation for a polyurethane adhesive suitable for bonding shoe uppers and shoe soles, *J.W. Becker; U.S. Patent 4,125,522; November 14, 1978; assigned to E.I. DuPont de Nemours and Company* bases the adhesive on an isocyanate-terminated prepolymer made from 1.2 to 2.0 mols of tolylenediisocyanate containing at least 65% of the 2,4-isomer, 1 mol of polytetramethyleneoxy glycol having a number average molecular weight of about 400 to 3,000, and a polyethyleneoxy compound having a number average molecular weight of about 310 to 4,000 and containing 7 to 30 $-CH_2CH_2O-$ groups and 1 or 2 hydroxyl groups, its $-CH_2CH_2O-$ groups providing 0.5 to 4% of the combined weight of the polyethyleneoxy compound and polytetramethyleneoxy glycol, while the polytetramethyleneoxy provides no less than 80% of the combined weight. A mixture of a methylenedianiline/sodium chloride complex with finely ground urea is used as the curing agent.

The preferred polyethyleneoxy compounds are the readily available and highly effective oxyethylated alcohols, phenols, and diols; for example, octylphenoxypolyethyleneoxy ethanols, nonylphenoxypolyethyleneoxy ethanols, polyethyleneoxy glycol, and oxyethylated polypropyleneoxy glycols. Suitable commercial products are available from several sources such as, for example, Pluronic (BASF Wyandotte Corp.) and Triton (Rohm and Haas). Other suitable polyethyleneoxy compounds can be made according to the process described in U.S. Patent 2,674,619.

The methylenedianiline/sodium chloride complex usually is sold as a dispersion in an inert liquid. Such dispersions can be prepared following the general procedure of U.S. Patent 3,876,604. Finely powdered urea is best added to methylenedianiline/sodium chloride complex as a dispersion. The urea dispersion can be made by ball-milling urea with an inert liquid (of U.S. Patent 3,876,604) until proper particle size (usually less than 20 microns) is obtained. The amount of urea used should be 2 to 5 parts per 100 parts by weight of complex. The proportion of the methylenedianiline/sodium chloride complex to the prepolymer is such that the amine $-NH_2$ to isocyanate $-NCO$ mol ratio is about 0.95-1.50:1, ratios of 1.00-1.30:1 being preferred.

The adhesive is used as a two-part system. The isocyanate-terminated prepolymer which contains the polyethyleneoxy compound is mixed with the curing agent shortly before use. The composition has at room temperature a pot life in excess of 8 hours; preferred compositions, in excess of 24 hours. The adhesive is applied by suitable means such as brushing, doctoring or transfer-coating onto one or both surfaces which are to be bonded; the adhesive surface or surfaces are heated to initiate reaction, and the parts joined in a press for at least 10 seconds. The bonded assembly is sufficiently strong to be subjected to the remaining steps of footwear finishing. The strength of the bond increases with time, approaching its maximum within several hours after bonding.

Example: To 100 g of an isocyanate-terminated prepolymer A containing 4.1% free isocyanate groups (prepared by a reaction of 1.0 mol polytetramethyleneoxy glycol of molecular weight 1,000 with 1.6 mols of 2,4-tolylenediisocyanate for 3 hours at 80°C) is added 3.25 g of octylphenoxypolyethyleneoxy ethanol having an average molecular weight of about 756 and a polyethyleneoxy block of 12 to 13 $-CH_2CH_2O-$ units. These proportions provide a $-CH_2CH_2O-$ content of 2.93% by weight and a polytetramethyleneoxy glycol content of 97.1% based on the total weight of polyethyleneoxy compound and polytetramethyleneoxy glycol. After 3 days at 25°C, the resulting reaction product (prepolymer B) contains 3.7% free isocyanate groups.

Both prepolymer A and prepolymer B are made into adhesives by combining with either curing agent C, consisting of a 1/1 dispersion of methylenedianiline/ sodium chloride complex in di-2-ethylhexyl-phthalate, or curing agent D which is prepared by adding 4.5 parts per 100 parts of curing agent C of a 50% dispersion of finely divided urea in di-2-ethylhexyl azelate:

     (a)   100 parts prepolymer A, 26.7 parts curing agent C.

     (b)   100 parts prepolymer B, 25.6 parts curing agent C.

     (c)   100 parts prepolymer A, 27.9 parts curing agent D.

     (d)   100 parts prepolymer B, 26.7 parts curing agent D.

All four adhesive compositions are still workable 24 hours after preparation. The four compositions are used as adhesives by coating 5 mils (0.13 mm) of the adhesive on roughed 0.3 cm thick slabs of cured styrene-butadiene copolymer. The slabs are immediately heated by radiant heat using a 220-volt adhesive activator so that the adhesive reaches a surface temperature of 73°C. Immediately after heating, the adhesive coating is placed in contact with the vinyl surface of a vinyl-coated fabric containing about 30% dioctyl phthalate as a plasticizer in the vinyl coating and the combined assembly is pressed at 0.41 MPa for 20 seconds. The combined pieces are then peel-tested as described in ASTM D-2558-69 at a jaw separation rate of 5 cm/min. The results are shown in the following table:

Peel Strength with Curing Agent C

	. . . After Release from Press. . . .	
	1 Minute	3 Hours
Prepolymer	. . . . (kN/m).	
A	<0.02	0.02
B	0.2	3.0

Peel Strength with Curing Agent D

A	0.2	1.6
B	1.1	5.8

The above results show the improved bonding performance of the modified prepolymer B with both curing agents. When prepolymer B is used with curing agent D, the resulting adhesive gives high-strength bonds.

Similar results are obtained using chemically equivalent amounts of nonylphenoxypolyethyleneoxy ethanols having number average molecular weights of 600 to 900.

Reinforcing Adhesives for Box-Toe Construction Resins

To obtain improved wear and reinforcement of shape, shoe manufacturers usually apply a thermoplastic stiffening resin, often called a box-toe resin, to the toe portion of the upper shoe.

For a thermoplastic resin to be an acceptable stiffener in this application it must satisfy the following requirements: The resin should have some adhesive character; it should have a low melt viscosity, preferably less than 150 poises at 190°C; the resin should set rapidly to prevent "welding" of stacked assemblages of the manufactured articles; and the resin must be stiff to impart and retain the desired shape but it must also have sufficient flexibility, even at low temperatures, to resist cracking upon impact and to snap back to its original shape. Heretofore, it has only been possible to obtain this difficult yet critical balance of properties by blending the thermoplastic polyamide resin with a minor amount of one or more other compositions including other polyamide resins.

M.L. Mitchell, III and H.J. Sharkey; U.S. Patent 4,122,229; October 24, 1978; assigned to Emery Industries, Inc. have discovered copolyamides derived from a mixture of a polymeric fatty acid and a short-chain dibasic acid and a mixture of a polyoxyalkylene diamine and a short-chain diamine are superior box-toe construction resins which satisfy all the aforementioned criteria for such adhesive compositions.

These copolyamide resins are obtained by the reaction of essentially stoichiometric amounts of a mixed acid component and a mixed amine component. A polymeric fatty acid obtained by the polymerization of an olefinically unsaturated monocarboxylic acid containing 16 to 20 carbon atoms is an essential component of the acid mixture. C_{36} dimer acids are especially useful. A short-chain saturated aliphatic dicarboxylic acid containing 7 to 12 carbon atoms, preferably azelaic acid or sebacic acid, is employed with the polymeric fatty acid. The equivalent ratio of polymeric fatty acid to short-chain dibasic acid ranges from 0.95-0.7:0.05-0.3. The mixed diamines consist of a short-chain diamine containing from 2 to 6 carbon atoms, preferably ethylenediamine, and a polyoxyalkylene diamine having a molecular weight from about 600 to 5,000. The equivalent ratio of short-chain diamine to polyoxyalkylene diamine, preferably having a molecular weight from 1,000 to 5,000, will range from about 0.92-0.995:0.08-0.005.

The resulting reinforcing copolyamide resins typically have an acid value and amine value less than 15, softening point in the range from 140° to 190°C, viscosity at 190°C less than 150 poises, compression strength greater than 85 grams with a brittleness temperature less than -10°C. These resins can be applied to a variety of substrates including leather and synthetic poromeric materials, woven and nonwoven fabrics, and a wide variety of polymeric materials and will readily adhere thereto. A 1- to 50-mil film of the copolymer on the substrate provides a tough resilient reinforcing coating on the substrate so that it can be shaped and otherwise molded to the desired configuration and will retain this shape during use.

These resins are particularly adaptable for use with leather, fabrics and vinyl polymers used in box-toe construction and impart greater stiffness to the substrate while maintaining flexibility.

Example: Azelaic acid, a polymerized fatty acid (75% C_{36} dimer acid, 20% trimer acid and 5% monocarboxylic acid), ethylenediamine and a polyoxypropylene diamine having an average molecular weight of about 2,000 were charged to a glass reactor as follows:

	Parts	Equivalent Ratio
Polymerized fatty acid	390.5	0.91
Azelaic acid	22.5	0.16
Ethylenediamine	43.2	0.96
Polyoxypropylene diamine	60.0	0.04

A slight excess of the acids was used to avoid excessive molecular weight build-up. The reactants were then heated under a nitrogen atmosphere to about 160°C and as the reaction progressed the temperature was steadily increased to a maximum temperature of 240°C. Water of condensation was removed by use of a condenser-trap arrangement and when about 95% of the theoretical amount of water was collected a vacuum of about 5 torrs was applied to remove the final traces of water and any other volatiles present.

Physical properties of the resin were as follows:

Acid value	4.4
Amine value	3.6
190°C viscosity (poise)	91.5
Softening point (°C)	158–162
Tensile strength (psi)	1,750
Elongation at break (%)	40
Young's modulus	12,100
Compression strength (g)	110
Brittleness temperature (°C)	–24

The resin exhibited good adhesive properties with a variety of materials. For example, the shear strength obtained in accordance with ASTM D-1002 with aluminum and steel was 1,100 psi. The molten resin was also applied to a variety of flexible substrates including cotton duck and leather. The resin exhibited good adhesion properties and imparted increased rigidity and resilience to these substrates without destroying their flexibility. A peel strength of 13 lb/in was obtained with cotton duck using ASTM D-1876. When this same test method was used to determine the peel strength of the reinforced leather specimen, failure of the leather occurred before adhesive failure.

Polyurethane Adhesives Having Uniform Molecular Weights

G. Falkenstein, O. Volkert and L. Mämpel; U.S. Patent 4,156,064; May 22, 1979; assigned to BASF Wyandotte Corporation have developed a process which makes possible the production of polyurethane adhesives with uniform molecular weights and reproducible mechanical properties in the absence of traditional polyurethane-dissolving organic solvents.

It has been found that polyurethanes with accurately defined molecular weight are obtained if instead of converting the polyol and diisocyanate at the calculated weight ratio and tempering the polyurethane until all NCO groups have disappeared, an excess of diisocyanate is used and the tempering process is terminated at the desired solution viscosity. The chronological increase in the solution viscosity, or, in other words, the chronological increase of the molecular weight of the polyurethane within certain limits is essentially independent of the NCO/OH ratio. At the normally applied temperatures of 120° to 125°C this process takes place so slowly that it can be followed conveniently by following the melting viscosity and interrupting at the appropriate point in time.

The polyaddition is thus carried out by using an excess of diisocyanate, the reaction is terminated upon reaching the desired molecular weight, and the still-existing excess isocyanate groups are reacted with alcohols, ammonia and/or amines. The process is characterized by the polyurethane-polyaddition products being changed into particles of 1 to 15 mm and then being treated with possibly substituted primary aliphatic mono- to trivalent alcohols, ammonia and/or aliphatic, cycloaliphatic or aromatic mono- or diamines in the liquid or vapor phase in the absence of organic solvents.

In detail, the polyurethane adhesives are obtained as follows: The polyol, the diisocyanate and, if applicable, the chain extenders, catalysts and other normally used materials are mixed at temperatures of approximately 40° to 100°C, preferably 50° to 80°C, within about 3 to 30, preferably 5 to 15, minutes. Subsequently, and as long as the material still flows, it is filled into molds and is tempered in customary tempering ovens at preferably 90° to 130°C until the polyurethanes have reached the desired molecular weight.

The progress of the polyaddition process is monitored by determining the solution or melting viscosity of samples removed from the product. At the preferred temperatures the viscosity (i.e., the molecular weight) of the polyurethane rises to the desired value within 2 to 24 hours (preferably within 5 to 10 hours). The viscosity increase is nearly independent of the applied NCO/OH-equivalency ratio. As soon as the polyurethane adhesive has reached the desired molecular weight, the product is crushed to a particle size of 1 to 15 (preferably 3 to 10) mm diameter by customary cutting, grinding or granulating equipment.

Following this process the material is treated with an alcohol, ammonia, and/or amines and/or mixtures of these compounds, as described below, at temperatures of 10° to 100°C (preferably 20° to 50°C) for about 3 to 120 (preferably 10 to 60) minutes. For this process the alcohols and/or amines may be used in liquid or gaseous form. The ammonia is applied either in the gas phase or, preferably, as an alcoholic solution. If liquids are used, the polyurethane particles are suspended in the alcohol, amine, the alcoholic ammonia and/or amine solution. After saturation of the isocyanate groups the product is separated from the liquid phase, preferably by filtration or centrifuging, and is subsequently dried. In the other case, the polyurethane particles are preferably placed in an alcohol, ammonia and/or amine atmosphere for 10 to 60 minutes. According to a preferred form of this process, the polyurethane particles are transported through the gas on a continuously operating conveyor.

The polyurethane adhesives produced have molecular weights of 40,000 to 60,000, the distribution of the molecular weight being very closely limited. The product is soluble in inert organic solvents such as acetone, methyl ethyl ketone, tetrahydrofuran, dioxane, butyl acetate, ethyl acetate, or dimethylformamide.

The polyurethane adhesives may be stored and shipped in the form of a granulate, and may be dissolved in the above-specified solvents. The viscosity of these adhesive solutions can be adjusted to the particular requirements of the gluing process or the materials to be glued by varying the polyurethane content.

Alcohols suitable for use in the abovedescribed process are primary aliphatic mono- to trivalent alcohols with 1 to 6 carbon atoms such as methanol, ethanol, ethylene glycol, glycerine, etc. Suitable amines are aliphatic or aromatic mono- or diamines such as ethylamine, butylamine, ethylene diamine, hexamethylene diamine, etc. Particularly well proven and therefore used on a preferential basis are ammonia, methanol, ethanol, butanol, methylamine and solutions of ammonia, methylamine and ethanolamine in methanol, ethanol, ethylene glycol and diethylene glycol. Suitable and simple to use are alcoholic amine solutions and preferably alcoholic solutions with an amine and/or ammonia content of 0.5 to 10, preferably approximately 1 to 5% by weight relative to the total weight of the solution.

High-strength adhesive bonds can be obtained with the polyurethanes manufactured according to this process. The adhesives are suitable for gluing numerous materials such as paper, cardboard, wood, metal and plastics. These adhesives are preferred for gluing rubber materials, polymers of vinyl chloride containing plasticizers to leather, particularly soles of these materials to the shoe uppers of leather or leather-like synthetic materials.

ADHESIVES FOR SKI MANUFACTURE

Containing an ABS Graft Polymer

H. Lehmann and H. Zondler; U.S. Patent 4,117,038; September 26, 1978; assigned to Ciba-Geigy Corporation have as their objective in this patent the provision of epoxy resin adhesives with adequate shelf lives which, using short curing times at 80° to 100°C, produce even stronger bonds than can epoxy resin mixtures of the prior art at curing temperatures of up to 100°C. The solution of this technical problem is of great importance especially in those fields in which plastics having relatively low softening ranges, such as ABS (acrylonitrile-butadiene-styrene) polymers, polyethylene and PVC, are to be bonded in as large a quantity as possible. The ski manufacturing industry is mentioned here as a particular example.

Accordingly, their formulation provides a storable, rapidly hardening epoxy resin adhesive which contains:

(a) a polyglycidyl compound containing on average more than one glycidyl group in the molecule and having a softening temperature between 40° and 90°C;

(b) a salt of 1 mol of ethylenediamine, 1 mol of bisphenol A and 2 g of N,N-dimethyl-1,3-diaminopropane;

(c) an ABS graft polymer; and

(d) a copolymer of ethylene, acrylic acid and acrylate having a melting range from 70° to 110°C;

such that the epoxy resin adhesive contains, per 1 equivalent of glycidyl groups, 0.5 to 1.5 equivalents of hydrogen atoms bonded to nitrogen of ethylenediamine and of N,N-dimethyl-1,3-diaminopropane, and per 100 parts of the polyglycidyl compound contains 3 to 30 parts by weight of the ABS graft polymer and 3 to 20 parts by weight of the copolymer of ethylene, acrylic acid and acrylate, and is in the form of a fine heterogeneous powder, optionally processed to tablets or granules, such that one category of particles contains the respective polyglycidyl compound and another category of particles contains the salt of 1 mol of ethylenediamine, 1 mol of bisphenol A and 2 g of N,N-dimethyl-1,3-diaminopropane, and the particles of a third category consist of the copolymer of ethylene, acrylic acid and acrylate.

These epoxy resin adhesives contain as polyglycidyl compound (a) preferably solid bisphenol A epoxy resins or glycidylated phenol or cresol novolaks.

The salt of 1 mol of ethylenediamine, 1 mol of bisphenol A and 2 g of N,N-dimethyl-1,3-diaminopropane contained in the adhesive is prepared in simple manner by fusing all the starting materials together.

Those adhesives which contain as copolymer (d) 87 to 91% by weight of ethylene, 3 to 5% by weight of acrylic acid and 6 to 8% by weight of acrylate are a preferred embodiment.

The principal use of these epoxy resin adhesives is in the field of ski construction. Chiefly powder, but also granulated, compositions are used. In producing bonds with these epoxy resin adhesives, it has proved advantageous to use coated fiber webs, woven materials, boards or sheets of plastics, glass, or the like as an intermediate layer between the surfaces to be bonded. In ski construction, this results in an elastic reinforcement of the laminate.

Example: A resin component is prepared as follows: 72 g of bisphenol A epoxy resin which is solid at room temperature and has an epoxide equivalent of 1,043 and 7 g of a liquid bisphenol A epoxy resin with an epoxide equivalent of 190 are fused together at 130°C and the melt is mixed with 21 g of heavy spar. After cooling, the solid mixture is ground in a beater mill to a powder having a particle size between 60 and 300 μ.

A hardener for the epoxy resin is prepared as follows: 1 mol of ethylenedi-amine, 1 mol of bisphenol A and 2 g of N,N-dimethyl-1,3-diaminopropane are heated to about 120°C until a homogeneous melt is formed. With stirring, 14 g of an ABS graft polymer in powder form (Novodur A-50) having a particle size smaller than 500 μ, a density of 1 g/ml and a powder density of 320 g/l, are then added. The resulting suspension is cooled to room temperature, whereupon the entire mixture solidifies. The solidified melt is ground in a beater mill to a powder having a particle size of less than 500 μ.

The resin component powder (100 g) is mixed intensively with 10 g of the hardener powder, 5 g of a copolymer of ethylene, acrylic acid and acrylate (Lupolen A 2910) having a particle size of less than 300 μ, and 0.5 g of a silicic acid powder prepared by hydrolysis of silicon tetrachloride in an oxyhydrogen flame (Aerosil 2431/380).

The finely powdered product thereby obtained is an epoxy resin adhesive of the process which has the following features:

> softening point (Kofler bench): 48° to 58°C;
>
> shelf life at 20° to 25°C: longer than 6 months;
>
> curing time at 80°C: 20 minutes; tensile shear strength (DIN 53 283), 8 to 10 N/mm^2; peel strength (British Standard DTD 5577), 3 to 5 N/mm;
>
> curing time at 90°C: 10 minutes; tensile shear strength, 15 to 17 N/mm^2;
>
> curing time at 100°C: 8 minutes; peel strength, 6 to 9 N/mm.

Containing a Substituted Pyrazine

H. Lehmann and H. Zondler; U.S. Patent 4,122,128; October 24, 1978; assigned to Ciba-Geigy Corporation have found another formulation which provides epoxy resin adhesives with short curing times at from 80° to 100°C. This adhesive contains:

(a) a polyglycidyl compound containing on average more than one glycidyl group in the molecule and having a softening temperature between 40° and 90°C;

(b) a 2,5-di-(ω-aminoalkyl-1')-pyrazine of the general formula:

$$H_2N(CH_2)_n \underset{N}{\overset{N}{\diagup}} \diagdown (CH_2)_n - NH_2,$$

where n is an integer from 3 to 6; and

   (c) a copolymer of ethylene, acrylic acid and acrylate having
        a melting range from 70° to 100°C

such that the epoxy resin contains, per 1 equivalent of glycidyl groups, 0.5 to
1.5 equivalent of hydrogen atoms bonded to nitrogen of the 2,5-di-(ω-amino-
alkyl-1')-pyrazine of the above formula, and, per 100 parts by weight of the
polyglycidyl compound, contains 3 to 20 parts by weight of the copolymer of
ethylene, acrylic acid and acrylate. It is in the form of a fine heterogeneous
powder, optionally compacted to tablets or granules, such that one category of
particles contains the respective polyglycidyl compound and another category
of particles contains the 2,5-di-(ω-aminoalkyl-1')-pyrazine of the formula, and
the particles of a third category consist of the copolymer of ethylene, acrylic
acid and acrylate.

A particularly suitable hardener for the adhesive is the 2,5-di-(ω-aminoalkyl-1')-
pyrazine of the above formula, where n is 5, namely 2,5-di-(5-aminopentyl-1')-
pyrazine. This diamine can be readily prepared in good yield by the hydrogenation of 2-nitromethane-perhydroazepine.

Example 1: Preparation of the Hardeners — 80 g of 2-nitromethane-perhydro-
azepine are dissolved in 800 ml of methanol and 80 ml of glacial acetic acid and
the solution is hydrogenated at 45°C under normal pressure in a hydrogenation
flask in the presence of 16 g of 10% palladium on charcoal. The uptake of hydrogen ceases after 3 to 4 hours. The catalyst is filtered off and, after addition of
2 equivalents of alcoholic hydrochloric acid, the filtrate is concentrated in
vacuo. The solid residue is recrystallized once from 100 ml of isopropanol and
then from a mixture of methanol and isopropanol, affording 55.5 g (67% of
theory) of 2,5-di-(5-aminopentyl-1')-pyrazine dihydrochloride with a melting
point of 305°C. The titration of the salt with NaOH confirms the presence of the
dihydrochloride.

For conversion into the free base, 150 g of the dihydrochloride are dissolved in
600 ml of 5 N sodium hydroxide solution. After extraction with chloroform, the
extract is dried over anhydrous potassium carbonate and concentrated by rotary
evaporation in vacuo at 30°C, affording 121.3 g of crystalline amine. Distillation
yields 101 g of pure 2,5-di-(5-aminopentyl-1')-pyrazine with a boiling point of
150°C at 0.02 torr and a melting point of 58° to 61°C.

Example 2: Epoxy Resin Adhesive — 72 g of a solid epoxy resin based on epichlorohydrin and bisphenol A and having an epoxide equivalent of 1,043 and 7 g
of a liquid bisphenol A epoxy resin having an epoxide equivalent of 190 are
fused together at 130°C and the melt is mixed with 21 g of heavy spar, cooled,
and ground to a powder having a granular size between 100 and 300 μ.

This resin powder (100 g) is mixed for 5 minutes at room temperature with
6.2 g of 2,5-di-(5-aminopentyl-1')-pyrazine, which also has a particle size of 100
to 300 μ, and with 0.5 g of Aerosil and 5 g of a copolymer powder of ethylene,
acrylic acid and acrylate (Lupolen A 2910). A single component adhesive powder having the following properties is obtained:

softening point (Kofler bench): $50°$ to $60°C$.
shelf life at $20°$ to $25°C$: 6 months
curing time at $90°C$: 10 to 12 minutes
curing time at $100°C$: 8 to 10 minutes
tensile shear strength (DIN 53283): 15 to 17 N/mm^2
peel strength (British DTD 5577): 5 to 7 N/mm

FOR DECALCOMANIAS

For Glass and Ceramic Decoration

The heat-release type of decalcomania includes a backing sheet which carries a heat-releasable wax layer, which is separated from the backing sheet by means of a so-called barrier layer. A design layer including a resinous carrier is printed on the wax layer and a layer of heat-activatable adhesive is disposed on the design layer.

In using the heat-release type of decalcomania, the article to be decorated is preheated and the decalcomania is applied to the article with the backing sheet up so that the layer of heat-activatable adhesive directly contacts the article. The heat of the article effects both a preliminary bonding of the design layer, via the heat-activatable adhesive, to the article, and also the release of the backing sheet including the barrier layer and a portion of the wax layer, from the design layer. Thereafter, the article is fired at a high temperature in the usual manner so that the design layer is melted and permanently fused to the article; the combustible organic ingredients including the remaining portion of the wax layer, the resinous carrier for the design layer as well as the heat-activatable adhesive layer are consumed during the firing.

H. Meade; U.S. Patent 4,068,033; January 10, 1978; assigned to Commercial Decal, Inc. has found that the disadvantages associated with the prior art heat-releasable decalcomanias which require preheating of the article to be decorated can be overcome by employing a special adhesive composition over the design layer or the protective layer for such design layer, which will allow preheating of the decalcomania so that it can be applied to a cold article or ceramic ware. This is possible through the use of an adhesive composition which has a relatively low melting point, for example, below $250°F$, which upon being heated to its softening point will soften and become tacky. The heated adhesive composition will remain tacky long enough to effect a preliminary bonding of the design layer to a cold ceramic ware while allowing release of the backing sheet including the barrier layer and a portion of the wax layer from the design layer.

The adhesive composition used comprises (a) an acrylic resin which functions as an adhesive base material, (b) a plasticizer or a tackifier for the acrylic resin which is capable of providing an adhesive which softens when subjected to heat and remains tacky at a temperature substantially below its melting point. Furthermore, the tackifier employed must be cleanly burned upon final firing without the deposition of a harmful residue. Such tackifier will include a polyethylene glycol, and/or a solid polyoxyalkylene derivative of propylene glycol or ethylenediamine. The adhesive composition also comprises (c) a major amount of an organic solvent. Preferably, a combination of tackifiers will be utilized, including a tackifier as noted above in combination with a second tackifier including one or more aromatic acid esters of monomeric and polymeric alkyl polyols.

The adhesive composition may also include an alkyd resin to modify the adhesive quality thereof and/or a cellulosic derivative which moderates the tackiness of the mixture of acrylic resin and tackifiers and enhances the cohesiveness thereof.

Thus there is provided a heat-release ceramic decalcomania comprising a decalcomania backing comprising a backing sheet (such as a paper sheet), a wax-impervious barrier coating disposed on the backing sheet, and a wax layer disposed on the barrier coating, and optionally a sealant or support layer disposed on the wax layer. A design layer is disposed on the wax layer or sealant layer, and a heat-activatable adhesive composition as described above disposed on the design layer.

The decalcomania is preheated to at least the softening point of the adhesive composition and is applied to an article to be decorated, with the backing sheet up, so that the layer of softened heat-activatable adhesive directly contacts the article and temporarily bonds the decalcomania to the article. The heat which activates the adhesive also causes the backing sheet, including the barrier layer and a portion of the wax, to be released from the design layer. Thereafter, the remaining assemblage is fired thereby causing the design layer to fuse and form a permanent bond with the article.

The following are preferred formulations of adhesive compositions for the decalcomanias:

Constituent	. . . . .Formulation.								
	A	B	C	D	E	F	G*	H	I
	. . . . (percent by weight).								
Acrylic resin									
Carboset 514A	—	16	11.5	18	—	19.5	12	—	—
Acryloid B-72	—	11	4.5	6.5	—	7	12	—	—
Acryloid B-48N	19	—	—	—	16.6	—	—	22.5	22.5
Polyol benzoate									
Benzoflex S-404	5.4	—	4.5	—	4.7	—	—	—	—
Benzoflex S-552	—	5.5	—	6.5	—	7	—	—	—
Polyethylene glycol									
Carbowax (4,000 or 6,000)	—	8	—	8	—	9	6	—	19
Polyoxyalkylene derivative									
Pluronic F-108	16	—	20.5	—	14.2	—	—	19	—

(continued)

	. . . . .Formulation.								
Constituent	A	B	C	D	E	F	G*	H	I
	. . . . (percent by weight)								
Cellulosic derivative									
Cellulose acetate butyr-									
ate (½ sec)	—	—	4.5	1.5	4.1	—	—	—	—
Polystyrene plasticizer									
Polystyrene 279V9	—	4	—	3	—	5	6	—	—
Alkyd resin									
Burnok 4040-Loms-60	—	4	2	4	—	5	—	—	—
Solvent	. . . . remainder up to 100%.								

*Includes 12% dicyclohexyl phthalate as a solid plasticizer.

A preferred solvent mixture is comprised of the following:

	Percent by Weight
Ethylene glycol monomethyl ether acetate	33
Ethylene glycol monomethyl ether	33
Diacetone alcohol	14
n-Butyl alcohol	20
Total	100

In another patent *assigned to Commercial Decal, Inc., R. Andrews; U.S. Patent 4,117,182; September 26, 1978* describes a heat-release layer for use in the same type of decalcomania described in the previous patent. This heat-release layer is disposed between the backing sheet and the design layer and is made from a normally solid straight chain, primary aliphatic oxyalkylated alcohol.

In a preferred embodiment, the heat-release layer has a molecular weight above 1,350 and also preferably has a melting point above 110°F. The release of the design layer from the backing sheet upon heat activation is considerably facilitated, while at the same time, by using this heat-release layer, after release from the backing sheet, the residue of the release layer which remains on the design layer does not interfere with subsequent firing of the decalcomania.

Figure 11.1 illustrates a heat-release decalcomania which includes a decalcomania backing 10 consisting of a paper sheet 12, with barrier layer 14, and covered by release layer 16. A sealant layer 18 of a cellulosic derivative is preferably disposed over the cellulosic derivative or sealant layer, which serves as an imprint receiving support for the design layer. Over the design layer 20 is disposed the layer 22 of heat-activatable adhesive material which serves as a temporary binder for securing the design layer to the article or ware to be decorated.

The heat-release layer comprises a normally solid straight chain, primary aliphatic oxyalkylated alcohol. The preferred compounds so employed include normally solid, straight chain, primary aliphatic oxyethylated alcohols marketed by BASF Wyandotte as Plurafac A-38 and most particularly Plurafac A-39.

The heat-activatable adhesive layer **22** used here is described in the previous patent by H. Meade.

Figure 11.1: Heat-Release Decalcomania

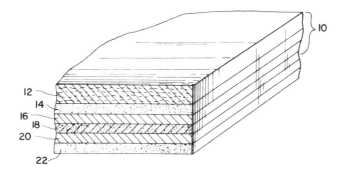

Source: U.S. Patent 4,117,182

ADDITIONAL APPLICATIONS

Adhesive Coated with Powdered Deliquescent Material for Wall Covering

Decorative wall coverings are usually applied either by having paste brushed onto them before application or providing the wall covering material with a coating of pressure sensitive adhesive which is then covered with a release paper or foil before it is rolled for selling. This prevents the adhesive from coming into contact with the other decorative surface of the sheet material thus preventing adhesion to that surface. However, the need to use a release paper or foil to prevent adhesion to the decorative surface considerably adds to the production costs and makes the application of the material to a surface difficult because of the need to remove the release paper or foil. Also, the intrinsic tackiness of pressure sensitive adhesives makes accurate positioning for pattern matching extremely difficult.

A. Burton and D.R. Reed; U.S. Patent 4,082,890; April 4, 1978; assigned to Imperial Chemical Industries Limited, England have provided a process for making a decorative sheet material for wall or shelf covering, etc. which comprises the application to the wrong side of the decorative material of a coating of a water-activatable adhesive; applying to the surface of the adhesive coating a dry coating of a deliquescent material; rolling up the sheet material to form a roll having a plurality of turns in which the coating of the deliquescent material contacts the decorative surface of the sheet material; and packaging the roll in such a manner that air, and hence moisture in the atmosphere, is prevented from contacting the rolled-up sheet material. Conveniently, the rolled-up sheet material is stored in a sealed bag which is impermeable to moisture vapor in the air.

It has been found advantageous to incorporate in the adhesive either a material which absorbs water or a material which reacts with water. The incorporation of such a material in the adhesive serves to remove from the activated adhesive any excess water which might be present and so assists in the development of a satisfactory bond. Such materials as plaster of Paris, Portland cement and other cementitious material have been found to be particularly useful for this purpose.

An adhesive which is soluble in, or activated by, water may be used in this process. Suitable adhesives are those based on starch, starch derivatives, cellulose derivatives, polyvinyl alcohol, polyvinyl acetate, polyacrylates and other natural and synthetic materials or blends thereof. In preference, however, the pressure sensitive adhesive ARG 746 (Samuel Jones Limited) is used. Desirably, the adhesive is suitably formulated to provide the desired degree of slip and final adhesion. Preferably the coating weight of the water-activatable adhesive will be in the range of 5 to 50 g/m^2, and more preferably in the range of 10 to 30 g/m^2.

The powdered coating which is applied over the surface of the adhesive layer may be based on any suitable deliquescent material, examples of which are lithium chloride, calcium chloride, magnesium chloride, ferric chloride, calcium nitrate. The critical factor in the selection of a suitable material is the rate at which the material absorbs water from the atmosphere, and this is related to the size of the particles of the material and the relative humidity of the atmosphere. In general, for most conditions of relative humidity which are likely to be encountered and with the above materials, a suitable average particle size can be selected below 200 μ. Usually the coating weight of the powdered deliquescent material will be up to 100% of the adhesive weight, and more preferably between 5 and 50% of the adhesive weight.

Example: A length of a printed and embossed wall covering based on a sheet of formed polyethylene was coated on its working surface with an adhesive composition comprising a commercially available adhesive, ARG 746. The applied coating weight was between 15 and 20 g/m^2. The coated wall covering was passed through an oven maintained at a temperature of 70°C and the solvent present in the adhesive evaporated off.

To the solvent-free, but still tacky, adhesive surface was applied a continuous coating of powdered (particle size less than 60 μ) anhydrous calcium chloride. This removed the tackiness from the adhesive coating and allowed the sheet to be rolled up without the adhesive adhering to the decorative surface. The rolled-up wall covering was stored in a moisture-vapor-impermeable bag. After a few days, the roll of wall covering was removed from the bag and cut to a suitable length for decorating a wall.

The length of wall covering was exposed to the atmosphere for between 10 and 15 minutes depending on the ambient conditions and this rendered the surface of the adhesive layer fluid and in a condition which allowed the wall covering to be positioned on the wall by sliding it about. When in position it could then be adhered to the wall by hand pressure. Within ten minutes to several hours, depending on the nature and absorbency of the wall, a satisfactory bond is achieved.

Adhesive Type-Correcting Material

J.N. Pierce; U.S. Patent 4,085,838; April 25, 1978; assigned to Minnesota Mining and Manufacturing Company has developed a type-correcting article, shaped somewhat like a pencil, which will contact and adhere to a character typed in a vertically cohesive ink on a piece of paper and can thereafter effect removal of the character from the substrate by quick withdrawal of the article from contact with the substrate, as with a jerking or snapping motion, whereby the ink character is "plucked" from the substrate. The articles can comprise the discriminating adhesive composition alone in a convenient form which may inherently define a body member or handle, or can comprise the adhesive in combination with associated supporting and/or applying means such as a sheet, stamp, key, handle or other body member.

A vertically cohesive ink is a specially formulated ink which is adapted to have a greater internal cohesion than external adhesion to the substrate to which it is applied. Inks having this characteristic generally are deposited as a thin, integral, coherent film on the typing substrate. This allows a typed ink character to be contacted on its surface by an adhesive means and plucked from the surface of the paper as an integral unit, i.e., without fracturing internally and without leaving ink remaining on the paper. Correctable typewriter ribbons employing vertically cohesive inks are commercially available. One such product is Correctable Film Ribbon (International Business Machines, Inc.).

The type-correction material must be made from a discriminating adhesive— one which has a greater affinity for vertically cohesive ink than for paper. This property results from a combination of factors including the chemical and physical natures of the adhesives. Thus, it has been found that in addition to having an adhesive affinity for ink, the adhesive must also be elastomeric and have greater internal cohesion than external adhesion to ink or paper.

The desired balance of physical properties can be achieved by maintaining the tensile, elongation and flexibility or hardness within certain ranges. It has been found that a solid, flexible, adhesive material having a tensile and elongation at break of greater than 11 kg/cm^2 tensile and 200% elongation (measured on an Instron tensile tester at room temperature and at an elongation rate of about 12.5 cm/min) is suitable. In addition, the Shore A hardness of a 1-cm thick slab of adhesive should be between 20 and 60.

Compositions which provide solid adhesives having the required discriminating properties can be obtained from blends of various classes of resins or polymers. A particularly preferred adhesive blend comprises the plasticized resins commonly referred to as plastisols, i.e., resins fused with a liquid plasticizer. Preferred plastisols comprise vinyl-containing resins such as polyvinyl chloride and vinyl chloride-vinyl acetate copolymers. These resins can be fused at elevated temperatures with liquid plasticizers so that a single-phase, solid, thermoplastic discriminating adhesive is obtained. Plasticizers which may be used are the conventional liquid monomeric and polymeric plasticizers known in the art. Typical monomeric plasticizers are the phthalate, adipate and sebacate esters as well as the aryl phosphate esters. Typical polymeric plasticizers are the acetylated polyesters such as the Morflex resins, and epoxy resins such as the DER, ERL and Epon resins commercially available. Other useful plasticizers are commercially available PA-3 and PA-5 (Eastman Chemicals) which plasticizers are particularly useful in combination with the polyvinyl chloride polymers and copolymers.

Adhesion to ink can be improved by the addition of minor amounts of mixtures of dimer acids and solid polyamide resins blended in a weight ratio of about 2:1 dimer acid:polyamide resin. The mixture is preferably added in amounts up to about 15% by weight of the total adhesive composition. The preferred dimer acids are aliphatic, dibasic acids produced by the polymerization of unsaturated fatty acids such as those commercially available from Emery Industries, Inc. under the trademark Empol. The polyamide resins preferred can be derived from the reaction of dimer acids with diamines. Useful polyamide resins are commercially available from Emery Industries, Inc. such as Emerz 1533 and 1540, and from General Mills such as Versamid 940.

Example 1: A type-correcting article was prepared by providing 40 g of a powdered vinyl chloride-vinyl acetate resin (PVC 74, Diamond Chemicals Co.) and homogeneously blending this resin with 60 g of a plasticizer (PA-5). The homogeneous mixture was then poured into a mold which was in the form of a long tapered glass tube about 11.5 cm in length (standard centrifuge tube) and heated to 125° to 130°C for about 10 minutes. At this temperature the resin melted and fused with the plasticizer to form a solid plastisol. This fused composition was cooled to room temperature and removed from the mold.

The elongated solid article was used to remove characters which had been typed onto paper with a correctable ribbon having a vertically cohesive ink, by pressing the tapered end of the article onto the freshly typed character, maintaining contact with the character briefly (about 1 second) and quickly jerking the article from the paper with a quick snapping action. In some cases it was necessary to repeat this step to completely remove the ink. The article was cleaned or "regenerated" after use by contacting the ink adhering to the article with a piece of pressure sensitive adhesive tape whereby the ink adhering to the tapered end of the article was transferred to the tape.

Example 2: A type-correcting article was prepared by melting an ethylene-vinyl acetate resin (Elvax 40, DuPont) in an oven at 150°C and the rounded eraser end of a wooden pencil was dipped into the melt and suspended vertically (eraser end down) in the 150°C oven long enough for the resin to develop a smooth protuberant shape. The pencil was then removed from the oven and cooled. The article was useful as a type-correcting stick when used as described in Example 1.

Adhesive Coating for Graphite Crucibles

Y_2O_3 because of its refractory nature and its inertness to molten metals is a desirable material for coating surfaces (especially graphite) which contact molten uranium during alloying and nuclear fuel fabrication processes. Eu_2O_3 and Gd_2O_3 are particularly useful as coatings or fabrication materials for radiation shielding and nuclear reactor control rods because of the exceptionally high thermal neutron absorption cross-section of naturally occuring Eu and Gd particularly [153]Eu and [157]Gd. Neodymium oxide is useful for its thermal and electrical insulating properties in high temperature applications.

The object of the process described by *C.E. Holcombe, Jr. and R.L. Swain, J.G. Banker and C.C. Edwards; U.S. Patent 4,087,573; May 2, 1978; assigned to U.S. Department of Energy* is to provide spontaneously hardening adhesive compositions comprising Y_2O_3, Nd_2O_3, Eu_2O_3 or Gd_2O_3 which may be applied as surface coatings and subsequently decomposed in place to the pure oxide which remains firmly bound to the surface.

It has been found that when powders of the four oxides, Y_2O_3, Eu_2O_3, Gd_2O_3 and Nd_2O_3 are contacted with dilute acids and agitated to form a suspension (either a slurry or a colloidal dispersion) that the resulting suspension, after a short fluid period (during which it may be applied to a surface or cast to a desired shape) and a short plastic period, will spontaneously harden into a rigid porous mass. This rigid material has a microstructure resembling that of plaster of Paris. While the composition of these materials is somewhat complex and undetermined, the materials lose weight upon heating until they are eventually decomposed into the initial oxides.

It was also found that these spontaneously hardening compositions during their fluid periods are adhesive and may be applied to a surface as a coating, in a like manner as paint. If the coated surface is heated to at least $1000°C$ for sufficient time to firmly bind the needles together, the composition decomposes to the original oxide which surprisingly remains firmly bound to the surface as a tough porous oxide coating.

The spontaneously hardening composition made from Y_2O_3 is particularly useful in a method for coating graphite crucibles used in uranium melting operations for nuclear fuel fabrication. Because Y_2O_3 isn't wet by molten uranium, the sintered Y_2O_3 coating of this process, though highly porous, is nevertheless effective as a crucible liner to prevent carbon contamination.

This Y_2O_3 coating is much tougher than other coatings and is more easily repairable. Of particular use in uranium melting operations is a coating composed of a mixture of Y_2O_3 and UO_2. This coating may be made by initially providing a mixed suspension of up to 2 mols UO_2 per mol of Y_2O_3 in dilute acid. Coatings of Eu_2O_3 and Gd_2O_3 for nuclear shielding or reactor control rods and Nd_2O_3 for thermal and electrical insulators may be applied in a like manner as the Y_2O_3 coating above.

In addition to coating applications, these self-hardening compositions provide a simple inexpensive fabrication technique for porous articles. The four oxides exhibiting this self-hardening behavior all have good insulating properties, so porous reticulated structures are very useful as electrical and thermal insulators and high temperature resistors.

The self-hardening compositions are prepared by contacting powders of Y_2O_3, Gd_2O_3, Eu_2O_3 or Nd_2O_3 with a dilute acid solution and agitating to bring the particles into a suspension (either a slurry or a colloidal dispersion) which soon spontaneously hardens. The agitation is only necessary to bring particles into suspension and the means of agitation is not critical. The particle size is not critical; however, finer particles result in a better article with a more strongly bound structure. The preferred particle diameter is less than 25 microns (equivalent area diameter) with 3 to 10 microns being optimum.

The preferred acid concentration for an easily workable mixture is 2.63 M. The preferred acids are nitric and hydrochloric, but other acids such as HBr, HI, $HClO_3$ etc. are suitable. It has been found that acids which have a strong affinity for water such as H_2SO_4, H_3PO_4 etc. do not form the self-hardening composition.

In order to form easily workable plaster-forming compositions, there must be sufficient liquid present to form a fluid slurry or colloidal dispersion. This may be easily determined by routine testing using various acid concentrations.

Example 1: 59 cc of 3.16 M HNO_3 was added slowly to 41 grams of Y_2O_3 powder having a mean agglomerate diameter of 3.5 microns with particles within the agglomerates being 100% less than 0.05 microns. The mixture was stirred until the powder became suspended in the solution. The resulting suspension, which resembled milk in appearance, was poured into a fluorocarbon-lubricated mold and allowed to harden. Twenty minutes were required for the suspension to set up into a rigid material.

The material was removed from the mold and dried at $50°$ to $60°C$ for 24 hours. The material was then sintered in argon at a heating rate of $200°C$/hr and held at $1100°C$ for 1 hour, followed by cooling at a similar rate. The sintering atmosphere is not critical and air may be used. The properties of the material as measured by a Mercury Intrusion Porosimeter are given in the following table.

	Before Sintering	Sintered
Percent porosity	57.8	70.1
Average pore diameter (microns)	1.9	2.6
Percent porosity less than 10 microns	99.0	97.3
Surface area (m^2/g)	11.8	6.0

Example 2: 100 cc of 2.01 M HCl was added slowly to 100 grams of Y_2O_3 powder having a mean agglomerate diameter of 3.5 microns with particles within the agglomerate 100% less than 0.05 microns. The mixture was stirred until the powder became suspended in the acid solution. The resulting milky suspension set up spontaneously into a plaster-like material identical in appearance to that of Example 1.

Cyanoacrylate Adhesives for Consumer Use

Liquid cyanoacrylate compositions have long been known in the art as excellent adhesives. Having achieved wide acceptance in industrial applications, cyanoacrylates have recently made enormous advances in sales in the consumer adhesive market. Various of the inherent characteristics of cyanoacrylates, which caused no particular difficulties in industrial automatic application equipment, have been detrimental to their unqualified acceptance in consumer use.

Chief among these characteristics are (1) the adhesives' relatively low viscosity, causing it to run or drip off the point of application to some undesired point, and (2) its extremely rapid cure speed when catalyzed by small amounts of moisture, as found in human skin.

E.R. Gleave; U.S. Patent 4,105,715; August 8, 1978; assigned to Loctite (Ireland) Limited, Ireland has formulated a thixotropic, curable adhesive composition comprising:

(a) a monomeric ester of 2-cyanoacrylic acid,

(b) to each 100 parts by weight of monomeric ester, about 5 to 100 parts by weight of a finely divided organic powder selected from the group consisting of: polycarbonates, polyvinylidene fluorides, polyethylenes and acrylic block copolymer resins containing saturated elastomer segments.

The composition also preferably contains one or more known acidic and free radical inhibitors, and optionally other functional additives for such purposes as improving thermal resistance, providing color, accelerating the cure reaction, providing crosslinking, etc.

By adjusting the quantity and kind of filler, these compositions may be prepared in various forms, for a loose jelly to a thick paste. In all forms, however, they have thixotropic properties.

Cyanoacrylate esters have the formula:

$$CH_2{=}C{-}COOR$$
$$\overset{\textstyle CN}{\underset{\textstyle |}{}}$$

and for the purposes of this formulation, R is most preferably a methyl or ethyl group. An anionic inhibitor, soluble in the ester of 2-cyanoacrylic acid should be used, such as soluble acidic gases like sulfur trioxide, nitric oxide, hydrogen fluoride; an organic sultone; or an organic sulfonic acid, preferably having a molecular weight less than about 400. To be optimally useful as a stabilizer in the adhesive compositions, the sulfonic acid should have a pKA value (dissociation constant in water) of less than about 2.8, and preferably less than about 1.5.

While not essential, the cyanoacrylate adhesive compositions of this process generally also contain an inhibitor of free radical polymerization. The most desirable of these inhibitors are of the phenolic type, such as quinone, hydroquinone, tert-butyl catechol, p-methoxyphenol, etc.

The above inhibitors may be used within wide ranges; but the following general guidelines are representative of common practice, all figures being weight percent of the adhesive composition: acidic gases, from about 0.001 to 0.06% by weight; sultones, from about 0.1 to 10% by weight; sulfonic acids, from about 0.0005 to 0.1% by weight; free radical inhibitors, from about 0.001 to 1%.

In addition to the monomer, the second required ingredient is a filler selected from: (a) polycarbonates, i.e., polyester of carbonic acid; (b) polyvinylidene fluoride (various grades); (c)) polyethylene; and (d) acrylic block copolymer resins containing saturated acrylic elastomer segments.

Preferred fillers of the type (d) include acrylic block copolymers made by polymerizing a mixture of methacrylate monomers in the presence of a latex of acrylate ester polymer.

Useful concentrations of the fillers are, respectively per 100 parts by weight of monomer,

   (a)   about 5 to 15 parts by weight, preferably about 8 to 12
         parts;

   (b)   about 35 to 100 parts by weight, preferably about 60 to
         80 parts;

   (c)   about 60 to 100 parts by weight, preferably about 70 to
         80 parts; and

   (d)   about 10 to 20 parts by weight, preferably about 12 to
         15 parts.

Example 1: A cyanoacrylate paste composition was prepared by mixing the following materials: ethyl cyanoacrylate, 100 parts by weight, containing 40 parts per million hydroxypropane sulfonic acid as inhibitor; and polyvinylidene fluoride powder, 65 parts by weight, approximately 5 microns particle size, Kynar 301 and supplied by Pennwalt (U.K.) Limited.

The resulting product was a smooth, white paste which was thixotropic and nondrip and could be easily spread. Adhesive bonds formed with this product cured rapidly and were only slightly less strong than similar bonds made with the corresponding unfilled (control) cyanoacrylate.

Example 2: Example 1 was repeated except that methyl cyanoacrylate was used. Equivalent results were obtained.

Example 3: Example 1 was repeated except that the polyvinylidene fluoride was replaced by an equivalent amount of either polycarbonate or polyethylene. Except for the color, equivalent results were obtained in both cases.

In another patent *E.R. Gleave; U.S. Patent 4,102,945; July 25, 1978; assigned to Loctite (Ireland) Limited, Ireland* describes a cyanoacrylate adhesive having improved peel strength. The compositions comprise:

(a) a monomeric ester of 2-cyanoacrylic acid,

(b) for each 100 parts by weight of monomeric ester, about 3 to 30 parts by weight of an organic material selected from the group consisting of: acrylonitrile-butadiene-styrene terpolymer, methacrylate-butadiene-styrene terpolymer, and vinylidene chloride-acrylonitrile copolymer containing at least 50 molar percent vinylidene chloride.

The composition also preferably contains one or more known acidic and free radical inhibitors, and optionally other functional additives for such purposes as improving thermal resistance, providing color, accelerating the cure reaction, providing crosslinking, etc.

Examples: The following is a description of a number of particular formulas. Parts quoted are by weight. Effective and preferred ranges have been included, but they are approximations only. Test adhesives were made by adding known amounts of preferred thickeners to suitably stabilized cyanoacrylate monomers, stirring the mixture and heating it sufficiently to obtain a uniform dispersion or solution of the thickener in the monomer. The peel strength of these adhesives was measured and compared to that of cyanoacrylate monomers thickened with polymethylmethacrylate, a well-known thickener of the prior art, designated below as PMM.

In these examples the abbreviated designations for the thickeners are to be understood as follows.

ABS No. 1: An ABS polymer of low molecular weight, low to medium elastomer content, and high styrene content; known as Blendex 211 (Borg-Warner).

ABS No. 2: An ABS polymer of medium molecular weight, medium elastomer content, and medium styrene content: known as Blendex 101 (Borg-Warner).

MBS No. 1 and 2: These are MBS polymers made by polymerizing a mixture of styrene and methacrylate monomers in the presence of a latex of polybutadiene. The ratios of the components are chosen so as to maximize certain properties of mixtures of these polymers with polyvinyl chloride, the properties being as follows. MBS No. 1, impact strength (only); Borg-Warner, Blendex 436. MBS No. 2, impact strength and clarity; Rohm & Haas, Paraloid KM 611.

VAC: A vinylidene chloride/acrylonitrile copolymer, wherein vinylidene chloride is the major component; known as Saran F 220 (Dow).

Ex.	Cyanoacrylate (100 parts)	Thickener	Quantity Used (parts)	Effective (parts)	Range Preferred (parts)	Peel Strength (kg/cm)
1	Methyl	None	–	–	–	0.11
2	Methyl	PMM	8	–	–	0.34
3	Methyl	ABS No. 1	10	5–50	10–30	0.58–0.88
4	Methyl	ABS No. 2	20	5–50	10–30	1.39
5	Methyl	MBS No. 1	20	5–50	10–30	0.43
6	Methyl	MBS No. 2	20	5–50	10–30	1.43
7	Methyl	VAC	10	2–20	4–15	1.32
8	Ethyl	None	–	–	–	0.15
9	Ethyl	PMM	8	–	–	0.23
10	Ethyl	ABS No. 1	20	5–50	10–20	0.87
11	Ethyl	ABS No. 2	20	5–50	10–30	1.36

Examples 1, 2, 8 and 9 relate to controls and are given for comparison purposes only. Each peel strength value given is the mean of from 10 to 20 determinations carried out on each of two pairs of bonded steel strips, i.e., it is the mean of from 20 to 40 determinations.

For Installing Artificial Turf

W.D. Emmons, S.S. Kim and D.A. Winey; U.S. Patent 4,115,169; September 19, 1978; assigned to Rohm and Haas Company have formulated compositions containing bis-ketimine esters of the general formula:

$$\underset{R}{\overset{R_1}{\diagdown}}C=N-X-O-\overset{\overset{O}{\|}}{C}-Y-\overset{\overset{O}{\|}}{C}-O-X-N=C\underset{R}{\overset{R_1}{\diagup}}$$

where, for example, R and R_1 are alkyl groups, X is alkylene, and Y is the noncarboxyl moiety of a dicarboxylic acid.

When such a compound is mixed with an acid polymeric anhydride, an adhesive is formed which has high solids, good storage stability, and is moisture-curable.

The adhesive composition is particularly advantageous in the glue down installation of outdoor carpeting on patios and around swimming pools, and for synthetic turfs on football fields, tennis courts and other recreational areas.

The polymeric anhydrides used are those which have two or more pendant anhydride groups. Preferred are those polyanhydrides prepared from maleic anhydride and one or more monomers selected from C_1-C_{18} alkyl acrylate or methacrylate, a polyunsaturated fatty acid derivative such as tung oil, an α-olefin such as 1-octene, 1-decene, 1-dodecene, 1-hexadecene, and the like, vinyl ethers such as methyl vinyl ether, butyl vinyl ether and the like, vinyl acetate, ethylene and styrene. Especially preferred is the polyanhydride prepared from polyunsaturated fatty acid esters and most preferably tung oil and maleic anhydride and the polyanhydride prepared from maleic anhydride C_3-C_6 acrylates and methacrylates and most preferably butyl acrylate and maleic anhydride.

Example: Preparation of Bis(Ketimino) Polyester — Diisobutyl ketone and ethanolamine, with toluene as diluent, are refluxed together for 12 hours then vacuum distilled. The product, $(iso\text{-}C_4H_9)_2C=NCH_2CH_2OH$ is distilled at $103°$ to $105°C$ 17mm. The diisobutyl ketiminoethanol is reacted with dimethyl adipate in an anhydrous system with toluene as diluent and under reduced pressure. Sodium methoxide is used as the catalyst. After 5 to 6 hours of reaction, during which a methanol/toluene azeotrope is taken off, the catalyst is decomposed with sodium bicarbonate and filtered off and the toluene is stripped, leaving bis(diisobutyl ketiminoethyl) adipate as the product.

Preparation of Polymeric Anhydrides — Maleic anhydride and tung oil are reacted in an exothermic reaction (after maleic anhydride is melted). Steam heat is then applied for 2 hours.

Preparation and Utilization of Carpet Adhesive — A small Brabender mixer is charged with 364 grams of the tung oil-maleic anhydride adduct prepared as above, and 454 grams of talc is mixed in for 20 minutes. Then over a period of 10 minutes, 180 grams of bis[2-(diisobutylketimino)ethyl] adipate prepared as above is added with mixing as dry nitrogen gas is passed over the ingredients to prevent moisture from initiating the curing reaction.

The above-prepared adhesive composition is spread on a concrete floor with a $1/16$ inch notched trowel. Nylon cut pile construction carpet is unrolled onto the spread adhesive and shifted into the desired position. By the next day the carpet is adhered to the floor. For the next five months the carpet is subjected to heavy pedestrian traffic and intermittent vacuuming. At the end of the five months, the carpet is shampooed with a Clarke jet cleaning machine. The shampooing causes no loosening or lifting of the carpet. Similarly good results are obtained with a like adhesive formulation containing no fillers.

A similarly efficacious carpet adhesive is formulated with bis[2-(diisobutylketimino)ethyl] terephthalate and the tung oil-maleic anhydride adduct used above.

Another good performing carpet adhesive is similarly formulated with bis[2-(methylisobutylketimino)ethyl] adipate and a polymeric anhydride of 95 parts by weight of butyl acrylate and 5 parts by weight of maleic anhydride.

In like manner, the tung oil-maleic anhydride adduct is mixed with the bis(ketimino) polyester prepared as described hereinabove to provide an adhesive formulation having good properties.

For Use in a Petrochemical Medium

R.A Veselovsky, T.E. Lipatova, R.A. Kimgir, J.K. Znachkov, D.M. Pyanykh, N.N. Shmanov, N.D. Trifonov and K.A. Zabela; U.S. Patent 4,119,681; Oct. 10, 1978 have formulated an adhesive which is suitable for repairing oil reservoirs and oil tanks, for sealing and mounting oil pipelines, for strengthening corroded ship structures, for bonding together natural stone and glass-fiber plastics in air, in water, and in a petrochemical medium.

The adhesive consists of a solution of an unsaturated polyester which is a product of interaction between carboxylic acids with polyols in an unsaturated monomer, and a polymerization agent, which contains also an urethane modifier, which is a product of interaction between aromatic diisocyanate and hydroxyl-containing polyesters, with fluorinated alcohol of the general formula $C_nF_{2n+1}OH$, where n=1-15, and with ethyleneglycol monomethacrylate. The ratio between the components in weight parts is as follows: a solution of unsaturated polyester in unsaturated monomer, 100; urethane modifier, 10-300; and polymerization agent, 0.1-15.

Introduction of the urethane modifier into the adhesive composition makes it possible to bond together surfaces contaminated with petrochemicals, to effect bonding in a petrochemical medium, to increase the elasticity of the hardened adhesive, to decrease the internal stress in the joint by more than 10 times, and to increase the adhesive strength and water resistance of the joints.

Example 1: An adhesive consists of the following components in weight parts: 100 parts 70% solution of oligodiethyleneglycol maleate in styrene; 50 parts of a product resulting from interacting 1 mol of urethane prepolymer, obtained from toluene diisocyanate and polydiethyleneglycol adipate, with 0.5 mol of perfluoropentyl alcohol and 0.5 mol of ethyleneglycol monomethacrylate; and 1 part α-methylbenzoyl alkyl ester.

To prepare this adhesive, 100 parts by weight of the 70% solution of oligodiethyleneglycol maleate in styrene are mixed with 50 parts by weight of the product of interacting 1 mol of urethane prepolymer, obtained from toluene diisocyanate and polydiethyleneglycol adipate, with 0.5 mol of perfluoropentyl alcohol and 0.5 mol of ethyleneglycol monomethacrylate, and with 1 part by weight of α-methylbenzoyl alkyl ester. Glass fabric was impregnated with the adhesive and put on ship structures, such as decks and partitions subjected to most intensive corrosive wear. Paint and rust were removed from the construction surface; spot corrosion was not removed. The section repaired was not treated with solvents.

The above adhesive was used for sealing a steel oil reservoir by coating leaky spots. Oil was not removed from the repaired sections. Air temperature during repairing was +15°C. The time of adhesive hardening was within the range of 10-30 minutes from the instant of pouring the adhesive out from a nontransparent container. The inspection of the reservoir a year after the repair showed that it was still sealed.

Example 2: An adhesive made by the same process as that described in Example 1 consists of the components in weight parts: 100 parts 70% solution of oligodiethyleneglycol maleate phthalate in styrene; 150 parts of a mixture of 1 weight part of a product of interacting allyl alcohol with 1,4-naphthalene diisocyanate and 5 weight parts of urethane prepolymer which is a product of interaction between polydiethyleneglycol adipate with 1,4-naphthylene diisocyanate; 4 parts methyl ethyl ketone peroxide; and 0.5 parts cobalt naphthenate.

The adhesive was used for sealing seams in fuel containers. Tests have shown that the containers were sealed.

For Installing Resin Bonded Rock Bolts in Mining Operations

A.C. Plaisted; U.S. Patent 4,136,134; January 23, 1979; assigned to Celtite, Inc. have formulated a fast-acting polyester cartridge catalyst for use in mining and other industries, particularly in cold climates.

There are in wide use polyester resin cartridges in the mining and other industries for installation of resin bonded rock bolts where speed of reinforcement is essential. It is known that in cold climates the reaction time between the resin and the catalyst is slowed and in many instances is excessively long. Under present practice it is the usual experience for a resin cartridge to have a gel time ranging from 30 seconds to 2 minutes which can be intolerably long. Available cartridges are exemplified in U.S. Patents 3,731,791 and 3,915,297.

Plaisted replaces the benzoyl peroxide catalyst previously used in the cartridges by a mixture of benzoyl peroxide and a halogen-substituted peroxide. By this means there is provided a cartridge which has been demonstrated to fulfill the need for a self-setting composition which will secure an anchor bolt in a rock or mine wall at a speed comparable to that required to install a mechanical bolt.

The resins used in these cartridges consist of solutions of unsaturated polyester resins in ethylenically unsaturated monomers, such as styrene or vinyl toluene, methyl methacrylate, etc. or mixtures of these materials. Suitable inhibitors are also included in the mixture such as tertiary butyl catechol or hydroquinone, together with promotors, such as tertiary amines, e.g., dimethyl aniline or dimethyl-p-toluidine. Inorganic fillers may also be incorporated into the resins to the extent of from 30 to 90%, preferable 70 to 80%, e.g., calcium carbonate, silica sand, talc, calcium sulfate, etc.

Catalyst pastes used heretofore, consist of benzoyl peroxide dispersed in a suitable plasticizer, such as dimethyl phthalate together with or without a filler, such as dolomite. The concentration of benzoyl peroxide is preferably in the range of 20 to 30%.

Cartridges are made by extrusion methods as described in U.S. Patent 3,731,791 with a resin mastic to catalyst paste ratio of 15:1.

In this formulation, two peroxides, namely (a) benzoyl peroxide and (b) halogen substituted benzoyl peroxide are combined. For example, 2,4-dichlorobenzoyl peroxide at a concentration of from 1 to 15% of the total paste formulation is quite effective.

Reaction of this foregoing catalyst blend with a highly promoted o-phthalic polyester resin results in significantly faster gel times which in some cases are over ten times the original reaction rate.

The unsaturated polyester resin based systems catalyzed with benzoyl peroxide pastes may be formulated within the following limits:

	Percent by Weight
70% unsaturated polyester resin	10–70
Inorganic filler	30–90
Dimethylaniline	0.1–0.5
Thixotropic additives	0–3
Benzoyl peroxide paste blend (23.5% benzoyl peroxide)	5–10

The benzoyl peroxide paste blends are prepared by dispersing the organic peroxide and some inorganic fillers in suitable inert plasticizers such as chlorinated paraffins, castor oil, phthalate esters, etc.

The preferred peroxide paste is made up from BPO powder together with a plasticizer system consisting of polypropylene glycol of molecular weight 400-450 and/or castor oil. The advantage of such a blend is twofold; (1) control of viscosity, and (2) very low solubility for benzoyl peroxide. A further preference in the paste formulation is for the use of the dihalogen BPO suspended in silicone oil (viscosity 1,000 cp) rather than dibutyl phthalate. There would appear to be marked improvements in the paste consistency. A typical formulation for the paste is:

	Percent by Weight
Benzoyl peroxide powder (35% on dicalcium phosphate)	50.0
2,4-Dichlorobenzoyl peroxide (50% in silicone oil)	12.0
Polypropylene glycol (MW 400)	38.0
	100.0

Cartridges were made with the above compositions. Size of cartridges is 22 mm diameter, and 12" long. An overhead hole 25 mm in diameter and 4' long was drilled into a shale/coal matrix and one of the above cartridges was inserted to the back of the hole.

A 6' length of 7/8" rebar attached to a drill was then spun into the cartridge at 500 rpm. The rebar bolt was fully home to the back of the hole in 10 seconds and the rotation discontinued. The very rapid reaction allowed the bolt to be fully supported within 20 seconds. After only 2 minutes, a hydraulic tensioning device was attached to the free end of the bolt and a direct tensile load applied at a rate of 1 ton every 15 seconds. The resin/rock interface yielded at 12 tons tensile load, which was considered more than adequate for that time interval from bolt insertion.

Adhesive Strip for Sanitary Napkin or Diaper

J.A. Collins and T.H. Quinn; U.S. Patent 4,136,699; January 30, 1979; assigned to H.B. Fuller Company have developed a pressure-sensitive adhesive (PSA) means for attaching a sanitary napkin or a diaper to a supporting garment. The adhesive used for such a purpose must meet a variety of exacting requirements. Collins and Quinn have found that heat-stable A-B-A block copolymers such as Shell Oil Co.'s Kraton G are useful as the hot-melt adhesive, provided tackifier and plasticizing oil are used.

Two different types of heat-stable A-B-A block copolymers have been found to meet the requirements of a PSA for absorbent articles. In one type (Kraton G) the A blocks comprise polystyrene and the B block is a rubbery poly(ethylene-butylene) center block. The other type is a teleblock copolymer comprising molecules having at least three branches radially branching out from a central hub, each of the branches having polystyrene blocks and a butadiene segment in the center.

The A-B-A polymer is a true elastomer and has an elongation at break well in excess of 200%, e.g., 500%. The elongation at break for the teleblock (radial) copolymer is in approximately the same range or slightly higher, e.g., 590%.

To ensure rubberiness or elastomeric behavior in the radial copolymer, the number of butadiene units should be greater than the number of styrene units. Shore A hardness for the teleblock (radial) copolymers can be in the range typical of true rubbers, e.g., above 60 or 70.

Plasticizing oils used as hydrocarbon process oils which are preferably low in aromatic content. For example, aromatic carbons should comprise less than 20% of the oil, while naphthenic carbons (i.e., carbons of cycloaliphatic compounds and the like) can range from about 25 to 60% and paraffinic carbons can range from about 35 to 75%.

Preferred tackifiers are of the type known as "hydrocarbon resins". In industrial practice "hydrocarbon resin" is a term relating to resins in the molecular weight range of a few hundred up to about 6,000 or 8,000 which are obtained or synthesized from rather basic hydrocarbonaceous materials such as petroleum, coal tar, turpentine, and the like. A good description of hydrocarbon resins can be found in Kirk-Othmer, *Encyclopedia of Chemical Technology,* Second Edition, Volume 11, Interscience, New York 1966, p 242.

Many of the hydrocarbon resins commercially available today are terpene resins, i.e., polymers with repeating isoprene (C_5H_8) or $C_{10}H_{16}$ units. These polymers can be natural or synthetic and can be copolymers (including terpolymers, etc.), since isoprene is an olefin which can be copolymerized with other olefins. Terpene-phenols are also produced.

It is preferred that the hydrocarbon resin (i.e., tackifier resin) component consist essentially of natural or synthetic polymers which are partially incompatible with the preferred A-B-A block copolymers. Such partial incompatibility is observed with a synthetic terpene resin having a softening point (ball and ring method) of about $80°$ to $115°C$, particularly the commercially available resin known as Wingtack 95.

These hot-melt PSA compositions can be formulated with techniques known in the art using heated mixers and the like. The rubbery copolymer and the oil can be blended together readily at moderately elevated temperatures (e.g., $150°$-$300°F$). The tackifier resin can be added to the copolymer-oil mixture. If a pigment is included in the PSA composition, it should be added to the copolymer/oil blend before the tackifier resin is introduced into the composition.

The resulting hot-melt PSA, once it is heated to the temperature where it will flow readily, can be applied to the outer covering layer of the absorbent structure or article by any of the techniques known in the art, including flow coating, roller coating, knife coating, or the like. The PSA can also be extruded into place by using a hot-melt extruder or die face.

As noted previously, hot-melt PSAs of this process can have a viscosity within the range of 500-15,000 cp, measured at $300°$-$400°F$ (e.g., $300°F$ or $150°C$) using a Brookfield Thermosel, spindle SC-4-37, 20 rpm. Viscosities less than 10,000 cp are preferred.

The following table sets forth broad and optimum amounts of the various components of PSAs of this process. Desired viscosities are also set forth. Although viscosity adjustments can be made by adding volatile solvents, it is preferred that PSA compositions of this process be essentially 100% "solids" (essentially free of volatiles). Parts per hundred of resin (phr) are given in weight.

	Broad	Optimum
Block copolymer		
wt %	10-20	13-16.5
phr	–	–
Terpene		
wt %	40-65	50-60
phr	250-500	300-400
Oil		
wt %	15-40	25-35
phr	120-400	150-300
Inert additives		
wt %	<25	<5
phr	0-100	<10
Brookfield viscosity, cp	<15,000*	500-8,000**

*at $300°$ to $400°F$.
**at $300°F$.

Preparation of Paper for Use in an Electrostatic Recording System

K. Nagai, R. Ishikawa, K. Imamura, F. Fujimura and Y. Shiokoshi; U.S. Patent 4,139,675; February 13, 1979; assigned to Jujo Paper Co.,Ltd., Japan describe paper for use in an electrostatic recording system which is capable of being recorded on by the usual electrostatic recording apparatus and which may then be adhesively applied to any desired surface by heating. This eliminates the necessity for a release paper backing on the paper.

The paper is a laminated one in which a photoconductive or dielectric recording layer is disposed on one surface of an electrically conductive base layer which serves as the support, and a heat-sensitive adhesion layer is disposed on the other side of the base layer.

The recording layer is made of zinc oxide and/or PVC, PVA, PVdC, acrylate and methacrylate resins, styrene-butadiene copolymers, melamine-formaldehyde resins or the like. The recording layer also contains filling material and/or a pigment.

The base layer is paper or synthetic film impregnated with a high molecular weight polymer quaternary ammonium salt or some electrolyte to make it electrically conductive.

For the heat-sensitive adhesion layer there is utilized a substance which develops the cohesive or adhesive property when heated. These substances include a large number of thermoplastic resins, for example, polyvinyl acetate resins; polyacrylate ester resins; polymethacrylate ester resins; polyvinyl chloride; polyvinylidene chloride and polyolefin and their copolymers; and thermoplastic prepolymers such as unsaturated polyester prepolymer and diallyl phthalate prepolymer.

It is desirable to mix into the base resin a tackifier which is solid at room temperature alone or to compound into the base resin a plasticizer which is solid at room temperature together with the tackifier or alone.

In order to render the adhesive layer electrically conductive, an electrolyte may be added to it in an amount of 10 to 50 parts by weight based on the total weight of solids of the adhesive layer. The paper is easily bonded to any substrate by, for example, use of a heated pressing roll which may be used at, e.g., 2 kg/cm^2 for about 3 seconds.

Example 1: Each surface of fine quality paper of 52 g/m^2 was impregnated with 2 g/m^2 of a quaternary ammonium salt series conductive polymer. On the front surface of conductive base paper thus obtained, 4 g/m^2 of the polymethacrylate resin (a substance forming the dielectric layer) was applied. On the back surface, there was applied 20 g/m^2 of a solution which was prepared by adding 50 parts of a particulate synthetic terpene resin to 80 parts of a solid content of a polyvinyl acetate-polyacrylate ester copolymer emulsion and 20 parts of a quaternary ammonium salt series conductive polymer as diluted in alcohol were added to the resultant mixture.

Then the paper was dried at 70°C. The recording paper had good electrostatic printing characteristics, and could be bonded to a paper by heating at 110°C.

Example 2: On the back surface of the same conductive base paper used in Example 1, a solution was applied in the same way as in the above example. The solution was made by mixing maleic acid, phthalic acid and ethylene glycol in proportions of 1:1:2, and reacted at 180°C for 5 hours to produce an unsaturated polyester prepolymer; 100 parts of hydrogenated rosin were added to 100 parts of the prepolymer. After the application, 4 g/m² of a polyvinyl chloride-vinyl acetate copolymer resin was applied in the form of a methyl ethyl ketone solution to the front surface. The recording paper thus obtained had a good electrostatic recording characteristic and exhibited a good thermal tackiness.

Grout Dressings

D.R. Burley; U.S. Patent 4,143,019; March 6, 1979; assigned to Tile Council of America, Inc. describes adhesive dressing compositions which are useful in restoring and improving adhesive joint surfaces such as grout surfaces in ceramic tile installations by easy, efficient methods. Such compositions form stain resistant, water repellent, washable coverings which adhere to most adhesive surfaces and further have properties of preferential adherability to certain adhesive surfaces compared with adjacent adherent surfaces. The compositions comprise a polymer in the form of an emulsion, an alkali-thickenable polymer, an alkaline material and water with other components including pigments, plasticizers and solvents.

These compositions have the particular advantage of being adherable to most adhesive joint surfaces while the compositions are easily removed from the adjacent surfaces of many adherends. This is due to the discovery of their selective adherability, or their ability to adhere and bond to certain materials and not to others. This property provides easy cleanup and simplified application methods of the composition. The workable consistency of the composition may be modified to allow it to be applied to surfaces by a variety of techniques.

The polymer in emulsion form must be alkali-stable and may be, for example, an acrylic polymer such as Rohm and Haas' Rhoplex AC-61, LC-40, and AC-707. Examples of useful alkali thickenable polymers include the Acrysol aqueous colloidal acrylics manufactured by Rohm and Haas as WS-50, WS-32, WS-24 and ASE-60. Other polymers suitable as the alkali thickenable polymer include Carboset (B.F. Goodrich Company).

Generally, the total weight of polymer solids in the uncured composition is in the range of 5 to 60% and preferably 10 to 40%. Of this polymer solid content, it is preferable to have at least 5% of the polymer solids be the emulsion polymers. The minimum amount of thickenable polymers present will depend upon the specific polymer used and the desired viscosity of the uncured composition.

The uncured dressing preferably contains at least about 20% by weight volatile components.

An alkaline agent or base is utilized to increase the pH of the composition. Monobasic water-soluble alkaline agents are preferred. Examples of such monobasic compounds include ammonium and alkali metal hydroxides and simple aliphatic amines.

Certain of the water-soluble solvents can prevent the formation of the film prematurely. Useful water-soluble solvents have a vapor pressure at room temperature lower than that of water and therefore tend to reduce the rate of evaporation of liquid constituents of the composition. These solvents can also affect the viscosity and rheology of the composition. The water-soluble solvent is chosen so as not to detrimentally affect the other ingredients in the dressing composition nor affect the physical stability of the dressing composition.

Among the solvents found most useful in the process are the short-chain diols such as ethylene glycol and propylene glycol. These ingredients are also referred to as antidrying aids. Not only do these solvents affect the film formation, but they also can function as pigment dispersers. Generally, ethylene glycol is the preferred diol although other diols may be used.

Example: The following composition was prepared as a batch on a Cowles Dissolver:

	Weight, g
Acrysol WS-50	2,000.00
Nopcocide N-96	80.00
TiO_2-RANC	80.00
Ethylene glycol	120.00
Diacetone alcohol	140.00
Butyl Cellosolve	140.00
Defoamer	0.50
NH_4OH to pH of 9.0	42.00
AC-61	2,000.00

This composition was used to dress cement grout joints which were discolored on a wall of glazed white wall tile. The composition was trowelled on the joints. The drying rate was slow enough to allow easy trowelling, but thin layers of the dressing which were allowed to dry were difficult to remove.

The appearance of the joints was improved. The white wall looked "new". When the composition dried, the application of permanent blue-black ink to the surface did not stain the dressing. The ink was wiped away with a damp cloth. Application of ink to untreated cement joints caused the formation of a spot which could not be removed by any cleansing method short of bleaching.

The above composition was tinted yellow with a yellow pigment dispersion. The pigment was just mixed in. The composition was applied to a wall of yellow tile with dirty joints. The effect was dramatic. The appearance of the whole wall was improved.

Similar compositions were prepared having blue pigment and a brown (iron oxide) pigment. In each case the joints on which the dressings were applied yielded dramatic improvement in the aesthetic properties of the wall. The compositions were also applied to a floor of ceramic mosaics with similar results obtained.

Forming of Endless Abrasive Belts

Coated abrasive belts are manufactured, in general, by adhesively joining together the free ends of strips of coated abrasive sheet material of a suitable length and width. The demands on an adhesive used in the formation of endless abrasive belts are severe, as a coated abrasive belt during use is subjected to relatively high tensile and flexural forces. Thus, any adhesive composition used in the formation of abrasive belt joints must be tough and durable, in addition to providing good adhesion (high peel strength) with the abrasive material, to prevent delamination of the belt joint upon being subjected to severe flexing conditions, especially when running over a small diameter contact roll. Moreover, the adhesive used in forming the joint must often be resistant to heat, water, lubricants, and other cutting fluids encountered in abrading operations.

J.F. Malloy; U.S. Patent 4,144,219; March 13, 1979; assigned to Norton Co. has discovered that a partially cured, heat-activatable, preformed polyurethane film can be used for this purpose which can be used over a relatively wide, as well as a relatively low, press-joining temperature yet will provide a joint of unexpectedly high strength. The adhesive films used in this formulation basically comprise the partially cured reaction product of a composition comprising in admixture:

> (1) a first active hydrogen containing component in the form of a hydroxyl terminated polyurethane polyester having a hydroxyl number of from 2 to 15,
>
> (2) a second active hydrogen containing component containing at least difunctional active hydrogen and having an active hydrogen equivalent weight in the range from about 27 to less than 500,
>
> (3) a component having available free isocyanate groups, and
>
> (4) a tackifier or reaction inhibitor such as a chlorinated aliphatic or aromatic hydrocarbon, or an alkylated polystyrene.

Use of such an adhesive film results in the optimum amount and composition of adhesive being used in every joint formed. It makes it unnecessary during belt manufacture for an operator to allow for a "dwell time" during which solvent is evaporated and whereby cure may be advanced to a degree resulting in cohesive strength satisfactory for joining. Most importantly, the adhesive film is the sole bonding member. A wet adhesive need not be applied at all to a surface mating with the adhesive film in preparation for joining.

Example: An adhesive composition was prepared by mixing the following ingredients.

Component	Parts by Weight
Bostik 7076* (United Shoe Machinery Corp.)	
(21% solids in acetone-toluene solvent)	88.8
Mondur** CB-75 (Mobay Chemical Co.)	
(75% solids in ethyl acetate)	4.5
Thylon*** D-406 (Thiokol Chemical Corp.)	
(70% solids in methyl ethyl ketone)	4.3
Paroil† 170HV (Dover Chemical Co.)	2.4††

    *A hydroxyl-terminated polyurethane polyester elastomer.
  **Polyurethane prepolymer-reaction product of trimethylol
      propane and toluene diisocyanate.
***Polyester-polyurethane isocyanate blocked prepolymer.
    †Chlorinated aliphatic hydrocarbon.
  ††10% by weight based on total solids excluding Paroil.

Mixing was conducted at room temperature, Mondur CB-75, Thylon D-406, and Paroil 170 HV being added to Bostik 7076 sequentially, a few minutes mixing occurring between each addition. Mixing was continued for about an hour after adding the chlorinated aliphatic hydrocarbon to provide a homogeneous mixture.

The above composition was coated onto a conventional silicone coated release carrier using a doctor blade with a 30 mil gap setting. The wet adhesive layer was allowed to air dry for 30 minutes at room temperature, i.e., 70°F 50% RH. Afterwards, the adhesive coated carrier was heated at 225°F for 40 minutes to form a partially cured heat-activatable adhesive film 3.0-3.5 mils thick. The film is observed to have a slight degree of finger tack and is, of course, extremely tacky with respect to itself.

The adhesive film-release carrier combination above-manufactured was slit into strips $13/16$ inches wide and wound into rolls of suitable length for the manufacture of abrasive belts.

A strip (6" x 24") of 120X Resinall Metalite abrasive cloth was prepared for formation of an endless abrasive belt. The abrasive surface of one free end of the abrasive strip was skived according to usual techniques thereby removing the abrasive material and the adhesive bond, leaving the upper portion of the backing member exposed. The skived surface ($3/8$" x 7.5") was at the angle of 55° with respect to the lengthwise direction of the abrasive strip and was slightly tapered, from trailing edge to leading edge thereof, at an angle of about 5° with respect to the horizontal. The mating surface of the other end of the abrasive strip was back rubbed with an abrasive belt, leaving the surface very slightly tapered and free of previously applied backsize.

A suitable length of the adhesive film-carrier combination was withdrawn from a roll thereof and was cut at a length and of a shape complementary to the skived surface area. The dried, partially cured, heat-activatable adhesive film was then positioned directly on the skived surface, the strip of abrasive material having been positioned previously on the bottom bar of a conventional platen-press.

The upper bar (platen) of the press was then brought into contact with the release liner, a pressure of about 386 psi being exerted on the adhesive film-carrier combination and skived abrasive end. This pressure was held for a period of about 3 seconds while the film-abrasive material was being heated solely by the bottom bar having a surface temperature of about 240°F. Thus, the adhesive film was softened and tackified in preparation for the subsequent joining operation.

Next, the pressure was released and the upper bar was withdrawn from contact with the release liner. The release liner was then removed from the heat activated, i.e., softened and tackified adhesive film. The bottom surface of the other end of the coated abrasive strip was then positioned in contact with the upper surface of the activated adhesive film.

The overlapped coated abrasive strip-adhesive film assembly was then again pressed while being heated only by the lower bar. A pressure of 386 lb/in^2 was maintained for 3 seconds (bar temperature 240°F) after which it was released and the upper bar was slightly withdrawn. The coated abrasive strip-film adhesive assembly was then removed from this press and was positioned in a similar press. The upper bar therein was brought into contact with the overlapped portions of the coated abrasive strip. A pressure of 8,600 lb/in^2, was maintained while the assembly was heated by both bars for about 20 seconds (bar temperature 240°F). The pressure was then released and the thus produced abrasive belt was removed from the press. On visual examination, the abrasive belt was observed to have a joint thickness not substantially greater than the thickness of the coated abrasive material per se.

On subjecting belt joints thus manufactured and conditioned for at least 24 hr at 70°F, 50% RH to severe and continuous flexing in the immediate area of the belt joint on a multihead oscillating flex fatigue tester, belt joints according to this process were found to have flexes which showed an improvement of at least 140% over the controls in which no Paroil was used.

Lamp Capping Cement

One of the operations of conventional lamp manufacture consists in cementing a cap, usually of metal such as brass or aluminum, to the lamp envelope or bulb in which the internal components such as a filament are already fitted. The resulting cemented connection must be heat resistant, especially in the case of incandescent filament lamps, since the temperature of the cap during burning of such lamps is in many cases above 200°C. A cement used in such lamps must therefore resist such temperatures for a long period, notably 1,000 to 2,000 hours, for the life of the lamp.

G.E. Coxon; U.S. Patent 4,145,332; March 20, 1979; assigned to Thorn Electrical Industries Limited, England has provided a heat-resistant thermosetting lamp capping cement comprising a resin component consisting essentially of a mixture of a condensation product of an aralkyl ether and a phenol combined with a crosslinkable polymeric substance, a heat resistant filler component and a solvent or diluent.

Condensation products of aralkyl ethers and phenol are available commercially. One example is Xylok (Albright and Wilson Ltd.) sold in a number of forms having different properties: Xylok 210 is ordinarily available as a 60% solution in methyl ethyl ketone (MEK) or 74° OP ethanol and the resin is cured by condensation with hexamine catalyst which is incorporated in the solution; Xylok 225 consists of a dry powder form of the resin to which is added 10-15% of hexamine agent (based on the weight of resin) to cure the resin; Xylok 234C is a flexible resin which is cured by an epoxy mechanism. By using Xylok component in the paste a thermally stable adhesive (up to 250°C) can be achieved which ensures efficient adhesion of the cap even at the end of life in the hottest fittings allowable by the British Standard.

The preferred crosslinkable polymer or resin is a carboxylated acrylic polymer, especially Alcolec (Allied Colloids Ltd.). This material has the form of a free flowing powder and imparts improved flow characteristics to the cement as well as acting as an adhesive. It is likely that this acrylic material crosslinks with the aralkyl ether/phenol condensation product to produce a more intractable polymer contributing to the stability of the cured paste.

As an alternative to a carboxylated acrylic polymer, a thermosetting condensation product such as phenol-formaldehyde, urea-formaldehyde or melamine-formaldehyde could be used with a similar result. A further alternative to the acrylic polymer is shellac.

Organic polar liquids such as ketones, esters, alcohols and ether solvents can be used as solvent, which is added to assist solubility of the resins and fillers. The preferred solvent, which is compatible with this capping cement and with accepted lamp making practice, is 74° OP ethanol or methylated spirit. To increase the drying time, solvent blends of ethanol with 2-ethoxyethanol (Cellosolve or Oxitol) can be used. Rosin can be included in the formulation to improve the solvent retention properties.

To ensure that the paste when fixed to the inside of the cap does not flow or sag when stored in a random manner in hoppers, an antislumping agent can be added to the mixtures. An addition of 1 to 2% by weight of fumed silica or alumina is preferably added to the solvent or resin solution for this purpose.

Preferred ranges of the polymeric components in the capping cements are: 1 to 15% aralkyl ether/phenol condensate (calculated as % solids out of the total dry solids) and 1 to 15% carboxylated acrylic polymer or alternative crosslinkable resin. The balance is filler whose preferred composition is marble dust and lithopone.

Example: A cement was manufactured using the following components: 45 g aralkyl ether/phenol condensate (Xylok 225); 50 g carboxylated acrylic polymer (Alcolec 860); 5 g hexamine; 30 g fumed alumina (Alon C); 870 g filler (marble dust and lithopone, 4:1); and 80 g ethanol (74° OP).

If longer drying times are required a solvent blend is used varying from 1:1 ethanol to Cellosolve to 1:10 ethanol to Cellosolve. Other suitable polar solvents can also be used.

In the production of this paste, the polymers, hexamine and filler are thoroughly blended dry together and added with mixing to a mixture of the alumina and solvent.

The performance of the capping paste in accordance with this process was evaluated in tests on 100 W single coil filament lamps in 60 mm diameter so-called Ribbon Pear Shape pearl glass envelopes, in a fitting which produces cap temperatures of 210°C.

Aluminum bayonet caps were pasted on conventional machines using the paste of the process and also conventional phenolic resin based pastes fortified with about 3% dry weight silicone. After testing by lamp operation to ultimate failure at about 1,000 hours in the hot fitting described, the adhesion of the cap at the end of life was measured and the adhesion of the experimental cement to the aluminum bayonet caps was found to be exceptionally good, better than that of the conventional phenolic resin fortified with silicon.

Company Index

The company names listed below are given exactly as they appear in the patents, despite name changes, mergers and acquisitions which have, at times, resulted in the revision of a company name.

Inventor Index

U.S. Patent Number Index

4,151,319 - 130	4,155,952 - 233	4,157,418 - 71
4,152,189 - 45	4,156,064 - 458	4,157,420 - 147
4,152,309 - 112	4,156,671 - 89	4,158,647 - 399
4,152,313 - 260	4,156,676 - 196	4,158,725 - 262
4,153,320 - 189	4,157,318 - 199	4,159,287 - 14
4,153,743 - 353	4,157,319 - 91	Reissue 29,548 - 324
4,154,774 - 397	4,157,328 - 383	Reissue 29,663 - 373
4,155,950 - 397	4,157,357 - 164	Reissue 29,699 - 137

Notice

Nothing contained in this Review shall be construed to constitute a permission or recommendation to practice any invention covered by any patent without a license from the patent owners. Further, neither the author nor the publisher assumes any liability with respect to the use of, or for damages resulting from the use of, any information, apparatus, method or process described in this Review.